GROUNDWATER HYDROLOGY

McGraw-Hill Series in Water Resources and Environmental Engineering

Ven Te Chow, Rolf Eliassen, and Ray K. Linsley
Consulting Editors

GROUNDWATER HYDROLOGY

Herman Bouwer

Director, U.S. Water Conservation Laboratory
Agricultural Research Service
United States Department of Agriculture
Phoenix, Arizona

and

Lecturer in Groundwater Hydrology
Geology Department
Arizona State University
Tempe, Arizona

McGraw-Hill Book Company

New York St. Louis San Francisco Auckland Bogotá Düsseldorf
Johannesburg London Madrid Mexico Montreal New Delhi Panama
Paris São Paulo Singapore Sydney Tokyo Toronto

GROUNDWATER HYDROLOGY

34567890DODO7832109

This book was set in Times Roman.
The editors were B. J. Clark and Susan Gamer;
the production supervisor was Leroy A. Young.
The drawings were reproduced by J & R Services, Inc.
R. R. Donnelley & Sons Company was printer and binder.

Library of Congress Cataloging in Publication Data

Bouwer, Herman.
    Groundwater hydrology.

    (McGraw-Hill series in water resources and environmental engineering)
    Includes bibliographical references and index.
    1. Water, Underground. I. Title.
GB1003.2.B68   551.4'9   77-24651
ISBN 0-07-006715-5

CONTENTS

PREFACE

This book is a basic groundwater hydrology text addressed to students and practitioners in hydrology, engineering, geology, agriculture, geography, ecology, and other disciplines concerned with groundwater. Groundwater hydrology is a broad field with many ramifications. Some books on the subject stress well-drilling technology. Others concentrate on mathematics of groundwater movement or groundwater geology. This book emphasizes environmental aspects of groundwater. Thus, in addition to the fundamentals of groundwater hydrology and well technology, the book includes topics like flow in the vadose zone as the link between surface and groundwater environments, surface-subsurface water relations, contamination of groundwater, land treatment and underground disposal of wastewater, geothermal power production, and land subsidence.

The topics are treated in a multidisciplinary way, using principles of hydrology, geology, hydrodynamics, various branches of engineering, and soil science. Yet the book is written so that the reader does not need a background in these disciplines. While elementary calculus is used in the derivation of some of the equations, the book as a whole does not require special knowledge of mathematics. Literature references have been selected to document the text and to provide the interested student with a bibliography for further study. The writing of a text of this nature was motivated by the author's teaching of groundwater courses at Arizona State University to graduate and undergraduate students with widely varying backgrounds and interests.

The author is indebted to the following persons for reviewing parts of the book and supplying valuable comments: C. R. Amerman, W. C. Bianchi, G. F. Briggs, M. D. Campbell, D. C. Helm, J. H. Lehr, B. E. Lofgren, S. W. Lohman, W. A. Pettyjohn, J. F. Poland, and T. A. Prickett. Their contributions to the quality of this book are gratefully acknowledged. Thanks are also due to the Geology Department of Arizona State University for the use of its facilities; to E. J. Bouwer for help on some of the engineering aspects; and to Jessie for typing the manuscript and for her patience.

Herman Bouwer

UNITS AND CONVERSIONS

Metric units are used in the book. Principal units and their English equivalents are:

Length

1 meter (m) = 39.37 inches = 3.28 feet = 1.09 yards
1 micrometer (μm) = 10^{-6} m = 10^{-3} mm = 0.000 039 37 inches
1 millimeter (mm) = 0.001 m = 0.039 37 inches
1 centimeter (cm) = 0.01 m = 0.393 7 inches = 0.032 8 feet
1 decimeter (dm) = 0.1 m = 3.937 inches = 0.328 feet
1 kilometer (km) = 1 000 m = 0.62 miles

Area

1 m^2 = 10.744 $feet^2$ = 1.196 $yards^2$
1 cm^2 = 0.155 $inches^2$
1 hectare (ha) = 10 000 m^2 = 2.471 acres
1 km^2 = 247 acres = 0.386 $miles^2$

Volume

1 m^3 = 1 000 liters = 264 U.S. gallons = 35.314 $feet^3$ = 1.307 9 $yards^3$
1 cm^3 = 0.061 $inches^3$
1 liter (l) = 1 000 cm^3 = 1 dm^3 = 0.264 U.S. gallons
1 million m^3 = 810.7 acre-feet
1 km^3 = 0.240 $miles^3$

Mass

1 kilogram (kg) = 1 000 grams = 2.205 pounds

1 gram (g) = 0.035 ounces = 0.002 205 pounds
1 milligram (mg) = 0.001 g
1 microgram (μg) = 10^{-6} g

Velocity and hydraulic conductivity

1 meter/day (m/day) = 3.28 feet/day = 1.64 inches/hour = 24.5 meinzers (at 15.56°C) = 1.204 darcys (at 20°C)
1 cm/s = 864 m/day = 1417 inches/hour

Transmissivity

1 m^2/day = 10.74 $feet^2$/day = 80.5 U.S. gallons per day per foot

Volume rate of flow

1 m^3/day = 0.011 57 1/s = 11.57 cm^3/s = 0.000 408 7 $feet^3$/s = 0.183 45 U.S. gallons/min (gpm)

Pressure

1 kg (force)/cm^2 = 14.223 pounds/$inch^2$
1 newton/$meter^2$ (N/m^2) = 0.000 145 pounds/$inch^2$
1 atmosphere (standard) = 1.033 kgf/cm^2 = 14.70 pounds/$inch^2$

GROUNDWATER HYDROLOGY

ONE

GROUNDWATER AND AQUIFERS

1.1 DEFINITION OF GROUNDWATER

Groundwater is that portion of the water beneath the surface of the earth that can be collected with wells, tunnels, or drainage galleries, or that flows naturally to the earth's surface via seeps or springs. Groundwater has been an important water resource throughout the ages. Old, dug wells can be found along the wadis of the Middle East, the cradle of our civilization. Some of the ancient tunnels or "ghanats" in Iran are still in use. Today, groundwater is a major source of water for many municipalities and industries and for irrigation, suburban homes, and farms. In the United States alone, groundwater is estimated to supply water for about half the population and about one-third of all irrigation water. Some three-fourths of the public water supply systems use groundwater, and groundwater is essentially the only water source for the roughly 35 million people with private systems (McCabe et al., 1970, and references therein). As with any natural resource, groundwater supplies are not unlimited. They must be wisely managed and protected against undue exploitation and contamination by pollutants or salt water.

Not all underground water is groundwater. If a hole is dug, moist or even saturated soil may be encountered. As long as this water does not seep freely into the hole, however, it is not groundwater. True groundwater is reached only when water begins to flow into the hole. Since the air in the hole is at atmospheric pressure, the pressure in the groundwater must be above atmospheric pressure if it is to flow freely into the hole. By the same token, the underground water that did not flow into the hole must be at less than atmospheric pressure. Thus, what distinguishes groundwater from the rest of the underground water is that its

Table 1.1 Estimated distribution of world's water

	Volume, 1 000 km³	Percentage of total water
Atmospheric water	13	0.001
Surface water		
Salt water in oceans	1 320 000	97.2
Salt water in lakes and inland seas	104	0.008
Fresh water in lakes	125	0.009
Fresh water in stream channels (average)	1.25	0.000 1
Fresh water in glaciers and icecaps	29 000	2.15
Water in the biomass	50	0.004
Subsurface water		
Vadose water	67	0.005
Groundwater within depth of 0.8 km	4 200	0.31
Groundwater between 0.8 and 4 km depth	4 200	0.31
Total (rounded)	1 360 000	100

Source: Adapted to metric system from Nace, 1960, and Feth, 1973.

pressure is greater than atmospheric pressure. Since such water moves freely under the force of gravity into wells, it is also called *free* water or *gravitational* water. Depths to groundwater may range from 1 m or less to 1 000 m or more. There are also places where groundwater has not been reached at all.

The zone between ground surface and the top of the groundwater is called the *vadose* zone or *unsaturated* zone. This zone still contains water (very little in dry climates), but the water is held to the soil particles or other underground material by capillary forces. Thus, while this water is still able to move within the vadose zone, it cannot move out of the zone into wells or other places that are exposed to atmospheric pressure.

The term *vadose zone* should be preferred over *unsaturated zone*. This is because portions of the vadose zone may actually be saturated, even though the pressure of the water is below atmospheric pressure. Examples of saturated regions in vadose zones are the capillary fringe above groundwater, rain-saturated topsoil, and saturated layers of clay or other fine materials that hold water more tightly than underlying coarser material (see Section 2.9). Thus, the zone above the groundwater should not be called the unsaturated zone. By the same token, it is not correct to refer blankly to the groundwater region as the saturated zone because air bubbles may remain entrapped in this zone and prevent complete saturation (see Section 2.9). For these reasons, the terms *vadose zone* and *groundwater zone* rather than *unsaturated* and *saturated zone* are used in this book. Atmospheric pressure is the dividing line between the two, with the pressure of vadose water being below atmospheric pressure and that of groundwater above atmospheric pressure.

Groundwater accounts for a major portion of the world's freshwater resources. Estimates of the global water supply show that groundwater represents about 0.6 percent of the world's total water (Table 1.1). Since much of the groundwater below a depth of 0.8 km is saline or costs too much to develop with present technology and economic conditions, the total volume of readily usable groundwater is about 4.2×10^6 km^3. This is much more than the 0.126×10^6 km^3 fresh water stored in lakes and streams. Next to glaciers and ice caps, groundwater reservoirs are the largest holding basins for fresh water in the world's hydrologic cycle.

1.2 AQUIFERS

Groundwater-bearing formations sufficiently permeable to transmit and yield water in usable quantities are called *aquifers*. By far the most common aquifer materials are unconsolidated sands and gravels, which occur in alluvial valleys, old stream beds covered by fine deposits (buried valleys), coastal plains, dunes, and glacial deposits. Sandstones also are good aquifer materials. Cavernous limestones with solution channels, caves, underground streams, and other karst developments (Legrand and Stringfield, 1973) can also be high-yielding aquifers. Other sedimentary rocks (shales, solid limestones, etc.) generally do not make good aquifers. Small water yields may be possible where these rocks are highly fractured. The same is true for granite, gneiss, and other crystalline or metamorphic rocks. Basalts, lavas, and other materials of volcanic origin can make excellent aquifers if they are sufficiently porous or fractured and, in the case of lava, if the vesicles are interconnected.

There are two types of aquifers: unconfined and confined. Unconfined aquifers (Figure 1.1) are like underground lakes in porous materials. There is no clay or other restricting material at the top of the groundwater, so that groundwater levels are free to rise or fall. The top of an unconfined aquifer is the *water table*, which is the plane where groundwater pressures are equal to atmospheric pressure. The water-table height corresponds to the equilibrium water level in a well penetrating the aquifer. Other terms for the water table are *free surface* or

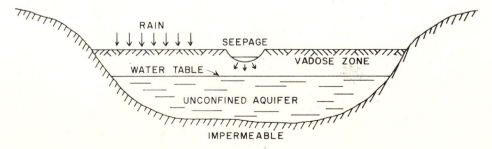

Figure 1.1 Unconfined aquifer in alluvial valley.

phreatic surface, after the Greek word *phrear*, "well." Above the water table is the vadose zone, where water pressures are less than atmospheric pressure. The air in vadose zones usually is continuous, so that unconfined aquifers tend to be open to the atmosphere. The lower boundary of unconfined aquifers is a layer of much less permeable material than the aquifer itself. Such "impermeable" layers may consist of clay or other fine-textured granular material, or of shale, solid limestone, igneous rock, or other bedrock. Unconfined aquifers are commonly found in alluvial valleys, coastal plains, dunes, and glacial deposits. They may range in thickness from a few meters or less to hundreds of meters or more. The principal source of groundwater in unconfined aquifers is precipitation that has infiltrated into the soil above the aquifers, either directly as it fell on the ground or indirectly via surface runoff and seepage from streams or lakes. In irrigated valleys, water not used by crops can percolate downward beyond the root zone and contribute to the groundwater. When groundwater flows from an unconfined aquifer into a pumped well, the water table drops and air moving through the vadose zone replaces water that has drained from the pores or other interstices in the upper aquifer material.

A confined aquifer is a layer of water-bearing material that is sandwiched between two layers of much less pervious material (Figure 1.2), like a sandy layer between two clay layers or sandstone between layers of shale or solid limestone. If the confining layers are essentially impermeable, they are called *aquicludes*. If they are sufficiently permeable to transmit water vertically to or from the confined aquifer, but not permeable enough to laterally transport water like an aquifer,

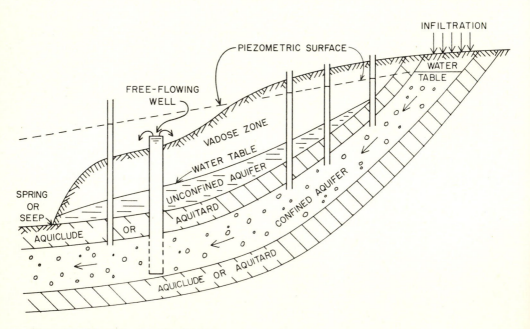

Figure 1.2 Schematic of confined aquifer.

they are called *aquitards*. An aquifer bound by one or two aquitards is called a *leaky* or *semiconfined* aquifer. In this book, the word *aquitard* refers to any layer in an aquifer or aquifer system that is much less permeable than the aquifers themselves, but not impermeable. Sometimes, usage of the term *aquitard* is restricted to fine-textured layers interbedded within one aquifer. Layers of low permeability forming boundaries of aquifers or separating various aquifers are then called *confining* layers. The source of water in confined aquifers is mostly rainfall that eventually moved through confining layers or infiltrated directly into the aquifer material at its outcrop (Figure 1.2).

Confined aquifers are completely filled with groundwater, and they do not have a free water table. The pressure condition in a confined aquifer is characterized by the *piezometric surface*, which is the surface obtained by connecting equilibrium water levels in tubes, or *piezometers*, penetrating the confined aquifer (Figure 1.2). If the piezometric surface is above the upper confining layer, the static water level in a well will be above the aquifer. Such a well is then called an *artesian* well, named after wells bored in Artois (a province in northern France) in the eighteenth century, where this phenomenon was observed. If the piezometric surface is above ground level, the confined aquifer will yield free-flowing wells (Figure 1.2). The "artesian" water can also move to ground surface through natural passages like faults or sinkholes, where it then produces springs. Artesian conditions occur where the confined aquifer is of lower elevation than its outcrop or other areas where recharge takes place.

When groundwater is pumped from a well in a confined aquifer, water is primarily yielded by compression of aquicludes and aquitards due to a lowering of the piezometric surface (see Chapter 9). Water may also seep into the aquifer through aquitards. Lowering of the water table at the outcrop may yield water by drainage of pore space, in the same way as unconfined aquifers.

Confined aquifers often occur as a succession of different aquifers separated

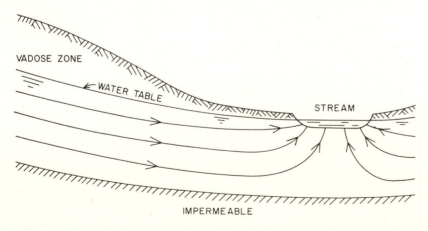

Figure 1.3 Drainage of groundwater into a stream.

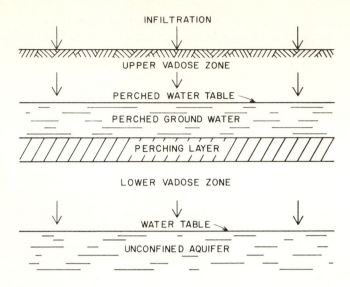

.Figure 1.4 Perched groundwater due to infiltration into soil with restricting layer.

by aquitards or aquicludes. The top aquifer may be unconfined. The permeability and water quality of the individual aquifers may differ considerably. Geologic formations that do not contain groundwater at all are called *aquifuges*.

Hillside seeps and springs occur where the aquifer and its lower impermeable boundary are exposed to the atmosphere at hillsides, canyons, etc. (Figure 1.2). Some aquifers, particularly unconfined aquifers, drain into bodies of surface water like streams, lakes, and oceans (Figure 1.3).

Temporary unconfined aquifers may form during periods of heavy infiltration into soil (spring thaw, periods of high rainfall, temporary flooding, etc.), if downward movement of infiltrated water is restricted by a deeper layer of relatively low permeability. The water will then "pond" on the restricting layer and form perched groundwater with a perched water table (Figure 1.4). Perching layers can be of large areal extent (essentially continuous), or they can be of restricted size, forming perching lenses. When infiltration at the surface has ceased, perched groundwater usually disappears as it moves down through the perching layer to underlying aquifers.

1.3 ORIGIN AND AGE OF GROUNDWATER

Atmospheric precipitation is the main source of fresh groundwater. The water may have infiltrated directly into the ground where it landed, or it may first have collected in streams and lakes via surface runoff and then seeped into the ground. For the conterminous United States, it is estimated that about 25 percent of the precipitation becomes groundwater (Nace, 1960). Groundwater of atmospheric origin that has been a recent part of the hydrologic cycle is called *meteoric* water. The term *recent* refers to geologic recency, which may cover tens of thousands of years.

Much older groundwater that is still of atmospheric origin but that may have been isolated from the hydrologic cycle for millions of years is called *connate* water. This water typically is groundwater that was already present in the geologic formation when it was formed, such as the water in which alluvial material was deposited (connate is derived from the Latin word *connatus*, which means "born together"). Connate water is often found in the lower parts of deep groundwater basins. It is normally of poor quality, especially if it has been in contact with salt deposits or other evaporites. Because of its long underground residence, connate water can have moved long distances, even though its velocity may be very small.

Juvenile water, also called *primary* water, is groundwater that has never been part of the hydrologic cycle. It was formed within the earth itself and is of volcanic or magmatic origin. Juvenile water can move up in the earth's crust with volcanic activity. It is, however, high in mineral content and probably insignificant as a water resource (McGuiness, 1963). *Magmatic* waters include volcanic water (derived from shallow magma) and plutonic water (derived from deep magma). *Metamorphic* water is water that was in rocks during the period of metamorphism. *Marine* water is water that has moved into aquifers from oceans.

The age of groundwater may range from a few years or less to tens of thousands of years or more. Assuming that 25 percent of the rainfall in the conterminous United States becomes groundwater, the volume of groundwater within a depth of 800 m is equivalent to the recharge of a 160-year period, which indicates the order of magnitude of the average groundwater age in the United States and other areas with similar climatic and geologic conditions (Nace, 1960). Much older meteoric groundwater has also been found. Using carbon 14 dating techniques (see Section 10.6), Pearson and White (1967) found groundwater ages in the confined Carizzo sand aquifer of southeast Texas that ranged from 5000 years at a distance of 22 km from the center of the outcrop to 30000 years at 60 km downgradient from the outcrop. Carbon 14 ages of groundwater from the Tucson basin in southern Arizona ranged from "modern" to 22000 years, with most of the samples falling in the range of 1000 to 4000 years (Wallick, 1973). The carbon 14 age of water from hot springs in Hot Springs, Arkansas, was 4600 years (Pearson et al., 1972).

Old meteoric water often occurs in arid areas where most of the groundwater was formed during previous climatic periods with higher rainfall. Thatcher et al. (1961) found carbon 14 ages of 20000 to more than 33000 years for groundwater sampled at depths of 382 to 1214 m from confined aquifers in Saudi Arabia. If the carbon 14 ages of more than 33000 years represent the true age of the groundwater, this water may then have fallen as rain during the early Wisconsin ice age. Other carbon 14 ages of groundwater in the Middle East are 10000 to 30000 years for groundwater from the Tiberias region in the Jordan Rift Valley (Carmi et al., 1971) and more than 31000 years for groundwater from the Suez Rift Valley (Münnich and Vogel, 1962). Recent groundwater, however, can also be present in arid areas. Using tritium-radiometry dating (see Section 10.6), Thatcher et al. (1961) found ages on the order of 10 years for groundwater in wadi gravels in the central region of the Arabian peninsula.

Carbon 14 groundwater ages in the Chalk of London Basin increased toward the central, confined part of the basin where it exceeded 25 000 years (Smith et al., 1976). In other European studies, Münnich and Vogel (1959) found carbon 14 ages that ranged from several thousand to more than 30 000 years for groundwater in German limestone aquifers.

1.4 DISTRIBUTION OF GROUNDWATER

Inventories of groundwater basins and aquifers have been made in many countries (see, for example, van der Leeden, 1975, and references therein). In classifying aquifers in the conterminous United States, 21 groundwater "provinces" were recognized by Meinzer (1923). These provinces were consolidated and rearranged into 10 groundwater regions by Thomas (1952). A brief description of these regions, condensed from the work by McGuinness (1963), follows.

1. *Western Mountain Ranges.* This region includes the Rocky Mountains and the mountains to the west. Precipitation and runoff from these mountain ranges are the main source of water for rivers and aquifers in the West. The aquifers may be in the thin mantle of rock and fracture zones that are found almost everywhere in the mountains, or they may occur in alluvial, intermontane valleys and in areas of glacial outwash. Also, sandstone and limestone aquifers may be found. One of the most productive wells on record is in a glacial outwash channel at Spokane, Washington.

2. *Alluvial Basins.* These are the alluvial valleys predominantly found in central and southern California, Nevada, Utah, Arizona, and New Mexico. A number of these valleys have excellent aquifers, which supply water for extensive irrigation and for municipalities. In many cases, groundwater is pumped out of these aquifers faster than the natural recharge rate, causing water tables to decline.

3. *Columbia Lava Plateau.* This is a high, generally dry plateau in Idaho, Washington, and Oregon, and it is bounded by the Cascade Range, the Rocky Mountains, and the Great Basin. It was formed by extrusive volcanic rocks, lavas, and basalts, which are 1 000 m or more deep and interbedded with or overlain by alluvial deposits. The lava deposits often are very porous and form excellent aquifers, which receive their water from local precipitation and runoff from the surrounding mountains. In many areas, the groundwater is at great depth. Groundwater is also found in interbedded, unconsolidated alluvial deposits.

4. *Colorado Plateau and Wyoming Basin.* These are areas that consist of sedimentary strata, chiefly interbedded sandstone and shale and mostly horizontal. Most aquifers are not highly productive and do not have much natural recharge for high-yielding wells. Small wells (about 1 to 10 m^3/day) are readily obtained, however. The most productive consolidated aquifer in this region is the Coconino Sandstone, which crops out on the north slope of the Mogollon Rim in Arizona.

5. *High Plains.* This is the area of the Texas Panhandle, western Kansas, and most of Nebraska that consists of extensive alluvium deposited on an eroded body of stratified sandstone, shale, and limestone. The alluvium forms a single stratigraphic unit called the *Ogallala formation*, which is more than 150 m thick in some places and which is the main source of groundwater in the area. Many productive wells have been installed, many for irrigation. The natural recharge of the Ogallala aquifer is restricted, however, due to low rainfall and tight overburden material (except where there are sand dunes). Thus, groundwater is being depleted and water-table levels are declining in many areas.

6. *Unglaciated Central Region.* This is a large, complex region mainly consisting of plains and plateaus underlain by sedimentary rocks. It extends from Texas into Iowa and through Oklahoma and Kansas, and also occurs in southeastern Colorado, eastern Montana, and parts of Wisconsin and Minnesota. The aquifers in most of the region are limestone and sandstone of low to moderate productivity. Many parts of the region have aquifers that are among the least productive in the United States. Salty groundwater is also common. The more productive aquifers include the limestone in the Edwards Plateau near San Antonio, Texas, the limestone and overlying alluvium in the Roswell basin and Carlsbad area of New Mexico, the limestone spring area of the Ozarks, the limestone in northern Alabama, Kentucky, and Tennessee, and some sandstone and glacial outwash areas in Wisconsin. Productive aquifers are also found in the alluvial valleys of major streams in the region.

7. *Glaciated Central Region.* This region takes in the area between the northern Rocky Mountains on the east, the Appalachians on the west, the Canadian border on the north, and the Unglaciated Central Region on the South, excluding the "driftless" area of Wisconsin, Minnesota, and Iowa. The region is hydrogeologically similar to the Unglaciated Central Region, except that it is covered by glacial drift, consisting mostly of fine-grained rock debris, but also of coarse sands and gravels that yield productive aquifers in many areas. The drift may locally be as much as 300 m thick. Where the drift is not as thick, high-yielding aquifers may be found in the underlying sedimentary rock.

8. *Unglaciated Appalachian Region.* This region extends from Alabama to Pennsylvania and New Jersey, and mainly consists of the unglaciated Appalachian Highlands, the Piedmont, the Blue Ridge, and the Valley and Ridge Province. The crystalline rocks of the Piedmont are sufficiently weathered and fractured to yield small but dependable wells in many places. The gneisses and schists are probably the best producers. Valleys tend to yield more productive wells than hilltops. The rocks of the Blue Ridge are hard and dense, and drilled wells commonly yield little water. The Valley and Ridge Province is somewhat erratic as a groundwater producer. Limestone rocks yield excellent wells if the driller is lucky enough to hit water-filled underground caverns. These rocks also are the source of many large springs. Alluvial valleys often consist of material that is too fine to yield producing wells.

9. *Glaciated Appalachian Region.* This region mainly covers New England and northeastern New York. It is primarily an extension of the Appalachian

region. The thickness of the glacial deposits varies widely. Typically, the hills are covered with a thin layer of till, whereas the valleys are underlain by thicker drift. Near the sea, the glacial deposits are covered or interbedded with marine sediment. The region as a whole does not contain productive aquifers. Most of the wells in the hills are drilled into the bedrock. In a few areas, the bedrock may yield some medium-capacity wells. The thicker glacial drift in the valleys is generally capable of yielding small amounts of water. The principal groundwater sources are unconsolidated sand and gravel deposits as may be found in outwash plains and channel fillings. The best place for a well is where these deposits have maximum thickness, areal extent, and permeability. It is often difficult, however, to determine where these optimum conditions occur.

10. *Atlantic and Gulf Coastal Plain.* This important region begins roughly at Cape Cod, covers the eastern portion of most of the Atlantic coast states, all of Florida, and the southern half of Alabama. It then expands northward to include all of Mississippi, the western part of Tennessee, the southern tips of Illinois and Missouri, and the southeastern section of Arkansas. It covers all of Louisiana and the southeastern part of Texas. The region consists of generally unconsolidated, but sometimes locally consolidated, sedimentary rocks on Precambrian, Paleozoic, and Mesozoic bedrocks. The sedimentary rocks mainly consist of clay, silt, sand, gravel, marl, and limestone. The thickness of these deposits increases toward the coast, where it may reach 10 000 m or more. The region contains many very productive aquifers, mainly consisting of sand and gravel. High-yielding limestone aquifers are found in the southeastern part of the region, including nearly all of Florida. One of the most productive aquifers of the United States is the Biscayne aquifer, which underlies the Miami area and consists of cavernous limestone and sand.

For more detailed information and descriptions of aquifers and groundwater for each state (including Hawaii and Alaska) and Puerto Rico, Virgin Islands, Guam, and Samoa, reference is made to McGuinness (1963). Local information about groundwater can be obtained from such sources as the U.S. Geological Survey (including many publications in the series *Water Supply Papers*), state geologists, state water resources departments, state land departments, consulting groundwater hydrologists, well drillers, etc.

PROBLEMS

1.1 Select a certain area and collect information about the local groundwater situation and aquifers, using the sources listed in Section 1.4.

1.2 Draw a schematic, vertical cross section through the earth's surface, showing a mountain, coastal plain, ocean, and underlying aquifers. Sketch in as many exchanges between atmospheric water, surface water, and underground water as you can think of, starting with rainfall in the mountains and ending with the flow of rivers and groundwater into the ocean. Complete the resulting *hydrologic cycle* by showing the return of evaporation from the ocean to the land via atmospheric precipitation.

REFERENCES

Carmi, I., Y. Noter, and R. Schlesinger, 1971. Rehovot radiocarbon measurements, 1. *Radiocarbon* **13:** 412–419.

Feth, J. H., 1973. Water facts and figures for planners and managers. *U.S. Geol. Survey Circular 601-1,* Washington, D.C., 30 pp.

Legrand, H. E., and V. T. Stringfield, 1973. Karst hydrology—a review. *J. Hydrol.* **20:** 97–120.

McCabe, L. J., J. M. Symons, R. D. Lee, and G. G. Robeck, 1970. Survey of community water supply systems. *J. Am. Water Works Assoc.* **62:** 670–687.

McGuinness, C. L., 1963. The role of groundwater in the national water situation. *U.S. Geol. Survey Water Supply Paper 1800,* Washington, D.C., 1121 pp.

Meinzer, O. E., 1923. The occurrence of groundwater in the United States, with a discussion of principles. *U.S. Geol. Survey Water Supply Paper 489,* Washington, D.C., 321 pp.

Münnich, K. O., and J. C. Vogel, 1959. C-14 Alterbestimmung von Süszwasser-Kalkablagerung. *Naturwissenschaften* **48:** 168–169.

Münnich, K. O., and J. C. Vogel, 1962. Untersuchungen an Pluvialen Wassern der Ost-Sahara. *Geol. Rundsch.* **52:** 611–624.

Nace, R. L., 1960. Water management, agriculture, and groundwater supplies. *U.S. Geol. Survey Circular 415,* Washington, D.C., 12 pp.

Pearson, F. J., Jr., M. S. Bedinger, and B. F. Jones, 1972. Carbon-14 ages of water from the Arkansas hot springs. *Proc. 8th Internat. Conf. on Radio Carbon Dating,* Wellington, N.Z., D19–D30.

Pearson, F. J., Jr., and D. E. White, 1967. Carbon 14 ages and flow rates of water in Carrizo Sand, Atascosa County, Texas. *Water Resour. Res.* **3:** 251–261.

Smith, D. B., R. A. Downing, R. A. Monkhouse, R. L. Otlet, and F. J. Pearson, 1976. The age of groundwater in the Chalk of London Basin. *Water Resour. Res.* **12:** 392–404.

Thatcher, L. M., M. Rubin, and G. Brown, 1961. Dating desert groundwater. *Science* **134**(3472): 105–106.

Thomas, H. E., 1952. Ground-water regions of the United States—their storage facilities. *U.S. 83d Congress, House Interior and Insular Affairs Comm., The Physical and Economic Foundation of Natural Resources* **3:** 78 pp.

van der Leeden, Frits, 1975. *Water Resources of the World.* Water Information Center, Port Washington, New York, 568 pp.

Wallick, E. I., 1973. "Isotopic and Chemical Considerations in Radiocarbon Dating of Groundwater within the Arid Tucson Basin, Arizona." Ph. D. thesis, Univ. of Arizona, Tucson, Dept. of Geosciences, 184 pp.

TWO

PHYSICAL PROPERTIES OF AQUIFERS AND VADOSE ZONES

Aquifers and vadose zones are porous media that store, transmit, and, in the case of aquifers, yield water. Clays and other unconsolidated materials may compress or expand, depending on whether groundwater is withdrawn or added. In order to quantify these processes, physical properties of aquifers and vadose zones have been expressed in a number of parameters, most of which will be presented in this chapter. A very important parameter is hydraulic conductivity, which is discussed in Chapter 3. Compression of unconsolidated underground materials is treated in Chapter 9.

2.1 PRESSURE HEAD

Positive-Pressure Heads and Water Tables

The pressure head of groundwater at a given point in an aquifer is the height to which water will rise in an open, vertical tube that is inserted down to that point (Figure 2.1). Such a tube is called a *piezometer*. The pressure head expresses the gage pressure of the water in terms of the height of a water column (gage pressure is the pressure with respect to atmospheric pressure). When the groundwater is at rest, or moving in a horizontal direction only, the pressure head at a given point in an aquifer is equal to the vertical distance of that point below the water table if the

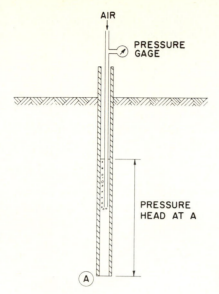

Figure 2.1 Piezometer for measuring pressure head, and bubble tube for measuring water level.

aquifer is unconfined, or below the piezometric surface if the aquifer is confined (see Section 3.7 for the effect of vertical flow).

The water level in a piezometer can be measured with a chalk-covered tape, an electric probe, a bell-shaped device that is lowered on a string until contact with the water is observed, or a bubble tube. The latter is a tube that is inserted into the piezometer until it is some distance below the water level (Figure 2.1). Air is pumped down the bubble tube, using a hand pump or slowly applying it with a cylinder of compressed air. This increases the air pressure in the bubble tube and forces the water level inside the tube down. When air starts bubbling up through the water in the piezometer, the air pressure (expressed in terms of water-pressure head) in the bubble tube equals the distance of the bottom of the bubble tube below the water level in the piezometer. Knowing the length of the bubble tube, the height of the water level in the piezometer can then be calculated. Continuous records of water levels in piezometers can be obtained with float-operated water-stage recorders or pressure transducers. Where pressure heads in aquifers change rapidly, piezometer diameters should be as small as possible to minimize the response time or lag.

Water levels in piezometers do not always indicate true heights of water tables or piezometric surfaces, even if vertical-flow components do not occur. Changes in barometric pressure, for example, are not always immediately transmitted to the groundwater because the air permeability of wet or saturated portions in the vadose zone may be very small. Barometric pressure changes are, however, directly transmitted to water levels in piezometers, which are open to the atmosphere. Thus, increases in barometric pressure may cause temporary drops in water levels in piezometers until the increased atmospheric pressure has been transmitted through the vadose zone to the groundwater. The position of the groundwater table itself is, of course, not affected by changes in atmospheric pressure. Decreases in barometric pressure can cause water levels in piezometers to rise. Each centimeter of change in barometric pressure (mercury scale) can thus cause the water level in a piezometer to temporarily deviate 13.6 cm from the true water table. Cyclic variations in barometric pressure are caused by the effects of the sun's transit on solar heating and, to a minor extent, on gravitational attraction. Because of this, atmospheric pressures tend to fluctuate with a resonance of 12 h, showing maximum values at 1000 and 2200 h and minimum values at 400 and 1600 h. Accordingly, water levels in piezometers tend to be lowest at 1000 and 2200 h and highest at 400 and 1600 h.

Temporary air-pressure increases in vadose zones and resulting discrepancies between groundwater-table positions and water levels in piezometers can also be caused by infiltration of water into the surface soil over a large area (rainfall, flooding, irrigation, etc.). The downward-moving wetting front (see Section 8.1) will then compress underlying air in the vadose zone (see also Section 2.9). When the increased air pressures are transmitted to the water table, water levels in piezometers will rise while the water-table position itself does not change. Dixon and Linden (1972) reported air-pressure increases of as much as 19 cm water below wetted zones in irrigated fields. They also found that differential air-pressure increases, resulting from nonuniform soil conditions and air escape routes, actually produced lateral flow of groundwater.

Water levels in piezometers can also give erroneous readings of water-table positions or piezometric surfaces if the bottom of the piezometer is in a layer of clay or other fine-textured material that is undergoing compression by increased loading, due, for example, to groundwater withdrawal. Restricted internal drainage of the compressing layers then produces excessive pressures in the pore water, which in turn causes the water level in the piezometer to be higher than the actual water table or piezometric surface (see Section 9.7).

True water-table fluctuations can be caused by transfer of water between the capillary fringe and the water table, resulting from changes in atmospheric pressure. An increase in barometric pressure, for example, can produce compression of entrapped air in the capillary fringe. The resulting increase in water content in the capillary fringe then is made up by flow from the groundwater, causing a drop in the water table. This mechanism produced daily fluctuations of 1.5 to 6 cm in summer and 0.5 and 1 cm in winter for a shallow water table in Utah (Turk, 1975).

Water tables or piezometric surfaces may also fluctuate under the influence of earth tides. Like the familiar ocean tides, earth tides are caused by bulging of the earth due to the gravitational attraction of the sun and moon. Earth tides produce an expansion (commonly called *dilatation*) of aquifers and other subsurface formations, resulting in groundwater declines. Water levels in wells reaching into confined aquifers have been observed to decline several centimeters, with a maximum of 17 cm (Bredehoeft, 1967). Conversely, the opposite action of bulging causes compression of aquifers, and groundwater levels will rise. For unconfined aquifers, the porosity must be small or the thickness of the aquifers must be large before earth tides can produce measurable groundwater fluctuations. Ocean tides affect groundwater levels in coastal aquifers. The amplitude of such tides is reduced and the phase is delayed as the tidal fluctuation is propagated inland through the aquifer (see Section 5.4).

Negative Pressure Heads

The pressure of water in vadose zones is negative, due to the surface tension of the water. This produces a negative pressure head which cannot be measured with open piezometers. Instead, a *tensiometer* is used, which consists of a vertical tube that is closed at the top and has a porous membrane (usually a ceramic plate or cup) at the bottom (Figure 2.2). The tensiometer is filled with water that is in contact with the water in the vadose zone through the pores of the membrane. At equilibrium, the water pressure inside the tensiometer, which can be measured with a pressure gage at the top, will be equal to that of the vadose water. If the water in the vadose zone is in equilibrium with the water table (hence, no vertical flow), the negative pressure head at a given point in the vadose zone will be equal to minus the height of that point above the water table. Because there is almost always some vertical flow in the vadose zone and equilibria take a long time to establish, negative pressure heads equal to minus the height above the water table may be found only relatively close to the water table.

The ceramic or other porous material at the bottom of the tensiometer is selected so that the pores are much smaller than those in the surrounding soil. The porous membrane will then remain saturated (even though the vadose zone at the contact point may be unsaturated), and air will not enter the tensiometer. Prior to installation, the porous bottom of the tensiometer is saturated (usually under vacuum). The tensiometer is then filled with water and installed to the desired depth, making sure that the porous bottom is in intimate contact with the soil of the vadose zone.

The pressure head of the water inside the tensiometer should be measured with a device that requires small volume displacement per unit change in pressure

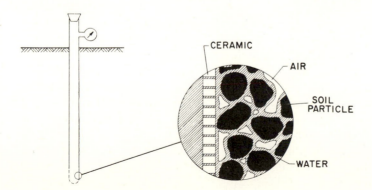

Figure 2.2 Tensiometer with vacuum gage for pressure measurement. The expanded section shows how the water inside the tensiometer is in contact through pores in the ceramic cup with the water in the unsaturated soil.

head. This will minimize the amount of water that has to move through the porous membrane of the tensiometer as the pressure head in the soil water changes, thus reducing the time lag between the measured negative pressure head and the actual pressure head of the water in the vadose zone. Suitable devices for measuring the pressure head in the tensiometer are small-bore mercury manometers, vacuum gages, or pressure transducers. If pressure transducers are used, continuous records of the measured pressure head can be obtained (Bianchi, 1962; Watson, 1967; and Rice, 1969).

At ambient temperature, water will reach its boiling point when the pressure head has dropped to about -1000 cm of water. Thus, the tensiometer as shown in Figure 2.2 cannot be used for pressure heads that are more negative than about -800 cm water because the water inside the tensiometer will " break." Pressure heads less (more negative!) than -800 cm water are usually not very important in subsurface hydrology. They can be measured, however, with special techniques commonly used in agronomic studies (Richards, 1965b). A recently developed technique consists of measuring the rate of dissipation of an electrically applied heat pulse in a ceramic block buried in the soil (Phene et al., 1971). Since the water content of the block decreases as the pressure head of the water in the surrounding soil decreases, and since the rate of the heat dissipation decreases as the water content of the block decreases, the apparatus can be calibrated to yield the pressure head of the water in the soil around the block.

In soil physics and agronomy, a negative pressure head of water in soil is often called *suction* or *tension*, which is positive. Thus, as the water content of a soil decreases and water is held more tightly by the soil, the pressure head of the water decreases (becomes more negative), but the suction or tension increases.

Tensiometers can also measure positive-pressure heads. They should be used, therefore, when both negative and positive pressure heads may occur. If only positive-pressure heads are expected, the simpler piezometer is preferable. A more detailed discussion of piezometers and tensiometers, including response time, is given by Bouwer and Jackson (1974).

2.2 TEXTURE OF UNCONSOLIDATED MATERIALS

The soils, sands, or gravels that occur in unconsolidated aquifers, aquicludes, aquitards, and vadose zones are classified according to the size of the individual particles. A commonly used classification scheme is the following system developed by the Soil Survey Staff of the U.S. Department of Agriculture (1951):

Clay	0–2 μm
Silt	2–50 μm
Very fine sand	50–100 μm
Fine sand	0.1–0.25 mm
Medium sand	0.25–0.5 mm
Coarse sand	0.5–1 mm
Very coarse sand	1–2 mm

Particles larger than 2 mm in size are classified as gravel, which is further differentiated as follows (ASTM D 2488; see, for example, Soil Conservation Service, 1973):

Fine gravel	0.6–1.9 cm
Coarse gravel	1.9–7.6 cm
Small cobbles	7.6–15.2 cm
Large cobbles	15.2–30.5 cm
Boulders	> 30.5 cm

According to ASTM D 2488, the fraction 2.0–4.76 mm is called *coarse sand*, which differs from the USDA classification.

The particle-size distribution is determined by washing a dispersed sample of the material through a 50-μm sieve. The material remaining on the sieve is dried and passed through various sieve sizes to obtain the weights of the different sand fractions. The size distribution of the particles that passed through the 50-μm sieve is determined by measuring the rate of settlement of these particles in a suspension, using Stokes' law (Day, 1965). This requires periodic measurement of the decrease in density at a given depth in the suspension as particles settle out. A hydrometer is commonly used for this purpose. Another method utilizes a pipette for periodically drawing a small sample from a certain depth in the suspension. The samples are then analyzed for suspended-solids content.

The results of the particle-size analysis, usually called *mechanical* analysis, are plotted on semilogarithmic paper to yield a distribution curve (Figure 2.3). The particle size is on the log-scale abscissa, whereas the "percentages smaller than" are on the regular-scale ordinate. These percentages express the sample fraction on a dry-weight basis. The right ordinate in Figure 2.3 shows the percentages greater than a given size, indicating the percentage by weight of the material that would be retained on a sieve with openings of that size. The sieve-opening size

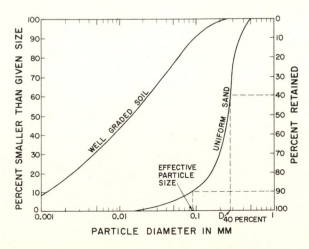

Figure 2.3 Particle-size distribution for a uniform sand and a well-graded soil.

retaining 90 percent of the material is called the *effective particle size* $D_{90\%}$ (Figure 2.3). The sieve size retaining 50 percent of the material is called the *average particle size* $D_{50\%}$.

Uniform materials consist of particles of predominantly one size. Such materials yield distribution curves with a rather abrupt reduction, almost like a step function (sand curve in Figure 2.3). Well-graded materials have particles of many different sizes, and their distribution curve is more gradual and S-shaped (soil curve in Figure 2.3). The uniformity of a granular material is described by the uniformity coefficient, which is the ratio between the sieve-opening sizes retaining 40 and 90 percent of the material, or $D_{40\%}/D_{90\%}$. Thus, the more uniform the material is, the lower its uniformity coefficient. The minimum value of the uniformity coefficient is 1, indicating that $D_{90\%} = D_{40\%}$ and that the material primarily consists of one particle size. Well-graded materials usually have uniformity coefficients of 5 to 10.

Depending on the relative amounts of sand, silt, and clay in a soil, the texture of the material is classified as sand, silt, loam, clay, or any of the various intermediate textural classes such as sandy loam, clay loam, silt loam, etc. The textural classification of a certain soil material can be evaluated from the soil-textural triangle (Figure 2.4) prepared by the Soil Survey Staff (1951). For example, a soil consisting of 60 percent sand, 30 percent silt, and 10 percent clay is a sandy loam. A soil with 10 percent sand, 60 percent silt, and 30 percent clay is a silty clay loam.

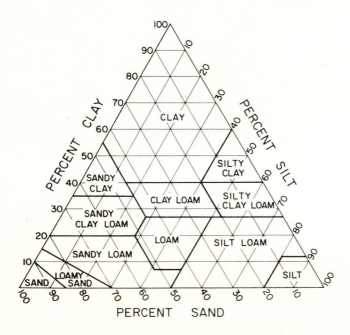

Figure 2.4 Triangular chart showing the percentages of sand, silt, and clay in the basic soil textural classes. (*From Soil Survey Staff*, 1951.)

2.3 STRUCTURE OF CLAY

If the soil contains appreciable amounts of clay, an important feature of the soil in addition to its texture is its structure (Baver et al., 1972). The structure of the soil depends on the arrangement of the clay particles. Depending on the type of ions that are adsorbed to the clay, clay particles can be dispersed as individual particles or they can be flocculated to form flocs and structural units (soil aggregates) that may be several millimeters in size. A soil with its clay in flocculated condition behaves like a coarser-textured soil than when its clay is dispersed. Whether a clay is dispersed or flocculated depends on how far the individual clay particles are separated from each other by the thickness of the layer of adsorbed cations surrounding each particle (Van Olphen, 1963). If the clay particles can be close together, the attractive van der Waals forces are dominant and the clay is flocculated. If the layer of adsorbed cations is thick and the clay particles are kept some distance apart, the repulsive electrostatic forces are dominant and the clay is dispersed.

Clay particles are negatively charged colloidal minerals, which can adsorb cations (H^+, Na^+, K^+, NH_4^+, Ca^{2+}, Mg^{2+}, etc.) from the soil water or, rather, the soil solution. The negatively charged surface of a clay particle and the surrounding mantle of adsorbed cations are called the *double layer*. If the predominant cation in the double layer is Na^+, the individual clay particles cannot come close together because the Na^+ ions are surrounded by water molecules or are hydrated, producing a thick double layer (Figure 2.5). Also, the monovalent Na^+ ions do not effectively mask the negative charges of the clay particles themselves. For Na^+ clays, therefore, the repulsive electrostatic forces between the negatively charged particles exceed the attractive van der Waals forces, causing the clay particles to exist as separate particles in a dispersed or deflocculated condition. This dispersion occurs already if 10 to 20 percent of the adsorbed cations consists of Na^+. Soils with Na clay have a poor "structure." They have a tendency to seal.

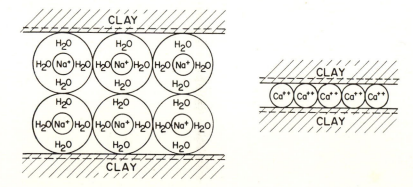

Figure 2.5 Schematic showing large distance between negatively charged clay particles with hydrated Na^+ ions, and small distance between clay particles with Ca^{2+} ions in their double layer.

They are low in permeability, are sticky and amorphous, and become hard upon drying.

If the cations in the double layer around the clay particles are mainly Ca^{2+} and Mg^{2+}, the clay particles can come much closer together (Figure 2.5). This is because Ca^{2+} and Mg^{2+} are not hydrated and are thus of smaller size. Also, these cations are divalent, producing better masking of the negative charges of the clay particles. Because of this, the clay particles can come so close together that they share Ca^{2+} ions in their double layers. These ions then act as a bond between the clay particles (Figure 2.5). Since the van der Waals forces increase very rapidly with decreasing distance between the clay particles, the attractive forces are dominant and the clay particles cling together to form flocs and aggregates. A soil with predominantly Ca or Mg clay behaves like a coarser-textured soil and has a "good" structure. Such soils are more permeable and friable than soils with dispersed clay.

Clay can be converted from a dispersed state to a flocculated condition by changing the adsorbed ions from Na^+ to Ca^{2+} or Mg^{2+}. This can be done through the process of cation exchange, adding a soluble Ca salt to the soil. The reverse, changing a flocculated clay to a dispersed clay, occurs when a Ca-Mg clay is brought into contact with dissolved Na salts, and Ca^{2+} and Mg^{2+} in the double layer are exchanged for Na^+. This is done when preparing a soil for mechanical analyses, so that soil aggregates and clay flocs are dispersed into individual particles. It is accomplished by adding a dispersant such as sodium metahexaphosphate to the soil suspension and stirring (Day, 1965).

A flocculated clay and associated "good" structure of the soil are preferred for agriculture. Thus, irrigation water should not contain too much Na^+ because this could cause the clay to become dispersed and the soil structure to deteriorate. An example of what can happen to a soil when the predominant ion in the double layer changes from Ca^{2+} to Na^+ is "the destruction of acres of prime farm land" caused by using the wrong neutralizer to correct the effects of an accidental acid spill from a derailed train in Canada (*Ground Water Newsletter*, 1974). Instead of using a lime or other calcium salt, sodium carbonate and sodium hydrochloride were applied to the land to raise the pH of the soil. This caused dispersion of the clay and associated deterioration of the soil structure. Conversely, Ca clays have been deliberately dispersed and sealed by applying soda ash to reduce seepage losses from ponds (Reginato et al., 1973).

Whether a clay is dispersed or flocculated also depends on the salt concentration of the soil solution. A high concentration of NaCl in the soil water, for example, could still cause a Na clay to be flocculated. The high ionic strength of the solution in that case compresses the double layer, allowing the clay particles to be sufficiently close to each other for the van der Waals forces to be dominant. For this reason, clay in a soil flooded with seawater will remain flocculated. Deflocculation will start only when seawater is leached out by rain, which causes the NaCl concentration in the soil solution to decrease. When this happens, lime or calcium chloride should be added to the soil to minimize deterioration of its structure.

Clay in saline aquifers will also be in flocculated condition. Dispersion of the clay may occur, however, when the saline groundwater is replaced by fresh water. This happened in a recharge project near Norfolk, Virginia, where fresh water was pumped through an injection well into a confined, brackish aquifer for temporary storage (Bouwer, 1974, and references therein). The reduction in salt concentration caused dispersion of the clay, which in turn resulted in a 75 percent reduction in the capacity of the recharge well. To prevent dispersion of clay in such a case, a solution of polyvalent cations—for example, zirconium or aluminum—should be pumped into the aquifer first. Such ions are tightly adsorbed by the clay particles and are not readily exchanged for Na^+ ions. For additional discussion of effects of ionic composition of soil solution on structure and hydraulic conductivity of soil, see Section 3.4.

2.4 POROSITY

The porosity of a soil or rock material is the percentage of the total volume of the material that is occupied by pores or interstices. These pores may be filled with water if the material is saturated, or with air and water if it is unsaturated. The porosity is determined on an undisturbed sample of the material. For soil or other unconsolidated material, such samples can be obtained with one of the various techniques developed for this purpose (Reeve, 1957; see also Section 6.2). If the material is consolidated, a drilled core sample is used. The total volume V_t of the undisturbed sample is determined. The sample is oven-dried to remove the water (24 h at 105°C is usually sufficient) and the dry weight W_d is determined. Dividing W_d by the density of the soil or rock material gives the volume V_s of the solid phase of the sample. The porosity n is then calculated as

$$n = \frac{V_t - V_s}{V_t} 100 \qquad (2.1)$$

The density of the solids depends on the mineral composition of the soil or rock. For soils and gravels, quartz is the predominant mineral, and a density of 2.65 g/cm^3 is commonly used for the density of the solid phase (Baver et al., 1972). Densities of rock materials, such as limestone and granite, are usually in the range of 2.7 to 2.8 g/cm^3. Basalt may have a density of close to 3 g/cm^3.

The porosity of a granular medium with spherical, uniform particles can be calculated as 47.6 percent if the spheres are packed loosely in a cubical array, and as 26.0 percent if the spheres are packed densely in a rhombohedral array. Uniform materials have a greater porosity than well-graded materials because in the latter, smaller particles occupy the voids between larger particles. Fine-textured soils tend to have a greater porosity than coarse-textured soils. The porosity of solid rock is very small. Fractured or weathered rock and porous materials like tuffs and lavas have higher porosities. Carbonate rock usually has some original porosity, including the porosity formed by the diagenetic change from calcite to dolomite, and secondary porosity due to solution of carbonate (solution channels, karst phenomena). Porosities of specific materials were listed by Davis (1969). Approximate ranges are:

	Porosity, percentage
Silts and clays	50–60
Fine sand	40–50
Medium sand	35–40
Coarse sand	25–35
Gravel	20–30
Sand and gravel mixes	10–30
Glacial till	25–45
Dense, solid rock	< 1
Fractured and weathered igneous rock	2–10
Permeable, recent basalt	2–5
Vesicular lava	10–50
Tuff	30
Sandstone	5–30
Carbonate rock with original and secondary porosity	10–20

2.5 VOID RATIO

The *void ratio* is the term more commonly used in soil mechanics to express the pore volume of the soil. It is defined as the ratio between the volume of the voids V_v and the volume of the solids V_s. Its symbol is usually e. Thus, $e = V_v/V_s$. The value of e varies from about 0.7 for dense sands and gravels to about 1.3 for unconsolidated clays. The relation between the porosity n and the void ratio is

$$n = \frac{e}{e+1} \tag{2.2}$$

2.6 BULK DENSITY

The bulk density is the density of the total soil or rock material, solids and voids, after drying. It is calculated as the dry weight W_d of a sample divided by its total volume V_t in undisturbed condition. The bulk density is equal to $(1 - n)\rho$, where n is the porosity expressed as a fraction and ρ is the density of the solid phase (2.65 g/cm^3 for most sands and soils).

2.7 WATER CONTENT

The water content of a soil or other porous material can be expressed per unit dry weight or per unit volume of the material. If it is expressed on a dry-weight basis, a sample of the material (a disturbed sample is sufficient) is weighed to obtain the wet weight W_w. The sample is then dried to obtain the dry weight W_d. The percentage water content on a dry-weight basis, or gravimetric water content, is calculated as $100(W_w - W_d)/W_d$, expressing all weights in grams.

The volumetric water content θ, which is the more useful parameter in subsurface hydrology, is determined by weighing and drying an undisturbed sample of the material. If the total volume of this sample is V_t cm^3, the weight before drying is W_w, and the weight after drying is W_d, the volumetric water content is calculated as $(W_w - W_d)/V_t$. This gives θ as a volume fraction. Multiplying by 100 yields θ as a percentage.

In addition to sampling and weighing techniques, the water content of subsurface materials can be measured indirectly. For example, a technique that has been used for many years consisted of burying small gypsum blocks in which two electrodes have been embedded. The electrodes are connected via a cable to the soil surface so that the electrical resistance of the gypsum block can be measured. Since this resistance changes with the water content of the block, which in turn changes with the water content of the surrounding soil, the soil's water content can be determined. The method requires calibration of the blocks for the soil in question to determine the relation between the electrical resistance of the block and the water content of the surrounding soil. For a more detailed description of this and other indirect techniques, reference is made to Gardner (1965).

The indirect methods are of special value where the water content of soil or other subsurface materials needs to be measured in a nondestructive manner over a certain period of time. A more recently developed technique is the neutron method, which requires an aluminum or steel access tube (usually 5 cm in diameter) placed vertically in the soil. The outside of the tube should be in good contact with the soil. A fast-neutron source is lowered into the tube (Figure 2.6) to the desired depth. When fast neutrons enter soil, they are moderated by hydrogen atoms and become slow neutrons. These slow neutrons are scattered in all directions. Since the main source of hydrogen in the soil is water, the density of the slow neutrons is a function of the amount of water in the soil. The density of the slow-neutron backscatter is measured by a slow-neutron detector, which is usually placed a few centimeters below the neutron source in a common probe (Figure 2.6). Calibration of the technique in soils of known water content is needed to relate the slow-neutron count to the water content. This can be done in the field by sampling the soil adjacent to a neutron access tube and determining its volumetric water content. Once calibrated, however, the neutron technique can be used over and over again to measure water content in vadose zones or aquifers in relation to depth and time. A recording neutron instrument has been constructed for neutron logging of wells (see Section 6.2).

The neutron method measures the water content of the soil in a sphere with a

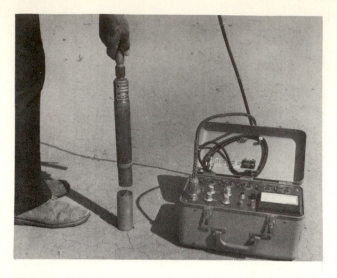

Figure 2.6 Probe (source and detector), scaler, and access tube for measuring water content of soil with neutron technique.

radius of about 15 cm around the neutron source. The size of this sphere is somewhat dependent on the water content of the soil (the drier the soil, the larger the sphere of influence). When the technique was first developed, the source of fast neutrons was ^{226}Ra mixed with beryllium. A 5-mCi source was commonly used. Nowadays, the preferred source is ^{241}Am, because it is safer (emits less gamma radiation) and can be used in stronger sources. The detector for the slow neutrons is usually a tube filled with lithium-enriched boron trifluoride gas. When hit by slow neutrons, short electric pulses are given off which can be counted by a scaler or ratemeter (Figure 2.6). The water content of the soil around the probe is then determined from the count rate.

The 30-cm sphere for which the neutron method gives some average water content is too large if it is important to measure water content in much smaller depth increments or closer to the surface. This requires greater resolution, which can be obtained with the gamma-ray technique. With this method, a probe with a fast gamma-ray source (usually ^{137}Cs) is lowered into an access tube. The fast gamma rays are attenuated by all matter, solid as well as water. Since water content is the only factor that changes in a soil or other subsurface material, a change in gamma-ray attenuation is associated with a change in water content. Field or laboratory calibration is necessary to convert gamma-ray attenuation into water-content data.

The attenuated gamma rays are measured with a detector (usually a sodium iodide scintillation crystal) which can be placed in a second access tube at the same depth as the gamma-ray source. The detector can be collumated so that water content can be measured in small depth increments—for example, 0.5 to 1 cm. The distance between the source and detector access tubes is on the order of

0.5 m. The detector can also be placed below the source, in a common probe, to measure the intensity of the backscattered gamma radiation, similar to the neutron technique. In this construction, the gamma-ray technique is used in well logging to obtain a density log of the various geologic formations (see Section 6.2).

Problems associated with the gamma-ray technique for water-content measurement are swelling and shrinking of soil upon wetting and drying, respectively, which cause errors because density changes are not due to water-content changes alone. Earlier problems with the technique were associated with electronic drift and temperature sensitivity in the instrumentation. At present, the gamma-ray technique is essentially a research tool not readily applicable to routine measurements of water content. This may change in the future, however, as the technique improves with continued research. For additional information regarding radiometric techniques for water-content measurements, reference is made to Gardner (1965) and Bouwer and Jackson (1974).

2.8 SATURATION PERCENTAGE

The saturation percentage is the percentage of the pore space that is occupied by water. A water-saturated soil has a saturation percentage of 100 percent. The saturation percentage is calculated as the volumetric water content divided by the porosity, or $100\theta/n$.

2.9 SOIL-WATER CHARACTERISTIC AND WATER IN THE VADOSE ZONE

The relation between water content and the negative-pressure head of the water for a given soil is called the *water characteristic*. This relation is usually expressed in graphical form. As discussed in Section 2.1, the negative pressure head in the vadose zone above an unconfined aquifer is numerically equal to the vertical distance above the water table if vertical flow does not occur. Thus, a plot of how the volumetric water content decreases with distance above the water table would be the water characteristic of the soil material in the vadose zone—assuming, of course, uniform soil and no vertical flow.

Examples of equilibrium water-content distributions in the vadose zone, and hence of soil-water characteristics, are shown in Figure 2.7, left. Pressure heads in this figure are equal to minus the distance above the water table. The negative pressure head whereby the water content undergoes its first major reduction and air becomes continuous in the pores is called the *air-entry value* of the soil. Air-entry values may vary from -5 to -30 cm water for coarse to medium sands, from -30 to -70 cm water for fine sands, and -70 to -200 cm water or less for dispersed clay soils. For materials with relatively uniform particle size, the water content usually decreases rather abruptly and significantly once the air-entry value is reached. These materials thus have a well-defined capillary fringe. The

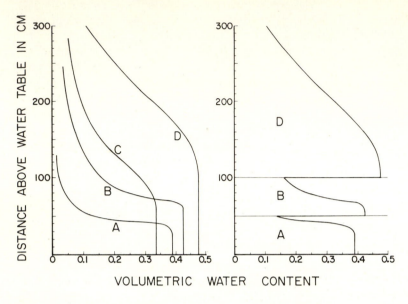

Figure 2.7 Schematic equilibrium water-content distribution above a water table (left) for a coarse uniform sand (A), a fine uniform sand (B), a well-graded fine sand (C), and a clay soil (D). The right plot shows the corresponding equilibrium water-content distribution in a soil profile consisting of layers of materials A, B, and D.

equilibrium height of this fringe is numerically about equal to the air-entry value. In well-graded materials, the water content decreases more gradually after the air-entry value is reached. The capillary fringe above the water table in such materials tends to be higher but with a less distinct top than in uniform materials (compare, for example, curves *A* and *C* in Figure 2.7, left).

If the vadose zone consists of layers of different soil materials, the negative pressure head at equilibrium conditions (no vertical flow) will still be equal to the vertical distance above the water table, regardless of the layering. The water content at each point, however, depends on the water characteristic of the particular soil material at that point. Thus, if the vadose zone consists of different soil layers, an irregular water-content distribution may occur above the water table. This is shown in Figure 2.7, right, where the different soil materials of the left graph are arranged in various layers above the water table. Some fine-textured layers in the vadose zone may actually be saturated, while coarser-textured materials above and below it may be unsaturated. Thus, the equilibrium water content in the vadose zone does not always decrease with increasing distance above the water table, and vadose zones should not be called unsaturated zones.

A saturated, fine-textured layer sandwiched between unsaturated materials acts as a barrier to air movement in the vadose zone. If such a layer occurs at a relatively shallow depth and heavy rainfall or flooding causes water to infiltrate at a high rate into the soil, the air in the soil between the downward-moving wetting front and the saturated layer cannot escape vertically and is compressed (Figure

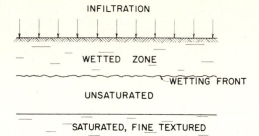

Figure 2.8 Infiltration into an unsaturated zone with restricted vertical escape of air.

2.8). The resulting pressure increase is transmitted to the water in the unsaturated zone beneath the advancing wet front and to the water in the underlying saturated layer, where it can increase the negative pressure heads to the extent that they become positive. The development of positive pressure heads of water in the vadose zone has surprised hydrologists carrying out piezometric or tensiometric surveys. Observed air-pressure increases below downward-moving wetting fronts were about 20 cm water in a border-irrigated field (Dixon and Linden, 1972) and 14 to 50 cm water in laboratory studies (Wilson and Luthin, 1963; Peck, 1965).

The equilibrium water-content distribution above a static water table depends on whether the water table was falling or rising before reaching a constant level (Figure 2.9). Specifically, the water content at a given negative pressure head is higher when this pressure head is reached by removing water from a saturated

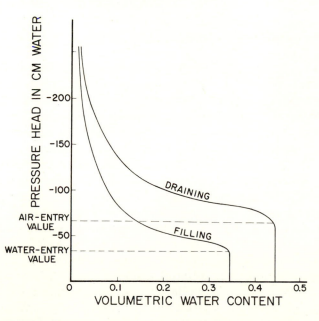

Figure 2.9 Schematic of water-content distribution above a water table after the water table was falling (soil pores drained) and rising (soil pores filled).

soil (as occurs above a falling water table) than when this pressure head is reached by adding water to an unsaturated soil (as occurs above a rising water table). This phenomenon is called *hysteresis*, and it is due to entrapped air that remains for some time in the soil pores after the soil has been wetted (see also Section 7.7). For a rising water table, the entrapped air will persist even if the water table has risen well above it. Adam et al. (1969) reported that 5 to 50 percent of the pore volume may contain air after the initial stage of wetting has been completed.

The pressure of the entrapped air below a water table is greater than atmospheric pressure because of the pressure of the groundwater surrounding the trapped air. Since the concentration of dissolved air in water increases with increasing pressure of the air with which the water is in contact, the concentration of the dissolved air increases with depth below the water table if entrapped air is present. The resulting upward gradient of the dissolved-air concentration then causes upward movement of dissolved air in the groundwater and eventual disappearance of the entrapped air. While this may take a long time (Adam et al., 1969; McWhorter et al., 1973), it explains why aquifers with old water are usually completely saturated. Near the soil surface, oxygen uptake by bacteria and other biologic activity may also contribute to the disappearance of entrapped air. Since entrapped air can occur below the water table, especially if the groundwater is relatively new, it is incorrect to call the groundwater zone the saturated zone.

When the pressure head of the water in a saturated soil decreases (becomes more negative), the negative pressure head whereby air first enters the soil and becomes essentially continuous in the soil pores is called the *air-entry* value. Conversely, when an unsaturated soil is wetted and the pressure head of the water in the soil increases (becomes less negative), the point where water has displaced most of the air and becomes essentially continuous in the pores is called the *water-entry* value or *air-exit* value (Figure 2.9). Since the water-entry value is about one-half the air-entry value for a number of granular materials (Bouwer, 1966), the capillary fringe above a rising water table is about one-half as high as that above a falling water table, all other conditions being the same. In soil physics, the water-content characteristic obtained by removing water from a wet soil is called the characteristic for *drying* or *desorption*. The curve obtained by adding water to a dry soil is called the characteristic for *wetting* or *sorption*.

Air entrapment below a rising water table causes the short-term storage capacity of an unconfined aquifer to be less than that indicated by the water yield of the aquifer when the water table was lowered. Wesseling (1958), for example, found that the water-table rise in a shallow, unconfined aquifer was twice the distance predicted on the basis of the water yielded by the soil when it was draining. Hysteresis effects can also be important in watershed hydrology (Royer and Vachaud, 1975) and in the study of irrigated soils (Watson et al., 1975).

The water in the vadose zone above an unconfined aquifer will seldom, if ever, be in complete equilibrium (no vertical flow) with the water table. Infiltration and downward movement of water from the soil surface, evaporation and water uptake by plants, and movement of the water table itself will almost always cause some vertical-flow components in the vadose zone. When vertical flow is present, the negative pressure head will not be equal to the distance above the water table and it will not be possible to determine the soil-water characteristic from the water-content distribution in the vadose zone. However, if negative pressure heads and the water content of the soil are measured simultaneously at a given depth (using tensiometers and the neutron method, for example), the relation between water content and negative pressure head can be obtained by taking measurements over a sufficient range in water content. If the natural changes in water content and pressure head are insufficient to yield the desired range, the soil may be wetted artificially and then allowed to drain.

Water-content characteristics can also be determined in the laboratory by placing undisturbed samples of the soil on a water-saturated ceramic plate or similar porous membrane. Different negative pressure heads in the soil water of the samples are then created by applying different water suctions to the bottom of each ceramic plate or by increasing air pressures above the samples (Richards, 1965a). When the water in the sample has reached equilibrium with the pressure head, the water content of the sample is determined. Water-content characteristics can be obtained in this manner for drying (desorption) and wetting (sorption) to include hysteresis effects. In soil physics, negative pressure heads are usually expressed as tension or suction, which are positive parameters, and water-content characteristics are plotted with the water content on the ordinate and the tension or suction on the abscissa (see Section 7.7).

Before the concept of water-content characteristics was developed, the water in the vadose zone was classified in such terms as *capillary* water, *funicular* water, *pellicular* water, and *hygroscopic* water (in order of decreasing water content). These terms are not very relevant, and they are seldom used in practice. Water content and pressure head are much more meaningful for describing the " condition " of water in the vadose zone, particularly as it relates to the movement of water in this zone (see Section 7.7).

2.10 STORAGE COEFFICIENT, SPECIFIC YIELD, AND FILLABLE POROSITY

The *storage coefficient* of an aquifer is defined as the volume of water yielded per unit horizontal area and per unit drop of water table (unconfined aquifers) or piezometric surface (confined aquifers). Thus, if an unconfined aquifer released 4 m^3 water for a water-table drop of 2 m over a horizontal area of 10 m^2, the storage coefficient is 0.2 or 20 percent. For unconfined aquifers, the storage coefficient can also be called *specific yield*, which is the volume of water released from a unit volume of saturated aquifer material drained by a falling water table.

Unconfined aquifers yield water to wells or other collection facilities because of drainage of pore space and air replacing water in the dewatered zone as the water table drops. The volume of water yielded by drainage of pores can be

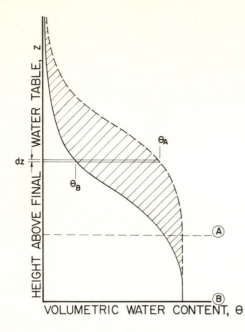

Figure 2.10 Determining specific yield from equilibrium water-content distributions above two water-table positions (A and B).

evaluated by plotting the equilibrium volumetric-water content above the water table at the beginning and at the end of a certain water-table drop (Figure 2.10). The vertical distance between the two water-content curves is the same as the distance between the two water-table positions. The amount of water yielded by the pores at each depth increment dz is equal to $(\theta_A - \theta_B)dz$, where θ_A and θ_B are the volumetric water contents (expressed as volume fraction) at that depth before and after the water-table drop. The total volume of water released by the pores is $\int (\theta_A - \theta_B)dz$, which is the hatched area in Figure 2.10. The specific yield is then calculated by dividing this integral (or hatched area) by the water-table drop. The specific yield of unconfined aquifers may range from only a few percent or less for fractured rock to as much as 20 to 30 percent for unconsolidated materials of relatively uniform particle size.

Estimates of specific yield can also be obtained by determining the difference between the volumetric water content below the water table and that above the water table after most of the water has drained from the pores. The latter is called *specific retention* and is similar to the term *field capacity* (Baver et al., 1972), used in irrigation to describe how much water a soil can retain against the force of gravity.

A third technique for estimating specific yield is to compare net groundwater withdrawals (difference between groundwater collected and groundwater replenished) with changes in water-table position. The net withdrawal per unit surface area and per unit water-table drop then is the specific yield.

In the field, specific yields or storage coefficients of aquifers are often obtained with pumping tests. A well is pumped at a constant rate, and the resulting drawdown of the water table or piezometric surface is measured with wells or piezometers at certain distances from the pumped well. These data yield both the transmissivity and the specific yield or storage coefficient (see Section 5.2).

If the water table drops at a relatively high rate, the drainage of pore space may not take place sufficiently fast to deliver the full specific yield (see Section 5.2). In that case, continued drainage of pore space will occur even if the water table has already receded to lower levels. Thus, the apparent specific yield is dependent upon the rate of fall of the water table, which in turn may vary with time and with distance from the well or other point where groundwater is collected from the aquifer. For a more detailed analysis of pore-space drainage under conditions of fast-dropping water tables and delayed yield, reference is made to Boulton (1955), Childs (1960 and 1969), Youngs (1969), and Streltsova (1973).

In confined aquifers, water is not yielded by drainage of pore space because there is no falling water table and the aquifer material remains saturated. The three main mechanisms whereby confined aquifers yield water are:

1. Consolidation or compression of the aquifer (particularly of interbedded clay or silt layers) and confining layers due to lowering of the piezometric surface (see Section 9.5)
2. Leakage from other aquifers (for example, an overlying unconfined aquifer) through aquitards
3. Drainage of pore space if the confined aquifer becomes an unconfined aquifer with a free water table at its outcrop (Figure 1.2)

The amount of water yielded by any of these mechanisms per unit drop in piezometric surface is much less than that yielded by drainage of pore space per unit drop of a water table. Hence, storage coefficients of confined aquifers are relatively small and often in the range of 0.01 to 0.000 05. Since water is essentially incompressible, expansion of water as it is brought from a pressurized condition in the aquifer to atmospheric pressure at ground surface contributes only very little to the value of the storage coefficient (see Section 9.5).

The term *storage coefficient* is not the best word to describe water release by confined and unconfined aquifers, because it implies storage while it should reflect yield. For unconfined aquifers, the terms *storage coefficient* and *specific yield* are interchangeable, with the latter term being preferred by some. The term *storage coefficient* would have been more appropriate in groundwater recharge, where it could describe the volume of water that can be stored per square meter of aquifer and per meter of rise in water table or piezometric surface. However, the term *storage coefficient* is so well established to describe the yield of aquifers (Lohman et al., 1972) that a change in its use is not warranted. Therefore, the amount of water that unconfined aquifers can store per unit rise in water table and per unit area will be called the *fillable porosity*, analogous to the term fillable porosity used for a rising water table in drained agricultural land (Bouwer and Jackson, 1974). Because of hysteresis, the fillable porosity will be less than the specific yield.

2.11 SAFE YIELD

The *safe yield* of an aquifer is the rate at which groundwater can be withdrawn without causing a long-term decline of the water table or piezometric surface. Thus, the safe yield is equal to the average replenishment rate of the aquifer. In coastal areas, safe yields may be limited by the danger of intrusion of seawater into the aquifer (see Section 11.2). Excessive groundwater pumping occurs so often, however, that the definition of *safe yield* has been stretched beyond the true hydrologic meaning of the term.

In areas of limited water resources, for example, groundwater may deliberately be "mined" because it is the cheapest and most readily available water resource for economic development. Once a thriving economy is established, more expensive schemes such as water importation, desalination, or wastewater reuse could then be financed. In that case, safe yield is defined as an economic safe yield, which is the rate at which groundwater can be withdrawn without danger of the wells drying up before an adequate tax base for the more expensive water supplies has been established. Another interpretation of safe yield could be the legal safe yield, which is the rate at which a well owner can pump groundwater without getting involved in legal action. In its broadest sense, therefore, safe yield can be considered as the rate at which groundwater can be withdrawn without producing undesirable effects.

Sooner or later, however, groundwater withdrawal in excess of the hydrological safe yield will produce undesirable effects. These effects may range from readily predictable consequences like declining groundwater levels and resulting increases in pumping lifts and costs and the need for deepening wells, to deterioration of well-water quality, saltwater intrusion, land subsidence, earth cracking, inducing flow of groundwater from adjacent areas where groundwater is used more conservatively, and eventual depletion of the groundwater resource.

PROBLEMS

2.1 A bubble tube is inserted to a depth of 30 m in a well. Air is pumped into the tube and the air pressure in the tube is measured as 0.5 kg/cm^2 when air begins to bubble up in the well water. What is the depth of the water level in the well?

2.2 As a class project, measure pressure heads below and above the water table in a transparent cylinder with uniform medium sand. The cylinder should be about 80 cm tall. A layer of gravel is placed on the bottom and a transparent tube with a diameter of about 2 cm (tube A in Figure 2.11) is installed so that it reaches into the gravel prior to filling the cylinder with sand. A record should be kept of the total dry weight of the sand going into the cylinder for use in a later test (see Problem 3.2). As the sand is placed in the cylinder, water is poured into tube A to saturate it and tensiometers are installed with their tips at vertical increments of 20 cm, starting about 10 cm above the bottom of the sand. The tensiometers may consist of 0.5-cm diameter tubing with a glass-wool or other porous plug at their bottom (Figure 2.11). The tensiometers are connected to a U-tube manometer arrangement to measure pressure heads (Figure 2.11), making sure that there is no air in the tubing. When the cylinder is completely filled and saturated to the top (water should be standing above the sand surface), water is siphoned out of tube A to produce successively lower, static water tables. Measuring the water-table depth as the depth of the water level in tube A, compare the positive and negative pressure heads indicated by the tensiometers with the distances of the tensiometer tips below and above the water

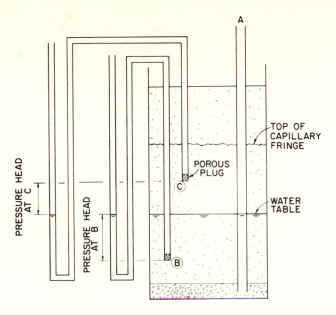

Figure 2.11 Cylinder filled with sand and tensiometers for measuring negative and positive pressure heads.

table. Observe the top of the capillary fringe when the water table has dropped far enough to produce air entry at the sand surface. Measure the height of the capillary fringe at static equilibrium. What is the air-entry value of the sand? Save the experimental setup for Problem 3.2.

2.3 The following results were obtained in mechanical analysis of three soil materials (all weights are expressed in grams of dry material):

	Sample		
	A	B	C
Total weight of sample	500	500	500
Weight of material retained on 53-μm sieve after washing sample through this sieve	500	395	25
Weight of material retained on sieves in a stack for dry-sieving of material not passing 53-μm sieve:			
on 2-mm sieve	0	0	0
on 1-mm sieve	105	0	0
on 0.5-mm sieve	245	100	0
on 0.3-mm sieve	150	130	0
on 0.106-mm sieve	0	85	0
on 0.075-mm sieve	0	60	15
on 0.053-mm sieve	0	20	10
The results of the hydrometer test of the material passing through the 53-μm sieve are:			
< 2 μm	0	20	150
< 1 μm	0	0	50

Calculate the "percentages greater than" and plot against particle size on semilogarithmic paper, as in Figure 2.3. For each sample, determine the effective particle size, the average particle size, the uniformity coefficient, and the textural class.

2.4 Using the parameters V_t, V_s, and V_v, derive Eq. (2.2).

2.5 An undisturbed core sample is obtained from a sandy material at a distance of about 50 cm above a water table. The core is 10.186 cm high and 5 cm in diameter (inside measurement). The net weight of the sample is 419 g before drying and 371 g after drying. Calculate the following parameters: water content by weight, volumetric water content, porosity, void ratio, saturation percentage, and bulk density.

2.6 Using the water-content characteristics in the left graph of Figure 2.7, determine the equilibrium water-content distribution above a water table if the vadose zone consists of 80 cm of material C, 60 cm of material A, 60 cm of material D, and 100 cm of material B (going upward from the water table).

2.7 The Salt River Valley in central Arizona is an alluvial valley with an unconfined aquifer system that is heavily pumped but receives little replenishment. The area is about 100 000 ha, the groundwater pumping about 500 million m^3 per year, and the water-table drop about 3 m per year. Assuming no replenishment, what is the specific yield of the aquifer material?

2.8 A rigid, noncompressible confined aquifer, sandwiched between two completely impermeable aquicludes, is recharged only at its outcrop, where it acts as an unconfined aquifer (like the confined aquifer of Figure 1.2). Water is pumped from the confined aquifer through numerous, evenly distributed wells. The average piezometric head of the confined aquifer drops twice as fast as the water table at the outcrop. The specific yield of the aquifer at the outcrop is 0.2. The confined aquifer covers an area that is 1 000 times as large as the unconfined portion at the outcrop. What is the average storage coefficient of the confined aquifer?

2.9 Apply the procedure of Figure 2.10 for determining specific yield to the curves in Figure 2.9, and determine the specific yield for a falling water table and the fillable porosity for a rising water table.

REFERENCES

Adam, K. M., G. L. Bloomsburg, and A. T. Corey, 1969. Diffusion of trapped gas from porous media. *Water Resour. Res.* **5**: 840–849.

Baver, L. D., W. H. Gardner, and W. R. Gardner, 1972. *Soil Physics*, 4th ed. John Wiley & Sons, New York, 498 pp.

Bianchi, W. C., 1962. Measuring soil moisture tension changes. *Agric. Eng.* **43**: 398–399.

Boulton, N. S., 1955. Unsteady radial flow to a pumped well allowing for delayed yield from storage. *Assoc. Int. Hydrol. Rome* **37**: 472–477.

Bouwer, H., 1966. Rapid field measurement of air-entry value and hydraulic conductivity of soil as significant parameters in flow system analysis. *Water Resour. Res.* **2**: 729–738.

Bouwer, H., 1974. What's new in deep-well injection? *Civ. Eng.* **44**(1): 58–61.

Bouwer, H., and R. D. Jackson, 1974. Determining soil properties. In *Drainage for Agriculture*, J. van Schilfgaarde (ed.), Agronomy Monograph No. 17, Am. Soc. Agron., Madison, Wis., pp. 611–672.

Bredehoeft, J. D., 1967. Response of well-aquifer systems to earth-tides. *J. Geophys. Res.* **72**(12): 3075–3087.

Childs, E. C., 1960. The nonsteady state of the water table in drained land. *J. Geophys. Res.* **65**: 780–782.

Childs, E. C., 1969. *An Introduction to the Physical Basis of Soil Water Phenomena.* John Wiley & Sons, New York, 493 pp.

Davis, S. N., 1969. Porosity and permeability of natural materials. In *Flow through Porous Media*, R. J. M. DeWiest (ed.), Academic Press, New York and London, pp. 53–89.

Day, P. R., 1965. Particle fractionation and particle-size analysis. In *Methods of Soil Analysis*, part I, C. A. Black (ed.), Agronomy Monograph No. 9, Am. Soc. Agron., Madison, Wis., 545–567.

Dixon, R. M., and D. R. Linden, 1972. Soil air pressure and water infiltration under border irrigation. *Proc. Soil Sci. Soc. Am.* **36:** 948–953.

Gardner, W. H., 1965. Water contents. In *Methods of Soil Analysis*, part I, C. A. Black (ed.), Agronomy Monograph No. 9, Am. Soc. Agron., Madison, Wis., pp. 82–127.

Ground Water Newsletter, 1974 (Jan. 29). Published by Water Information Center, N. P. Gillies (ed.), **3**(2): 1.

Lohman, S. W., et al., 1972. Definitions of selected ground-water terms—revisions and conceptual refinements. *U.S. Geol. Survey Water Supply Paper 1988*, 21 pp.

McWhorter, D. B., A. T. Corey, and K. M. Adam, 1973. The elimination of trapped gas from porous media by diffusion. *Soil Sci.* **116:** 18–25.

Peck, A. J., 1965. Moisture profile development and air compression during water uptake by bounded porous bodies: 3. Vertical columns. *Soil Sci.* **100:** 44–51.

Phene, C. J., S. L. Rawlins, and G. J. Hoffman, 1971. Measuring soil matric potential in situ by sensing heat dissipation within a porous body: II. Experimental results. *Proc. Soil Sci. Soc. Am.* **35:** 225–229.

Reeve, R. C., 1957. The measurement of permeability in the laboratory. In *Drainage of Agricultural Lands*, J. N. Luthin (ed.), Agronomy Monograph No. 7, Am. Soc. Agron., Madison, Wis., pp. 414–419.

Reginato, R. J., F. S. Nakayama, and J. B. Miller, 1973. Reducing seepage from stock tanks with uncompacted sodium-treated soils. *J. Soil Water Conserv.* **28:** 214–215.

Rice, R. C., 1969. A fast-response field tensiometer system. *Trans. Am. Soc. Agric. Eng.* **10:** 80–83.

Richards, L. A., 1965a. Physical condition of water in soil. In *Methods of Soil Analysis*, part I, C. A. Black (ed.), Agronomy Monograph No. 9, Am. Soc. Agron., Madison, Wis., pp. 128–152.

Richards, L. A., 1965b. Soil suction measurements with tensiometer. In "*Methods of Soil Analysis*," part I, C. A. Black (ed.), Agronomy Monograph No. 9, Am. Soc. Agron., Madison, Wis., pp. 153–163.

Royer, J. M., and G. Vachaud, 1975. Field determination of hysteresis in soil-water characteristics. *Proc. Soil Sci. Soc. Am.* **39:** 221–223.

Soil Conservation Service, U.S. Dept. of Agriculture, 1973. *Drainage of Agricultural Land*. Water Information Center, Inc., Port Washington, N.Y., 430 pp.

Soil Survey Staff, U.S. Dept. of Agriculture, 1951. *Soil Survey Manual*. U.S. Dept. of Agriculture Handbook No. 18, 503 pp.

Streltsova, T. D., 1973. On the leakage assumption applied to equations of groundwater flow. *J. Hydrol.* **20:** 237–254.

Turk, L. J., 1975. Diurnal fluctuations of water tables induced by atmospheric pressure changes. *J. Hydrol.* **26:** 1–16.

Van Olphen, H., 1963. *An Introduction to Clay Colloid Chemistry*. John Wiley & Sons, New York and London, 301 pp.

Watson, K. K., 1967. A recording field tensiometer with rapid response characteristics. *J. Hydrol.* **5:** 33–39.

Watson, K. K., R. J. Reginato, and R. D. Jackson, 1975. Soil water hysteresis in a field soil. *Proc. Soil Sci. Soc. Am.* **39:** 242–246.

Wesseling, J., 1958. The relation between rainfall, drain discharge, and depth of the water table in tile-drained land. *Techn. Bull. No. 2*, Institute of Land and Water Management Research, Wageningen, Netherlands, 60 pp.

Wilson, L. G., and J. N. Luthin, 1963. Effect of air flow ahead of the wetting front on infiltration. *Soil Sci.* **96:** 136–143.

Youngs, E. G., 1969. Unconfined aquifers and the concept of the specific yield. *Bull. Int. Assoc. Sci. Hydrol.* **14**(2): 191–196.

THREE

DARCY'S LAW AND HYDRAULIC CONDUCTIVITY

Underground water is almost always in motion, although the velocities may be very small. Groundwater primarily moves in horizontal or lateral directions at velocities that generally range from 1 to 500 m/year. Water movement in vadose zones and through restricting layers like perching layers or aquitards is mostly vertical (upward or downward). In this chapter, only the flow of groundwater as such (positive-pressure flow) will be considered. Water movement in the vadose zone is treated in Section 7.7.

3.1 DARCY'S LAW

In analyzing the subsurface movement of water, the actual tortuous paths of the water molecules as they flow through pores, cracks, and crevices of the soil or other aquifer material are taken as smooth paths as if the water molecules moved right through the solid particles. The resulting smooth lines of travel of the water molecules are called *streamlines*.

Figure 3.1 shows a system of linear, parallel streamlines below a water table in a vertical cross section of an aquifer parallel to the direction of flow. Since the streamlines are straight and parallel, the flow of water does not change with distance. This is called *uniform flow*, in contrast with nonuniform flow, where the flow changes with distance and streamlines may curve, diverge, or converge (as, for example, in flow of groundwater to wells, flow of groundwater to or from streams, and other two-dimensional, three-dimensional, or axisymmetric flow

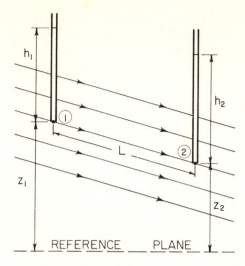

Figure 3.1 Vertical cross section of groundwater flow with linear, parallel streamlines.

systems). The one-dimensional flow in Figure 3.1 is also assumed to be steady, meaning that the flow does not change with time. If the flow changes with time (as, for example, in connection with a rising groundwater table below a recharge basin or with a falling water table around a pumped well), the flow is called *nonsteady* or *transient*.

If piezometers are placed at two points on a streamline (points 1 and 2 in Figure 3.1), the velocity of the groundwater in that streamline can be calculated with the equation

$$v = K \frac{(h_1 + z_1) - (h_2 + z_2)}{L} \tag{3.1}$$

where v = Darcy velocity of water (length/time)
h_1 = pressure head at point 1 (length)
z_1 = elevation head at point 1 (length)
h_2 = pressure head at point 2 (length)
z_2 = elevation head at point 2 (length)
L = distance of flow between points 1 and 2 as measured along streamline (length)
K = hydraulic conductivity of soil or aquifer material (length/time)

Equation (3.1) or modifications thereof are called *Darcy's equation*, after the French hydrologist Henri Darcy (1856), who discovered that velocity was proportional to hydraulic gradient. The Darcy velocity is not the real macroscopic velocity of the water, but the velocity as if the water were moving through the entire cross-sectional area normal to the flow, solids as well as pores. The pressure head h at a given point in the flow system is the height to which water will rise in a piezometer inserted down to that point. The elevation head of a given point is the vertical distance of that point above an arbitrary, horizontal reference plane (dashed line in Figure 3.1). The sum of pressure head and elevation head at a given point in the flow system is called the *total head H*. Thus, $h_1 + z_1$ in Eq. (3.1) is the total head H_1 at point 1, and $h_2 + z_2$ is the total head H_2 at point 2. The distance *L* between points 1 and 2 must be measured along the streamline on which the points are located. The ratio $(H_1 - H_2)/L$ is called the *hydraulic gradient* of the flow.

Darcy's law basically states that *v* is directly proportional to hydraulic gradient [Eq. (3.1)]. The factor of proportionality *K* is a property of the soil or rock material, and it is called the *hydraulic conductivity*. Thus, *K* is the Darcy velocity at unit gradient. Since the hydraulic gradient is dimensionless, *K* has the same dimension as *v*, or length divided by time. A convenient unit for *K* is meters per day. The value of *K* depends on the size and number of pores in the soil or aquifer material. Orders of magnitudes for *K* of granular materials are:

Clay soils (surface)	0.01–0.2 m/day
Deep clay beds	10^{-8}–10^{-2} m/day
Loam soils (surface)	0.1–1 m/day
Fine sand	1–5 m/day
Medium sand	5–20 m/day
Coarse sand	20–100 m/day
Gravel	100–1 000 m/day
Sand and gravel mixes	5–100 m/day
Clay, sand, and gravel mixes (till)	0.001–0.1 m/day

The hydraulic conductivity of sandstone is considerably less than that of unconsolidated sand with the same grain sizes, due to cementation and higher density of the sandstone. *K* values of other consolidated materials depend entirely on the secondary porosity of the rock (fractures, weathering, solution channels in carbonate rock, etc.). General ranges of *K* for consolidated materials are:

Sandstone	0.001–1 m/day
Carbonate rock with secondary porosity	0.01–1 m/day
Shale	10^{-7} m/day
Dense, solid rock	$< 10^{-5}$ m/day
Fractured or weathered rock (aquifers)	0.001–10 m/day
Fractured or weathered rock (core samples)	almost 0–300 m/day
Volcanic rock	almost 0–1 000 m/day

Specific values of *K* for various consolidated and unconsolidated materials were given by Davis (1969).

Since *v* in Eq. (3.1) is the velocity as if the water were moving through the entire porous material, solids as well as pores, the volume rate of flow through a given cross-sectional area perpendicular to the flow is simply calculated as

$$Q = vA \tag{3.2}$$

where Q = volume rate of flow (length^3/time)

v = Darcy velocity (length/time)

A = area normal to flow direction (length^2)

For example, if the Darcy velocity in an aquifer is 0.1 m/day and the aquifer normal to the flow direction is 10 m thick and 1 000 m wide, the flow in the aquifer is 1 000 m^3/day.

Because water movement occurs through the pores and cracks of the aquifer material only, the actual or macroscopic velocity of the water is greater than the Darcy velocity. Assuming that the flow occurs through a bundle of straight capillary tubes in the direction of the streamlines (Figure 3.2), the relation between the actual velocity v_m and the Darcy velocity v is

$$v_m = \frac{A}{A_{cap}} v \tag{3.3}$$

where A = total area normal to the flow direction

A_{cap} = sum of the cross-sectional areas of the capillary tubes

The ratio A_{cap}/A, however, is the porosity n, so that Eq. (3.3) can be written as

$$v_m = \frac{v}{n} \tag{3.4}$$

Equation (3.4) theoretically is correct only for the straight-capillary-tube model of Figure 3.2. However, Eq. (3.4) also tends to give reasonable estimates of the actual velocity in soil or aquifer materials. This has been demonstrated for small-scale systems where water moved only a few centimeters in a matter of minutes (Bouwer and Rice, 1968) and for large-scale systems where water moved as much as 50 km in more than 20 000 years (Pearson and White, 1967).

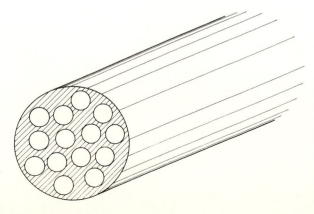

Figure 3.2 Bundle of parallel capillary tubes.

3.2 LABORATORY DETERMINATION OF HYDRAULIC CONDUCTIVITY

Hydraulic conductivity can be determined in the laboratory by placing a cylindrical sample of a particular material in a device that maintains a flow of water through the sample and enables measurement of the flow rate and the head loss across the sample. These devices, called *permeameters*, are divided into constant-head permeameters where constant pressure heads are maintained at both the inflow and outflow ends of the sample, and falling-head permeameters where constant pressure head is maintained only at the outflow end and a variable or falling pressure head is used at the inflow end (Figure 3.3). The falling pressure head is obtained by letting the water level drop in a standpipe that has a much smaller diameter than the sample. Constant pressure heads are maintained by letting the water spill over the top of a cylinder. The outflow from such a cylinder can be collected by another reservoir and discharged into a graduated cylinder for measurement (Figure 3.3, left). Falling-head permeameters are used for materials of relatively low hydraulic conductivity, whereas constant head permeameters are suitable for measuring K of permeable materials like sands and gravels.

Applying Darcy's equation [Eq. (3.1)] to the constant head permeameter and solving for K yields

$$K = LQ/H\pi R^2 \qquad (3.5)$$

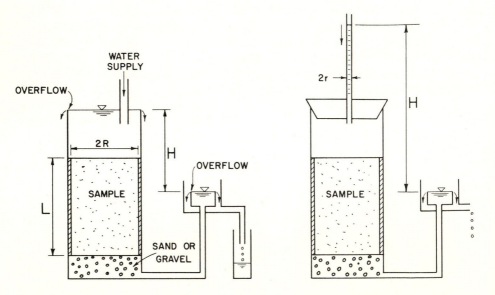

Figure 3.3 Constant-head permeameter (left) and falling-head permeameter (right) for measuring K of sample.

where K = hydraulic conductivity of sample

$\quad\quad H$ = total-head loss across sample (measured as vertical distance between constant water level above sample and water level in constant-level outflow cylinder; Figure 3.3, left)

$\quad\quad L$ = height of sample

$\quad\quad Q$ = volume rate of flow (measured in graduated cylinder below outflow device)

$\quad\quad R$ = radius of sample in permeameter

For the falling-head permeameter, the volume rate of flow Q can be expressed on the basis of the rate of fall dH/dt of the water level in the narrow standpipe as $Q = \pi r^2\, dH/dt$ (see Figure 3.3, right, for meaning of symbols), or on the basis of the flow through the sample with Darcy's equation as $Q = K\pi R^2 H/L$. Equating the two expressions for Q, integrating, and solving for K yields

$$K = \frac{Lr^2}{tR^2}\ln\frac{H_1}{H_2} \tag{3.6}$$

where H_1 and $H_2 = H$ values at beginning and end of a certain time period t

$\quad\quad r$ = radius of standpipe

$\quad\quad t$ = time required for water level in standpipe to drop from H_1 to H_2

The time required to complete a falling-head permeameter test can be reduced by increasing H and decreasing r.

The sand or gravel beneath the sample in the permeameter should be much coarser than the sample itself, so that head losses other than those in the sample are negligible (sometimes the sample is placed on a screen in the permeameter). The samples should be saturated prior to measuring K. Complete saturation may be obtained only by saturating under vacuum. Because a change in the particle arrangement and pore sizes in a porous material can drastically alter its K value, the sample should be as close to its natural condition as possible. Special techniques for taking such samples have been developed (see Sections 5.7 and 6.2.2). Since truly undisturbed samples of subsurface materials are difficult if not impossible to obtain, in-place determination with the field techniques described in Chapter 5 tends to give much more reliable data for natural materials than laboratory measurements.

3.3 VALIDITY OF DARCY'S LAW

Darcy's equation is valid only for laminar flow. With this type of flow, velocities are relatively small and water molecules travel in smooth paths more or less parallel to the solid boundaries of the pores (in capillary tubes with uniform diameters, water molecules move exactly parallel to the tube walls). Laminar flow is governed by the viscous forces of the fluid, so that head losses vary linearly with velocity, as in Darcy's equation. If the velocities increase, a point is reached whereby the inertial forces in the water become dominant. Water molecules then travel in irregular paths (even in a straight tube), forming eddies, swirls, and other turbulences as can be observed in rapidly flowing streams. In this type of flow, which is called *turbulent* flow, head losses vary exponentially with the velocity of the fluid (for example, with the second power of velocity for fully developed turbulent flow in open channels or pipes).

Underground movement of water occurs almost always as laminar flow. Turbulent flow may develop where the pores and the hydraulic gradients are both large. This could happen in the immediate vicinity of pumped wells (for example, in gravel packs or developed zones; see Sections 4.5 and 6.3.3) or in very porous formations (basalts, cavernous limestones), particularly near springs or seeps where the flow is concentrated. Sometimes water may flow in large, subterranean channels, as in cavernous limestone. Such flow can no longer be considered porous-media flow and should be treated as open-channel or pipe flow.

In fluid mechanics, the type flow is characterized by the Reynolds number N_R. This is a dimensionless number that expresses the ratio between inertia and viscous forces in the fluid, as follows:

$$N_R = \frac{\rho v D}{\mu} \tag{3.7}$$

where v = velocity of fluid
$\quad\quad D$ = characteristic dimension of conduit (pipe diameter in case of a full-flowing pipe)
$\quad\quad \rho$ = density of fluid (g/cm^3)
$\quad\quad \mu$ = viscosity of fluid (P or g/cm·s)

In engineering applications, the ratio μ/ρ is often expressed as one parameter, called the kinematic viscosity v with dimension length/time^2 (normally cm/s^2). Pipe flow is laminar when N_R is less than 2 100. If N_R exceeds 2 100, a gradual transition to full turbulent flow occurs as N_R increases. For flow through granular media, v is taken as the Darcy velocity and D as the average particle size $D_{50\%}$. Turbulent flow in porous media has been observed to start at N_R values of 60 to 700 (Lindquist, 1933; Schneebeli, 1955; Hubbert, 1956). Some departure of Darcian flow has been observed at much lower values of N_R (Lindquist, 1933), probably because inertia forces in the tortuous paths of porous-media flow already have an effect even if the flow regime is still laminar. For most cases of underground water movement, N_R will be less than 1 and Darcy's law is valid.

Another situation where Darcy's law may not be valid is where water flows through dense clays. The pores in such materials can be so small that the water molecules in the pores are influenced by the double-layer effects of the clay particles. Because water molecules are polar, water near the electrically charged clay particles then has a more crystalline or "icelike" structure, which causes the viscosity to be higher than that of "free" water. Under such conditions, small hydraulic gradients may not be sufficient to produce water movement, giving rise to threshold gradients and a nonlinearity between flow rate and hydraulic gradient, in contrast with Darcy's law (Swartzendruber, 1969). Other phenomena that cause flow through dense clays to deviate from Darcian flow include movement and rearrangement of clay particles due to frictional drag by the flowing water, development of electrokinetic streaming potentials, and electroosmotic counterflow (Kutilek, 1972; Elnaggar et al., 1974).

Except for the flow immediately around pumped wells, non-Darcian behavior of underground water movement is seldom if ever considered in groundwater hydrology. This is because (1) non-Darcian flow is rare (Sunada, 1969), (2) very accurate solutions of underground flow systems cannot be obtained anyway (because of the heterogeneity of aquifers and inaccurate knowledge of geometry and boundary conditions), and (3) analysis of non-Darcian flow is complicated and involves the use of hydraulic-conductivity values that are dependent on N_R (Volker, 1975).

3.4 FACTORS AFFECTING HYDRAULIC CONDUCTIVITY

The hydraulic conductivity of a certain soil or aquifer is affected by temperature, ionic composition of the water, and presence of entrapped air. The effect of temperature on K is due to the effect of temperature on water viscosity. The higher the temperature, the lower the viscosity of the water will be and the easier it will be for the water to move through the pores of a soil or aquifer. This will then be reflected in an increase in K. The effect of viscosity on K is linear. A 50 percent reduction in viscosity, for example, will double the value of K. Values of K normally are expressed at 20°C. K values at other temperatures can be calculated as

$$K_t = \frac{\mu_{20}}{\mu_t} K_{20} \tag{3.8}$$

where $K_t = K$ at temperature t
μ_t = absolute viscosity of water at temperature t (P or g/cm·s)
μ_{20} = absolute viscosity of water at 20°C
$K_{20} = K$ at 20°C

The relation between μ and water temperature is shown in Figure 3.4.

The ionic composition of the water has an effect on K if the porous material contains clay and if the cations in the water are not yet in equilibrium with the cations in the double layer of the clay particles. The resulting ion exchange may then change the condition of the clay particles from a flocculated to a dispersed state (see Section 2.3), which in turn causes a decrease in K, or vice versa. This phenomenon is mainly of concern where foreign water is added to the ground, as in irrigation, groundwater recharge, and land disposal or deep-well injection of waste liquids.

The effect on K of the cations in water passing through a soil can be quite

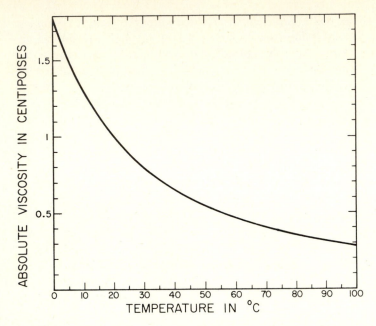

Figure 3.4 Relation between absolute viscosity of water and temperature.

large, as shown in Figure 3.5 for Pachappa sandy loam (a California soil). The SAR values in this figure (taken from McNeal, 1968) are the sodium adsorption ratios, which are calculated as

$$SAR = \frac{Na}{\sqrt{(Ca + Mg)/2}} \qquad (3.9)$$

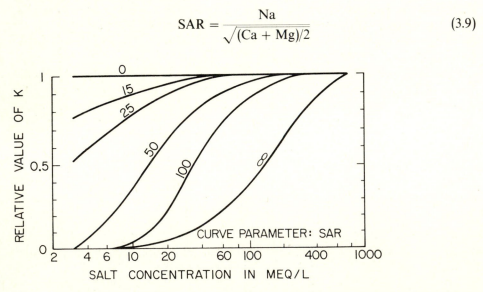

Figure 3.5 Effect of SAR and salt concentration of soil solution on hydraulic conductivity of Pachappa sandy loam. (*Redrawn from McNeal, 1968.*)

from the concentrations of Na^+, Ca^{2+}, and Mg^{2+} in meq/l. The curves show that, for a given salt concentration, K decreases with increasing Na content as expressed by the increasing SAR values. For a given SAR value, K increases with increasing concentration of the solution. This is in accordance with the effects of Na^+ concentration and total salt concentration in soil water on the flocculation-dispersion status of the clay (see Section 2.3).

Changes in salt concentration and ionic composition of underground water can take place when such water moves through dense clay layers. The pores in these clays can be so small that the larger ions in the water are actually filtered out and held back. Heavy soil, clays, and shales can thus behave as selective, geologic membranes that produce salt-sieving effects (White, 1965; Berry, 1969; Kharaka, 1971; and Groenevelt and Bolt, 1972). At high hydraulic gradients, monovalent cations were retarded more than divalent cations. At lower hydraulic gradients, however, Ca was retarded more than Na (Kharaka, 1973). The retardation sequence for monovalent cations is $Li < Na < NH_4 < K < Rb < Cs$, and that for divalent cations is $Mg < Ca < Sr < Ba$. Since a change in concentration and composition of cations in water moving through a clay layer can alter the flocculation-dispersion status of the clay, salt sieving can produce changes in the hydraulic conductivity of the clay layers or other fine materials involved (Hardcastle and Mitchell, 1976).

Entrapped air in soil or aquifer material physically blocks pores and causes K to be less than when the material is completely saturated. Entrapped air may occur after a rise of the water table (see Section 2.9). For sands and relatively fine textured materials, K after wetting the material from an unsaturated condition to one where the pore water is at positive pressure may be only about one-half the K value at complete saturation (Bouwer, 1966). Entrapped air may also accumulate in aquifers invaded by foreign water that is of lower temperature than the native groundwater. After entering the aquifer, the temperature of the foreign water will then increase, which can cause some of its dissolved air to go out of solution and to accumulate in the pores of the aquifer material. This phenomenon, known as *air blocking* or *air binding*, has been observed to reduce K when water that was not de-aired was injected with recharge wells into warmer aquifers (see Section 8.3.3).

3.5 INTRINSIC PERMEABILITY AND HYDRAULIC-CONDUCTIVITY UNITS

Equation 3.8 shows that the product $K\mu$ is constant. This is the basis for the intrinsic permeability k, which was developed to express the permeability of a porous medium in a parameter that is a property of the medium only and independent of the density and viscosity of the fluid. The intrinsic permeability is defined as

$$k = \frac{K\mu}{\rho g} \tag{3.10}$$

where K = hydraulic conductivity (length/time)
  μ = absolute viscosity of fluid (P)
  ρ = density of fluid
  g = acceleration due to gravity

Expressing μ in g/cm·s, ρ in g/cm³, g in cm/s², and K in cm/s, Eq. (3.10) shows that k is expressed in cm². Hence, the dimension of k is length squared. The customary unit of k is the darcy, which is 0.987×10^{-8} cm². The intrinsic permeability is seldom used in groundwater hydrology. It is mainly applied in the petroleum and natural gas industries, which deal with underground flow of fluids of different viscosity and density, liquids as well as gases.

The preferred unit of hydraulic conductivity is meters per day. This unit gives a reasonable range of values (from essentially zero for virtually impermeable materials to as much as 1000 m/day or more for very permeable media). The unit also is readily visualized, and it is compatible with other metric units (length, surface, volume, discharge, etc.) used in hydrology and engineering.

English units of hydraulic conductivity include inches per hour or per day, feet per day, and U.S. gallons per day through a cross-sectional area of 1 ft² and per unit hydraulic gradient (gpd/ft²). The latter unit is standardized to 60°F and is called the *meinzer*, in honor of the American hydrologist O. E. Meinzer. The meinzer has also been expressed as U.S. gallons per day per cross section of 1 ft × 1 mi and per gradient of 1 ft/mi. Sometimes this unit has erroneously been identified as a separate unit, called the field coefficient of hydraulic conductivity. However, the multiplication of the cross-sectional area in the meinzer by a mile and the division of the gradient by a mile cancel each other and do not change the unit.

Conversion of the various units of hydraulic conductivity and of the intrinsic permeability to meters per day is as follows:

$$1 \text{ darcy} = 0.831 \text{ m/day (at } 20° \text{ C)}$$

$$1 \text{ cm/s} = 864 \text{ m/day}$$

$$1 \text{ in/day} = 0.0254 \text{ m/day}$$

$$1 \text{ ft/day} = 0.3048 \text{ m/day}$$

$$1 \text{ meinzer} = 0.0408 \text{ m/day (at } 60° \text{ F or } 15.56° \text{ C)}$$

3.6 AQUIFER FLOW AND TRANSMISSIVITY

Flow in Aquifers

Equation (3.1) can be used to calculate one-dimensional flow in aquifers, such as the unconfined aquifer in the alluvial, intermontane valley of Figure 3.6. The aquifer in this figure is primarily recharged through an alluvial fan at the head of the valley. Groundwater flow in the valley is essentially in a longitudinal

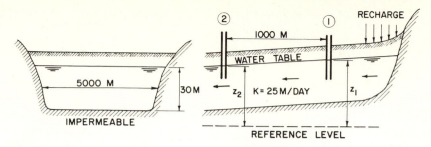

Figure 3.6 Transverse (left) and longitudinal (right) cross section of unconfined aquifer in inter-montane valley.

direction, parallel to the boundaries of the valley (as in a large trough). The aquifer consists of sand and gravel layers with an average hydraulic conductivity of 25 m/day. Two piezometers are installed a distance of 1 000 m apart in the center line of the valley (Figure 3.6). The water level in piezometer 1 is 0.4 m higher than that in piezometer 2, indicating that the water table at 1 is 0.4 m higher than at 2. The average height of the aquifer between points 1 and 2 is 30 m, and the average width of the aquifer is 5 000 m. What is the volume rate of groundwater flow?

Assuming parallel, linear flow and no other additions or withdrawals of water to or from the aquifer (no other "sources" and "sinks"), the flow in the aquifer can be calculated by applying Darcy's equation to any one streamline, for example the top streamline which is at the water table. The pressure head at the water table is zero. Applying Eq. (3.1) to the streamline at the water table between points 1 and 2 yields

$$v = 25\frac{(0 + z_1) - (0 + z_2)}{1\,000} \tag{3.11}$$

where z_1 and z_2 are the elevation heads of the water table at points 1 and 2 above an arbitrary reference level. Since $z_1 - z_2 = 0.4$ m (as indicated by the water levels in the piezometers), v is calculated as 0.01 m/day. Because the streamlines are considered parallel, all other streamlines below the water table have the same hydraulic gradient and, hence, the same Darcy velocity. Thus, multiplying v by the cross-sectional area of the aquifer normal to the flow direction yields the total flow Q of the aquifer, or $Q = 30 \times 5\,000 \times 0.01 = 1\,500$ m³/day. If this flow were representative of the average flow conditions in the aquifer over a long period of time, the safe yield of the aquifer would be 1 500 m³/day.

The hydraulic gradient in Eq. (3.11) is $(z_1 - z_2)/1\,000$, which is the slope i of the water table. Thus, if the flow is in essentially a horizontal direction, the flow in an unconfined aquifer can be calculated by using the slope of the water table as the hydraulic gradient. The same is true for confined aquifers, where the slope of the piezometric surface is used. Thus, Eq. (3.1) becomes $v = Ki$, where i is the slope of the water table or piezometric surface. To obtain the volume rate of flow Q in the aquifer, v is multiplied by the cross-sectional area of the aquifer normal to the flow, or

$$Q = WDKi \qquad (3.12)$$

where $W =$ width of aquifer (5000 m in Figure 3.6)
$D =$ height of aquifer (30 m in Figure 3.6)

Transmissivity

The product DK as it appears in Eq. (3.12) is often combined into one parameter called the *transmissivity* of the aquifer, symbol T. KD has also been called *transmissibility*. Transmissivity, however, is semantically the more correct term, since it is the aquifer that is transmissive while the water itself is transmissible (Lohman et al., 1972). The transmissivity of confined or unconfined aquifers usually is evaluated from pumping tests on wells (Chapter 5). The dimension of T is length^2/time—for example, m^2/day. If K is expressed in meinzer units, the unit of T is U.S. gallons per day per foot, or gpd/ft (1 m^2/day $= 80.5$ gpd/ft). Using water-table slope i and transmissivity T, the flow rate in an aquifer can thus be calculated with the simple equation

$$Q = WTi \qquad (3.13)$$

where W is the width of the aquifer normal to the direction of the flow.

3.7 DUPUIT-FORCHHEIMER ASSUMPTION AND APPLICATION OF DARCY'S LAW TO SIMPLE FLOW SYSTEMS

Darcy's law, expressed by Eq. (3.1) for one-dimensional flow, can be used to solve various simple systems of lateral or vertical groundwater flow. Some systems have both vertical and horizontal flow components. These systems must be simplified so that the flow is in one direction only before Eq. (3.1) is applicable. Vertical-flow components are often neglected where groundwater moves primarily in a lateral direction. The flow is then considered to be purely horizontal and also to be uniformly distributed with depth. These assumptions, collectively called the *Dupuit-Forchheimer assumption* (De Wiest, 1965), will be used in some of the following examples of flow systems that can be solved with Eq. (3.1).

Seepage from Open Channels

Figure 3.7 shows a stream in a floodplain. The soil is considered uniform throughout, including around the wetted perimeter of the stream. Thus, the water in the channel is in direct hydraulic contact with the groundwater in the floodplain (see Section 8.2). A horizontal, impermeable layer occurs at a relatively small distance

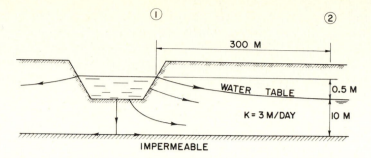

Figure 3.7 Seepage from stream in unconfined aquifer with impermeable layer at relatively shallow depth.

below the stream bottom. The water level in a piezometer at a distance of 300 m from the channel is 0.5 m below the water level in the stream. The impermeable layer is 10 m below the water level in the piezometer. Assuming that K of the soil is 3 m/day, how much are the seepage losses from the stream?

In solving this flow system, vertical-flow components which are concentrated below the channel (the streamlines starting at the channel bottom are initially vertically downward) are ignored and all flow is considered to be in a horizontal direction only. Also, the flow is considered to be evenly distributed with depth below the water table. These assumptions, called the horizontal-flow or Dupuit-Forchheimer assumption, were first used by Dupuit (1863) and later by Forchheimer (1901). Proof that the Dupuit-Forchheimer assumption can yield exact solutions for flow below mildly sloping water tables was given by Charny (as discussed by Polubarinova-Kochina, 1962).

In solving the flow system of Figure 3.7, Darcy's equation will be applied to the flow below the water table from the edge of the channel (section 1) to the piezometer location (section 2). Applying Darcy's equation to the top streamline (along the water table) between sections 1 and 2 and using the impermeable layer as reference plane for the elevation heads yields

$$v = 3 \frac{(0 + 10.5) - (0 + 10)}{300}$$

or $v = 0.005$ m/day. According to the Dupuit-Forchheimer assumption, all the water below the water table will move at this velocity. In calculating the volume rate of flow, the height of the flow system is taken as the average height of the water table above the impermeable layer between sections 1 and 2, or 10.25 m. The width of the flow system will be taken as 1 m, so that the seepage will be expressed per unit length of stream. Thus, the cross-sectional area is $1 \times 10.25 = 10.25$ m^2 and the seepage is $0.005 \times 10.25 = 0.05125$ m^3/day per m length of stream. Assuming that the conditions on the other side of the channel are the same (in other words, the flow system is symmetrical), the total seepage from the stream is 0.1025 m^3/day per meter length of stream.

The Dupuit-Forchheimer theory loses accuracy in this case if the depth of the impermeable layer increases, because of the increased importance of vertical flow. In comparing seepage rates based on the Dupuit-Forchheimer theory with solutions obtained with an electrical resistance network analog, which takes vertical-flow components into account (see Section 7.3), Bouwer (1969) found that the Dupuit-Forchheimer theory gave reasonably accurate seepage values if the distance of the impermeable layer below the stream bottom was not more than twice the width of the water level in the stream (see Section 8.2).

Subsurface Runoff

Figure 3.8 shows a hillside with relatively shallow soil draining into a stream. The slope of the land is 2 percent, the soil is a sandy loam with $K = 2.5$ m/day, and impermeable bedrock occurs at a depth of 6 m for the entire slope. To reduce pollution of the stream, effluent from a sewage treatment plant which discharges directly into the stream will be sprinkled on land at some distance from the stream. After infiltration, the effluent will then move downhill as subsurface runoff and drain back into the stream. The underground flow and associated "land filtration" considerably improve the quality of the effluent (Section 11.4), so that the pollution load on the stream will be greatly reduced. The system must be designed and operated to avoid surface runoff of the effluent. If the application rate of the sprinklers is 2 cm/day, what will be the maximum width W of the area that can be sprinkled at any one time?

Maximum underground flow is obtained when the soil between the disposal field and the channel is completely saturated and the water table coincides with the surface of the soil. The transmissivity of the saturated soil will then be $2.5 \times 6 = 15$ m²/day. Applying Eq. (3.13), the maximum underground flow q per unit length of system (normal to the cross section of Figure 3.8) is

$$q = 1 \times 15 \times 0.02$$

$$= 0.3 \text{ m}^3/\text{day per m}$$

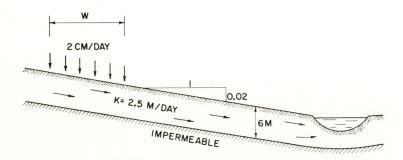

Figure 3.8 Land application of sewage effluent and subsurface runoff to stream.

At an infiltration rate of 2 cm/day over an area of width W, the rate of flow per unit length of system is $0.02W$ m³/day per m. Thus, the maximum value of W whereby surface runoff does not occur is $0.3/0.02 = 15$ m. If the soil should receive effluent only one day per week, the disposal field can be 105 m wide and consist of 7 parallel 15-m-wide strips that are irrigated in turn on a 7-day cycle.

Uniform Infiltration and Drainage to a Stream

A system of precipitation and drainage of infiltrated water to a stream via an unconfined aquifer with a horizontal impermeable barrier is shown in Figure 3.9. Assuming a uniform infiltration rate P and steady-state conditions, how high is the equilibrium water-table position at the top of the hill?

Using the Dupuit-Forchheimer theory, the velocity v_x of the groundwater at a distance x from the top of the hill is

$$v_x = -K \frac{dh}{dx}$$

where dh/dx is the slope of the water table at that point. Since h decreases with increasing x, dh/dx is negative. Thus, a minus sign must be placed in front of dh/dx in Darcy's equation so that v will be positive in the direction of decreasing h. The flow q_x per unit length of system at distance x is

$$q_x = -Kh \frac{dh}{dx}$$

However, q_x is also equal to Px. Thus we have

$$Px = -Kh \frac{dh}{dx}$$

or

$$Kh\, dh = -Px\, dx$$

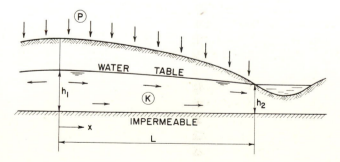

Figure 3.9 Infiltration of rainfall and drainage to stream.

Integrating this equation between the top of the hill and the edge of the stream yields

$$K(h_1^2 - h_2^2) = PL^2$$

or

$$h_1 = \sqrt{h_2^2 + \frac{PL^2}{K}}$$

where h_1 = height of water table above impermeable layer at top of hill
$\quad\; h_2$ = height of water level in stream above impermeable layer
$\quad\; L$ = distance between top of hill and stream
$\quad\; P$ = infiltration rate (length/time)
$\quad\; K$ = hydraulic conductivity of soil

Error in Water-Table Measurement by Piezometer due to Vertical Flow

An unconfined aquifer in glacial till is recharged at the top (by infiltration of rainfall). The water then flows vertically down through an aquitard and into a leaky, confined aquifer (Figure 3.10). A piezometer is installed at a depth of 10 m below the water table. If the infiltration rate is 1.5 cm/day and K of the unconfined aquifer is 0.2 m/day, how much will the water level in the piezometer differ from the true water table?

   The pressure head at point 1, which is on the water table, is zero. The pressure head at point 2 is equal to h, the height of the water level in the piezometer. Taking the bottom of the piezometer tube as the horizontal reference level, the elevation head at point 1 is 10 m and that at point 2 is zero. Applying Eq. (3.1) to the vertical flow from point 1 to point 2 then yields

$$0.015 = 0.2 \frac{(0 + 10) - (h + 0)}{10}$$

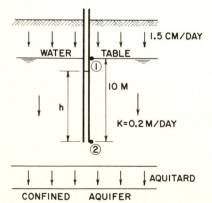

1.5 CM/DAY

WATER TABLE

10 M

h

K=0.2 M/DAY

AQUITARD

CONFINED AQUIFER

Figure 3.10 Piezometer in unconfined aquifer with vertically downward flow.

from which h is calculated as 9.25 m. Thus, the water level in the piezometer is 0.75 m below the water table in the aquifer.

This shows that water levels in piezometers do not yield true water-table positions if vertical flow occurs in the aquifer. The higher the vertical-flow rate is in relation to the hydraulic conductivity, the greater the error will be. To minimize the error, piezometers should penetrate the aquifer as little as possible. If the water table fluctuates, several piezometers may be installed at different depths in the zone of fluctuation. The water level in the shallowest piezometer that contains water can then be used as the best estimate of the water-table position.

Recharge Rate of Leaky Aquifer

The piezometric surface of a confined aquifer is 1.5 m below the water table of an overlying unconfined aquifer. The two aquifers are separated by an aquitard of 0.5 m thickness (Figure 3.11). The water table in the unconfined aquifer is 20 m above the top of the aquitard and is in equilibrium with downward flow due to infiltration from above. K of the unconfined aquifer is 0.8 m/day and K of the aquitard is 0.1 m/day. What is the flow rate from the unconfined aquifer to the confined aquifer?

In order to calculate the flow through the aquitard, the pressure head at point 2 must be known. However, the only known pressure heads are at point 1 (the water table) and at point 3 (the piezometric surface). To solve this problem, Eq. (3.1) will be applied between points 1 and 2, and between 2 and 3. Since the flow rate from 1 to 2 is the same as from 2 to 3, this yields two equations with two unknowns (the downward velocity v and the pressure head h_2 at point 2), which can both be solved. Taking the bottom of the aquitard as the reference level for the elevation heads, the resulting equations are:

Between points 1 and 2: $\qquad v = 0.8 \dfrac{(0 + 20.5) - (h_2 + 0.5)}{20}$

Between points 2 and 3: $\qquad v = 0.1 \dfrac{(h_2 + 0.5) - (19 + 0)}{0.5}$

Solving these equations yields $h_2 = 18.75$ m and $v = 0.05$ m/day.

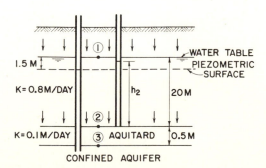

Figure 3.11 Vertically downward flow from unconfined aquifer through aquitard to confined aquifer.

If K of the aquitard had been much smaller than K of the unconfined aquifer, h_2 would have been very close to 20 m. Thus, the water-table position in the unconfined aquifer could have been used in that case to approximate the pressure head at point 2.

Height of Perched Water Table

Heavy infiltration (due to spring snow melt, for example) of 3 cm/day causes a perched water table to form above a slowly permeable, flow-restricting layer. The restricting layer is at a depth of 2 m, it is 0.4 m thick, and it has a K of 0.01 m/day (Figure 3.12). The material above the restricting layer is a silt loam with $K = 0.12$ m/day. Coarse sands and gravels occur below the restricting layer. After flowing through the restricting layer, the water moves as unsaturated flow through the sand and gravel to an unconfined aquifer. What is the height z of the perched water table above the top of the restricting layer?

This flow system will be solved by applying Darcy's equation to the flow in the perched groundwater and to the flow through the restricting layer. The pressure head h_3 at the bottom of the restricting layer will be negative and will depend on the unsaturated hydraulic conductivity of the underlying material (see Section 7.7). Since this material is coarse (sand and gravel), h_3 will be only slightly negative and can be taken as zero. The value h_2 of the pressure head at point 2 can be calculated by applying Eq. (3.1) to points 2 and 3. Taking the reference level for the elevation heads at the bottom of the restricting layer, this yields

$$0.03 = 0.01 \frac{(h_2 + 0.4) - (0 + 0)}{0.4}$$

from which h_2 is calculated as 0.8 m. The height z of the perched water table is calculated by applying Darcy's equation to the flow between points 1 and 2, or

$$0.03 = 0.12 \frac{(0 + z + 0.4) - (0.8 + 0.4)}{z}$$

which yields $z = 1.067$ m for the height of the perched water table.

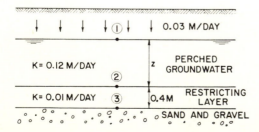

K= 0.12 M/DAY   0.03 M/DAY   PERCHED GROUNDWATER   K= 0.01 M/DAY   0.4M   RESTRICTING LAYER   SAND AND GRAVEL

Figure 3.12 Perched groundwater above restricting layer during period of infiltration.

Effect of River Stage on Water Table in Floodplain

A floodplain is underlain by a very permeable gravel layer that is in direct hydraulic connection with a leveed river (Figure 3.13). Impermeable bedrock occurs below the gravel layer. The hydraulic conductivity of the gravel layer is so much higher than that of the overlying soil in the floodplain that the piezometric surface in the entire gravel layer can, at all times, be taken as equal to the water-surface elevation in the stream. A flood causes a sudden rise in the water depth of the river from H_1 to H_2. How fast will the water table rise in the soil above the gravel layer?

To calculate the water-table rise in the floodplain, K and the fillable porosity f of the soil above the gravel layer must be known. The fillable porosity is the difference between the volumetric water content above and below the rising water table. The relation between the upward Darcy velocity v in the soil and the rate of rise dz/dt of the water table is

$$\frac{dz}{dt} = \frac{v}{f}$$

or

$$v = f\frac{dz}{dt} \tag{3.14}$$

When the water table is a given height above the gravel layer, v can also be expressed with Darcy's equation as

$$v = K\frac{H - z}{z} \tag{3.15}$$

where $H = H_2 - H_1$ or the rise of the water level in the stream (Figure 3.13). Equating Eqs. (3.14) and (3.15) yields

$$f\frac{dz}{dt} = K\frac{H - z}{z}$$

which can be integrated to

$$-z - H \ln (H - z) = \frac{Kt}{f} + C \tag{3.16}$$

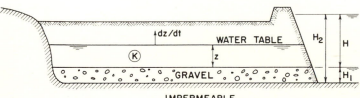

Figure 3.13 Rising water table in floodplain due to high water level in river.

where C is the integration constant. Since $z = 0$ when $t = 0$, C is equal to $-H \ln H$. Substituting this value of C in Eq. (3.16) and rearranging terms yields

$$\frac{Kt}{f} = H \ln \frac{H}{H - z} - z \qquad (3.17)$$

Thus, for given values of K, f, and H, the time t in which the water table in the floodplain rises to a certain height z above the permeable gravel layer can be computed.

The examples in this section show how Darcy's equation can be used to solve a variety of problems of one-dimensional underground water movement. Solutions of more complex flow systems, including flow to wells and the use of model and analog techniques, are presented in Chapters 4 and 7.

3.8 ANISOTROPY

Individual particles of granular subsurface materials are seldom spherical. When deposited under water, the particles usually come to rest on their flat side. Particles deposited in flowing water may be tilted slightly upward in the direction of flow and overlap somewhat (Figure 3.14). This arrangement, called *imbrication*, can often be observed on gravel deposits (Figure 3.15).

The path of water molecules flowing through imbricated material is more tortuous in vertical than in horizontal directions. Consequently, the hydraulic conductivity K_z in a vertical direction will be less than K_x in a horizontal direction. It is not unusual to find K_z values that are only one-fifth or one-tenth of K_x. This phenomenon, called *anisotropy*, is the rule rather than the exception for (undisturbed) alluvial deposits. If K is the same in all directions, the material is called *isotropic*. Such conditions can be approached in soils consisting of near-round particles, or in the laboratory with glass beads or mixed soils packed dry or damp.

Anisotropy is caused not only by particle orientation, but also by layering of materials with different K values, even though each layer itself may be isotropic. For example, an aquifer consisting of separate, horizontal sand and gravel layers will behave like an anisotropic medium because the resistance to vertical flow, where all the water has to move through both sand and gravel layers, will be more than the resistance to horizontal flow, where most of the water can move through the gravel layers only. If K of the individual sand and gravel layers is known, K_z and K_x of the entire packet of sand and gravel layers can be computed, as shown next.

STREAM   FLOW

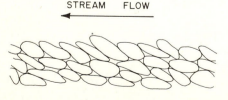

Figure 3.14 Imbrication of gravel deposited by stream.

Figure 3.15 Imbrication of gravel (lower one-third in right half of picture) in north wall of gravel pit, Salt River bed and 27th Avenue, Phoenix, Arizona. Stream flow at time of deposit was from right to left. Person is pointing to a layer of black, manganese-oxide-coated gravel indicative of past water-table position.

An aquifer or other underground formation consisting of n horizontal isotropic layers of different thickness z and with different K values is shown in Figure 3.16. If there is horizontal flow through the system, the hydraulic gradient i is the same in each layer (if i were not the same, pressure-head differences would exist along the interfaces between the layers, which is an impossibility for horizontal flow). The flow q_1 in the top layer per unit width of the system can be expressed as

$$q_1 = iK_1z_1 \tag{3.18}$$

The horizontal flow in the other layers can be expressed similarly (for example, $q_n = iK_nz_n$). Summing the q values of each layer to get the total horizontal flow q_x per unit width of the layered system yields

$$q_x = i(K_1z_1 + K_2z_2 + \cdots + K_nz_n) \tag{3.19}$$

If the layered system of Figure 3.16 is considered as one homogeneous medium, q_x can also be expressed as

$$q_x = iK_xZ \tag{3.20}$$

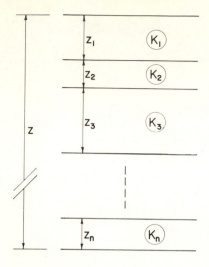

Figure 3.16 System of horizontal, isotropic layers with different thicknesses and K values.

where K_x is the average hydraulic conductivity of the medium in a horizontal direction and Z is the height of the entire system. Equating Eqs. (3.19) and (3.20) and solving for K_x yields

$$K_x = \frac{K_1 z_1 + K_2 z_2 + \cdots + K_n z_n}{Z} \tag{3.21}$$

If the layers are equally thick, Eq. (3.21) reduces to

$$K_x = \frac{K_1 + K_2 + \cdots + K_n}{n} \tag{3.22}$$

where n is the number of layers.

If there is vertical flow through the system of Figure 3.16, the flow q per unit horizontal area can be expressed for the top layer as

$$q = K_1 \frac{\Delta H_1}{z_1} \tag{3.23}$$

where ΔH_1 is the total head loss in the first layer. Solving this equation for ΔH_1 yields

$$\Delta H_1 = \frac{z_1}{K_1} q \tag{3.24}$$

Since q is the same for all layers, similar expressions can be written for H across the other layers (for example, $\Delta H_n = z_n q/K_n$). The total-head loss ΔH_t for the vertical flow through all the layers of the system can be calculated as the sum of the head losses in each layer, or

$$\Delta H_t = \left(\frac{z_1}{K_1} + \frac{z_2}{K_2} + \cdots + \frac{z_n}{K_n} \right) q \tag{3.25}$$

If the system of Figure 3.16 is considered as one homogeneous medium, the vertical flow q can be expressed as

$$q = K_z \frac{\Delta H_t}{Z} \tag{3.26}$$

where K_z is the average hydraulic conductivity of the medium in a vertical direction. Solving Eq. (3.26) for ΔH_t, equating the resulting expression to Eq. (3.25), and solving for K_z yields

$$K_z = \frac{Z}{\dfrac{z_1}{K_1} + \dfrac{z_2}{K_2} + \cdots + \dfrac{z_n}{K_n}} \tag{3.27}$$

If the layers are equally thick, this equation reduces to

$$K_z = \frac{n}{\dfrac{1}{K_1} + \dfrac{1}{K_2} + \cdots + \dfrac{1}{K_n}} \tag{3.28}$$

where n is the number of layers.

Figure 3.17 Microstratification in sand layer of same gravel pit as in Figure 3.15.

Figure 3.18 Sand and gravel layers of different thickness and texture in wall of same gravel pit as in Figure 3.15. The top layer is about 1.5 m thick and consists of a loam.

Equations (3.21) and (3.22) express what is known as the arithmetic mean of a group of numbers, whereas Eqs. (3.27) and (3.28) express the harmonic mean. Inspection of these equations shows that the harmonic mean K_z is much more influenced by a small number in the group than the arithmetic mean K_x. As a matter of fact, the harmonic mean of a certain group is always smaller than the arithmetic mean. Thus, K_z of a horizontally layered medium is always less than K_x when such a medium is treated as a homogeneous, anisotropic system.

The thickness of the different layers in alluvial deposits may vary from a few millimeters or less (Figure 3.17) to several decimeters or more (Figure 3.18). The microstratification in Figure 3.17 is in a fairly uniform sand layer about 50 cm thick and causes the sand layer itself to be anisotropic. The directional hydraulic conductivities of an aquifer with sand and gravel layers similar to those in Figure 3.18 were $K_z = 5.4$ m/day and $K_x = 86.0$ m/day (Bouwer, 1970). These values were determined from the rise of the water table below recharge basins about 15 km downstream from the gravel pit in which Figure 3.18 was taken. Weeks (1969) found K_x/K_z ratios of 2 to 20 for glacial-outwash aquifers in central Wisconsin.

When applying Darcy's equation to an anisotropic system, the hydraulic conductivity must be taken in accordance with the direction of flow (K_x for horizontal flow, K_z for vertical flow). The K value in other directions is obtained from an ellipsoid with $\sqrt{K_x}$ and $\sqrt{K_z}$ as principal axes. The theory for the ellipsoid of directional permeability was given by Muskat (1937); it states that the square root of directional K, when plotted in all directions from a certain point in an anisotropic medium, forms the ellipsoid $x^2/\sqrt{K_x} + z^2/\sqrt{K_z} = 1$. A derivation of this equation is presented in Section 7.6.

In most cases, alluvial deposits are considered anisotropic in two directions: vertical and horizontal. However, on a large scale, aquifers and groundwater basins deposited by flowing water may also exhibit anisotropy in the horizontal plane itself, because K_x tends to be greater in the downstream direction than perpendicular thereto. This results from the fact that gravel layers, buried valleys, and similar coarse strata tend to be more continuous in the direction of stream flow at the time they were formed than normal thereto. Such aquifers then have three-dimensional anisotropy with principal K axes in the vertical direction, the horizontal direction parallel to past prevailing stream flows, and the horizontal direction at a right angle to these flows.

PROBLEMS

3.1 Apply Eq. (3.1) to the inflow and outflow ends of the sample in constant-head and falling-head permeameters and derive Eqs. (3.5) and (3.6).

3.2 Use the device of Figure 2.11 as a permeameter to determine K of the sand in the cylinder. Pour water in tube A until water starts spilling over the rim of the cylinder containing the sand. Perform a constant-head test by adding just enough water to tube A to maintain the water level at the top of that tube without causing water to spill over the edge. This maintains a constant pressure head at the inflow end (bottom) of the sand. The overflow from the cylinder itself provides the constant pressure head at the outflow end (top) of the sand. The value of Q in Eq. (3.5) is determined from the volume of water that had to be added to tube A during a certain period of time. During this period of time, also measure water-pressure heads in the sand with the piezometers (installed as tensiometers) and calculate the hydraulic gradient of the upward flow. Calculate K by dividing this gradient into the measured Q. Stop adding water to tube A and measure the rate of fall of the water level in this tube. Calculate K with the falling-head Eq. (3.6), and compare the values of K obtained with the three different techniques.

Add water to tube A and again maintain the water level at the top of this tube. Put a soluble dye like fluorescein, rhodamine, food coloring, or methylene blue in the inflow water and observe and measure the rate of rise of the dye in the sand. Compare this rate with the macroscopic velocity calculated from the Darcy velocity and the porosity of the sand. The porosity of the sand is calculated from the dry weight of the sand in the cylinder and the total volume of the sand column, assuming a density of 2.65 g/cm^3 for the sand particles as such.

3.3 Assuming that the porosity of the aquifer in Figure 3.6 is 20 percent, how long would it take for the water in the aquifer to travel from the head of the valley to a point 20 km downstream? Assuming an average water use of 300 l per person per day, how large a population could the aquifer of Figure 3.6 support without overdraft of the groundwater?

3.4 A confined aquifer has a transmissivity of 40 m^2/day. The slope of the piezometric surface is 0.25 m per kilometer. How much water per day flows through the aquifer per kilometer width of the aquifer?

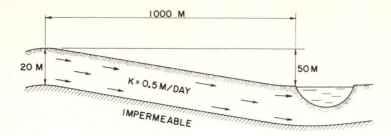

Figure 3.19 Subsurface flow to stream.

3.5 A hillside drains to a stream as shown in Figure 3.19. Using the parameters given in the figure, calculate the maximum seepage into the stream per unit length of stream (when the entire soil is saturated). What is the maximum runoff rate at the base of the hillside if it rains steadily at 0.5 cm/day? Express this maximum surface runoff as a percentage of the rainfall rate.

3.6 Calculate the difference between the water level in the piezometer and the water table in the aquifer of Figure 3.10 for downward-flow rates of 0.1, 1, and 5 cm/day and piezometer depths of 1, 5, and 10 m below the water table. Plot the results to show how the difference between the water level in the piezometer and the true water-table position is affected by vertical-flow rate and depth of piezometer below water table.

3.7 Taking $K = 0.5$ m/day, $f = 0.1$, and $H = 10$ m, calculate t with Eq. (3.17) for z values of 1, 2, 3, 4, 5, 6, 7, and 8 m. Plot a graph with z on the ordinate and t on the abscissa. Determine dz/dt from the slope of the resulting curve at z values of 2, 4, and 6 m. Multiply each value of dz/dt by f to obtain the Darcy velocity v [Eq. (3.14)]. Compare the resulting values with v calculated with Eq. (3.15).

3.8 Two piezometers are installed in the soil above the gravel layer in the system of Figure 3.13. One piezometer reaches to a distance of 2 m above the gravel layer and the other to a distance of 4 m (Figure 3.20). If the water depth in the deep piezometer is 4 m and that in the shallow piezometer 1 m, how high is the water table above the top of the gravel layer, and what is the rate of rise of the water table at that height? The value of K of the soil above the gravel layer is 0.5 m/day. The volumetric water content of this soil is 0.35 below the water table and 0.25 above the water table.

3.9 A confined aquifer is recharged from an overlying unconfined aquifer through an aquitard. From pumped-well data, it is estimated that the recharge rate is 2.92 cm/year. If the average piezometric surface of the confined aquifer is 10 m below the water table in the unconfined aquifer and the aquitard is 1 m thick, what is K of the aquitard in meters per day? This problem shows that considerable recharge can take place through an "essentially impermeable" aquitard. For example, if a well in the confined aquifer pumps 5000 m³/day, how many square kilometers of recharge area are required to sustain the flow to the well at the recharge rate of 2.92 cm/year?

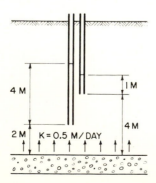

Figure 3.20 Upward flow from gravel layer into overlying soil.

3.10 A perching layer with a K value of 0.01 m/day supports a perched groundwater body 1 m high. The perching layer is 0.5 m thick and the downward-flow rate through the perching layer is 0.038 m/day. What is the pressure head at the bottom of the perching layer? Assume that K of the material above the perching layer is so large compared to K of the perching layer itself that the perching water table can be used to indicate the pressure head at the top of the perching layer.

3.11 Water is flowing through coarse sand with an average particle size of 0.75 mm. Assuming that turbulent flow starts at $N_R = 10$, what is the upper velocity limit for laminar flow (take the absolute viscosity of the water as 1 cP and the density as 1 g/cm^3)? If K of the sand is 20 m/day, what hydraulic gradient would be necessary to produce this velocity limit? Could such a gradient ever occur in practice?

3.12 Five horizontal and differently textured layers separate a confined aquifer from ground surface. Each layer is anisotropic itself. The thickness and directional K values of each layer are as follows:

Layer	Thickness, m	K_x, m/day	K_z, m/day
1 (top)	1.5	0.5	0.1
2	2	3	0.5
3	0.3	0.1	0.05
4	4	2	0.8
5	1	0.01	0.002

If the piezometric surface in the confined aquifer is at the bottom of layer 5, what is the recharge rate (per unit area) of the confined aquifer, assuming that all layers are saturated, that the infiltration rate at the surface is sufficient to maintain any downward flow, and that no water is standing above the surface? If the five layers are underlain by impermeable bedrock instead of a confined aquifer and the packet of layers slopes 0.1 percent, what is the lateral flow per unit width through the layers?

REFERENCES

Berry, F. A. F., 1969. Relative factors influencing membrane filtration effects in geologic environments. *Chem. Geol.* **4:** 295–301.

Bouwer, H., 1966. Rapid field measurement of air entry value and hydraulic conductivity of soil as significant parameters in flow system analysis. *Water Resour. Res.* **2:** 729–738.

Bouwer, H., 1969. Theory of seepage from open channels. In *Advances in Hydroscience*, vol. 5, V. T. Chow (ed.), Academic Press, New York and London, pp. 121–172.

Bouwer, H., 1970. Groundwater recharge design for renovating waste water. *J. Sanit. Eng. Div., Am. Soc. Civ. Eng.* **96**(SA1): 59–74.

Bouwer, H., and R. C. Rice, 1968. A salt penetration technique for seepage measurement. *J. Irrig. Drain. Div., Am. Soc. Civ. Eng.* **94**(IR4): 481–492.

Darcy, H., 1856. Les fontaines publiques de la ville de Dijon, pp. 570, 590–594, V. Dalmont, Paris.

Davis, S. N., 1969. Porosity and permeability of actual materials. In "Flow Through Porous Media," R. J. M. DeWiest (ed.), Academic Press, New York and London, pp. 53–89.

DeWiest, R. J. M., 1965. History of the Dupuit-Forchheimer assumptions in groundwater hydraulics. *Trans. Am. Soc. Agric. Eng.* **8:** 508–509.

Dupuit, J., 1863. Etudes théoriques et pratiques sur le mouvement des eaux dans les canaux découverts et à travers les terrains permeables. Dunod, 2d ed., Paris.

Elnaggar, H. A., G. M. Karadi, and R. J. Krizek, 1974. Non-Darcian flow in clay soils. In *Flow: Its Measurement and Control in Science and Industry*, vol. 1, R. B. Dowdell and H. W. Stoll (eds.), Instrument Society of America, Pittsburgh, pp. 53–61.

Forchheimer, P., 1901. Hydraulik. *Encyklopadie der Mathematischen Wissenschaften* **4**: 20. B. G. Teubner, Leipzig.

Groenevelt, P. H., and G. H. Bolt, 1972. Permiselective properties of porous materials as calculated from diffuse double-layer theory. In *Fundamentals of Transport Phenomena in Porous Media*, International Association for Hydraulic Research, Elsevier Publishing Company, Amsterdam, pp. 241–258.

Hardcastle, J. H., and J. K. Mitchell, 1976. Water quality and aquitard permeability. *J. Irrig. Drain. Div., Am. Soc. Civ. Eng.* **102**(IR2): 205–220.

Hubbert, M. K., 1956. Darcy's law and the field equations of underground fluids. *Trans. Am. Inst. Min. Metall. Pet. Eng.* **27**: 222–239.

Kharaka, Y. K., 1971. "Simultaneous flow of water and solutes through geological membranes: Experimental and field investigations." Ph.D. dissertation, Univ. of California, Berkeley, 274 pp.

Kharaka, Y. K., 1973. Retention of dissolved constituents of waste by geologic membranes. Proc. 2nd Internat. Symp. Underground Waste Management and Artificial Recharge. *Am. Assoc. Pet. Geol.* **1**: 420–435.

Kutilek, M., 1972. Non-Darcian flow of water in soils—laminar region. In *Fundamentals of Transport Phenomena in Porous Media*, International Association for Hydraulic Research, Elsevier Publishing Company, Amsterdam, pp. 327–340.

Lindquist, E., 1933. "On the flow of water through porous soil." Premier Congrès des grand barrages. pp. 81–101, Stockholm.

Lohman, S. W., et al., 1972. Definitions of selected ground-water terms—revisions and conceptual refinements. *U.S. Geol. Survey Water Supply Paper 1988*, 21 pp.

McNeal, B. L., 1968. Prediction of the effect of mixed-salt solutions on the soil hydraulic conductivity. *Proc. Soil Sci. Soc. Am.* **32**: 190–193.

Muskat, M., 1937. *The Flow of Homogeneous Fluids through Porous Media*, McGraw-Hill Book Co., New York. (Reprinted 1946, J. W. Edwards, Ann Arbor, Mich.)

Pearson, F. J., Jr., and D. E. White, 1967. Carbon-14 ages and flow rates of water in Carrizo Sand, Atascosa County, Texas. *Water Resour. Res.* **3**: 251–261.

Polubarinova-Kochina, P. Y., 1962. Theory of groundwater movement. Translated from the Russian by R. J. M. DeWiest, Princeton University Press, Princeton, N.J.

Schneebeli, G., 1955. Expériences sur la limite de validité de la loi de Darcy et l'apparition de la turbulence dans un écoulement de filtration. *Houille Blanche* **10**: 141–149.

Sunada, D. K., 1969. Laminar and turbulent flow of water through porous media. *Final Report, CER68-69DK533*, Colorado State University, Fort Collins.

Swartzendruber, D., 1969. The flow of water in unsaturated soils. In *Flow through Porous Media*, R. J. M. DeWiest (ed.), Academic Press, Inc., New York and London, pp. 215–292.

Volker, R. E., 1975. Solutions for unconfined non-Darcy seepage. *J. Irrig. Drain. Div., Am. Soc. Civ. Eng.* **101**(IR1): 53–65.

Weeks, E. P., 1969. Determining the ratio of horizontal to vertical permeability by aquifer-test analysis. *Water Resour. Res.* **5**: 196–214.

White, D. E., 1965. Saline waters of sedimentary rocks. In *Fluids in Subsurface Environments*, A. Young and J. E. Galley (eds.), *Am. Soc. Petrol. Geol. Mem.* **4**: 342–366.

FOUR

WELL-FLOW SYSTEMS

4.1 INTRODUCTION

Various equations have been developed to relate well flow to drawdown of piezometric surface (or water table), transmissivity, and storage coefficient (or specific yield) of aquifers. Prediction of well discharge is necessary for proper selection of the pump and power unit and of pumping depth (see Section 6.4). In some areas, there is enough local experience to predict fairly accurately the performance of a new well. In other places, the expected yield of a well must be calculated from the hydraulic properties of the aquifer.

Equations for calculating well discharge have been derived for steady-state flow and for nonsteady or transient flow. The steady state is an equilibrium condition whereby no changes occur with time. It will seldom if ever occur in practice, but it may be approached after prolonged pumping of the well when piezometric surfaces or water tables decline at very slow rates. Transient-flow equations include the factor time, and they enable the calculation of the drop in piezometric surface or water table in relation to time since pumping began. The derivation of equations relating well discharge to water-level drawdown and hydraulic properties of aquifers is generally based on the following assumptions (unless stated otherwise for a given situation):

1. The well is pumped, or flows, at a constant rate.
2. The well fully penetrates the aquifer and is screened, perforated, or otherwise open for the entire height of the aquifer.
3. The aquifer is homogeneous, isotropic, horizontal, and of infinite horizontal extent.

4. Water is released from storage in the aquifer or other underground material in immediate response to a drop in water table or piezometric surface.

The hydraulic properties of the aquifer (transmissivity and storage coefficient or specific yield) used to predict well discharge normally are evaluated from the decline of the piezometric surface or water table in the aquifer in response to test-pumping of the well (Chapter 5).

4.2 STEADY FLOW

Confined Aquifers

The flow system around a pumped or flowing well in a confined aquifer can be analyzed with the Dupuit-Forchheimer assumption of horizontal flow. At steady state, the flow in the aquifer comes from far (theoretically infinitely far) away, so that the flow across an imaginary cylindrical surface in the aquifer at radius r from the pumped well is the same as the flow Q from the well (Figure 4.1). Since the surface area of the cylinder is $2\pi rD$, this flow can be expressed with Darcy's equation as

$$Q = K2\pi rD \, (dh/dr) \tag{4.1}$$

where Q = flow from well (volume/time)
K = hydraulic conductivity of aquifer
r = radial distance from well center
D = height of aquifer
dh/dr = hydraulic gradient (slope of piezometric head h at distance r from pumped well)

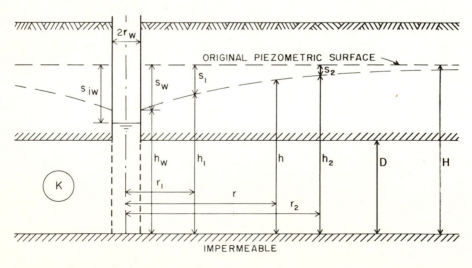

Figure 4.1 Geometry and symbols for pumped well in confined aquifer.

The piezometric head h is expressed with respect to the bottom of the aquifer (Figure 4.1). Equation (4.1) can be written as

$$Q \; dr/r = 2\pi KD \; dh \tag{4.2}$$

which after integrating between two points at different distances from the well (r_2, h_2 and r_1, h_1) yields

$$Q = \frac{2\pi KD(h_2 - h_1)}{\ln (r_2/r_1)} \tag{4.3}$$

or:

$$Q = \frac{2\pi T(h_2 - h_1)}{\ln (r_2/r_1)} \tag{4.4}$$

where T is the transmissivity of the aquifer. This equation is known as the Thiem equation, after the father-son team of Adolph and Gunther Thiem that developed this equation late in the nineteenth century (see Lohman, 1972).

Equation (4.4) makes it possible to calculate Q in relation to h at two different distances from the well. By taking r_2 large, so that h_2 approaches the original or static height H of the piezometric surface (Figure 4.1) and taking r_1 as the well radius r_w, the relation between Q and the height h_w of the piezometric surface at the well can be calculated. For an ideal well in a confined aquifer, the height of the water level in the well will then be equal to h_w. Normally, however, there are additional friction losses as water enters the well through screens or slots in the casing, so that the water depth inside the well tends to be less than h_w (see Section 4.5). The r value where withdrawal of water from the well causes an insignificant decline of the piezometric surface or water table is called the *radius of influence* of the well.

Unconfined Aquifers

For unconfined aquifers, the factor D in Eq. (4.1) is replaced by the height h of the water table above the lower boundary of the aquifer (Figure 4.2), yielding

$$Q = 2\pi rhK \; (dh/dr) \tag{4.5}$$

Separating variables and integrating between r_2, h_2 and r_1, h_1 then gives

$$Q = \frac{\pi K(h_2^2 - h_1^2)}{\ln (r_2/r_1)} \tag{4.6}$$

The term $h_2^2 - h_1^2$ in this equation can be written as $2(h_2 - h_1)(h_2 + h_1)/2$, where $(h_2 + h_1)/2$ is the average height of the aquifer between r_2 and r_1. The product $K(h_2 + h_1)/2$ then represents the average transmissivity T_h between r_2 and r_1. When T_h is substituted for $K(h_2 + h_1)/2$, Eq. (4.6) becomes identical to Eq. (4.4). Thus, the Thiem equation also applies to unconfined aquifers, as long as the effect of water-table drop on transmissivity is taken into account.

For unconfined aquifers, the height h_{iw} of the water level inside a pumped well is always less than the height h_w of the water table adjacent to the well, even if head

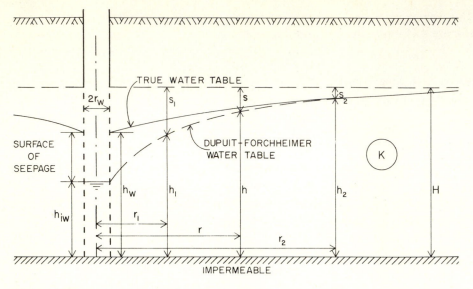

Figure 4.2 Geometry and symbols for pumped well in unconfined aquifer.

losses due to entry of water into the well are ignored. This is because of a surface of seepage or seepage face that develops when water freely drains out of a wall of saturated material that is exposed to the atmosphere (Figure 4.2). The water table in the saturated material will then intersect the exposed wall at some distance above the free-water level at the other side of the wall, forming a surface of seepage where water moves out of the saturated material and then down along the saturated face to the free-water surface. If Eq. (4.6) is used to calculate Q, h_{iw} should be used for h at r_w, while taking r_2 large enough so that h_2 can be taken as H. Theoretical and experimental investigations have shown that the use of h_{iw} for h at r_w in Eq. (4.6) yields Q values that are within 1 or 2 percent of the true values (Hantush, 1964, and references therein). These statements apply only to ideal wells, where there are no head losses as water enters the well through slots or screens. Where such head losses occur, the height of the water level inside the well will be less than h_{iw}.

Because of the surface of seepage at the well and the occurrence of vertical-flow components in the vicinity of the well, Eq. (4.6), which is based on the assumption of horizontal flow (Dupuit-Forchheimer assumption), does not yield an accurate prediction of the water-table height near the well. However, at $r > 1.5H$, the effects of the vertical flow and of the seepage surface have become negligible and Eq. (4.6) yields essentially correct estimates of the water-table position. The water-table height h_w at r_w (ignoring well losses) can be calculated with the following equation for the height of the seepage surface:

$$h_w - h_{iw} = \frac{(h_2 - h_{iw})[1 - (h_{iw}/h_2)^{2.4}]}{(1 + 5r_w/h_2)[1 + 0.02 \ln (r_2/r_w)]} \qquad (4.7)$$

where r_2 must be taken as $500r_w$. This equation was developed by Hall (1955) on the basis of numerical and experimental solutions. Other equations for the height of the seepage face were developed by Boulton (1951). Petersen (1957) presented a graph relating h_w/r_w to Q/Kr_w^2 for different values of h_{iw}/r_w and h_{115}/r_w, where h_{115} is h at $115r_w$. Hall (1955) also presented equations for the water-table position between the well and a distance of $500r_w$ from the well.

Unconfined Aquifers with Uniform Recharge

True equilibrium well-flow conditions may be approached in unconfined aquifers that are replenished from above by excess rainfall, deep percolation from irrigation, or other water seeping down in the vadose zone. Assuming a uniform recharge rate v (as may occur in maritime climates with long periods of low-intensity rain and in large irrigated areas during the irrigation season), the flow Q from the well can be calculated as

$$Q = \pi r_v^2 v \tag{4.8}$$

where r_v is the radius of influence of the well (radius of recharge area supplying Q). The flow q across an imaginary vertical cylinder with radius r ($r < r_v$, Figure 4.3) can be expressed as

$$q = Q - \pi r^2 v \tag{4.9}$$

The flow q can also be expressed with Darcy's equation as

$$q = 2\pi r h K \ (dh/dr) \tag{4.10}$$

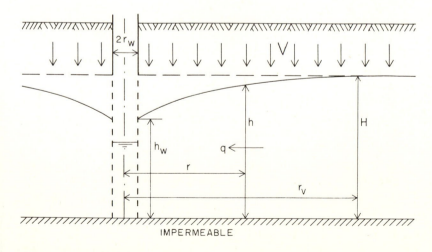

Figure 4.3 Pumped well in unconfined aquifer in equilibrium with vertical recharge.

Equating Eqs. (4.9) and (4.10), substituting Eq. (4.8) for Q, and separating variables gives

$$\left(\frac{r_v^2}{r} - r\right) dr = 2\frac{K}{v} h\, dh \tag{4.11}$$

Integrating this equation between r_v, H and r_w, h_{iw} then yields

$$r_v^2 \ln\frac{r_v}{r_w} - \frac{1}{2}(r_v^2 - r_w^2) = \frac{K}{v}(H^2 - h_{iw}^2) \tag{4.12}$$

Substituting $Q/\pi v$ for r_v^2 before the ln term and multiplying the equation by v/K gives

$$\frac{Q}{\pi K}\ln\frac{r_v}{r_w} - \frac{v}{2K}(r_v^2 - r_w^2) = H^2 - h_{iw}^2 \tag{4.13}$$

If v is known, r_v can be computed for a given Q and h_{iw} can be calculated with Eq. (4.13). The equation can also be used to estimate v for an existing well if the other parameters are known, or to estimate r_v if v and the other parameters are known.

Leaky Aquifers

Confined aquifers may be recharged from above if the upper confining layer is semipermeable. Such aquifers are then called *leaky* aquifers or *semiconfined* aquifers. The upper confining layer may connect the confined aquifer to an overlying unconfined aquifer (Figure 4.4, left), or the upper confining layer itself may have a water table (Figure 4.4, right). The latter situation may occur in low-lying alluvial valleys or plains, where sands and gravels are covered by loams and clays with a free-water table.

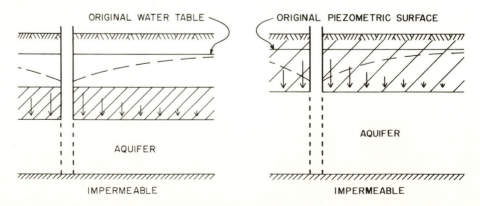

Figure 4.4 Vertical recharge of leaky aquifer from unconfined aquifer above upper confining layer (left), or with water table in upper confining layer (right).

At static conditions, the water table will coincide with the piezometric surface of the leaky aquifer for both cases illustrated in Figure 4.4. When the well is pumped, however, the piezometric surface drops below the water table, causing water to move downward through the upper confining layer and into the leaky aquifer. The assumption is usually made that the water table in the unconfined aquifer or in the upper confining layer itself remains unchanged, which will be valid during the first stages of pumping. Since the downward flow is proportional to the vertical difference between the water table and the piezometric surface (see Section 3.7), the recharge rate of the leaky aquifer decreases with increasing r (Figure 4.4). This must be taken into account in mathematical analyses of the flow system. The same applies to leaky aquifers that are recharged from below. Under those conditions, the lower confining layer is an aquitard which separates the aquifer from an underlying aquifer with higher piezometric surface. Equations relating Q to drawdown of piezometric surface and hydraulic properties of leaky aquifers are presented in Section 5.2.4.

4.3 TRANSIENT FLOW

Equations developed for transient well flow normally show how the drawdown s of the piezometric surface or water table ($s = H - h$) is related to time of pumping the well. In addition to the well and aquifer parameters considered so far, the equations also contain the storage coefficient (or specific yield) S of the aquifer and the time of pumping. Since S is the volume of water released from storage per unit horizontal area and per unit drop of piezometric surface or water table, the rate $\delta V/\delta t$ at which a certain volume V is released from storage over an aquifer area A can be calculated as

$$\frac{\delta V}{\delta t} = -\frac{\delta h}{\delta t} SA \tag{4.14}$$

where V = volume of water released per horizontal area A of aquifer
$\quad\quad h$ = height of piezometric surface or water table above lower boundary of aquifer
$\quad\quad S$ = storage coefficient or specific yield
$\quad\quad A$ = area of aquifer to which $\delta h/\delta t$ applies
$\quad\quad t$ = time

The minus sign in Eq. (4.14) is introduced because $\delta h/\delta t$ is negative (h decreases with increasing t).

For an annular area of infinitesimal radial thickness dr at radius r from the pumped well (Figure 4.5), the volume rate of water release can thus be expressed as

$$\frac{\delta V}{\delta t} = -\frac{\delta h}{\delta t} S2\pi r \, dr \tag{4.15}$$

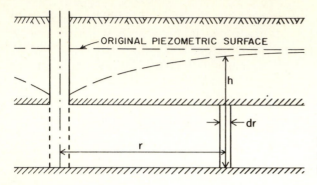

Figure 4.5 Geometry and symbols for derivation of basic equation for transient well flow.

This volume rate of water release is equal to the increase in flow q as it passes through the annular area. Since the rate of increase in q can be expressed as $-(\delta q/\delta r)$ (the minus sign is necessary because q increases with decreasing r), the increase in q through the annular area is $-(\delta q/\delta r)\, dr$. Substituting this term for $\delta V/\delta t$ in Eq. (4.15) yields

$$\frac{\delta q}{\delta r}\, dr = \frac{\delta h}{\delta t}\, S 2\pi r\, dr \tag{4.16}$$

or

$$\frac{\delta q}{\delta r} = \frac{\delta h}{\delta t}\, S 2\pi r \tag{4.17}$$

For confined aquifers, the flow q toward the well at distance r can also be expressed with Darcy's equation as

$$q = 2\pi r T \frac{\delta h}{\delta r} \tag{4.18}$$

which after differentiating with respect to r yields

$$\frac{\delta q}{\delta r} = 2\pi T \frac{r\,(\delta h/\delta r)}{\delta r}$$

or

$$\frac{\delta q}{\delta r} = 2\pi T \left(\frac{\delta h}{\delta r} + r \frac{\delta^2 h}{\delta r^2} \right) \tag{4.19}$$

Combining Eqs. (4.17) and (4.19) then gives

$$2\pi S r \frac{\delta h}{\delta t} = 2\pi T \left(\frac{\delta h}{\delta r} + r \frac{\delta^2 h}{\delta r^2} \right) \tag{4.20}$$

which can be simplified to

$$\frac{1}{r}\frac{\delta h}{\delta r} + \frac{\delta^2 h}{\delta r^2} = \frac{S}{T}\frac{\delta h}{\delta t} \tag{4.21}$$

This is the basic equation for transient flow to a well, essentially following the derivation by Stallman (see Lohman, 1972). The equation can also be used for unconfined aquifers if T can be considered constant (drawdown near well small compared to height of aquifer). Equation (4.21) is an expression of the Laplace equation for transient flow (Section 7.1) in radial coordinates.

Confined Aquifers

A solution of Eq. (4.21) was developed by Theis (1935) for a well of infinitesimally small diameter in a confined aquifer, using heat-flow theory as analogy. The resulting equation is

$$s = \frac{Q}{4\pi T} \int_u^\infty \left(\frac{e^{-u}}{u}\right) du \qquad (4.22)$$

where s is the drawdown $H - h$ of the piezometric surface (Figure 4.1), Q is the constant well discharge, and

$$u = \frac{r^2 S}{4Tt} \qquad (4.23)$$

The solution of Eq. (4.22) is

$$s = \frac{Q}{4\pi T} \left| -0.577\,216 - \ln u + u - \frac{u^2}{2 \cdot 2!} + \frac{u^3}{3 \cdot 3!} - \cdots \right| \qquad (4.24)$$

The function between brackets in Eq. (4.24) is called the well function $W(u)$, and it is listed in Table 4.1 for different values of u (taken from Ferris et al., 1962). Both u and $W(u)$ are dimensionless. Equation (4.24) can thus be written as

$$s = \frac{QW(u)}{4\pi T} \qquad (4.25)$$

To calculate s versus t at a given distance from the well, u is calculated with Eq. (4.23), $W(u)$ is found from Table 4.1, and s is computed with Eq. (4.25).

   Example. The fall of the piezometric surface at distances of 100 m and 200 m from a pumped well will be calculated for a confined aquifer with $T = 1\,000$ m^2/day and $S = 0.000\,1$. The well is pumped for 10 days at a rate of $1\,000$ m^3/day. A number of t values is selected (Table 4.2, first column), u is calculated according to Eq. (4.23), the corresponding values of $W(u)$ are found from Table 4.1, and s is calculated with Eq. (4.25). The resulting s values are plotted downward against time in Figure 4.6 to produce hydrographs of the piezometric surface at 100 m and 200 m, as may be observed with piezometers or observation wells. The piezometric surfaces initially decline rapidly, but then fall at a slower rate as pumping is continued. After about 2 days of pumping, the curves become virtually parallel. When s is plotted against $\log t$, essentially straight lines are obtained except when t is very small (less than 0.002 days in this example).

Table 4.1 Values of $W(u)$ for different values of u

u \ N	N	$N \times 10^{-1}$	$N \times 10^{-2}$	$N \times 10^{-3}$	$N \times 10^{-4}$	$N \times 10^{-5}$	$N \times 10^{-6}$	$N \times 10^{-7}$	$N \times 10^{-8}$	$N \times 10^{-9}$	$N \times 10^{-10}$	$N \times 10^{-11}$	$N \times 10^{-12}$	$N \times 10^{-13}$	$N \times 10^{-14}$	$N \times 10^{-15}$
1	0.219	1.82	4.04	6.33	8.63	10.9	13.2	15.5	17.8	20.1	22.4	24.8	27.1	29.4	31.7	34.0
1.2	0.158	1.66	3.86	6.15	8.45	10.8	13.1	15.4	17.7	20.0	22.3	24.6	26.9	29.2	31.5	33.8
1.5	0.100	1.46	3.64	5.93	8.23	10.5	12.8	15.1	17.4	19.7	22.0	24.3	26.6	29.0	31.3	33.6
2	0.0489	1.22	3.35	5.64	7.94	10.2	12.5	14.8	17.2	19.5	21.8	24.1	26.4	28.7	31.0	33.3
2.2	0.0372	1.14	3.26	5.54	7.84	10.1	12.4	14.6	17.1	19.4	21.7	24.0	26.3	28.6	30.9	33.2
2.5	0.0249	1.04	3.14	5.42	7.72	10.0	12.3	14.6	16.9	19.2	21.5	23.8	26.1	28.4	30.7	33.0
3	0.0130	0.906	2.96	5.23	7.53	9.84	12.1	14.4	16.7	19.0	21.3	23.6	26.0	28.3	30.6	32.9
3.2	0.0101	0.858	2.90	5.17	7.47	9.77	12.1	14.4	16.7	19.0	21.3	23.6	25.9	28.2	30.5	32.8
3.5	0.00697	0.794	2.81	5.08	7.38	9.68	12.0	14.3	16.6	18.9	21.2	23.5	25.8	28.1	30.4	32.7
4	0.00378	0.702	2.68	4.95	7.25	9.55	11.9	14.2	16.5	18.8	21.1	23.4	25.7	28.0	30.3	32.6
4.2	0.00300	0.670	2.63	4.90	7.20	9.50	11.8	14.1	16.4	18.7	21.0	23.3	25.6	27.9	30.2	32.5
4.5	0.00207	0.625	2.57	4.83	7.13	9.43	11.7	14.0	16.3	18.6	20.9	23.2	25.5	27.9	30.2	32.5
5	0.00115	0.560	2.47	4.73	7.02	9.33	11.6	13.9	16.2	18.5	20.8	23.1	25.4	27.7	30.0	32.4
5.2	0.000909	0.536	2.43	4.69	6.98	9.29	11.6	13.9	16.2	18.5	20.8	23.1	25.4	27.7	30.0	32.3
5.5	0.000641	0.503	2.38	4.63	6.93	9.23	11.5	13.8	16.1	18.4	20.7	23.0	25.3	27.7	30.0	32.3
6	0.000360	0.454	2.30	4.54	6.84	9.14	11.4	13.7	16.1	18.4	20.7	23.0	25.3	27.6	29.9	32.2
6.2	0.000286	0.437	2.26	4.51	6.81	9.11	11.4	13.7	16.0	18.3	20.6	22.9	25.2	27.5	29.8	32.1
6.5	0.000203	0.411	2.22	4.47	6.76	9.06	11.4	13.7	16.0	18.3	20.6	22.9	25.2	27.5	29.8	32.1
7	0.000115	0.374	2.15	4.39	6.69	8.99	11.3	13.6	15.9	18.2	20.5	22.8	25.1	27.4	29.7	32.0
7.2	0.0000922	0.360	2.12	4.36	6.66	8.96	11.3	13.6	15.9	18.2	20.5	22.8	25.1	27.4	29.7	32.0
7.5	0.0000658	0.340	2.09	4.32	6.62	8.92	11.2	13.5	15.8	18.1	20.4	22.7	25.0	27.3	29.6	32.0
8	0.0000377	0.311	2.03	4.26	6.55	8.86	11.2	13.5	15.8	18.1	20.4	22.7	25.0	27.3	29.6	31.9
8.2	0.0000301	0.300	2.00	4.23	6.53	8.83	11.1	13.4	15.7	18.0	20.3	22.6	24.9	27.3	29.6	31.9
8.5	0.0000216	0.284	1.97	4.20	6.49	8.80	11.1	13.4	15.7	18.0	20.3	22.6	24.9	27.2	29.5	31.8
9	0.0000124	0.260	1.92	4.14	6.44	8.74	11.0	13.3	15.6	17.9	20.3	22.6	24.9	27.2	29.5	31.8
9.2	0.00000999	0.251	1.90	4.12	6.41	8.72	11.0	13.3	15.6	17.9	20.2	22.5	24.8	27.1	29.4	31.7
9.5	0.00000718	0.239	1.87	4.09	6.38	8.68	11.0	13.3	15.6	17.9	20.2	22.5	24.8	27.1	29.4	31.7
10	0.00000415	0.219	1.82	4.04	6.33	8.63	10.9	13.2	15.5	17.8	20.1	22.4	24.8	27.1	29.4	31.7

Source: From Ferris et al., 1962. Reference to the original table is made for more detail and more significant figures.

Table 4.2 Calculation of s in relation to t, for example

t, days	r = 100 m			r = 200 m		
	u	W(u)	s, m	u	W(u)	s, m
0.001	0.25	1.044	0.083	1	0.219	0.017
0.005	0.05	2.468	0.196	0.2	1.223	0.097
0.01	0.025	3.136	0.249	0.1	1.823	0.145
0.05	0.005	4.726	0.376	0.02	3.355	0.267
0.1	0.002 5	5.417	0.431	0.01	4.038	0.322
0.5	0.000 5	7.024	0.559	0.002	5.639	0.449
1	0.000 25	7.717	0.614	0.001	6.331	0.504
5	0.000 05	9.326	0.742	0.000 2	7.940	0.632
10	0.000 025	10.019	0.797	0.000 1	8.633	0.687

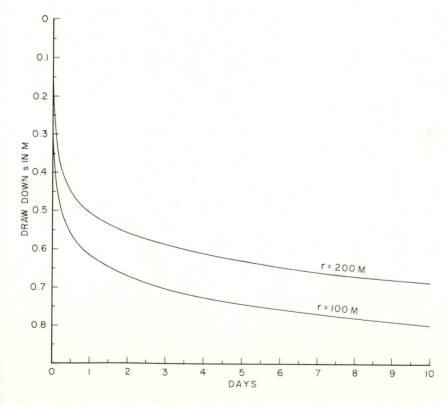

Figure 4.6 Fall of piezometric surface in observation wells at 100 m and 200 m from pumped well in example.

Simplified Solution

A simplified solution of Eq. (4.24) was developed by Cooper and Jacob (1946), who found that if u is small (for example < 0.01), only the first two terms of the series between brackets are significant. Equation (4.24) can thus be reduced to [substituting Eq. (4.23) for u and bringing the constant under the ln term]

$$s = \frac{Q}{4\pi T} \ln \frac{2.25Tt}{r^2 S} \qquad (4.26)$$

This equation enables the direct calculation of s in relation to r and t for given values of Q, T, and S.

Unconfined Aquifers

Exact solution of Eq. (4.21) for unconfined aquifers is difficult because T changes with t and r as the water table declines during pumping. Also, vertical-flow components may be significant near the well, invalidating the Dupuit-Forchheimer assumption. If s is small compared to H, however, the Theis and Jacob solutions of Eq. (4.21) can also be used for unconfined aquifers. For larger drawdowns, Boulton (1954) presented a solution which is valid if the water depth in the well exceeds $0.5H$ (Figure 4.2). Boulton's equation is

$$s = \frac{Q}{2\pi KH}(1 + C_k)V(t', r') \qquad (4.27)$$

where C_k is a correction factor and $V(t', r')$ is Boulton's well function of t' and r' defined as

$$t' = \frac{Kt}{SH} \qquad (4.28)$$

and

$$r' = \frac{r}{H} \qquad (4.29)$$

The other parameters are as earlier defined. Values of $V(t', r')$ are shown in Table 4.3 for different values of t' and r' (from Boulton, 1954). Reference to the original table is made for expressions for $V(t', r')$ when $t' < 0.01$.

The correction factor C_k varies from about -0.30 to 0.16. When t' is between 0.05 and 5, C_k can be taken as zero with an error of less than 6 percent. When $t' < 0.05$, C_k depends on a number of factors (see Boulton, 1954, or Hantush, 1964). Such small t' values occur only during the first stages of pumping, and the associated s values are usually of minor practical interest. When $t' > 5$, C_k is a function of r' and can be obtained by drawing a curve through the following data points (taken from a graph by Boulton, 1954):

r'	0.03	0.04	0.06	0.08	0.1	0.2	0.4	0.6	0.8	1	2	4
C_k	-0.27	-0.24	-0.19	-0.16	-0.13	-0.05	0.02	0.05	0.05	0.05	0.03	0

Table 4.3 Values of the function $V(t', r')$ for different values of t' and r'

t' \ r'	0.001	0.002	0.003	0.004	0.005	0.006	0.007	0.008	0.009	0.01	0.02	0.03	0.04	0.05	0.06	0.07	0.08	0.09
0.01	2.99	2.30	1.90	1.64	1.42	1.28	1.15	1.04	0.95	0.875	0.474	0.322	0.240	0.192	0.158	0.135	0.118	0.104
0.02	3.68	2.97	2.58	2.30	2.09	1.92	1.76	1.64	1.52	1.42	0.860	0.610	0.468	0.378	0.316	0.270	0.236	0.210
0.03	4.08	3.40	3.00	2.70	2.46	2.28	2.13	2.00	1.88	1.79	1.18	0.860	0.675	0.555	0.465	0.400	0.350	0.310
0.04	4.35	3.68	3.26	2.98	2.75	2.58	2.42	2.29	2.17	2.06	1.42	1.07	0.850	0.710	0.600	0.525	0.460	0.410
0.05	4.58	3.90	3.49	3.20	2.96	2.79	2.64	2.50	2.38	2.28	1.60	1.24	1.010	0.850	0.725	0.630	0.560	0.500
0.06	4.76	4.06	3.65	3.36	3.15	2.96	2.80	2.68	2.56	2.45	1.78	1.40	1.15	0.970	0.840	0.735	0.650	0.585
0.07	4.92	4.20	3.80	3.51	3.30	3.12	2.96	2.82	2.70	2.60	1.91	1.54	1.28	1.09	0.950	0.835	0.740	0.670
0.08	5.08	4.34	3.94	3.65	3.42	3.24	3.09	2.95	2.84	2.72	2.04	1.65	1.39	1.20	1.04	0.925	0.825	0.750
0.09	5.18	4.47	4.05	3.75	3.54	3.35	3.20	3.05	2.95	2.84	2.14	1.75	1.50	1.29	1.14	1.02	0.910	0.825
0.1	5.24	4.54	4.14	3.85	3.63	3.45	3.30	3.15	3.04	2.94	2.25	1.85	1.58	1.38	1.22	1.09	0.985	0.890
0.2	5.85	5.15	4.78	4.50	4.28	4.10	3.93	3.80	3.66	3.56	2.87	2.46	2.20	1.98	1.80	1.65	1.52	1.42
0.3	6.24	5.50	5.12	4.85	4.61	4.43	4.28	4.14	4.01	3.90	3.24	2.84	2.54	2.32	2.14	1.98	1.85	1.74
0.4	6.45	5.75	5.35	5.08	4.85	4.67	4.50	4.38	4.26	4.15	3.46	3.05	2.76	2.54	2.36	2.20	2.07	1.96
0.5	6.65	6.00	5.58	5.25	5.00	4.85	4.70	4.55	4.45	4.30	3.65	3.24	2.95	2.72	2.52	2.38	2.24	2.14
0.6	6.75	6.10	5.65	5.40	5.15	4.98	4.82	4.68	4.56	4.45	3.76	3.37	3.09	2.85	2.67	2.50	2.38	2.26
0.7	6.88	6.20	5.80	5.50	5.25	5.08	4.92	4.80	4.68	4.55	3.90	3.50	3.20	2.99	2.80	2.64	2.50	2.38
0.8	7.00	6.25	5.85	5.60	5.35	5.20	5.00	4.90	4.80	4.65	3.96	3.55	3.26	3.05	2.86	2.71	2.58	2.46
0.9	7.10	6.35	6.00	5.70	5.50	5.30	5.12	5.00	4.90	4.75	4.05	3.65	3.36	3.15	2.96	2.80	2.66	2.55
1	7.14	6.45	6.05	5.75	5.55	5.35	5.20	5.05	4.95	4.83	4.10	3.74	3.45	3.22	3.04	2.90	2.75	2.64
2	7.60	6.88	6.45	6.15	5.92	5.75	5.60	5.50	5.35	5.25	4.59	4.18	3.90	3.68	3.50	3.34	3.20	3.09
3	7.85	7.15	6.70	6.45	6.20	6.00	5.85	5.75	5.60	5.50	4.82	4.42	4.12	3.90	3.72	3.57	3.45	3.31
4	8.00	7.28	6.85	6.58	6.35	6.15	6.00	5.90	5.75	5.70	4.95	4.55	4.26	4.04	3.86	3.70	3.59	3.46
5	8.15	7.35	7.00	6.65	6.50	6.25	6.10	6.00	5.85	5.80	5.05	4.68	4.40	4.19	4.00	3.85	3.71	3.60
6	8.20	7.50	7.10	6.75	6.55	6.35	6.20	6.10	5.95	5.85	5.20	4.78	4.50	4.26	4.09	3.92	3.80	3.69
7	8.25	7.55	7.15	6.85	6.62	6.40	6.30	6.20	6.05	5.95	5.25	4.85	4.58	4.35	4.18	4.00	3.90	3.78
8	8.30	7.60	7.20	6.90	6.70	6.50	6.35	6.25	6.10	6.05	5.30	4.92	4.65	4.40	4.25	4.10	3.95	3.82
9	8.32	7.65	7.25	7.00	6.75	6.55	6.40	6.30	6.15	6.10	5.35	5.00	4.70	4.49	4.30	4.15	4.00	3.90
10	8.35	7.75	7.35	7.05	6.80	6.60	6.45	6.35	6.20	6.14	5.40	5.02	4.80	4.52	4.35	4.19	4.05	3.92

(continued on page 78)

Table 4.3 (Continued)

t' \ r'	0.1	0.2	0.3	0.4	0.5	0.6	0.7	0.8	0.9	1	2	3	4	5
0.01	0.093	0.0430	0.0264	0.0180	0.0132	0.0100	0.0078	0.0062	0.0049	0.0040	0.00057	0.00015		
0.02	0.187	0.0865	0.0530	0.0365	0.0268	0.0205	0.0160	0.0125	0.0100	0.0081	0.00118	0.00020		
0.03	0.278	0.130	0.0800	0.0550	0.0405	0.0310	0.0240	0.0190	0.0150	0.0122	0.00184	0.00032		
0.04	0.368	0.174	0.107	0.0735	0.0540	0.0415	0.0322	0.0255	0.0202	0.0165	0.00244	0.00043		
0.05	0.450	0.215	0.133	0.0920	0.0675	0.0520	0.0400	0.0320	0.0255	0.0206	0.00305	0.00055		
0.06	0.530	0.257	0.160	0.110	0.0810	0.0610	0.0478	0.0380	0.0305	0.0250	0.00365	0.00065		
0.07	0.610	0.298	0.186	0.130	0.0950	0.0725	0.0565	0.0450	0.0360	0.0292	0.00430	0.00078		
0.08	0.680	0.340	0.214	0.148	0.108	0.0825	0.0645	0.0510	0.0412	0.0336	0.00500	0.00090		
0.09	0.750	0.378	0.236	0.164	0.122	0.0930	0.0730	0.0585	0.0470	0.0380	0.00570	0.00105		
0.1	0.815	0.415	0.260	0.180	0.134	0.103	0.0805	0.0640	0.0515	0.0420	0.00635	0.00118		
0.2	1.32	0.750	0.500	0.359	0.268	0.208	0.165	0.132	0.107	0.0880	0.0145	0.00278		
0.3	1.64	1.02	0.700	0.515	0.392	0.308	0.246	0.200	0.164	0.135	0.0238	0.00490		
0.4	1.86	1.22	0.870	0.650	0.510	0.405	0.328	0.268	0.220	0.182	0.0350	0.00750	0.00160	0.00038
0.5	2.03	1.37	1.00	0.770	0.610	0.490	0.400	0.330	0.275	0.230	0.0450	0.0104	0.00240	0.00056
0.6	2.16	1.49	1.12	0.875	0.700	0.570	0.468	0.390	0.325	0.276	0.0580	0.0138	0.00320	0.00080
0.7	2.28	1.60	1.22	0.965	0.775	0.640	0.525	0.445	0.375	0.320	0.0715	0.0175	0.00425	0.00108
0.8	2.36	1.69	1.30	1.04	0.850	0.715	0.600	0.500	0.425	0.364	0.0840	0.0212	0.00525	0.00140
0.9	2.45	1.75	1.38	1.11	0.920	0.775	0.650	0.550	0.475	0.404	0.0980	0.0260	0.00630	0.00165
1	2.54	1.85	1.45	1.18	0.975	0.825	0.700	0.595	0.510	0.444	0.113	0.0310	0.00840	0.00235
2	2.97	2.29	1.88	1.60	1.38	1.22	1.07	0.950	0.840	0.750	0.259	0.0950	0.0330	0.0115
3	3.20	2.50	2.10	1.82	1.60	1.42	1.28	1.15	1.05	0.960	0.388	0.165	0.0700	0.0275
4	3.36	2.66	2.25	1.97	1.75	1.58	1.42	1.30	1.20	1.10	0.495	0.235	0.112	0.0535
5	3.49	2.78	2.38	2.09	1.87	1.69	1.54	1.42	1.30	1.21	0.580	0.300	0.150	0.0715
6	3.59	2.90	2.47	2.18	1.95	1.78	1.65	1.52	1.40	1.30	0.660	0.360	0.195	0.0990
7	3.66	2.96	2.55	2.25	2.04	1.85	1.70	1.58	1.48	1.38	0.730	0.415	0.230	0.125
8	3.74	3.00	2.60	2.32	2.11	1.94	1.79	1.66	1.55	1.44	0.790	0.465	0.272	0.155
9	3.80	3.09	2.67	2.39	2.17	2.00	1.85	1.72	1.60	1.50	0.850	0.515	0.307	0.182
10	3.84	3.12	2.74	2.45	2.24	2.05	1.90	1.77	1.65	1.55	0.890	0.550	0.340	0.210

Note: For $t' > 5$, $V(t', r')$ is about equal to $0.5W[(r')^2/4t]$, which is the well function in Table 4.1.

Source: From Boulton, 1954.

The height h_{iw} of the water level in the well (taking into account the surface of seepage but neglecting well losses) can be calculated from the equation

$$h_{iw}^2 = H^2 - \frac{Q}{\pi K} \ln\left(1.5\sqrt{\frac{Kt}{Sr_w}}\right) \qquad (4.30)$$

which is valid if $Kt/SH > 5$ (Boulton, 1954). If $0.05 < Kt/SH < 5$, h_{iw} is calculated as

$$h_{iw} = H - \frac{Q}{2\pi KH}\left(m + \ln\frac{H}{r_w}\right) \qquad (4.31)$$

where m is a function of Kt/SH and can be obtained from a curve plotted through the following points (Boulton, 1954):

Kt/SH	0.05	0.2	1	5
m	-0.043	0.087	0.512	1.288

The range $Kt/SH < 0.05$ will usually be of minor practical significance. Hantush (1964) has shown how h_{iw} can be calculated with Eq. (4.27) for this case.

Leaky Aquifers and Special Conditions

Equations relating drawdown of piezometric surface to time of pumping for wells in leaky aquifers are presented in Section 5.2.4.

In addition to the "standard" aquifer conditions considered so far, flow systems around pumped wells have also been analyzed for aquifers that are sloping, wedge-shaped, two-layered, anisotropic, or bound at some distance by a solid boundary (rock outcrop) or recharging boundary (stream). Other solutions have been developed for wells not flowing at a constant rate and wells not fully penetrating the aquifer. Literature references for these cases are listed in Section 5.2.6.

4.4 PARTIALLY PENETRATING WELLS

Wells that do not tap the aquifer for its entire height (Figure 4.7) require more drawdown for a given Q than do fully penetrating wells. For confined aquifers, the additional drawdown can be represented by including a dimensionless term s_p in Eq. (4.26), as follows:

$$s_{wp} = \frac{Q}{4\pi T}\left(\ln\frac{2.25Tt}{r^2S} + 2s_p\right) \qquad (4.32)$$

where s_{wp} is the drawdown of the piezometric surface at the well. An expression for s_{wp} was developed by Hantush (1962, 1964). This expression, which contains an infinite series and an integral of exponential functions, was solved by Visocky (1970) for a limited number of parameters. Earlier, a solution by Nisle (1958) of the pressure-head distribution around a partially penetrating well was used by

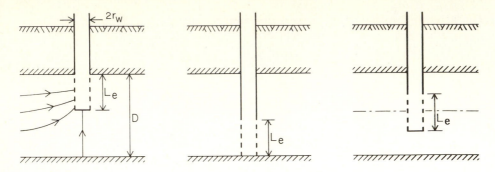

Figure 4.7 Partially penetrating well in confined aquifer with open section at top of aquifer (left), at bottom of aquifer (center), and in center of aquifer (right).

Brons and Marting (1961) to evaluate s_p. Sternberg (1973) then found that the resulting s_p values, though approximate, agreed quite well (at least for practical purposes) with Visocky's solutions of Hantush's expression. Graphs, adapted from Brons and Marting, were presented by Sternberg for ready evaluation of s_p. One of the graphs, showing s_p as a function of D/r_w for different values of L_e/D, is presented as Figure 4.8 (L_e is the length of the open portion of the well; see Figure 4.7.) For the other graph, which has s_p plotted against L_e/D for different values of D/r_w, reference is made to Sternberg (1973). After s_p is evaluated from Figure 4.8, s_{wp} can be calculated with Eq. (4.32).

The performance of a partially penetrating well normally is expressed as an efficiency, defined as the ratio of the flow Q_p from the partially penetrating well at a given drawdown to the flow Q that would be yielded at the same drawdown if the well were completely penetrating. To develop an expression for Q_p/Q, the equilibrium discharge Q of a fully penetrating well is described with Eq. (4.4) as

$$Q = \frac{2\pi T s_w}{\ln{(r/r_w)}} \tag{4.33}$$

where r is the radius of influence of the well (r value where h is essentially equal to H, Figure 4.1). According to Eq. (4.32), the extra drawdown at the well to produce the same flow Q with a partially penetrating well is $Qs_p/2\pi T$, yielding a total drawdown of $s_w + Qs_p/2\pi T$. If the drawdown for the partially penetrating well were the same as that for the fully penetrating well, i.e., s_w, the flow Q_p from the partially penetrating well would be proportionally less, or

$$Q_p = \frac{s_w}{s_w + Qs_p/2\pi T}Q \tag{4.34}$$

Dividing this equation by Q and multiplying the top and bottom of the right-hand side by $2\pi T/Q$ then yields

$$\frac{Q_p}{Q} = \frac{s_w 2\pi T/Q}{s_w 2\pi T/Q + s_p} \tag{4.35}$$

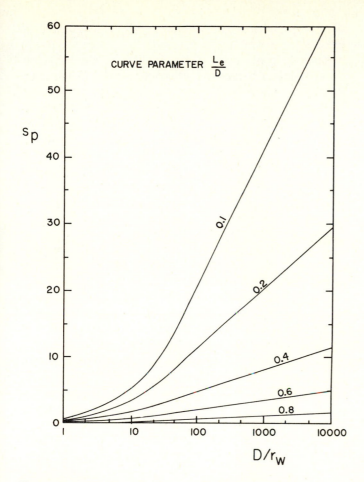

Figure 4.8 Graph of s_p versus D/r_w for different values of L_e/D (on the curves). (*From Sternberg, 1973.*)

However, Eq. (4.33) shows that $s_w 2\pi T/Q$ is equal to $\ln r/r_w$, reducing Eq. (4.35) to

$$\frac{Q_p}{Q} = \frac{\ln (r/r_w)}{\ln (r/r_w) + s_p} \qquad (4.36)$$

The ratio Q_p/Q for a partially penetrating well always exceeds the penetration ratio L_e/D of the well, particularly if r_w is relatively large. This is because vertical-flow components contribute additional inflow to the well (Figure 4.7, left). Because of symmetry, Eq. (4.36) and Figure 4.8 apply equally to wells having their perforated or otherwise open section at the top or at the bottom of the aquifer (Figure 4.7, left and center). If the open section of the well is in the center of the aquifer (Figure 4.7, right), vertical-flow components occur at both the top and bottom of the section. To account for this, the aquifer is split midway along the symmetry line and s_p is evaluated for one-half the length of the open section. Since the half-sections of the well are symmetrical, the resulting Q_p/Q value then represents the efficiency ratio for the entire well.

Example. Assuming a partially penetrating well as in Figure 4.7, left, with $D = 100$ m, $L_e = 10$ m, $r_w = 0.2$ m, and a radius of influence of 500 m, Figure 4.8 shows that for $D/r_w = 500$ and $L_e/D = 0.1$, $s_p = 35.2$. Equation (4.36) then yields $Q_p/Q = 0.18$. The same ratio applies if the open section had been in the lower part of the aquifer (Figure 4.7, center). If the open section is halfway down the aquifer (Figure 4.7, right), the geometry parameters are evaluated for the upper or lower half of the open section and aquifer, yielding $L_e/D = 5/50 = 0.1$ and $D/r_w = 50/0.2 = 250$. Figure 4.8 then gives $s_p = 28$, which when substituted into Eq. (4.36) yields $Q_p/Q = 0.22$. As expected, this efficiency is higher than that for a well with its entrance section at the top or bottom of the aquifer. Both ratios of Q_p/Q exceed the L_e/D ratio of 0.1.

Other equations for calculating Q_p/Q for partially penetrating wells in confined aquifers have been developed by Kozeny (1933), using an analytic treatment and experimental data by Muskat (1937); by De Glee (1930; see also Todd, 1959); and by Huisman (1972). Kirkham (1959) presented an exact computer solution which, however, is not readily amenable for routine or practical application. Li et al. (1954) and Franke (1967) studied flow to partially penetrating wells with electric analogs.

The s_p values in Figure 4.8 probably also give reasonable estimates of Q_p/Q for wells in unconfined aquifers, particularly if the drawdown is small compared to the aquifer height and the well has been pumped for some time. For the early stages of pumping, Q_p/Q can be estimated with equations developed for the slug test to evaluate K of aquifers (Bouwer and Rice, 1976), as discussed in Section 5.3.1. For this purpose, the Thiem equation is modified to

$$Q_p = 2\pi K L_e \frac{H - h_w}{\ln (R_e/r_w)} \tag{4.37}$$

where R_e is the effective radius of the flow system over which the head difference $H - h_w$ is dissipated. Values of R_e were determined with a resistance-network analog, and the results were expressed in empirical equations relating $\ln (R_e/r_w)$ to the geometry of the well and aquifer system [Eqs. (5.42) and (5.43)]. To calculate Q_p/Q for initial stages of pumping, Q_p is determined with Eq. (4.37) using $\ln (R_e/r_w)$ as calculated for the particular L_e value of the well. The value of Q is then computed similarly, calculating $\ln (R_e/r_w)$ for full penetration of the well $(L_e = H)$.

A special type of partially penetrating well is obtained when the intake portion of the well is divided over several screen sections separated by lengths of solid casing. Selim and Kirkham (1974) found that the flow into such a well is greater than that into a well with the same total length of screen in one continuous section. The authors give an example showing that five sections of 1.5-m screen evenly distributed across an aquifer with a height of 15 m give 25 percent more flow into the well than one screen of 7.5-m length in the lower half of the aquifer.

4.5 WELL LOSSES

To determine the total lift for pumping water from a well, the distance s_{iw} between the static piezometric surface or water table and the water level in the well (Figure 4.1) must be known. The value of s_{iw} normally is taken as the sum of the draw-down s_w in the aquifer at the well, and the total-head loss s_e incurred when water moves into the well through the developed zone or gravel envelope and the screen or slots in the casing. The loss s_w is called the *formation* loss because it occurs in the aquifer, while s_e is called the *well* loss.

Since the flow in the aquifer is laminar, s_w will vary linearly with Q, as indicated by the various well-flow equations [for example, Eq. (4.4) or Eq. (4.24)]. Of course s_w also varies with time, but if pumping is continued long enough so that s_w changes only very little, the "final" s_w will be directly proportional to the pumping rate (ignoring the effect of drawdown on transmissivity if the aquifer is unconfined). The flow through the gravel envelope or developed zone outside the well casing and through the screen or other openings in the casing, as well as the flow inside the well itself, however, will usually be turbulent. Thus, s_e can be expected to vary with some power of Q, and the total head loss s_{iw} can be expressed as

$$s_{iw} = C_f Q + C_w Q^n \qquad (4.38)$$

where C_f is the formation constant relating Q to s_w, C_w is the well-loss constant relating Q^n to s_e, and n is the exponent due to turbulence. Jacob (1947) suggested that $n = 2$, while Rorabaugh (1953) concluded that a value of 2.5 may be more appropriate. In actual tests, Lennox (1966) found n values as high as 3.5. Values of n less than 2 may also occur—for example, if Q is relatively low and full tur-bulence has not yet developed in the entire well-entry flow. For very low values of Q, the flow may even be laminar throughout the system, in which case C_w will be zero. The value of C_f could be calculated from one of the well-flow equations. For example, Eq. (4.4) shows that C_f could be estimated as $\ln (r_2/r_w)/2\pi T$, where r_2 is the radius of influence of the well.

The best way to determine C_f, C_w, and n for a given well is by experiment. This can be done with the step-drawdown test, where s_{iw} is measured for successively increasing values of Q. The well is pumped at a certain Q until s_{iw} changes only relatively little; Q is then increased, and s_{iw} is measured after the same time interval used for the first flow rate. This process is repeated until s_{iw} is known for about four or five different Q values. The depth of the water level in the well during pumping is usually measured with the bubble-tube technique (Section 2.1).

Jacob (1947) developed equations for evaluating C_f and C_w from the increase in s_{iw} due to an increase in Q, assuming $n = 2$. Rorabaugh (1953) presented a graphical procedure where n is evaluated from the test results themselves, as are C_f and C_w. For this purpose, Eq. (4.38) is written as

$$\frac{s_{iw}}{Q} - C_f = C_w Q^{n-1} \qquad (4.39)$$

Taking the logarithm of both sides, this equation becomes

$$\log\left(\frac{s_{iw}}{Q} - C_f\right) = \log C_w + (n-1)\log Q \qquad (4.40)$$

which shows that a plot of $s_{iw}/Q - C_f$ versus Q on double-logarithmic paper should yield a straight line with slope $n-1$ and intercept C_w when $s_{iw}/Q - C_f = 1$. Since C_f is not known, however, such a plot cannot be constructed. The procedure then is to assume different values of C_f and plotting $s_{iw}/Q - C_f$ versus Q on double-logarithmic paper until a straight line is obtained. Usually, the first C_f value is taken as zero, which yields a concave curve as shown in Figure 4.9. Then, C_f is successively increased until a straight line is obtained (if C_f is too large, a concave curve on the other side of the straight line will be obtained, as in Figure 4.9). The slope of the straight line is equal to $n-1$, which yields n. To evaluate C_w, the line can be extended until $s_{iw}/Q - C_f = 1$, or C_w can be calculated by substituting C_f, n, and a certain combination of the measured s_{iw} and Q into Eq. (4.38).

Example. The following values of Q and s_{iw} are obtained in a hypothetical step-drawdown test:

Q (m³/day)	1 000	2 000	4 000	6 000	8 000
s_{iw} (m)	4.56	10.74	29.48	58.26	98.41

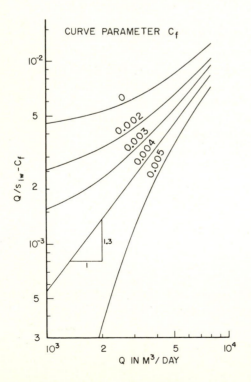

CURVE PARAMETER C_f

Figure 4.9 Double-logarithmic plot of $Q/s_{iw} - C_f$ versus Q for different values of C_f (on the curves) to obtain C_f yielding straight line.

On the basis of these data, $Q/s_{iw} - C_f$ is plotted against Q on double-logarithmic paper, first taking $C_f = 0$ and then larger values for C_f until a straight line is obtained. As shown in Figure 4.9, this is the case if $C_f = 0.004$. The slope of the resulting line is 1.3, yielding $n = 2.3$. Taking an arbitrary combination of Q and s_{iw}—for example, 4000 m^3/day and 29.48 m—and substituting these values along with $n = 2.3$ and $C_f = 0.004$ into Eq. (4.38) then yields $C_w = 7 \times 10^{-8}$. Using these values of C_f, C_w, and n, the formation losses and well losses for the Q values in this example are calculated as follows:

Q (m^3/day)	1000	2000	4000	6000	8000
s_w (m)	4	8	16	24	32
s_e (m)	0.556	2.74	13.48	34.26	66.41

Because well losses vary with Q raised to a power of about 2 or 3, s_e increases rapidly with increasing Q until at high Q values most of s_{iw} may consist of well losses. This shows the importance of having properly constructed wells with sufficient screen or slot area and sufficient radius r_w to keep well losses to a minimum. While r_w has only a minor effect on the formation loss s_w [as shown by well-flow equations such as Eq. (4.4)], it has a very significant effect on the well-entry loss s_e because a larger r_w means lower entry velocities. Since s_e varies with the nth power of the entry velocity, considerable reductions in s_e can be obtained by increasing r_w. For example, doubling r_w will reduce the entry velocity by 50 percent, which will reduce entry losses by 75 percent if $n = 2$, and by 87.5 percent if $n = 3$.

The step-drawdown test gives information regarding the relation between Q and s_{iw} of a given well, which is important in selecting the optimum pump and depth of pumping (Section 6.4.3). The test also shows how much head loss occurs in the aquifer, and how much in and around the well. Excessive well losses indicate poor design and construction, poor development of the well, or deterioration of the screen. Also, the C_f value yielded by the test can be used to estimate T of the aquifer, using the appropriate well-flow equation relating s_w to Q.

In addition to the graphical trial-and-error procedure of Rorabaugh (1953), a method for direct analysis of step-drawdown data using type curves was developed by Sheahan (1971). A computer program for processing the data was presented by Labadie and Helweg (1975). For additional details and field application of the technique, reference is made to Lennox (1966).

4.6 SPECIFIC CAPACITY

The specific capacity of a well is the well flow per unit drop of water level in the well, or Q/s_{iw}. Specific capacity is not a constant. First, since s_{iw} generally continues to increase during pumping (even though eventually very slowly), the specific capacity of a well decreases with continued pumping. Second, since well loss varies with Q^n where n may be 2 or 3 more, the specific capacity of a certain well will decrease with increasing Q. Third, s_{iw} may increase faster than Q for unconfined aquifers because the drawdown of the water table reduces the transmissivity. However, specific capacity still is a useful concept because it describes the productivity of both aquifer and well in a single parameter. A decline in specific capacity may indicate deterioration of the well screen or perforations, or declining S or T values in aquifer(s) due to declining water tables or piezometric surfaces, for example.

Sometimes, specific capacities of wells in a certain aquifer system are used to determine the transmissivity distribution of the aquifer. The relation between T and specific capacity is then determined for a few wells (using pumping tests, Chapter 5), after which T for the other wells is inferred from their specific capacities. This may be a valid procedure if the wells are of similar depth and construction and situated in the same aquifer system. The resulting relations, however, may not be valid for other wells in different groundwater systems. Specific capacities of wells in one part of an aquifer have been divided by aquifer thickness to give a parameter that can be used to predict yields of wells in relation to aquifer thickness for other parts of the aquifer—assuming, of course, that the aquifer material itself is uniform (Walton and Neill, 1963). Summers (1972) found substantially different specific capacities in wells located close together in crystalline rocks.

4.7. INFILTRATION GALLERIES AND RADIAL COLLECTOR WELLS

Horizontal wells or infiltration galleries are mostly installed below or adjacent to streams or other surface water to collect groundwater that derives primarily from stream seepage (see Section 6.3.1). When located in a permeable layer between two less permeable, confining layers, the yield of the gallery can be estimated from equations describing the discharge of a horizontal drain in a confined aquifer (see, for example, Ferris et al., 1962, and Lohman, 1972).

A special type of horizontal well is the radial collector well, which consists of a number (usually 4 to 16) of horizontal collectors extending radially and symmetrically from a central caisson (Figure 4.10; see also Section 6.3.1). The flow system around such a well is very complex, and the simplest approach is to treat the system as flow to a vertical well with a radius that is about 75 to 85 percent of the lateral extent of the collectors (Mikels and Klaer, 1956). Analytic solutions of the flow system around radial collector wells were presented by Hantush and Papadopulos (1962); see also Hantush (1964) and Walton (1970). Empirical equations to calculate the flow from radial collector wells were reported by Milojevic (1963). Some equations, however, must be used with caution because they can yield well-discharge values which, when substituted in steady-state equations for flow to vertical wells [Eqs. (4.4) and (4.6)], produce radii of equivalent vertical wells that exceed the lateral extent of the collectors.

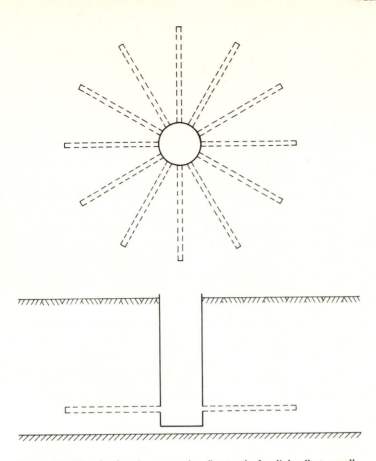

Figure 4.10 Plan (top) and cross section (bottom) of radial collector well.

PROBLEMS

4.1 Using Eq. (4.3) or Eq. (4.6), show that doubling the well diameter generally decreases the drawdown in the aquifer by not more than 10 percent if the well flow is kept constant, or increases the well flow by not more than 10 percent if the drawdown in the aquifer is kept constant (use r_w values in the range of 0.1 to 1 m and r_2 values from a few hundred to a few thousand meters).

4.2 A well with a radius of 0.5 m, including gravel envelope and developed zone, completely penetrates an unconfined aquifer with $K = 30$ m/day and H = 50 m. The well is pumped so that the water level in the well remains at 40 m above the bottom. Assuming that pumping has essentially no effect on water-table height at $r = 500$ m and that well losses are zero, what is the steady-state well discharge?

4.3 Calculate h between $r = 1.5H$ and $r = 500$ m (include calculation of h for $r = 250$ m) and plot h versus r to obtain the water-table position at $r > 1.5H$ for the well and aquifer of Problem 4.2.

4.4 Taking h calculated at $r = 250$ m in Problem 4.3, calculate the height of the seepage surface.

4.5 Calculate h_w in aquifer at well and sketch water table between r_w and $r = 1.5H$ in water-table plot of Problem 4.3.

4.6 For the above conditions, what is C_f in Eq. (4.38)?

4.7 Using the data in Problem 4.2 and assuming $S = 0.15$, calculate with Eq. (4.30) how fast the water level in the well drops after pumping is started (calculate h_{iw} for t values of 1, 5, 10, and 100 days).

4.8 If well losses are taken into account and $n = 2.5$ and $C_w = 4.788 \times 10^{-10}$, what is the total final drawdown of the water level in the well of Problem 4.2, and what is the specific capacity?

4.9 Calculate the drawdown in an observation well 100 m from a pumped well which completely penetrates an unconfined aquifer. The aquifer properties are $K = 20$ m/day, $H = 100$ m, and $S = 0.16$. The well is pumped at 5027 m³/day. Calculate s with Eq. (4.27) for $t = 0.08, 0.32, 0.8, 3.2, 8, 32,$ and 80 days. Compare the results with the s-vs.-t data calculated in Problem 5.1 with the Theis solution for the same conditions. Why do the results agree so well?

4.10 Calculate the specific capacity of the well in the example of Section 4.5 (Well Losses) for the five different Q values. Why does the specific capacity decrease with increasing Q?

REFERENCES

Boulton, N. S., 1951. The flow pattern near a gravity well in a uniform water-bearing medium. *J. Inst. Civ. Eng.* (London) **36:** 534–550.

Boulton, N. S., 1954. The drawdown of water table under non-steady conditions near a pumped well in an unconfined formation. *Proc. Inst. Civ. Eng. (London)* **3,** part III: 564–579.

Bouwer, H., and R. C. Rice, 1976. A slug test for determining hydraulic conductivity of unconfined aquifers with completely or partially penetrating wells. *Water Resour. Res.* **12:** 423–428.

Brons, F., and V. E. Marting, 1961. The effect of restricted fluid entry on well productivity. *Pet. Technol.* **13**(2): 172–174.

Cooper, H. H., Jr., and C. E. Jacob, 1946. A generalized graphical method for evaluating formation constants and summarizing well-field history. *Trans. Am. Geoph. Union* **27:** 526–534.

De Glee, G. J., 1930. "Over grondwaterstromingen bij wateronttrekking door middel van putten." Doctoral dissertation, Techn. Univ. of Delft. (Printed by J. Waltman, Delft, Netherlands, 175 pp.)

Ferris, J. G., D. B. Knowles, R. H. Brown, and R. W. Stallman, 1962. Theory of aquifer tests. *U.S. Geol. Survey Water Supply Paper 1536-E,* pp. 69–174.

Franke, O. L., 1967. Steady-state discharge to a partially penetrating artesian well: an electrolyte-tank model study. *Ground Water* **5**(1): 29–34.

Hall, H. P., 1955. An investigation of steady flow toward a gravity well. *Houille Blanche* **10:** 8–35.

Hantush, M. S., 1962. Aquifer tests on partially penetrating wells. *Trans. Am. Soc. Civ. Eng.* **127,** part 1: 284–308.

Hantush, M. S., 1964. Hydraulics of wells. In *Advances in Hydroscience,* V. T. Chow (ed.), Academic Press, New York and London, vol. 1, pp. 281–432.

Hantush, M. S., and I. S. Papadopulos, 1962. Flow of ground water to collector wells. *J. Hydraul. Div., Am. Soc. Civ. Eng.* **88**(HY5): 221–244.

Huisman, L., 1972. *Ground Water Recovery.* Winchester Press, New York.

Jacob, C. E., 1947. Drawdown test to determine effective radius of artesian well. *Trans. Am. Soc. Civ. Eng.* **112:** 1047–1070.

Kirkham, D., 1959. Exact theory of flow into a partially penetrating well. *J. Geophys. Res.* **64:** 1317–1327.

Kozeny, J., 1933. Theorie und Berechnung der Brunnen. *Wasserkraft und Wasserwirtschaft* **28:** 101–105.

Labadie, J. W., and O. J. Helweg, 1975. Step-drawdown test analysis by computer. *Ground Water* **13:** 438–444.

Lennox, D. H., 1966. Analysis and application of step-drawdown test. *J. Hydraul. Div., Proc. Am. Soc. Civ. Eng.* **92**(HY6): 25–48.

Li, W. H., P. Bock, and G. Benton, 1954. A new formula for flow into partially penetrating wells in aquifers. *Trans. Am. Geophys. Un.* **35:** 806–811.

Lohman, S. W., 1972. Groundwater hydraulics. *U.S. Geol. Survey Prof. Paper 708,* 70 pp.

Mikels, F. C., and F. H. Klaer, Jr., 1956. Application of groundwater hydraulics to the development of water supplies by induced infiltration. In *Symposia Darcy* (*Dijon, France*), vol. 2, pp. 232–242. Internat. Assoc. Scientific Hydrology, Publ. 41.

Milojevic, M., 1963. Radial collector wells adjacent to the river bank. *J. Hydraul. Div., Am. Soc. Civ. Eng.* **89**(HY6): 133–151.

Muskat, M., 1937. *The Flow of Homogeneous Fluids Through Porous Media.* McGraw-Hill, New York. (Second printing by J. W. Edwards, Ann Arbor, Mich.)

Nisle, R. G., 1958. The effect of partial penetration on pressure buildup in oil wells. *Trans. Am. Inst. Min. Metall. Pet. Eng.* **213**: 85–90.

Petersen, D. F., 1957. Hydraulics of wells. *Trans. Am. Soc. Civ. Eng.* **122**: 502–517.

Rorabaugh, M. I., 1953. Graphical and theoretical analysis of step-drawdown test of artesian well. *Proc. Am. Soc. Civ. Eng.* **79,** separate no. 362, 23 pp.

Selim, S. M., and D. Kirkham, 1974. Screen theory for wells and soil drainpipes. *Water Resour. Res.* **10**: 1019–1030.

Sheahan, N. T., 1971. Type-curve solution of step-drawdown test. *Ground Water* **9**(1): 25–29.

Sternberg, Y. M., 1973. Efficiency of partially penetrating wells. *Ground Water* **11**(3): 5–8.

Summers, W. K., 1972. Specific capacities of wells in crystalline rocks. *Ground Water* **10**(6): 37–47.

Theis, C. V., 1935. The relation between the lowering of the piezometric surface and the rate and duration of discharge of a well using groundwater storage. *Trans. Am. Geophys. Un.* **16**: 519–524.

Todd, D. K., 1959. *Ground Water Hydrology.* John Wiley & Sons, New York.

Visocky, A. P., 1970. Values of $W(u, r/m, \gamma)$ presented in W. C. Walton, *Groundwater Resources Evaluation*, McGraw-Hill Book Co., New York, Table 3.3, p. 140.

Walton W. C., 1970. *Groundwater Resources Evaluation.* McGraw-Hill Book Co., New York, 664 pp.

Walton, W. C., and J. C. Neill, 1963. Statistical analysis of specific-capacity data for a dolomite aquifer. *J. Geophys. Res.* **68**: 2251–2262.

FIVE

MEASUREMENT OF HYDRAULIC CONDUCTIVITY, TRANSMISSIVITY, SPECIFIC YIELD, AND STORAGE COEFFICIENT

5.1 INTRODUCTION

Hydraulic conductivity K, transmissivity T, and storage coefficient S (specific yield for unconfined aquifers) are the hydraulic properties of aquifers and soil materials that determine how fast water moves into, through, and out of subsurface materials, and how piezometric surfaces or water tables are affected. Much of the success in predicting underground water movement depends on how accurately the pertinent hydraulic parameters can be evaluated. To this end, numerous techniques have been developed—most of them in the last 40 years—to evaluate K, T, or S of aquifers and other materials below a water table. Some techniques enable the measurement of K in the vadose zone. Knowledge of flow and hydraulic properties of this zone is important because the vadose zone is the link between surface water (and the surface environment in general) and groundwater.

Pumped-well techniques traditionally play a prominent role in evaluating hydraulic properties of aquifers and semiconfining layers. The results are used for predicting well yields, positions of water tables or piezometric surfaces, and recharge rates of aquifers; they are also used for developing optimum schemes of groundwater management. Other techniques, such as the auger-hole and piezometer methods, have been developed to measure K of soil profiles where the water table is close to ground surface, or of shallow aquifers where pumped wells are not available. The methods for measuring K in vadose zones require artificial wetting of a small part of the vadose zone in which K is then measured. The results of K measurements at relatively small depths are used to analyze or predict surface-subsurface water relations (infiltration, subsurface runoff, etc.), to design drainage systems and other agricultural water management schemes, to estimate seepage from open channels or reservoirs, to predict groundwater contamination from waste disposal sites, etc.

The methods described in this chapter all measure K or T at or near saturation, whether below a water table or in an artificially wetted part of the vadose zone. Unsaturated flow and hydraulic conductivity of unsaturated materials are discussed in Section 7.7.

5.2 PUMPED-WELL TECHNIQUES

5.2.1 Basic Aspects

With the pumped-well techniques, often simply called *pumping tests*, hydraulic properties of aquifers are determined by pumping a well at a constant rate and observing the drawdown of the piezometric surface or water table in observation wells at some distance from the pumped well. Two types of tests are used: steady-state and nonsteady or transient-state tests. With the steady-state tests, pumping is continued sufficiently long for the water levels in the observation wells to approach equilibrium (true equilibrium will seldom if ever be reached). The equilibrium drawdown then enables calculation of T. With the transient pumping tests, water-level drops in observation wells are measured in relation to time, which then yields not only T but also S. Transient pumping tests are more common than steady-state tests.

The procedures for calculating T and S from pumping-test data are based on the following assumptions, unless differently stated:

1. The aquifer is homogeneous, isotropic, and of infinite horizontal extent.
2. The flow in the aquifer is in a horizontal direction only (Dupuit-Forchheimer flow).
3. Water is released from storage in the aquifer in immediate response to a drop in piezometric surface or water table (for transient methods only). The case of delayed yield is treated separately.
4. There is no flow in the aquifer other than the flow caused by pumping the well.
5. The well is pumped at a constant rate.
6. The volume of water that was inside the well and that is removed by pumping is negligibly small compared to the volume coming out of the aquifer (for transient methods only).
7. The well completely penetrates the aquifer and is screened, perforated, or otherwise open for the entire height of the aquifer.

5.2.2 Observation Wells

Observation wells for measuring drop of piezometric surface or water table in response to pumping may consist of existing wells or of piezometers especially installed for this purpose. At least three observation wells at different distances from the pumped well are desired, so that results can be averaged and obviously erroneous data can be disregarded. Observation wells may be located 10 to 100 m from the pumped well. For thick aquifers, distances of 100 to 300 m may be desirable. Lohman (1972) mentions that a good arrangement consists of a pair of observation wells at distances of one, two, and four times the thickness of the aquifer from the pumped well. Each pair consists of a shallow well reaching just into the aquifer and a deep well extending to the bottom of the aquifer. For unconfined aquifers, observation wells should be at a distance of at least 1.5 times the aquifer thickness from the pumped well to avoid errors due to vertical-flow components in the vicinity of the well (see Section 4.2). Financial constraints and the availability of existing wells at usable but less than ideal locations often force a compromise between what is theoretically desirable and what is practical.

If there is reason to suspect that water levels in observation wells are affected by factors other than pumping the well, efforts should be made to correct the drawdowns for these effects, especially if the drawdowns are relatively small. One possible source of error is a change in barometric pressure (see Section 2.1). To correct for this, the response of the water level in the observation well to barometric pressure should be observed for some period prior to pumping, so that the appropriate correction can be applied when there is a change in barometric pressure during the pumping test. Other sources of error include effects of tides or other changes in surface-water levels on groundwater tables or piezometric surfaces, earth tides (see Section 2.1), pumping of other wells in the aquifer, and recharge or depletion of groundwater. If these factors are expected to significantly affect water levels in observation wells during pumping tests, groundwater levels should be observed for some time prior to pumping, so that trends can be extrapolated to the pumping period for correction of the observed drawdowns.

To minimize the possibility that water levels in observation wells are affected by compression of air in the vadose zone (Section 2.1), pumping tests should be carried out when there is no heavy rainfall, flooding, or irrigation in the area. If the observation well or piezometer inadvertently reaches into a fine-textured, readily compressible layer within the aquifer, the water level in the piezometer may actually rise in response to pumping the well. This is because declining water tables or piezometric surfaces can cause compression of such a layer, which in turn can produce temporary increases in pore-water pressure (Section 9.7). To avoid such reverse water-level reactions, care must be taken that observation wells reach into true aquifer material.

5.2.3 Confined Aquifers

Procedures for evaluating T and S of confined aquifers from pumping-test data can be divided into steady-state and transient-state methods.

Steady-state methods. The evaluation of T with the steady-state approach is based on the Thiem equation. Substituting the drawdown s, defined as

$$s = H - h \tag{5.1}$$

(Figure 4.1) into Eq. (4.4) and solving for T yields

$$T = \frac{Q \ln (r_2/r_1)}{2\pi(s_1 - s_2)} \tag{5.2}$$

This equation enables calculation of T from the pumping rate Q and the equilibrium drawdowns s_1 and s_2 measured in two observation wells at distances of r_1 and r_2 from the pumped well, respectively.

Theoretically, the water levels in the observation wells will never reach equilibrium. They may approach the equilibrium position sufficiently closely, however, to yield reasonably accurate estimates of T. A factor in favor of the steady-state test is that, after a while, the water levels in the observation wells drop at about the same rate (Figure 4.6), so that $s_1 - s_2$ will already be essentially constant while the water levels in the observation wells are still falling. For example, Figure 4.6 shows that the difference of 0.11 m between s_1 and s_2 after 9 days' pumping is already reached after 1 day even though equilibrium conditions are not even approached in such a short time.

After T has been calculated with Eq. (5.2), S can be determined in a relatively simple manner with the transient-flow equations (4.23) and (4.25) if the drawdown s in one of the observation wells is measured at a certain time t. Since T, Q, and s are then known, $W(u)$ can be calculated with Eq. (4.25) and the corresponding value of u can be found from Table 4.1. Substituting this u value into Eq. (4.23) then enables calculation of S. Other procedures for calculating S were given by Lohman (1972).

Example. The curves in Figure 4.6 indicate that $s_1 - s_2$ remains essentially constant at 0.11 m after 1 day of pumping. Substituting this value into Eq. (5.2)—along with $Q = 1\,000$ m³/day, $r_1 = 100$ m, and $r_2 = 200$ m—yields $T = 1\,003$ m²/day. Taking an arbitrary point on the s-vs.-t curve for $r_1 = 100$ m in Figure 4.6—for example, $s = 0.615$ m and $t = 1$ day—and substituting $s = 0.615$ m, $T = 1\,003$ m²/day, and $Q = 1\,000$ m³/day into Eq. (4.25) yields $W(u) = 7.751$. Table 4.1 shows that the corresponding value of u is 0.000 242, which when substituted into Eq. (4.23) along with $T = 1\,003$ m²/day, $r = 100$ m, and $t = 1$ day, yields $S = 0.000\,097$. The values of T and S calculated in this manner agree closely with the "true" values of $1\,000$ m²/day and $0.000\,1$, respectively, used in calculating the s-vs.-t curves in Figure 4.6.

Transient-state methods. The calculation of T and S from the rate of drawdown of water levels in observation wells due to pumping a test well at constant rate Q is based on the Theis equation [Eq. (4.22)] and its solution [Eq. (4.24)]. Solving Eq. (4.25) for T yields

$$T = \frac{Q}{4\pi s} W(u) \tag{5.3}$$

while solving Eq. (4.23) for S gives

$$S = \frac{4Tu}{r^2/t} \tag{5.4}$$

Since u and $W(u)$ are both functions of T and S, Eqs. (5.3) and (5.4) cannot be solved directly, and special procedures, discussed in the following paragraphs, must be employed.

Theis solution. A graphical procedure for evaluating T and S was developed by Theis (1935); see also Lohman (1972). If Eq. (5.3) is written as

$$\log s = \log \frac{Q}{4\pi T} + \log W(u) \tag{5.5}$$

and Eq. (5.4) as

$$\log \frac{r^2}{t} = \log \frac{4T}{S} + \log (u) \tag{5.6}$$

it can readily be seen that, since $Q/4\pi T$ and $4T/S$ are constant for a given test, the relation between $\log s$ and $\log r^2/t$ must be similar to the relation between $\log W(u)$ and $\log u$. Thus, if s is plotted against r^2/t and $W(u)$ against u on the same double-logarithmic graph paper, the resulting curves will be of the same shape but horizontally and vertically offset by the constants $Q/4\pi T$ and $4T/S$. If each curve is plotted on a separate sheet, the curves can be made to match by placing one graph on top of the other and moving it horizontally and vertically (keeping the coordinate axes parallel) until the curves coincide. An arbitrary point on the matching curves is then selected, and the coordinates of this matching point are read on both graphs. This yields related values of s, r^2/t, u, and $W(u)$, which are used to calculate T and S with Eqs. (5.3) and (5.4).

Example. The Theis solution will be applied to the drawdown data in Table 4.2, which were calculated for the following conditions:

$$Q = 1\,000 \text{ m}^3/\text{day}$$

$$r_1 = 100 \text{ m}$$

$$r_2 = 200 \text{ m}$$

$$T = 1\,000 \text{ m}^2/\text{day}$$

$$S = 0.000\,1$$

Values of r^2/t for the s-vs.-t data of both observation wells are calculated in Table 5.1 and plotted against s on double-logarithmic paper (s on the ordinate) in Figure 5.1. The next step is to construct a curve of $W(u)$ versus u on a sheet of the same logarithmic paper, using the data of Table 4.1. The resulting curve is shown in Figure 5.2. The u range for this curve, i.e., 10^{-4} to 1, was selected because this was the range where the shape of the $W(u)$-vs.-u curve best resembled the shape of the data curve in Figure 5.1. Figure 5.1 is now placed on Figure 5.2 and moved until

Table 5.1 Calculation of r^2/t for s-vs.-t data from Table 4.2

	$r = 100$ m		$r = 200$ m	
t, days	s, m	r^2/t, m^2/day	s, m	r^2/t, m^2/day
0.001	0.083	10^7	0.017	4×10^7
0.005	0.196	2×10^6	0.097	8×10^6
0.01	0.249	10^6	0.145	4×10^6
0.05	0.376	2×10^5	0.267	8×10^5
0.1	0.431	10^5	0.322	4×10^5
0.5	0.559	2×10^4	0.449	8×10^4
1	0.614	10^4	0.504	4×10^4
5	0.742	2×10^3	0.632	8×10^3
10	0.797	10^3	0.687	4×10^3

the two curves coincide while keeping the coordinate axes parallel (Figure 5.3). A match point is arbitrarily selected, and the coordinates of the point are read on both graphs, yielding

$$s = 0.167 \text{ m}$$

$$r^2/t = 3 \times 10^6 \text{ m}^2/\text{day}$$

$$W(u) = 2.1$$

$$u = 8 \times 10^{-2}$$

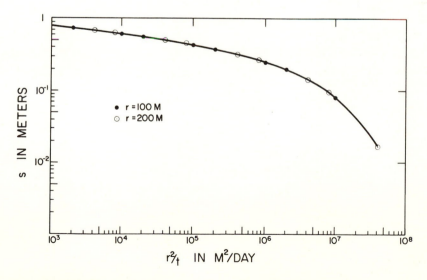

Figure 5.1 Logarithmic plot of s versus r^2/t for data in Table 5.1.

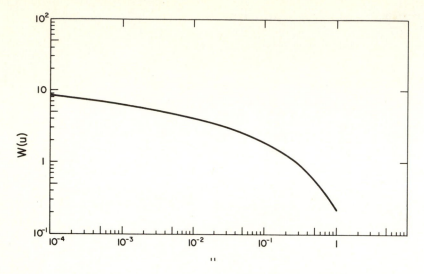

Figure 5.2 Logarithmic plot of $W(u)$ versus u, taking data from Table 4.1.

Substituting the above values of s and $W(u)$ in Eq. (5.3) yields $T = 1\,000.7$ m^2/day, which when substituted along with the values of u and r^2/t in Eq. (5.4) yields $S = 0.000\,107$. These results agree with $T = 1\,000$ m^2/day and $S = 0.000\,1$ used in the calculation of the s-vs.-t data in Table 4.2.

Once the curves are made to coincide, it is not necessary to select the match point on the curves. A more practical approach is to take some convenient point on one graph [for example, $W(u) = 1$ and $u = 10^{-4}$ for Figure 5.3] and read the corresponding coordinates on the other graph to obtain related values of $W(u)$, u, s, and r^2/t.

Chow solution. A direct solution of Eqs. (5.3) and (5.4) that eliminates the curve matching of the Theis procedure was developed by Chow (1952), who introduced the function

$$F(u) = \frac{W(u)\,e^u}{2.3} \tag{5.7}$$

to relate $W(u)$ and u to a certain combination of s and t. The relation between $F(u)$, $W(u)$, and u is shown in Figure 5.4 with the u scale on the curve. $F(u)$ is calculated from s and t of a given observation well by plotting s on the ordinate versus $\log t$ on the abscissa of semilogarithmic paper (Figure 5.5). A point A is selected on the curve, and the slope of the tangent to the curve at A is determined as the drawdown difference Δs_A for one log cycle of time. The value of $F(u)$ is then calculated as

$$F(u) = \frac{s_A}{\Delta s_A} \tag{5.8}$$

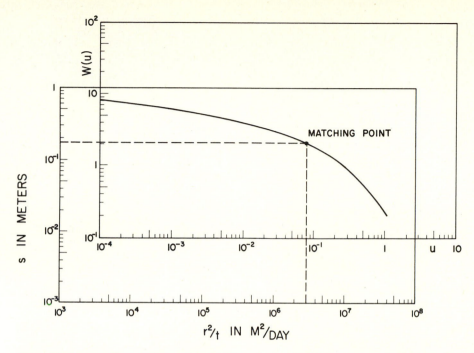

Figure 5.3 Superposition of Figures 5.1 and 5.2 to obtain coinciding curves, and selection of match point.

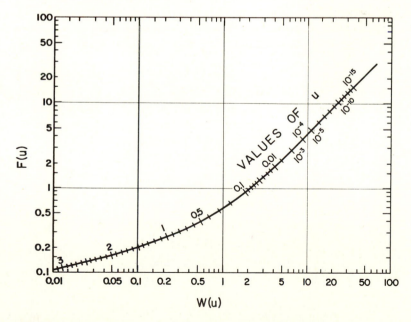

Figure 5.4 Relation between $F(u)$, $W(u)$, and u. (*From Chow, 1952.*)

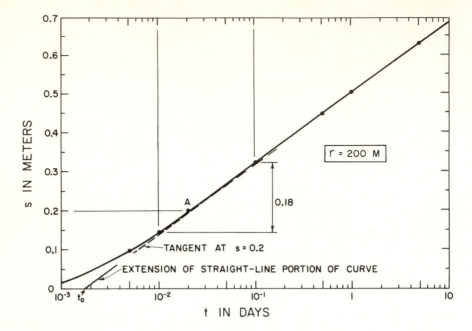

Figure 5.5 Plot of s versus log t for data in Table 5.1, $r = 200$ m.

where s_A is the drawdown at point A. The values of u and $W(u)$ corresponding to $F(u)$ are determined from Figure 5.4, after which T and S are calculated with Eqs. (5.3) and (5.4), respectively. If $F(u) > 2$, Eq. (5.7) reduces to $W(u) = 2.3F(u)$, which enables direct calculation of $W(u)$. The corresponding value of u is then found from Table 4.1, which in turn enables calculation of T and S with Eqs. (5.3) and (5.4).

Example. The s-vs.-t data for the observation well at $r = 200$ m in Table 4.2 are plotted on semilogarithmic paper in Figure 5.5. Selecting point A at $s = 0.2$ m, the tangent to the data curve at this point has a slope that gives an increase in s of 0.18 m for each log cycle of t. Thus, $s_A = 0.2$ m and $\Delta s_A = 0.18$ m, which when substituted into Eq. (5.8) yields $F(u) = 1.11$. Figure 5.4 shows that if $F(u) = 1.11$, then $u = 0.065$ and $W(u) = 2.2$. Substituting the values for s_A and $W(u)$ in Eq. (5.3) yields $T = 875$ m²/day, which when substituted with the values for u, r, and t in Eq. (5.4) yields $S = 0.000\,114$. These values are within 15 percent of the values of T and S used to calculate the s-vs.-t data.

Cooper-Jacob solution. The solution of Eq. (4.24) by Cooper and Jacob (1946)— shown as Eq. (4.26) in Chapter 4—enables direct calculation of T by plotting s versus log t (as done in Figure 5.5) and noting the time t_0 where the extended straight-line portion of the data curve intersects the abscissa. Since $s = 0$ at that point, Eq. (4.26) reduces to

$$0 = \frac{2.3Q}{4\pi T} \log \frac{2.25Tt_0}{r^2 S} \tag{5.9}$$

Because $2.3Q/4\pi T \neq 0$, the log term in this equation must be zero so that

$$\frac{2.25Tt_0}{r^2S} = 1 \qquad (5.10)$$

or
$$S = \frac{2.25Tt_0}{r^2} \qquad (5.11)$$

Since the log term in Eq. (5.9) is zero at t_0, the log term must be equal to one at $10t_0$. The s value at time $10t_0$ is equal to Δs, which is the increase in s per log cycle of t for the straight-line portion of the data curve (Figure 5.5). Substituting these values into Eq. (5.9) yields

$$\Delta s = \frac{2.3Q}{4\pi T} \qquad (5.12)$$

or
$$T = \frac{2.3Q}{4\pi \, \Delta s} \qquad (5.13)$$

This equation enables the calculation of T, which when substituted along with t_0 and r in Eq. (5.11) yields S.

As stated in Section 4.3, the Cooper-Jacob solution is valid if $u \leq 0.01$. If r is relatively small, this condition normally will be met after about 1 h of pumping for confined aquifers and about 12 h for unconfined aquifers (Kruseman and de Ridder, 1970). Pumping for 12 h should not be objectionable, because long pumping times may be desirable anyway, especially for unconfined aquifers where there is the possibility of delayed yield (see Section 5.2.5).

Example. Extension of the straight-line portion of the data curve in Figure 5.5 to the abscissa yields $t_0 = 1.6 \times 10^{-3}$ days. The slope Δs of the straight line is 0.181. Substituting these values into Eq. (5.13) yields $T = 1011$ m^2/day, after which S is calculated with Eq. (5.11) as 0.000 091. These data agree with the values of 1 000 m^2/day and 0.000 1 for T and S, respectively, used to calculate the s-vs.-t data.

Recovery test. When a pumping test is completed and the pump is stopped, the water levels in the observation wells will rise. Theis (1935) presented an equation for the residual drawdown s', which for small values of $r^2S/4Tt'$ reduces to

$$s' = \frac{2.3Q}{4\pi T} \log \frac{t}{t'} \qquad (5.14)$$

where t is the time since pumping was started and t' is the time since pumping was stopped (t is total time of pumping plus t'). To solve this equation, s' is plotted against t/t' on semilogarithmic paper. A straight line is fitted through the data points. The slope of this line is equal to $2.3Q/4\pi T$ and also to the change $\Delta s'$ in s' per unit log cycle. Thus, T can be calculated as

$$T = \frac{2.3Q}{4\pi \, \Delta s'} \qquad (5.15)$$

The recovery test, which can be used as a check on T calculated from drawdown data during pumping, does not yield S (see also Lohman, 1972).

5.2.4 Leaky Aquifers

Confined aquifers that receive water through overlying material often occur in alluvial valleys or plains, where deeper sand and gravel strata are covered by an overburden of loams or other fine-textured material in which the water table is located (Figure 5.6). The sand and gravel layers can then be classified as leaky or semiconfined aquifers and the overburden as aquitards or semiconfining layers. Pumping water from a well in the aquifer causes a downward flow from the overburden to the aquifer, which at any point is proportional to the vertical difference between the water table in the overburden and the piezometric surface of the aquifer. The assumption is usually made that the water table in the overburden is not affected by the pumping, so that the downward flow is proportional to the drop in piezometric surface. This assumption is valid during the first stages of pumping the well.

The downward flow or leakage rate v is given by the equation (Section 3.7)

$$v = K_a \frac{s}{L_a} \tag{5.16}$$

where K_a = hydraulic conductivity of upper confining layer (Figure 5.6)
$\quad\quad L_a$ = height of water table above bottom of confining layer
$\quad\quad s$ = vertical distance between water table in confining layer and piezometric surface of aquifer (drawdown if water table and piezometric surface were at same level prior to pumping)

The ratio L_a/K_a normally is combined into a single parameter called the *hydraulic resistance* R_a of the confining layer. R_a has the dimension of time (days). The

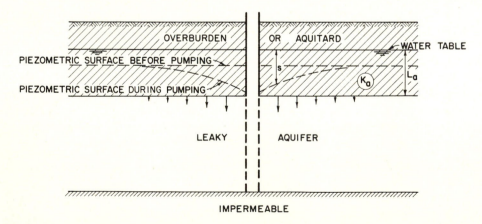

Figure 5.6 Pumped well in leaky aquifer with water table in semiconfining layer.

square root of the product of R_a and T of the aquifer is called the *leakage factor L*, so that $L = \sqrt{R_a T}$ with the dimension of length (meters).

In addition to downward flow through the upper confining layer, water is also released from storage by compression of fine-textured layers within the aquifer and, to a much lesser extent, by compression of the aquifer itself due to the decline in piezometric surface caused by pumping (Chapter 9). Several procedures have been developed, steady-state as well as transient-state methods, to evaluate hydraulic properties of the aquifer and the overlying aquitard from pumping-test data.

Steady-state methods. The steady-state methods include curve-matching procedures and simplified methods that enable direct calculation of hydraulic properties.

De Glee-Hantush-Jacob solution. De Glee (1930, 1951) and Hantush and Jacob (1955) presented the following equation for the final drop s_f of the piezometric surface at a distance r from the pumped well when steady-state flow is reached:

$$s_f = \frac{Q}{2\pi T} K_0\left(\frac{r}{L}\right) \tag{5.17}$$

where s_f = equilibrium drawdown at distance r from well
$\quad L = \sqrt{TR_a}$
$\quad K_0(r/L)$ = modified Bessel function of the second kind and zero order (Hankel function)

and Q and T are as defined earlier. Values of $K_0(r/L)$ in relation to r/L are presented in Table 5.2 (taken from Hantush, 1956).

Values of s_f can be obtained by pumping for long periods (which tends to invalidate the assumption of negligible water-table drop around the well), or by pumping for shorter periods and extrapolating s to large t. To solve Eq. (5.17) for T, s_f must be measured in several observation wells so that s_f can be plotted against r on double-logarithmic paper (s_f on the ordinate). On another sheet of the same logarithmic paper, $K_0(r/L)$ is plotted on the ordinate against r/L. One graph is placed on top of the other and moved until most of the experimental points fall on the $K_0(r/L)$-vs.-r/L curve, keeping the coordinate axes parallel. A match point is selected, and the coordinates of this point are read on both graphs. The resulting values of s_f and $K_0(r/L)$ are then substituted into Eq. (5.17) for calculation of T. Since $L = \sqrt{TR_a}$, $R_a = L^2/T = [1/(r/L)^2](r^2/T)$. Substituting T and the values of r and r/L of the matching point into this equation then yields R_a, which after substitution for L_a/K_a in Eq. (5.16) enables calculation of the recharge rate of the aquifer.

Hantush solution. A direct solution, not requiring the matching procedure, was developed by Hantush (1956, 1964a), who found that if $r/L < 0.05$, Eq. (5.17) can be approximated by

$$s_f = \frac{2.3Q}{2\pi T} \log \frac{1.12L}{r} \tag{5.18}$$

Table 5.2 Values of $K_0(x)$ and $e^x K_0(x)$ for different values of x

x	$K_0(x)$	$e^x K_0(x)$	x	$K_0(x)$	$e^x K_0(x)$	x	$K_0(x)$	$e^x K_0(x)$
0.010	4.72	4.77	0.10	2.43	2.68	1.0	0.421	1.14
12	4.54	4.59	12	2.25	2.53	1.2	0.318	1.06
14	4.38	4.45	14	2.10	2.41	1.4	0.244	0.988
16	4.25	4.32	16	1.97	2.31	1.6	0.188	0.931
18	4.13	4.21	18	1.85	2.22	1.8	0.146	0.883
0.020	4.03	4.11	0.20	1.75	2.14	2.0	0.114	0.842
22	3.93	4.02	22	1.66	2.07	2.2	0.0893	0.806
24	3.85	3.94	24	1.58	2.01	2.4	0.0702	0.774
26	3.77	3.87	26	1.51	1.95	2.6	0.0554	0.746
28	3.69	3.80	28	1.44	1.90	2.8	0.0438	0.721
0.030	3.62	3.73	0.30	1.37	1.85	3.0	0.0347	0.698
32	3.56	3.68	32	1.31	1.81	3.2	0.0276	0.677
34	3.50	3.62	34	1.26	1.77	3.4	0.0220	0.658
36	3.44	3.57	36	1.21	1.73	3.6	0.0175	0.640
38	3.39	3.52	38	1.16	1.70	3.8	0.0140	0.624
0.040	3.34	3.47	0.40	1.11	1.66	4.0	0.0112	0.609
42	3.29	3.43	42	1.07	1.63	4.2	0.0089	0.595
44	3.24	3.39	44	1.03	1.60	4.4	0.0071	0.582
46	3.20	3.35	46	0.994	1.58	4.6	0.0057	0.570
48	3.16	3.31	48	0.958	1.55	4.8	0.0046	0.559
0.050	3.11	3.27	0.50	0.924	1.52	5.0	0.0037	0.548
52	3.08	3.24	52	0.892	1.50			
54	3.04	3.21	54	0.861	1.48			
56	3.00	3.17	56	0.832	1.46			
58	2.97	3.14	58	0.804	1.44			
0.060	2.93	3.11	0.60	0.777	1.42			
62	2.90	3.09	62	0.752	1.40			
64	2.87	3.06	64	0.728	1.38			
66	2.84	3.03	66	0.704	1.36			
68	2.81	3.01	68	0.682	1.35			
0.070	2.78	2.98	0.70	0.660	1.33			
72	2.75	2.96	72	0.640	1.32			
74	2.72	2.93	74	0.620	1.30			
76	2.70	2.91	76	0.601	1.28			
78	2.67	2.89	78	0.583	1.27			
0.080	2.65	2.87	0.80	0.565	1.26			
82	2.62	2.85	82	0.548	1.24			
84	2.60	2.83	84	0.532	1.23			
86	2.58	2.81	86	0.516	1.22			
88	2.55	2.79	88	0.501	1.21			
0.090	2.53	2.77	0.90	0.487	1.20			
92	2.51	2.75	92	0.473	1.19			
94	2.49	2.73	94	0.459	1.18			
96	2.47	2.72	96	0.446	1.16			
98	2.45	2.70	98	0.443	1.15			
0.100	2.43	2.68	1.00	0.421	1.14			

Source: From Hantush, 1956. Reference to the original article is made for more extensive data and expression of the functions in more significant figures.

To solve this equation for T, s_f is plotted against $\log r$ on semilogarithmic paper. The data points will form a straight line where $r/L < 0.05$. The change Δs_f in s_f per unit log cycle is then the slope of this line, which according to Eq. (5.18) is equal to $2.3Q/2\pi T$. Thus, T can be calculated as

$$T = \frac{2.3Q}{2\pi \, \Delta s_f} \tag{5.19}$$

Extending the straight line to the abscissa yields the intercept r_0 where $s_f = 0$. When these values are substituted into Eq. (5.18), the log term must be zero, yielding

$$L = \frac{r_0}{1.12} \tag{5.20}$$

Substituting $\sqrt{TR_a}$ for L in this equation then enables calculation of R_a as

$$R_a = \frac{r_0^2}{1.25T} \tag{5.21}$$

Transient-state methods. Hantush and Jacob (1955) showed that the drawdown of the water level in a piezometer reaching into a leaky aquifer at some distance from a pumped well is described by the equation

$$s = \frac{Q}{4\pi T} \, W(u, r/L) \tag{5.22}$$

where

$$u = \frac{r^2 S}{4Tt} \tag{5.23}$$

Equation (5.22) is similar to Eq. (5.3) for the confined aquifer, except that the well function contains the additional term r/L. Values of $W(u, r/L)$ for different values of u and r/L, taken from Hantush (1956), are shown in Table 5.3.

Walton solution. The solution of Eq. (5.22) by Walton (1962) is similar to the Theis method for solving Eq. (5.3). For the leaky aquifer, a number of type curves are plotted on double-logarithmic paper with $W(u, r/L)$ on the ordinate, u on the abscissa, and r/L on the curves (one curve for each r/L value). On another sheet of the same logarithmic paper, s is plotted against t/r^2 for the observation wells or piezometers used in the test. The data curve is superimposed on the type curves and moved until the best fitting type curve is found (keeping the coordinate axes parallel). A match point is selected, and its coordinates are read on both graphs. The resulting values of $W(u, r/L)$ and s are substituted into Eq. (5.22) for calculation of T. Substituting the values of u and t/r^2 of the match point and the calculated T into Eq. (5.23) then yields S. The R_a value of the upper confining layer is calculated as L^2/T, where L is obtained from the r/L value of the type curve showing the best fit with the data curve.

Table 5.3 Values of $W(u, r/L)$ for different values of u and r/L

u \ r/L	0.002	0.004	0.006	0.008	0.01	0.02	0.04	0.06	0.08	0.1	0.2	0.4	0.6	0.8	1	2	4	6	8
0	12.7	11.3	10.5	9.89	9.44	8.06	6.67	5.87	5.29	4.85	3.51	2.23	1.55	1.13	0.842	0.228	0.0223	0.0025	0.0003
0.000002	12.1	11.2	10.5	9.89	9.44														
4	11.6	11.1	10.4	9.88	9.44														
6	11.3	10.9	10.4	9.87	9.44														
8	11.0	10.7	10.3	9.84	9.43														
0.00001	10.8	10.6	10.2	9.80	9.42	8.06													
2	10.2	10.1	9.84	9.58	9.30	8.06													
4	9.52	9.45	9.34	9.19	9.01	8.03													
6	9.13	9.08	9.00	8.89	8.77	7.98	6.67												
8	8.84	8.81	8.75	8.67	8.57	7.91	6.67												
0.0001	8.62	8.59	8.55	8.48	8.40	7.84	6.67												
2	7.94	7.92	7.90	7.86	7.82	7.50	6.62	5.86	5.29										
4	7.24	7.24	7.22	7.21	7.19	7.01	6.45	5.83	5.29										
6	6.84	6.84	6.83	6.82	6.80	6.68	6.27	5.77	5.27	4.85									
8	6.55	6.55	6.54	6.53	6.52	6.43	6.11	5.69	5.25	4.84									
0.001	6.33	6.33	6.32	6.32	6.31	6.23	5.97	5.61	5.21	4.83									
2	5.64	5.64	5.63	5.63	5.63	5.59	5.45	5.24	4.98	4.71	3.50								
4	4.95	4.95	4.95	4.94	4.94	4.92	4.85	4.74	4.59	4.42	3.48								
6	4.54				4.54	4.53	4.48	4.41	4.30	4.18	3.43								
8	4.26				4.26	4.25	4.21	4.15	4.08	3.98	3.36	2.23							
0.01	4.04				4.04	4.03	4.00	3.95	3.89	3.81	3.29	2.23							
2	3.35				3.35	3.35	3.34	3.31	3.28	3.24	2.95	2.18	1.55	1.13					
4	2.68				2.68	2.68	2.67	2.66	2.65	2.63	2.48	2.02	1.52	1.13					
6	2.30				2.30	2.29	2.29	2.28	2.27	2.26	2.17	1.85	1.46	1.11	0.839				
8	2.03					2.03	2.02	2.02	2.01	2.00	1.94	1.69	1.39	1.08	0.832				
0.1	1.82						1.82	1.82	1.81	1.80	1.75	1.56	1.31	1.05	0.819	0.228			
2	1.22						1.22	1.22	1.22	1.22	1.19	1.11	0.996	0.857	0.715	0.227			
4	0.702						0.702	0.702	0.701	0.700	0.693	0.665	0.621	0.565	0.502	0.210			
6	0.454						0.454	0.454	0.454	0.453	0.450	0.436	0.415	0.387	0.354	0.177	0.0222		
8	0.311						0.311	0.310	0.310	0.310	0.308	0.301	0.289	0.273	0.254	0.144	0.0218		
1	0.219									0.219	0.218	0.213	0.206	0.197	0.185	0.114	0.0207	0.0025	
2	0.049										0.049	0.048	0.047	0.046	0.044	0.034	0.011	0.0021	0.0002
4	0.0038											0.0038	0.0037	0.0037	0.0036	0.0031	0.0016	0.0006	0
6	0.0004														0.0004	0.0003	0.0002	0.0001	0
8	0																		

Source: From Hantush, 1956. Reference to the original article is made for more extensive tables and expression of $W(u, r/L)$ in more significant figures (see also Hantush, 1964).

Hantush solutions. Hantush (1956) developed a procedure for calculating T, S, and R_a from transient-pumping-test data which utilizes the halfway point or inflection point P_i on a curve relating s to $\log t$. The inflection point is defined as the point where the drawdown s_i is one-half the final or equilibrium drawdown s_f. The equation for s_i is

$$s_i = \frac{Q}{4\pi T} K_0(r/L) \tag{5.24}$$

where $K_0(r/L)$ is the modified Bessel function of the second kind and zero order (Table 5.2). Furthermore, the u value at the inflection point was found to be equal to $r/2L$, so that Eq. (5.23) yields

$$\frac{r}{2L} = \frac{r^2 S}{4Tt_i} \tag{5.25}$$

where t_i is t at the inflection point. The slope of the curve at the inflection point, expressed as the change Δs_i in s per unit log cycle of t, is

$$\Delta s_i = \frac{2.3Q}{4\pi T} e^{-r/L} \tag{5.26}$$

Solving this equation for r yields

$$r = 2.3L\left(\log \frac{2.3Q}{4\pi T} - \log \Delta s_i\right) \tag{5.27}$$

Finally, the ratio between s_i and Δs_i was derived as

$$2.3 \frac{s_i}{\Delta s_i} = e^{r/L} K_0(r/L) \tag{5.28}$$

Values of the function $e^{r/L}K_0(r/L)$ for different values of r/L are given in Table 5.2.

To determine T, S, and R_a from pumping-test data, s for a given observation well or piezometer is plotted against $\log t$ on semilogarithmic paper (s on ordinate) and the best-fitting curve is drawn. The value of s_f is estimated or "judged," extrapolating the curve to large t if necessary. The inflection point P_i is located at $s_i = \frac{1}{2}s_f$, and the corresponding value of t_i is read from the graph. The tangent to the curve at point P_i is drawn, and Δs_i is determined. The values of s_i and Δs_i are substituted into Eq. (5.28) for calculation of $e^{r/L}K_0(r/L)$, for which the corresponding value of r/L is determined from Table 5.2. Using this r/L value, $K_0(r/L)$ is evaluated from Table 5.2, after which T is calculated with Eq. (5.24) and S with Eq. (5.25). The r value of the piezometer is then used to calculate L from r/L, which enables the calculation of R_a as L^2/T.

Another solution by Hantush (1956) requires s-vs.-t data for more than one piezometer or observation well. Values of s are plotted against $\log t$ (s on the ordinate) on semilogarithmic paper for each piezometer, and the slope of the straight-line portion of each curve through the data points is evaluated as the change Δs in s per log cycle of t. The next step is to plot r against $\log \Delta s$ on semilogarithmic paper, draw the best-fitting straight line, and evaluate the slope of this line as the change Δr in r per log cycle of Δs. The straight line is extended to the abscissa, yielding the intercept Δs_0 where $r = 0$. From the values of Δr and Δs_0, L can be calculated as

$$L = \frac{\Delta r}{2.3} \tag{5.29}$$

and T as

$$T = \frac{2.3Q}{4\pi\,\Delta s_0} \tag{5.30}$$

after which R_a is calculated as L^2/T. To calculate S, the values of Q, T, and $K_0(r/L)$ for the ratio r/L are substituted into Eq. (5.24) for calculation of s_i. The corresponding values of t_i are determined from the plots of s versus log t, after which S is calculated with Eq. (5.25).

A third procedure by Hantush (1964a) is based on a simplified solution of Eq. (5.22), which is valid if $t > 4t_i$ and $(Tt/SL) > 2r$.

5.2.5 Unconfined Aquifers

Procedures for determining T and S of unconfined aquifers from pumping-test data can be divided into steady-state and transient-state methods.

Steady-state methods. The calculation of T of unconfined aquifers from equilibrium drawdown data of the water table is based on the Thiem equation, similar to the evaluation of T for confined aquifers (see Section 5.2.3). Substituting $H - s_1$ for h_1 and $H - s_2$ for h_2 in Eq. (4.6) yields

$$Q = \frac{\pi K(2H - s_1 - s_2)(s_1 - s_2)}{\ln\,(r_2/r_1)} \tag{5.31}$$

The average height of the aquifer between r_1 and r_2 is $(h_1 + h_2)/2$, which can be written as $(2H - s_1 - s_2)/2$. The transmissivity T_h for this average height thus is equal to $K(2H - s_1 - s_2)/2$, and the term $K(2H - s_1 - s_2)$ in Eq. (5.31) is equal to $2T_h$. Substituting $2T_h$ into Eq. (5.31) and solving for T_h yields

$$T_h = \frac{Q \ln\,(r_2/r_1)}{2\pi(s_1 - s_2)} \tag{5.32}$$

The transmissivity T of the aquifer at full thickness H is then calculated by multiplying T_h by the ratio of H to the average height of the aquifer between r_1 and r_2 during pumping, or

$$T = \frac{2H}{2H - s_1 - s_2}\,T_h \tag{5.33}$$

This equation is identical to Jacob's procedure for calculating T when s is not small compared with H [see Lohman, 1972, Eq. (33)].

When applying the steady-state method, pumping does not have to be continued until true steady-state conditions are approached, because $s_1 - s_2$ will reach an essentially constant value before s_1 and s_2 will have become constant individually, similar to confined aquifers (see Section 5.2.3).

Once T has been calculated with Eq. (5.32), the specific yield can be calculated in the same manner as discussed for the storage coefficient of confined aquifers (see Section 5.2.3). A given point on the s-vs.-t curve from one of the observation wells is selected, so that $W(u)$ can be calculated with Eq. (5.3). The corresponding u value is then obtained from Table 4.1, after which S is calculated with Eq. (5.4).

Transient-state methods. Equations (5.3) and (5.4) and the solutions thereto (Theis, Chow, Jacob) can also be used to calculate T and S of unconfined aquifers, provided of course that the basic assumptions (see Section 5.2.1) are met. The term S for unconfined aquifers then stands for specific yield. The T value obtained for unconfined aquifers applies to the average height $\bar{h}$ of the water table between the observation wells during pumping. To obtain T for the aquifer at original height H, the T value calculated from pumping-test data must be multiplied by $H/\bar{h}$ [see Eq. (5.33)].

Delayed yield. One of the assumptions underlying Eqs. (5.3) and (5.4) is that the release of water from storage in the aquifer is in immediate response to the drop of the water table. For unconfined aquifers, this often is not true. For example, the rate of fall of the water table may be faster than the rate at which pore water is released. Some pore water will drain immediately when the water table is lowered. However, it will take some time before all the pore water that eventually will be released will actually have reached the falling water table. The rate of pore-water release is also affected by the rate at which air can move into the zone immediately above the water table to occupy the space left by the draining water. When such air movement is restricted (for example, by wet soil layers higher in the vadose zone), negative air pressures will develop, which will cause a delay in the release of pore water in response to a water-table drop. Delayed yield is not restricted to unconfined aquifers. It can also occur with leaky aquifers that receive water from upper confining layers with a free water table.

When a pumping test is performed in an unconfined aquifer with delayed yield, the water level in an observation well initially will drop relatively fast; then it will drop at a slower rate and almost reach a constant level for a while, after which it begins to drop faster again. This produces a sigmoid drawdown curve, shown schematically in Figure 5.7. The initial fast drop, which may occur only during the first few minutes of pumping, corresponds to releases of water other than by free drainage of pore space. These releases may be caused by compression of fine-textured material within the aquifer, due to a decline in water-table level (see Chapter 9), or by expansion of air in the vadose zone or air entrapped below the water table. Since water is virtually incompressible, decompression of the water itself has a negligible effect on yield (see Chapter 9). Values of S evaluated

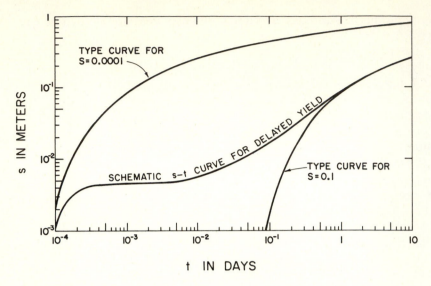

Figure 5.7 Type curves of s versus t calculated for $S = 0.000\,1$ and $S = 0.1$, and schematic s-vs.-t curve for pumping test in unconfined aquifer with delayed yield ($Q = 1\,000$ m^3/day, $T = 1\,000$ m^2/day, and $r = 100$ m).

from the early s values are often on the order of 0.000 05 to 0.01, which is typical of confined aquifers. As the water table continues to drop due to pumping, however, more and more water is yielded by drainage of pore space above the water table. This causes the s-vs.-t relation to curve upward and follow the pattern that would be obtained for the full specific yield more typical of unconfined aquifers (0.05 to 0.25, for example). This behavior is schematically shown in Figure 5.7, where the s-vs.-t curve for $S = 0.000\,1$ was taken from Figure 4.5 for $r = 100$ m, and the s-vs.-t curve for $S = 0.1$ was calculated with Eq. (4.24), using the same values for Q, T, and r (i.e., 1 000 m^3/day, 1 000 m^2/day, and 100 m, respectively). The s-vs.-t curve for delayed yield was then drawn so that the initial part of the curve would lie closely to the type curve for $S = 0.000\,1$, and the final part closely to the type curve for $S = 0.1$, in accordance with observed patterns (Boulton, 1963; Prickett, 1965).

Boulton's solution. A procedure for evaluating S and T from s-vs.-t data showing the sigmoid relation indicative of delayed yield was presented by Boulton (1963). The yield of the aquifer was considered to consist of a term S_A describing the initial, small release, and a term S_Y describing the final release or true specific yield of the unconfined aquifer. The release S_Y is delayed in time according to the empirical formula $\alpha S_Y \exp\left[-\alpha(t - t_A)\right]$, where t_A is the time that S_A prevails ($t > t_A$). The term α is an empirical constant which is characteristic of the aquifer in question. It is normally taken in its reciprocal form $1/\alpha$, where it has the dimension of time and is called the *delay index*. Also, a term B was introduced, defined as

$$B = \sqrt{\frac{T}{\alpha S_Y}} \tag{5.34}$$

The delay index is related via r/B to the time t_{wt} whereby delayed yield ceases to affect drawdown and the aquifer begins to perform as an unconfined aquifer with full yield. A graph relating αt_{wt} to r/B is shown in Figure 5.8.

Boulton's solution for the first part of the s-vs.-t curve, before the flat portion, is given by the equation

$$s = \frac{Q}{4\pi T} W(u_A, r/B) \tag{5.35}$$

where

$$u_A = \frac{r^2 S_A}{4Tt} \tag{5.36}$$

The equation for the third segment of the s-vs.-t curve, following the flat portion, is

$$s = \frac{Q}{4\pi T} W(u_Y, r/B) \tag{5.37}$$

where

$$u_Y = \frac{r^2 S_Y}{4Tt} \tag{5.38}$$

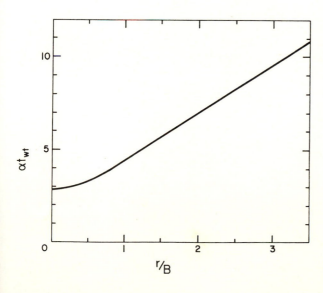

Figure 5.8 Relation between αt_{wt} and r/B. (*From Boulton, 1963.*)

Theoretically, these equations are valid only if S_Y/S_A approaches infinity. In practice, this means that S_Y/S_A should be greater than 100. When this is true, Boulton's solution yields an essentially horizontal line for the flat, second segment of the s-vs.-t curve that forms the transition between the initial and final rise (Figure 5.7). The value of s for the horizontal part of the s-vs.-t curve is given by the equation

$$s = \frac{Q}{2\pi T} K_0(r/B) \tag{5.39}$$

where $K_0(r/B)$ is the modified Bessel function of second kind and zero order (Table 5.2). If $(S_Y/S_A) < 100$, the second segment of the s-vs.-t curve is not horizontal, but reasonably accurate solutions can still be obtained with the procedure.

The functions $W(u_A, r/B)$ and $W(u_Y, r/B)$ are called *Boulton's well functions.*

Table 5.4A Values of $W(u_A, r/B)$, abbreviated as W_A, for different values of $1/u_A$ and r/B

r/B	$1/u_A$	W_A	r/B	$1/u_A$	W_A	r/B	$1/u_A$	W_A
0.01	10	1.82	0.4	1	0.213	1.5	0.5	0.039
	100	4.04		2	0.534		1	0.151
	10^3	6.31		5	1.11		1.25	0.199
	5×10^3	7.82		10	1.56		2	0.301
	10^4	8.40		50	2.18		5	0.413
	10^5	9.42		100	2.22		10	0.427
0.01	10^6	9.44	0.4	10^3	2.23	1.5	20	0.428
0.1	10	1.80	0.6	1	0.206	2	0.333	0.010
	50	3.24		2	0.504		0.5	0.033
	100	3.81		5	0.996		1	0.114
	200	4.30		10	1.31		1.25	0.144
	500	4.71		20	1.49		2	0.194
	10^3	4.83		50	1.55		5	0.227
0.1	10^4	4.85	0.6	100	1.55	2	10	0.228
0.2	5	1.19	0.8	0.5	0.046	2.5	0.5	0.027
	10	1.75		1	0.197		1	0.080
	50	2.95		2	0.466		1.25	0.096
	100	3.29		5	0.857		2	0.117
	500	3.50		10	1.05		5	0.125
0.2	10^3	3.51		20	1.12	2.5	10	0.125
			0.8	50	1.13			
0.316	1	0.216	1	0.5	0.044	3	0.5	0.021
	2	0.544		1	0.185		1	0.053
	5	1.15		2	0.421		1.25	0.061
	10	1.65		5	0.715		2	0.068
	50	2.50		10	0.819		5	0.070
	100	2.62		20	0.841	3	10	0.070
0.316	10^3	2.65	1	50	0.842			

Source: From Boulton, 1963.

Table 5.4B Values of $W(u_Y, r/B)$, abbreviated as W_Y, for different values of $1/u_Y$ and r/B

r/B	$1/u_Y$	W_Y	r/B	$1/u_Y$	W_Y	r/B	$1/u_Y$	W_Y
0.01	400	9.45	0.4	0.1	2.23	1.5	0.0711	0.444
	4×10^3	9.54		1	2.26		0.355	0.509
	4×10^4	10.2		5	2.40		0.711	0.587
	4×10^5	12.3		10	2.55		2.67	0.963
0.01	4×10^6	14.6		37.5	3.20	1.5	7.11	1.57
			0.4	100	4.05			
0.1	4	4.86	0.6	0.444	1.59	2	0.04	0.239
	40	4.95		2.22	1.71		0.2	0.283
	400	5.64		4.44	1.84		0.4	0.337
	4×10^3	7.72		16.7	2.45		1.5	0.614
0.1	4×10^4	10.0	0.6	44.4	3.26	2	4	1.11
0.2	0.4	3.51	0.8	0.025	1.13	2.5	0.0256	0.132
	4	3.54		0.25	1.16		0.128	0.162
	20	3.69		1.25	1.26		0.256	0.199
	40	3.85		2.5	1.39		0.96	0.399
	150	4.55		9.37	1.94	2.5	2.56	0.798
0.2	400	5.42	0.8	25	2.70			
0.316	0.4	2.66	1	0.04	0.844	3	0.0178	0.0743
	4	2.74		0.4	0.901		0.0889	0.0939
	40	3.38		4	1.36		0.178	0.119
	400	5.42	1	40	3.14		0.667	0.262
0.316	4×10^3	7.72				3	1.78	0.577

Source: From Boulton, 1963.

They are tabulated in Table 5.4A and B and shown graphically in Figure 5.9 in relation to $1/u_A$ and $1/u_Y$. The left portion of Figure 5.9 shows the type curves (type A curves) for the first segment of the s-vs.-t curve when yield is delayed. The right portion shows the type Y curves for the part of the s-vs.-t curve when full specific yield is obtained.

The solution process consists of plotting s and t for a given observation well or piezometer on double-logarithmic paper (s on the ordinate). Then, type curves are plotted on a different sheet of the same paper, as in Figure 5.9. The graphs are superimposed and moved so that most of the s-vs.-t data of the first segment of the curve fall on one of the type A curves, keeping the coordinate axes parallel. The r/B value of the type A curve giving the best fit is noted. A match point is selected on the coinciding curves, and the coordinates of the match point are read on both graphs. This yields corresponding values of s, t, $1/u_A$, and $W(u_A, r/B)$, from which T is calculated with Eq. (5.35) and S_A with Eq. (5.36).

The s-vs.-t curve is again placed on the type curves, such as in Figure 5.9, but now the graph is moved until the best fit is obtained between the field data at large t (third segment of data curve) and the type Y curve with the same r/B value as the type A curve that best fitted the initial s-vs.-t data. A matching point is again

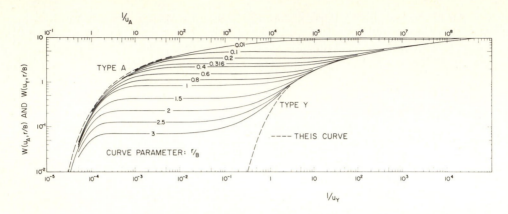

Figure 5.9 Logarithmic plot of $W(u_A, r/B)$ versus $1/u_A$ (top scale), and $W(u_Y, r/B)$ versus $1/u_Y$ (bottom scale) for data in Table 5.4. (*From Boulton, 1963.*)

selected and the coordinates read on both graphs. The resulting values of s, t, $1/u_Y$, and $W(u_Y, r/B)$ are then used to calculate T with Eq. (5.37) and S_Y with Eq. (5.38). This T value should agree with the one calculated earlier with Eq. (5.35).

The r/B value from the best-fitting type curve is used to calculate B, which is then substituted into Eq. (5.34) for calculation of α. Next, αt_{wt} is evaluated from Figure 5.8 for the same r/B value. Knowing α and αt_{wt}, t_{wt} can be calculated; t_{wt} indicates the time that yield is no longer delayed and the time that the s-vs.-t curve should merge with one of the type Y curves, which are the Theis curves for full release of pore water.

The value of t_{wt} for a given piezometer or observation well depends on r, T, S_Y, and α. Prickett (1965) showed that the delay index $1/\alpha$ is about 10 min if the material in which pore drainage takes place is coarse sand, 60 min for medium sand, 200 min for fine sand, 1 000 min for very fine sand, and 2 000 min for silt. Thus, for an aquifer consisting of fine sand with $1/\alpha = 288$ min and having a T value of 100 m²/day and $S_Y = 0.1$, t_{wt} for a piezometer at $r = 42.4$ m can be calculated as 1.9 days. This is the length of time that a pumping test should continue before the drawdown in the piezometer is no longer affected by delayed yield. Kroszynski and Dagan (1975) concluded that restricted flow above the water table affects drawdown only in shallow observation wells close to the pumped well and at the beginning of pumping only. From a practical standpoint, S_Y is the most important value for specific yield. Pumping tests in unconfined aquifers should always be continued long enough so that the third segment of the s-vs.-t curve is adequately obtained.

Further refinements in analyzing pumping-test data from unconfined aquifers with delayed yield, including anisotropy of aquifers and partial penetration of wells, were developed by Dagan (1967) and Neuman (1975). In an earlier paper, Neuman (1972) concluded that unsaturated flow above the water table had little effect on delayed yield, and analyzed delayed water-table response on the basis of two storage coefficients: a small S associated with compression of the aquifer, and a large S associated with drainage of pore space due to a falling water table. The transition from the small to the large S followed from the analytical solution. Moench and Prickett (1972) considered delayed yield as a situation of an aquifer that initially is confined but changes to an unconfined aquifer as the piezometric surface drops below the upper confining layer. Ehlig and Halepaska (1976) obtained numerical solutions of well-flow systems in which both T and S were taken as functions of total head. Drawdown curves were then calculated for delayed yield and leakage in confined-unconfined aquifers. The authors concluded that the conditions producing delayed yield or delayed water-table response are not unique and that it is, therefore, difficult to determine aquifer properties from such pumping-test data.

5.2.6 Finite Aquifers, Anisotropy, Special Cases

The theory of the pumping test and its application to basic aquifer conditions discussed in this chapter has been extended and refined to cover several complicating conditions found in practice [see, for example, Ferris et al. (1962), Bentall (1963), Hantush (1964a), and Walton (1962, 1970)]. These conditions include aquifers of finite lateral extent due to a solid boundary (for example, a mountain range) or a recharging boundary (a stream or lake). Procedures to calculate T and S of such aquifers from pumping-test data are summarized by Kruseman and de Ridder (1970) and Ferris et al. (1962); see also Hantush (1959b). Other special conditions for which procedures for analyzing pumping-test data have been developed include wells with constant drawdown like flowing wells in artesian, leaky aquifers (Hantush, 1959a), sloping or wedge-shaped aquifers (Hantush, 1964a), anisotropic aquifers (Hantush, 1966; Hantush and Thomas, 1966; Weeks, 1969; and Neuman, 1975), partial penetration of the pumped well (Dagan, 1967), varying pumping rates (Cooper and Jacob, 1946; Hantush, 1964a; Aron and Scott, 1965; Sternberg, 1967), large-diameter wells with nonnegligible storage of water inside well (Papadopulos and Cooper, 1967), and two-layered aquifers (Huisman and Kemperman, 1951; Neuman and Witherspoon, 1969). Boulton and Streltsova (1975) presented theory and equations for analyzing pumping-test data that take into account compressibility and anisotropy of the main aquifer, partial penetration of the pumped well, depth of observation wells, leaky-aquifer conditions, and flow in saturated and unsaturated zones above the water table.

5.3 RATE-OF-RISE TECHNIQUES

With the rate-of-rise tests, local values of K of aquifers or other subsurface materials are determined from the rate of rise of the water level in a well or similar hole after this level was abruptly lowered by the sudden removal of a certain volume of water from the well. This removal can be accomplished with a bucket or bailer. Another technique consists of submerging a closed cylinder or other solid body in the well, letting the water level reach equilibrium, and then quickly pulling it out.

Enough water must be removed or displaced to lower the water level in the well 10 to 50 cm.

Main advantages of rate-of-rise tests are that pumping is not needed, observation wells are not required, and tests can be completed in a short time. Rate-of-rise tests can be used on wells after construction to get a preliminary estimate of aquifer conditions. They are also useful where continuous pumping at constant rate is difficult, where observation wells are not available, where there is interference from other wells, or where there are other disturbances that conflict with the basic conditions required for pumping tests. Disadvantages of the techniques are that K is measured on a relatively small portion of the aquifer and that S normally cannot be evaluated.

When rate-of-rise tests are carried out on wells, they are usually called *slug tests* because a slug of water is removed. Other rate-of-rise tests are the auger-hole and piezometer techniques, where K is calculated from the water-level rises in holes especially constructed for K measurement.

5.3.1 Slug Test

Cooper et al. (1967) obtained a solution of Eq. (4.21) to calculate T and S of confined aquifers from the rate of rise of the water level in a fully penetrating well after a sudden removal of a slug of water. Type curves were prepared so that T and S could be evaluated by matching field data with type curves, similar to the Theis procedure for pumping tests. The S value obtained with this technique may not be reliable because the shape of the type curves is rather insensitive to S (Lohman, 1972).

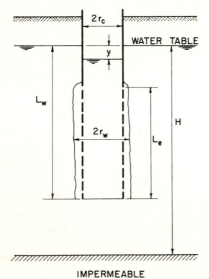

Figure 5.10 Geometry and symbols of partially penetrating, partially perforated well in unconfined aquifer with gravel pack or developed zone around perforated section.

A slug-test procedure applicable to fully or partially penetrating wells in unconfined aquifers was developed by Bouwer and Rice (1976). The procedure is based on the Thiem equation (4.3) and assumes negligible drawdown of the water table around the well and no flow above the water table. The term $h_2 - h_1$ in Eq. (4.3) then represents the distance y of the water level in the well below the water table (Figure 5.10). The rate of rise dy/dt of the water level after removal of water is expressed as

$$\frac{dy}{dt} = -\frac{Q}{\pi r_c^2} \tag{5.40}$$

where r_c is the radius of the well section where the water level is rising and Q is the flow of groundwater into the well. The minus sign in Eq. (5.40) is introduced because y decreases with increasing t, so that dy/dt is negative. Substituting the Thiem equation (4.3) for Q in Eq. (5.40), integrating, and solving for K yields

$$K = \frac{r_c^2 \ln (R_e/r_w)}{2L_e} \frac{1}{t} \ln \frac{y_0}{y_t} \tag{5.41}$$

where R_e = effective radial distance over which the head difference y is dissipated
$\quad r_w$ = radial distance between well center and undisturbed aquifer (r_c plus thickness of gravel envelope or developed zone outside casing)
$\quad L_e$ = height of perforated, screened, uncased, or otherwise open section of well through which groundwater enters
$\quad y_0$ = y at time zero
$\quad y_t$ = y at time t
$\quad t$ = time since y_0

The effective radius R_e is essentially the effective value of r_2 to be used in Eq. (4.3) so that it gives the correct value of Q (the Thiem equation was developed for horizontal flow only and as such cannot be used to calculate Q for the system of Figure 5.10). Values of R_e were experimentally determined with a resistance network analog for different values of r_w, L_e, L_w and H (see Figure 5.10 for meaning of symbols). The following empirical equation was then developed to relate R_e to the geometry and boundary conditions of the system

$$\ln \frac{R_e}{r_w} = \frac{1}{\dfrac{1.1}{\ln (L_w/r_w)} + \dfrac{A + B \ln [(H - L_w)/r_w]}{(L_e/r_w)}} \tag{5.42}$$

where A and B are dimensionless parameters shown in Figure 5.11 in relation to L_e/r_w. If H is much larger than L_w, a further increase in H has little effect on the flow system and, hence, on R_e. The analog analyses indicated that the effective upper limit of $\ln [(H - L_w)/r_w]$ is 6. Thus, if $H - L_w$ is so large that $\ln [(H - L_w)/r_w] > 6$, a value of 6 should still be used for this term in Eq. (5.42), including the theoretical case of $H = \infty$. If $H = L_w$ (well penetrating to bottom of

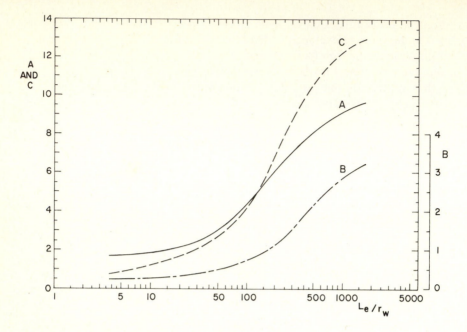

Figure 5.11 Curves relating coefficients A, B, and C to L_e/r_w.

aquifer), the term $\ln\left[(H - L_w)/r_w\right]$ in Eq. (5.42) cannot be used. For this situation, the equation for $\ln\left(R_e/r_w\right)$ is

$$\ln\frac{R_e}{r_w} = \cfrac{1}{\cfrac{1.1}{\ln\left(L_w/r_w\right)} + \cfrac{C}{\left(L_e/r_w\right)}} \qquad (5.43)$$

where C is a dimensionless coefficient shown in Figure 5.11 as a function of L_e/r_w. The value of $\ln\left(R_e/r_w\right)$ calculated with Eqs. (5.42) and (5.43) is within 10 percent of the analog value if $L_e > 0.4L_w$ and within 25 percent if $L_e < 0.2L_w$.

Since K, r_c, R_e, r_w, and L_e are constant for a given well, $1/t \ln\left(y_0/y_t\right)$ must also be constant, as indicated by Eq. (5.41). Thus, when the observed values of y are plotted against t on semilogarithmic paper (y on the log scale), the data points should form a straight line. This is exemplified in Figure 5.12, showing data from a slug test on a well in the Salt River bed west of Phoenix, Arizona (see Problem 5.7). The data begin to deviate from a straight line at small y, probably because of measurement error. The straight-line portion of the points should be used to evaluate $1/t \ln\left(y_0/y_t\right)$ for calculation of K.

The time $t_{90\%}$, necessary for the water level in the well to rise 90 percent of the distance back to the equilibrium level, is given by the equation (Bouwer and Rice, 1976)

$$t_{90\%} = 0.0527\frac{r_c^2}{KL_e}\ln\frac{R_e}{r_w} \qquad (5.44)$$

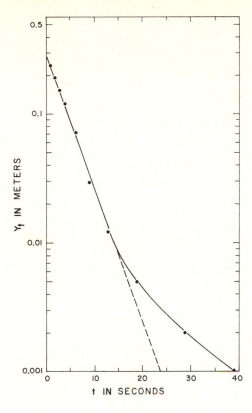

Figure 5.12 Plot of y versus t for slug test on East Well.

If K and/or L_e are relatively large, $t_{90\%}$ may be only a few seconds. Such fast water-level rises can be measured with sensitive pressure transducers and fast strip-chart recorders or x-vs.-y plotters to record the transducer output.

Although streamlines in flow systems around slug-tested wells contain both vertical and horizontal portions, most of the head loss is dissipated in a horizontal direction (Bouwer and Rice, 1976). Thus, K yielded by the slug test primarily reflects K in horizontal direction. The portion of the aquifer on which K is measured is approximately a cylinder with a radius of about R_e and a height slightly larger than L_e. The T value of the aquifer is obtained by multiplying K by H, assuming of course that the aquifer is uniform.

Since the water table in the aquifer was held at a constant level and taken as a plane source of water in the analog evaluations of R_e, the slug test of Bouwer and Rice can also be used to estimate K of confined aquifers that receive most of their water from the upper confining layer, through leakage or compression.

5.3.2 Auger-Hole Method

The auger-hole method is similar to the slug method for wells, but the water-level rise is measured in an unlined, cylindrical hole, usually 10 to 20 cm in diameter, 1 to several meters deep, and dug with an auger or other tool for the express purpose of measuring K. In unstable soils, the hole may have to be lined with a screen to prevent caving of the wall. After the hole is dug and the water level in the hole has reached equilibrium with the water table, a volume of water is removed (usually with a bailer) and the rate of rise of the water level is measured with a float and tape, an electric probe, or other technique (Figure 5.13).

Numerous equations and charts have been developed to calculate K from the rate of water-level rise in the hole (Bouwer and Jackson, 1974, and references therein). A recent solution by Boast and Kirkham (1971) is based on an exact mathematical analysis of the flow system around the well. The distance of the lower boundary below the hole bottom was taken as a variable, and this boundary was considered as either an impermeable or an infinitely permeable layer to represent deeper soil material that is either much less or much more permeable than the soil around the auger hole (Figure 5.13). In addition, the usual assumption of no drawdown of the water table around the hole, no flow above the water table, and homogeneous soil were made. Boast and Kirkham reduced the results of their analyses to the equation

$$K = C_a \frac{(dy/dt)}{864} \tag{5.45}$$

and presented a table listing C_a for different values of L_w/r_w, y/L_w, and $(H - L_w)/L_w$ for both the impermeable and infinitely permeable lower boundary (Table 5.5). The factor y in y/L_w is the value of y at which dy/dt is evaluated. Usually, dy/dt should be determined for the early portion of the water-level rise. Values of C_a for geometry factors between those listed in the table can be determined by interpolation. Equation (5.45) is dimensionally correct, so that K is

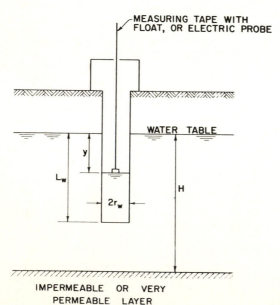

IMPERMEABLE OR VERY
PERMEABLE LAYER

Figure 5.13 Geometry and symbols for auger-hole method.

Table 5.5 Values of C_a for Eq. (5.45) for auger hole underlain by impermeable or infinitely permeable material

L_w/r_w	y/L_w	$(H-L_w)/L_w$ for impermeable layer								$H-L_w$	$(H-L_w)/L_w$ for infinitely permeable layer			
		0	0.05	0.1	0.2	0.5	1	2	5	∞	5	2	1	0.5
1	1	447	423	404	375	323	286	264	255	254	252	241	213	166
	0.75	469	450	434	408	360	324	303	292	291	289	278	248	198
	0.5	555	537	522	497	449	411	386	380	379	377	359	324	264
2	1	186	176	167	154	134	123	118	116	115	115	113	106	91
	0.75	196	187	180	168	149	138	133	131	131	130	128	121	106
	0.5	234	225	218	2 7	188	175	169	167	167	166	164	156	139
5	1	51.9	48.6	46.2	42.8	38.7	36.9	36.1		35.8		35.5	34.6	32.4
	0.75	54.8	52.0	49.9	46.8		41.0	40.2		40.0		39.6	38.6	36.3
	0.5	66.1	63.4	61.3	58.1	53.9	51.9	51.0		50.7		50.3	49.2	46.6
10	1	18.1	16.9	16.1	15.1	14.1	13.6	13.4		13.4		13.3	13.1	12.6
	0.75	19.1	18.1	17.4	16.5	15.5	15.0	14.8		14.8		14.7	14.5	14.0
	0.5	23.3	22.3	21.5	20.6	19.5	19.0	18.8		18.7		18.6	18.4	17.8
20	1	5.91	5.53	5.30	5.06	4.81	4.70	4.66		4.64		4.62	4.58	4.46
	0.75	6.27	5.94	5.73	5.50	5.25	5.15	5.10		5.08		5.07	5.02	4.89
	0.5	7.67	7.34	7.12	6.88	6.60	6.48	6.43		6.41		6.39	6.34	6.19
50	1	1.25	1.18	1.14	1.11	1.07	1.05			1.04			1.03	1.02
	0.75	1.33	1.27	1.23	1.20	1.16	1.14			1.13			1.12	1.11
	0.5	1.64	1.57	1.54	1.50	1.46	1.44			1.43			1.42	1.39
100	1	0.37	0.35	0.34	0.34	0.33	0.32			0.32			0.32	0.31
	0.75	0.40	0.38	0.37	0.36	0.35	0.35			0.35			0.34	0.34
	0.5	0.49	0.47	0.46	0.45	0.44	0.44			0.44			0.43	0.43

Source: From Boast and Kirkham, 1971.

obtained in the same units as dy/dt. In the original equation, Boast and Kirkham expressed dy/dt in centimeters per second and K in meters per day [hence the factor 1/864 in Eq. (5.45) to make it dimensionally correct and to enable use of the original C_a values].

An equation identical to Eq. (5.45) was also developed by Ernst (Bouwer and Jackson, 1974, and references therein), who evaluated the C_a factor with relaxation techniques (see Section 7.1.4) and presented both empirical equations and nomographs relating C_a to the geometry of the auger-hole system (Bouwer and Jackson, 1974).

Variations on the auger-hole technique are the two-well, four-well, or multiple-well methods. These are steady-state techniques whereby two or more auger holes are installed relatively close together (1 to several meters apart). Water is pumped at constant rate from one hole to the next until steady state is essentially reached. The value of K between the auger holes is then calculated from the pumping rate and the equilibrium difference between the water levels in the holes (Bouwer and Jackson, 1974, and references therein).

The auger-hole technique enables measurement of K in unconfined or perching aquifers relatively close to ground surface. The K value obtained is essentially a point measurement (the soil volume sampled for K is about $0.4L_w$ m^3) and reflects mostly the hydraulic conductivity in a horizontal direction (except when L_w is relatively small).

5.3.3 Piezometer Method

With the piezometer technique, a rigid pipe is driven (removing the soil below the pipe with an auger) or jetted into the soil. When the pipe has reached the desired depth, a cavity is augered out below the bottom of the pipe (Figure 5.14). After the water level in the pipe has reached equilibrium, a slug of water is rapidly removed and the subsequent rate of rise of the water level is measured. The K value of the soil around the cavity is calculated as

$$K = \frac{\pi r_w^2}{A_p t} \ln \frac{y_0}{y_t} \tag{5.46}$$

where A_p is a factor depending on the shape of the cavity and the depth of the lower boundary (Luthin and Kirkham, 1949). Youngs (1968) evaluated A_p, which has the dimension of length, for different values of L_c, L_w, H, and r_w, using an electric analog. The lower boundary was again taken as impermeable and infinitely permeable. The results were expressed in dimensionless ratios (Table 5.6).

The diameter of the piezometer pipe and cavity normally are relatively small (5 to 10 cm), although large values are also possible. The height L_c of the cavity usually varies between 10 and 30 cm. If $L_c = 0$, the piezometer becomes a cased auger hole, in which case the technique is called the *tube method* (Bouwer and Jackson, 1974, and references). Piezometers normally are capable of measuring K at greater depths than the auger-hole method. In unstable soils, the piezometer cavity must be screened to prevent caving. This can be accomplished with a perforated, screened section at the bottom of the pipe.

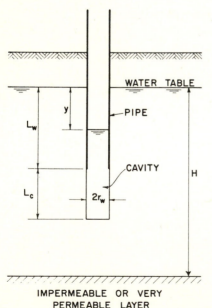

IMPERMEABLE OR VERY
PERMEABLE LAYER

Figure 5.14 Geometry and symbols for piezometer method.

Table 5.6 Values of A_p/r_w for piezometer method for cylindrical cavities

L_c/r_w	L_w/r_w	$(H - L_w - L_c)/r_w$ for impermeable layer							$(H - L_w - L_c)/r_w$ for infinitely permeable layer						
		∞	8.0	4.0	2.0	1.0	0.5	0	∞	8.0	4.0	2.0	1.0	0.5	0
0	20	5.6	5.5	5.3	5.0	4.4	3.6	0	5.6	5.6	5.8	6.3	7.4	10.2	∞
	16	5.6	5.5	5.3	5.0	4.4	3.6	0	5.6	5.6	5.8	6.4	7.5	10.3	∞
	12	5.6	5.5	5.4	5.1	4.5	3.7	0	5.6	5.7	5.9	6.5	7.6	10.4	∞
	8	5.7	5.6	5.5	5.2	4.6	3.8	0	5.7	5.7	5.9	6.6	7.7	10.5	∞
	4	5.8	5.7	5.6	5.4	4.8	3.9	0	5.8	5.8	6.0	6.7	7.9	10.7	∞
0.5	20	8.7	8.6	8.3	7.7	7.0	6.2	4.8	8.7	8.9	9.4	10.3	12.2	15.2	∞
	16	8.8	8.7	8.4	7.8	7.0	6.2	4.8	8.8	9.0	9.4	10.3	12.2	15.2	∞
	12	8.9	8.8	8.5	8.0	7.1	6.3	4.8	8.9	9.1	9.5	10.4	12.2	15.3	∞
	8	9.0	9.0	8.7	8.2	7.2	6.4	4.9	9.0	9.3	9.6	10.5	12.3	15.3	∞
	4	9.5	9.4	9.0	8.6	7.5	6.5	5.0	9.5	9.6	9.8	10.6	12.4	15.4	∞
1.0	20	10.6	10.4	10.0	9.3	8.4	7.6	6.3	10.6	11.0	11.6	12.8	14.9	19.0	∞
	16	10.7	10.5	10.1	9.4	8.5	7.7	6.4	10.7	11.1	11.6	12.8	14.9	19.0	∞
	12	10.8	10.6	10.2	9.5	8.6	7.8	6.5	10.8	11.1	11.7	12.8	14.9	19.0	∞
	8	11.0	10.9	10.5	9.8	8.9	8.0	6.7	11.0	11.2	11.8	12.9	14.9	19.0	∞
	4	11.5	11.4	11.2	10.5	9.7	8.8	7.3	11.5	11.6	12.1	13.1	15.0	19.0	∞
2.0	20	13.8	13.5	12.8	11.9	10.9	10.1	9.1	13.8	14.1	15.0	16.5	19.0	23.0	∞
	16	13.9	13.6	13.0	12.1	11.0	10.2	9.2	13.9	14.3	15.1	16.6	19.1	23.1	∞
	12	14.0	13.7	13.2	12.3	11.2	10.4	9.4	14.0	14.4	15.2	16.7	19.2	23.2	∞
	8	14.3	14.1	13.6	12.7	11.5	10.7	9.6	14.3	14.8	15.5	17.0	19.4	23.3	∞
	4	15.0	14.9	14.5	13.7	12.6	11.7	10.5	15.0	15.4	16.0	17.6	20.1	23.8	∞
4.0	20	18.6	18.0	17.3	16.3	15.3	14.6	13.6	18.6	19.8	20.8	22.7	25.5	29.9	∞
	16	19.0	18.4	17.6	16.6	15.6	14.8	13.8	19.0	20.0	20.9	22.8	25.6	29.9	∞
	12	19.4	18.8	18.0	17.1	16.0	15.1	14.1	19.4	20.3	21.2	23.0	25.8	30.0	∞
	8	19.8	19.4	18.7	17.6	16.4	15.5	14.5	19.8	20.6	21.4	23.3	26.0	30.2	∞
	4	21.0	20.5	20.0	19.1	17.8	17.0	15.8	21.0	21.5	22.2	24.1	26.8	31.5	∞
8.0	20	26.9	23.0	25.5	24.0	23.0	22.2	21.4	26.9	29.6	30.6	32.9	36.1	40.6	∞
	16	27.4	26.3	25.8	24.4	23.4	22.7	21.9	27.4	29.8	30.8	33.1	36.2	40.7	∞
	12	28.3	27.2	26.4	25.1	24.1	23.4	22.6	28.3	30.0	31.0	33.3	36.4	40.8	∞
	8	29.1	28.2	27.4	26.1	25.1	24.4	23.4	29.1	30.3	31.2	33.8	36.9	41.0	∞
	4	30.8	30.2	29.6	28.0	26.9	25.7	24.5	30.8	31.5	32.8	35.0	38.4	43.0	∞

Source: From Youngs, 1968.

The volume of soil on which K is measured with the piezometer method is smaller (usually 1 to a few cubic decimeters) than with other techniques. This is a disadvantage if the objective is to measure K of extensive soil or aquifer formations, but an advantage if great resolution is required (for example, when measuring K of thin, individual layers).

5.4 FLOW-SYSTEM TECHNIQUES

Subsurface formations are inherently heterogeneous and anisotropic. If K or T of such formations must be measured to predict the performance of a certain underground flow system (flow rates, response of water table or piezometric surface, etc.), it is very important that the region on which K is measured and the predominant flow direction therein are as much as possible the same as those for the system to be predicted. Thus, in selecting a technique for measuring K or T, the first choice should always be a method whereby K or T is evaluated from a similar flow system in the same region. Such a system may already be present, or it may be created. If, for example, T of an aquifer must be measured to predict the yield of a well, it is best to evaluate T with a pumping test on the same well or on a similar well in the same aquifer, because the flow system then duplicates that of the well. If wells are not present and the aquifer is shallow, a method that measures K predominantly in a horizontal direction (auger holes, or piezometers with large cavity heights) may be the next choice. Because these methods measure K on a small portion of the aquifer, replicate measurements at different sites are necessary to obtain a representative average (see Section 5.8).

In river valleys or coastal areas, aquifers may be hydraulically connected to rivers, oceans, or other bodies of surface water. In those situations, it may be possible to evaluate T and S for large aquifer sections from the response of the water table or piezometric surface in the aquifer to fluctuations in the surface-water level (tidal movement, floods, etc.). Such fluctuations are delayed in phase and reduced in amplitude when propagated in the aquifer (Steggewentz, 1933). Based on this, Ferris (1951) developed a method for calculating the ratio T/S, called the *hydraulic diffusivity*, of the aquifer from water-level measurements in observation wells at various distances from the river bank or shoreline. Type curves were generated for each well, and the hydraulic diffusivity was obtained from the type curve that best matched the observed water-level response in the well. The equations for these type curves and other procedures for evaluating T/S (Pinder et al., 1969, and references therein) normally assume that the fluctuations of the surface-water level can be described by a step function, a steady-state periodic sinusoid, or a linear equation. A procedure for calculating T/S for a flood-stage hydrograph of any shape was developed by Pinder et al. (1969). Carr and Van Der Kamp showed how T and S can be solved individually.

Once T has been determined at several locations in an aquifer, the spatial distribution of T over the entire aquifer can be inferred from contours of groundwater levels and accretion or depletion of water to or from the aquifer (see Section

7.5). Such information is needed for modeling aquifers to determine, for example, optimum schemes for groundwater exploitation.

In spreading water for groundwater recharge, it is often desirable to predict how high the water table will rise below the recharge basins or other infiltration facilities. The flow in the underlying groundwater mounds initially will be mainly downward, particularly if the aquifer is thick, and will then become more horizontal. This makes it difficult to obtain a meaningful K value with pumping tests or other conventional methods for measuring K. The best solution in this situation is to use a similar recharge system in the same area (on a smaller scale if it has to be especially constructed) and evaluate K from the rise of the mound in response to a known recharge rate (see Section 8.3). Bouwer (1970), for example, determined K of an unconfined aquifer in both horizontal and vertical directions from the infiltration rate in recharge basins and the rise of water levels in two piezometers installed at different depths in the center of the recharge area.

Sometimes, restricting or semiconfining layers occur at relatively small depth, and perching water tables could rise relatively close to ground surface during periods of high infiltration rates. To predict the height of such water tables for a given infiltration rate, K in a vertical direction must be measured for the restricting layer. A piezometer with zero cavity height (tube method) could be used for this purpose if enough replicate measurements are made to obtain a representative average. A better approach, however, would be to create a perched water table with an infiltration flow system using, for example, a sprinkler system over a relatively large area. The vertical K value of the restricting layer can then be calculated from the infiltration rate and the height of the perched water table (see Section 3.7).

Agricultural land with excessively high water tables frequently is drained by installing parallel, horizontal drain tubes. To find the proper depth and spacing of the drains, K of the soil profile must be known in a primarily horizontal direction (see Section 8.4.3). While the auger-hole method is commonly used for this purpose, a more representative K value could be obtained if a drain already existed or an experimental drain could be installed. K could then be evaluated from measurements of the drain flow and the water-table position (Hoffman and Schwab, 1964).

In practice, it is not always possible to evaluate hydraulic properties of subsurface materials from flow systems that are of a similar type and occur in a similar region as the flow system to be predicted. Measurements of K must then be made using techniques like those discussed in this chapter.

5.5 HYDRAULIC-CONDUCTIVITY MEASUREMENT IN THE VADOSE ZONE

The methods for K measurement discussed so far all require that the material on which K is to be measured is in the groundwater, where water pressures are above atmospheric. However, it may also be desirable to know K of soil materials in the vadose zone. Such information is needed, for example, to evaluate the suitability of sites for groundwater recharge projects, to predict seepage from proposed irrigation canals or storage reservoirs, to determine ultimate drainage requirements of new irrigation projects where water tables are expected to rise, to assess the potential for groundwater contamination below waste disposal areas, to calculate infiltration rates, to analyze surface-subsurface water relationships, or to determine the suitability of sites for land treatment of sewage or other wastewater.

The principle of measuring K of soil or other material that is in the vadose zone, and hence not saturated with water of greater-than-atmospheric pressure at the time of the measurement, is to artificially wet a portion of the soil and to evaluate K from a flow system created in the wetted zone. Since artificial wetting seldom gives complete saturation (entrapped air is difficult to avoid), the resulting K value will be less than K at saturation. Limited experimental data indicate that K after artificial wetting may be about one-half the K value at complete saturation (Bouwer, 1966).

5.5.1 Air-Entry Permeameter

The simplest and quickest technique for measuring K in the absence of a water table is the air-entry permeameter (Bouwer, 1966). The device essentially consists of a metal cylinder of about 30 cm in diameter that is driven 10 to 20 cm into the soil (Figure 5.15). The soil is covered with a layer of coarse sand, and a disk is placed in the center to avoid soil erosion when water is applied. A lid assembly with standpipe and reservoir is clamped to the cylinder, water is poured into the reservoir, and the supply valve at the base of the standpipe is opened while water is continually added to the reservoir. After the cylinder is filled with water, the

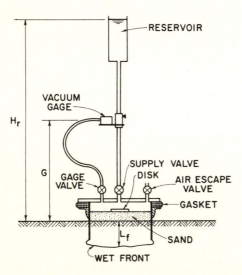

Figure 5.15 Schematic of air-entry permeameter.

air-escape valve is closed and the reservoir is kept full while water infiltrates into the soil. When the wet front is expected to have reached a depth of about 10 cm, water is no longer added to the reservoir, a few time and water-level readings are taken to determine the rate of water-level fall in the reservoir, and the supply valve is closed (a few trial tests may be needed to determine how long infiltration should continue before the wet front has reached the desired depth). After closing the valve, the wet front no longer advances in the soil. Also, the pressure of the aboveground water inside the cylinder will decrease to negative values and reach a minimum when the air-entry value of the wetted zone is reached. At this point, air will start moving up through the wetted zone and eventually will bubble up through the soil surface, increasing the pressure of the aboveground water inside the cylinder. As soon as minimum pressure is observed on the vacuum gage, it is measured, the cylinder is immediately removed, and the height L_f of the wetted zone is measured. This can be done visually, using a shovel, or by pushing a rod down and observing the depth where the penetration resistance increases.

Referring the minimum pressure in the aboveground water to the elevation of the wet front yields the air-entry value P_a of the wetted zone [$P_a = P_{min} + G + L_f$, where P_{min} is the minimum-pressure head measured with the vacuum gage and $G + L_f$ is the height of the gage above the wet front (see Figure 5.15)]. The water-entry value, which is the pressure head at the wet front while it was moving down, is then taken as $\frac{1}{2}P_a$ (see Figure 2.9). Since the pressure head at the top and bottom of the wetted zone during infiltration, the height of the wetted zone, and the infiltration rate prior to closing the supply valve are known, K of the wetted zone can be calculated with Darcy's equation as

$$K = \frac{L_f(dH_r/dt)(r_r/r_c)^2}{H_r + L_f - 0.5P_a} \tag{5.47}$$

where L_f = height of wetted zone at end of test
dH_r/dt = rate of fall of water level in reservoir just before closing supply valve
H_r = height of water level in reservoir above soil just before closing supply valve
r_r = radius of reservoir
r_c = radius of cylinder
P_a = air-entry value of wetted zone (expressed as a negative water-pressure head)

The air-entry permeameter is a surface device, at least in the construction shown in Figure 5.15. To use the technique for K measurement at greater depths, pits or trenches must be dug. Also, the method requires a relatively dry soil for easy detection of the wet front. If the soil is already moist, small tensiometers or other sensors can be placed in the soil below the cylinder to indicate the arrival of the wet front (Topp and Binns, 1976). The air-entry permeameter measures K in vertical direction. About 10 l of water are required per test, and tests are usually completed in $\frac{1}{2}$ to 1 h.

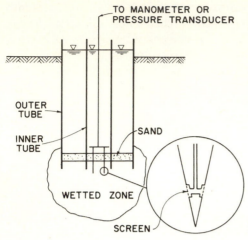

Figure 5.16 Schematic of infiltration-gradient technique.

5.5.2 Infiltration-Gradient Techniques

The infiltration-gradient technique in principle is similar to the air-entry permeameter, but the vertical gradient in the wetted zone is directly measured with tensiometers or piezometers. To ensure vertical flow in the wetted zone, two concentric cylinders are used. The infiltration rate and gradient are then measured for the inner cylinder, keeping the water levels in both cylinders the same (Figure 5.16). If the water depth in the cylinders is relatively large, the pressures of the water in the upper part of the wetted zone will be positive and the gradient can be measured with small, fast-reacting piezometers of the type described by Rice (1967). The piezometers are pushed down a small distance at the time, so that the pressure head is measured in relation to depth below the bottom of the hole. A plot of pressure head against depth should then yield a straight line (at least for the top part of the wetted zone where the flow is vertical), from which the gradient can be evaluated (Bouwer, 1964; Bouwer and Rice, 1967). Knowing gradient and infiltration rate, K of the wetted zone can be calculated with Darcy's equation.

The infiltration-gradient technique measures K in a vertical direction. If the method is used to determine K at or near the surface, short cylinders are used and the water depth may not be sufficient to produce positive-pressure heads in the wetted zone. In that case, the gradient is measured with tensiometers (Bouwer and Jackson, 1974, and references therein). The technique requires 10 to 200 l water per test (depending on size of cylinders) and can usually be completed in 1 to 4 h (depending on the depth of the hole).

5.5.3 Double-Tube Method

The double-tube method enables the measurement of K at depths of 0.2 to 3 m or more without need for tensiometers or piezometers (Bouwer, 1961 and 1962). Detection of a wet front is not required, so that the technique works in wet as well

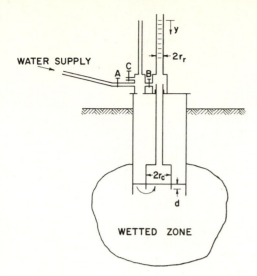

WATER SUPPLY

WETTED ZONE

Figure 5.17 Schematic of double-tube apparatus.

as in dry soils (including soils below a water table). The device consists of two concentric tubes placed in an auger hole of which the bottom is carefully cleaned to expose an "undisturbed" surface, which is then covered by a protective layer of sand (Figure 5.17). The diameter of the outer tube should be at least twice that of the inner tube (Bouwer and Rice, 1967). Convenient diameters are 25 and 12.5 cm, respectively. The outer tube is installed first. It may penetrate about 5 cm into the hole bottom. After filling the outer tube with water, the inner tube, which is thin-walled and has a beveled edge, is placed centrally in the hole and pushed about 2 cm into the hole bottom. The outer tube is covered by a lid with two standpipes, one connected to the inner tube and the other to the outer tube, or rather to the annular space between the two tubes (Figure 5.17). Water levels are maintained at the top of the standpipes to rapidly wet the soil beneath the tubes. After a while (usually 0.5 to 3 h, depending on soil and hole depth), the water supply to the inner tube is stopped (by closing valve B), and time and water-level (t and y) readings are taken of the falling water level in the standpipe on the inner tube, while the water level in the outer-tube standpipe stays at the top. Valve B is opened, and the water level is again maintained at the top of the standpipe on the inner tube. After 10 to 20 min, the water supply to the inner tube is stopped again and time and water-level readings are again taken of the falling water level in the inner-tube standpipe, but this time the water level in the standpipe on the outer tube is adjusted by manipulating valve C at the bottom of the standpipe so that the water level falls at the same rate as that in the inner-tube standpipe. The rate of water-level fall in the inner-tube standpipe is now greater than during the first measurement because a flow component from the outer tube to the inner tube through the wetted soil—caused by the pressure difference between the outer and inner tubes—is no longer present (the flow component is indicated by an arrow in Figure 5.17). Thus, a curve of water-level drop y versus t for the second set of

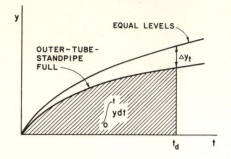

Figure 5.18 Curves of water-level drop versus time in standpipe on inner tube when water level in outer-tube standpipe was held at top of pipe (lower curve) and when allowed to fall at same rate as water level in inner-tube standpipe (upper curve).

measurements (equal levels in standpipes) will be above that for the first set (outer-tube standpipe kept full), as shown in Figure 5.18. The K value is calculated from these curves as

$$K = \frac{r_r^2}{F_f r_c} \frac{\Delta y_t}{\int_0^t y \, dt} \tag{5.48}$$

where r_r = radius of standpipe on inner tube
   r_c = radius of inner tube
   Δy_t = vertical difference between curves in Figure 5.18 at time t_d
   F_f = dimensionless factor
   $\int_0^t y \, dt$ = area under lower y-vs.-t curve between $t = 0$ and $t = t_d$

The time t_d for which Δy_t is read from the graph and for which the area under the lower curve is evaluated should be selected as small as possible while still yielding a sufficiently accurate value of Δy_t. The factor F_f is dependent on the geometry of the system, and it was evaluated with a resistance network analog as a function of r_c/d (Figure 5.19), where d is the penetration of the inner tube into the hole bottom. The curve in this figure applies to soil that is uniform to great depth,

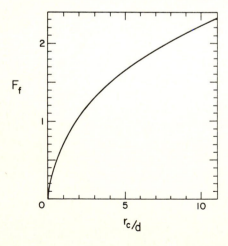

Figure 5.19 Plot of F_f in relation to r_c/d for double-tube method if soil is uniform to a depth of at least $2r_c$ below hole bottom.

which in practice means a depth of at least $2r_c$ below the hole bottom. Values of F_f were also determined for the case where impermeable or infinitely permeable material occurs at depths less than $2r_c$ below the hole bottom (Bouwer, 1961; Bouwer and Jackson, 1974). A simplified procedure for calculating K, based on replacing the y-vs.-t curves as in Figure 5.18 by straight lines to eliminate the integral, was developed by Bouwer and Rice (1964).

The K value obtained with the double-tube method is affected by K in both horizontal and vertical direction, but it is closest to K in vertical direction. The soil region sampled for K is approximately a cylinder with a radius somewhat larger than r_c and a height of about $2r_c$. Tests usually take 3 to 6 h to complete, and several hundred liters of water may be required per test.

5.5.4 Well Pump-in Technique

With the well pump-in technique, which is also called the *reverse auger-hole* method, a well or auger hole is dug and filled with water to a certain depth L_w, which should not be less than 10 times the hole radius r_w (Figure 5.20). Water is added to the hole to maintain this water depth until the outflow Q from the hole into the soil has become essentially constant. If the depth S_i of the impermeable layer below the hole bottom is greater than $2L_w$, K of the wetted zone can be calculated as

$$K = \frac{Q}{2\pi L_w^2}\left\{\ln\left[\frac{L_w}{r_w} + \sqrt{\left(\frac{L_w^2}{r_w^2} - 1\right)}\right] - 1\right\} \tag{5.49}$$

If $S_i < 2L_w$, the equation for K is

$$K = \frac{3Q\,\ln\dfrac{L_w}{r_w}}{\pi L_w(3L_w + 2S_i)} \tag{5.50}$$

These equations were developed by Zangar (Bouwer and Jackson, 1974, and references therein).

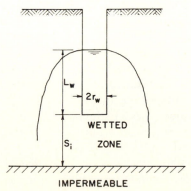

IMPERMEABLE

Figure 5.20 Schematic of well pump-in method.

The value of K obtained with the well pump-in technique mostly reflects K in horizontal direction. Sealing and other disturbances of the soil at the hole wall may cause underestimation of the true K value. The volume of the soil region sampled for K is approximately the same as that for the auger-hole method, i.e., about $0.4L_w^3$. Water may have to be added to the hole for several days before Q approaches a constant value. The water requirements are about 1 m³ per test. Lining the hole with perforated casing or filling it with gravel may be necessary in unstable soils. The method can be used in gravelly or stony materials, which normally is not possible for the other techniques for measuring K in the absence of a water table.

5.6 MEASURING ANISOTROPY OF SOILS AND AQUIFERS

Anisotropy of subsurface material can be evaluated by combining techniques that measure K in different directions. In isotropic soils, these methods theoretically yield the same K value. Different K values indicate anisotropy, provided of course that the soil is otherwise homogeneous. To minimize the effect of local nonuniformities, the different techniques should measure K on essentially the same body of soil or aquifer material, so that the predominant flow direction is the only factor that is different.

Measurements of K in predominantly vertical and horizontal directions can be obtained with the piezometer method by first using a cavity of zero height and then deepening the cavity to a depth of about 4 or 5 times the cavity diameter below the piezometer tube (Maasland, 1957). Talsma (1960), however, found that the hydraulic conductivity K_z in a vertical direction showed much greater variation with depth than the hydraulic conductivity K_x in a horizontal direction. Thus, increasing the height of the piezometer cavity could yield a different K value merely by including soil with different K_z values in the flow system around the cavity. This would lead to erroneous conclusions regarding anisotropy.

Measurement of K in different directions but on essentially the same soil region can be accomplished by first measuring K with the double-tube method (yielding a value affected by both K_z and K_x) and then measuring K_z of the same soil by applying the infiltration-gradient technique to the same installation. Based on the theory of flow in anisotropic media, K_x can then be calculated (Bouwer, 1964; Bouwer and Rice, 1967).

Anisotropy in the horizontal plane can be evaluated with the two-well method, which measures mostly K_x in the direction of the line connecting the wells. If a well or auger hole is dug on each corner of a square, K_x in two mutually perpendicular directions can be evaluated by applying the two-well method in succession to each pair of diagonally opposite holes. Estimates of three-dimensional anisotropy can then be obtained by measuring K_z in the same area with, for example, the infiltration-gradient method.

The above methods give directional K values for relatively small volumes of soil or aquifer material (" point " measurements) and at relatively small depths. To obtain K in different directions for aquifers as a whole, some of the recent refinements in analyzing pumping-test data could be used (see references in Section 5.2.6). Other techniques include simulating certain flow systems in a model and varying K_z and K_x in the model until the simulated system produces the same cause-and-effect relations as observed for the actual system. This technique was employed by Bouwer (1970) to evaluate K_z and K_x of an unconfined aquifer from the response of the water levels in two piezometers to infiltration from ground-water recharge basins. Stallman (1963) developed a procedure, based on electric analog analyses, to determine anisotropy of unconfined aquifers from drawdown measurements at five points. Weeks (1969) evaluated K_z and K_x from drawdown measurements around a pumped well in a glacial-outwash aquifer.

5.7 CORE SAMPLING AND PERMEAMETER TECHNIQUE

The oldest technique for measuring K of subsurface materials is to collect a sample, place it in a cylinder or "permeameter" in the laboratory (Klute, 1965; Reeve, 1957), let water flow through the cylinder, and calculate K with Darcy's equation from the observed flow rate and head loss across the sample (see Section 3.2). Needless to say, this disturbed-sample permeameter technique yielded reliable results only for uniform sands or other coarse materials consisting of relatively round particles. Better results are obtained with an "undisturbed" sample, collected by pushing, driving, or drilling (for rock) a metal cylinder into the material (Campbell and Lehr, 1974; Reeve, 1957; see also Section 6.2.2). The cylinder with the sample inside is then taken to the laboratory for determination of K with the permeameter principle. Disadvantages of the technique are disturbance of the material (truly undisturbed samples of unconsolidated materials are almost impossible to obtain, no matter how sophisticated the core sampling technique is), the small size of the sample (the diameter is usually 5 to 10 cm and the height 5 to 50 cm), and the possibility of leakage flow between the sample and the cylinder wall. Core samples are usually taken in vertical bore holes so that the technique yields K_z. Horizontal samples for measuring K_x can be obtained by pushing the sampler horizontally into walls of pits, trenches, etc.

5.8 PLANNING AND INTERPRETING HYDRAULIC-CONDUCTIVITY MEASUREMENTS

Careful planning of hydraulic-conductivity measurements and rational interpretation of the results are needed where the similar-flow-system approach is not possible and "point" measurement techniques must be resorted to. Because of the heterogeneity of soils and aquifers, the locations and depths of K measurements and the number of replications must be carefully selected. This requires detailed knowledge of the various soils, soil layers, and stratigraphy of the area so that each layer or soil type will be adequately covered by the measurements.

Using technically sound techniques, such as those described in this chapter, and careful field procedures, reasonably accurate K data can be obtained. Variations in the results of replicate tests are then real and attributable to differences in soil or aquifer materials, and not to test errors. It is not uncommon to find that replicate tests in seemingly uniform material yield K values that differ by an order of magnitude, especially if the material has a low permeability. Deeper strata tend to be of more uniform K than surface soils, where K is affected by root activity and soil structure and where K may also change with time. For uniform surface soils, 10 to 20 replicate point measurements may be required to produce standard deviations that are less than 20 percent of the mean (Bouwer and Jackson, 1974, and references therein).

When K measurements have been performed to characterize a certain underground region, the complex situation usually reflected by raw K data must be simplified to obtain a manageable system amenable to analysis. Sometimes the K data will indicate distinct layering or zones of different K. In that case, the K data for each layer can be averaged and the system can be treated as flow in a layered or otherwise nonuniform medium. It may also be possible to combine several layers and consider the resulting medium as homogeneous. If the flow through such layers is mainly horizontal (or parallel to the layers), the arithmetic mean $\bar{K}_a$ of K of the individual layers [Eq. 3.21] should be used. If the flow is mainly vertical (or normal to the layers), the harmonic mean $\bar{K}_h$ should be taken [Eq. 3.27].

If the individual K data reveal a heterogeneous material with random K distribution, the average K can be expected to be between the arithmetic and harmonic means. This is because the average K along a streamline will be equal to the harmonic mean of the K values on that streamline, but the average K normal to streamlines will be closer to the arithmetic mean (Bouwer, 1969). Thus, for a random K distribution, the average K of the medium should be expressed by a

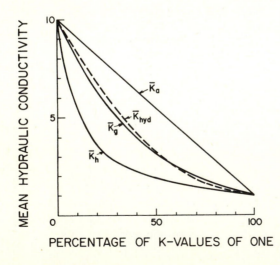

Figure 5.21 Hydraulic, arithmetic, geometric, and harmonic means for different distributions of K values of 1 and 10. (*From Bouwer, 1969.*)

mean that is between the arithmetic and harmonic means of the individual values. An example of such a mean is the geometric mean $\bar{K}_g$, calculated as

$$\overline{K_g} = \sqrt[n]{K_1 K_2 K_3 \cdots K_n} \tag{5.51}$$

Studies with an electric resistance network analog (Bouwer, 1969) showed that the geometric mean gave a better estimate of the true or hydraulic mean $\bar{K}_{hyd}$ of K values in randomly heterogeneous media than the arithmetic and harmonic means (Figure 5.21). The media were obtained by randomly distributing K values of 1 and 10 over a square area. The relative proportion of the 1-and-10 values was varied, and $\bar{K}_{hyd}$ for each medium was calculated from flow rate and total head loss.

PROBLEMS

5.1 A completely penetrating well in an unconfined aquifer is pumped at a rate of $5\,027$ m³/day. The aquifer constants are $T = 2\,000$ m²/day and $S = 0.16$. Calculate the drawdown s in observation wells, 50, 100, and 200 m from the pumped well after 0.01, 0.025, 0.1, 0.25, 1, 2.5, 10, 25, and 100 days of pumping.

Calculate the difference between s for the wells at 50 and 100 m from the pumped well, and plot this difference against log t for the above t values. Determine the t value whereby the s difference has reached 90 percent of the "final" difference when $t = 100$ days. Repeat for the s difference between the wells at 100 and 200 m, and that between the wells at 50 and 200 m from the pumped well. What do the results indicate regarding the best location for observation wells to determine T with the steady-state method?

5.2 Assume that the calculated s-vs.-t data for the observation wells in Problem 5.1 are s-vs.-t data observed in a pumping test to determine T and S of the aquifer. The calculated values of T and S should then agree with the "known" values of $2\,000$ m²/day and 0.16, respectively. Using the difference between s in the 50- and 100-m observation wells after 10 days pumping, calculate T with the steady-state method ($Q = 5\,027$ m³/day). Use the resulting T value and the "measured" s value in the 100-m observation well after 1 day pumping to calculate S with Eqs. (5.3) and (5.4).

5.3 Plot a curve of s (on the ordinate) versus r^2/t on double-logarithmic paper (5×5 cycles) for the s-vs.-t data calculated for all three observation wells in Problem 5.1. Calculate T and S according to the Theis solution (select the match point at $u = 0.2$).

5.4 Plot s versus log t on semilogarithmic paper for the 100-m observation well, using the s-vs.-t data calculated in Problem 5.1. Calculate T and S according to the Chow method (draw the tangent to the curve at $s = 0.2$ m).

5.5 Using the same data curve as in Problem 5.14, calculate T and S with the Cooper-Jacob method.

5.6 Superimpose the schematic data curve of s versus t in Figure 5.7 on the type curves of Figure 5.9 and determine the r/B value of the type curve showing the best fit. Select the match point for the first segment (A portion) of the s-vs.-t curve at $s = 0.004$ m and for the third segment (Y portion) at $s = 0.135$ m. Read the coordinates of the two match points on both graphs and calculate T and S for the A portion (yield delayed) and for the Y portion (yield no longer delayed) of the data curve ($r = 100$ m and $Q = 1\,000$ m³/day).

Calculate s for the horizontal portion of the s-vs.-t curve in Figure 5.7 with Eq. (5.39) and compare with the s value on the graph (use average of calculated T_A and T_Y for T).

Calculate t_{wt} (use the average of T_A and T_Y for T and the calculated value of S_Y) and compare with t where the s-vs.-t curve in Figure 5.7 joins the type curve for $S = 0.1$ and delayed yield no longer occurs.

5.7 Using the y-vs.-t data in Figure 5.12, calculate K for a slug test on a well with $r_c = 0.076$ m, $r_w = 0.12$ m, $L_w = 5.5$ m, $L_e = 4.56$ m, and $H = 80$ m.

5.8 Derive Eq. (5.47) for evaluating K with the air-entry permeameter. Calculate K for the following data: $r_r = 5$ cm, $r_c = 15$ cm, $H_r = 80$ cm, and $G = 50$ cm. The water level in the reservoir fell 1 cm in 10 s just before closing the supply valve, the minimum-pressure head measured with the gage was -90 cm water, and the depth of the wet front in the soil was 15 cm at the conclusion of the test.

5.9 K is determined with the infiltration-gradient technique. Piezometer readings in the wetted zone showed that the pressure head decreased 4 cm per cm depth at an infiltration rate of 0.174 cm/min for the inner cylinder. Calculate K.

5.10 The following time and water-level readings were obtained in measuring K with the double-tube method:

y, cm	Outer-tube standpipe full, s	Equal levels, s
0	0	0
5	13	12
10	26	24
15	41	38
20	57	52
25	73	65

Calculate K if the diameters were 12.7 cm for the inner tube, 25.4 cm for the outer tube, and 2.54 cm for the standpipe on the inner tube. Also, the inner tube penetrated the hole bottom 2.3 cm, and the soil was uniform to a depth of at least 30 cm below the hole bottom.

REFERENCES

Aron, G., and V. H. Scott, 1965. Simplified solutions for decreasing flow in wells. *Proc. Am. Soc. Civ. Eng.* **91**(HY 5): 1–12.

Bentall, R., 1963. Methods of determining permeability, transmissibility, and drawdown. *U.S. Geol. Survey Water Supply Paper 1536-I*, pp. 243–341.

Boast, C. W., and D. Kirkham, 1971. Auger hole seepage theory. *Soil Sci. Soc. Am. Proc.* **35**: 365–374.

Boulton, N. S., 1963. Analysis of data from non-equilibrium pumping tests allowing for delayed yield from storage. *Proc. Inst. Civ. Eng.* **26**: 469–482.

Boulton, N. S., and T. D. Streltsova, 1975. New equations for determining the formation constants of an aquifer from pumping test data. *Water Resour. Res.* **11**: 148–153.

Bouwer, H., 1961. A double-tube method for measuring hydraulic conductivity of soil in situ above a water table. *Soil Sci. Soc. Am. Proc.* **25**: 334–339.

Bouwer, H., 1962. Field determination of hydraulic conductivity above a water table with the double-tube method. *Soil Sci. Soc. Am. Proc.* **26**: 330–335.

Bouwer, H., 1964. Measuring horizontal and vertical hydraulic conductivity of soil with the double-tube method. *Soil Sci. Soc. Am. Proc.* **28**: 19–23.

Bouwer, H., 1966. Rapid field measurement of air-entry value and hydraulic conductivity of soil as significant parameters in flow system analysis. *Water Resour. Res.* **2**: 729–738.

Bouwer, H., 1969. Planning and interpreting soil permeability measurements. *J. Irrig. Drain. Div., Am. Soc. Civil Eng.* **95**(IR 3): 391–402.

Bouwer, H., 1970. Groundwater recharge design for renovating waste water. *J. Sanit. Eng. Div., Am. Soc. Civ. Eng.* **96**(SA 1): 59–74.

Bouwer, H., and R. D. Jackson, 1974. Determining soil properties. In *Drainage for Agriculture*. J. van Schilfgaarde (ed.), Agronomy Monograph No. 17, Am. Soc. Agron., Madison, Wis., pp. 611–672.

Bouwer, H., and R. C. Rice, 1964. Simplified procedure for calculation of hydraulic conductivity with the double-tube method. *Soil Sci. Soc. Am. Proc.* **28**: 133–134.

Bouwer, H., and R. C. Rice, 1967. Modified tube diameters for the double-tube apparatus. *Soil Sci. Soc. Am. Proc.* **31**: 437–439.

Bouwer, H., and R. C. Rice, 1976. A slug test for determining hydraulic conductivity of unconfined aquifers with completely or partially penetrating wells. *Water Resour. Res.* **12**: 423–428.

Campbell, M. D., and J. H. Lehr, 1974. *Water Well Technology.* McGraw-Hill, New York, 681 pp.

Carr. P. A., and Van Der Kamp, G. S., 1969. Determining aquifer characteristics by the tidal method. *Water Resour. Res.* **5**: 1023–1031.

Chow, V. T., 1952. On the determination of transmissivity and storage coefficients from pumping test data. *Trans. Am. Geoph. Union* **33**: 397–404.

Cooper, H. H., Jr., J. D. Bredehoeft, and I. S. Papadopulos, 1967. Response of a finite-diameter well to an instantaneous charge of water. *Water Resour. Res.* **3**: 263–269.

Cooper, H. H., Jr., and C. E. Jacob, 1946. A generalized graphical method for evaluating formation constants and summarizing well-field history. *Trans. Am. Geoph. Union* **27**: 526–534.

Dagan, G., 1967. A method of determining the permeability and effective porosity of unconfined anisotropic aquifers. *Water Resour. Res.* **3**: 1059–1071.

De Glee, G. J., 1930. "Over grondwaterstromingen bij wateronttrekking door middel van putten." Doctoral dissertation, Techn. Univ., Delft, The Netherlands, printed by J. Waltman, 175 pp.

De Glee, G. J., 1951. Berekeningsmethoden voor de winning van grondwater. In *Drinkwatervoorziening 3e Vacantie cursus*, pp. 38–80. Moorman's periodieke pers, The Hague, Netherlands.

Ehlig, C., and J. C. Halepaska, 1976. A numerical study of confined-unconfined aquifers including effects of delayed yield and leakage. *Water Resources Res.* **12**: 1175–1833.

Ferris, J. G., 1951. Cyclic fluctuations of water level as a basis for determining aquifer transmissibility. *Assemblée Generale de Bruxelles, Assoc. Int. Hydrol. Sci.* **2**: 148–155. (Also in *U.S. Geol. Survey Water Supply Papers 1536 E*, pp. 132–135, 1962; and *1536 I*, pp. 305–318, 1963.)

Ferris, J. G., D. B. Knowles, R. H. Brown, and R. W. Stallman, 1962. Theory of aquifer tests. *U.S. Geol. Survey Water Supply Paper 1536-E*, pp. 69–174.

Hantush, M. S., 1956. Analysis of data from pumping tests in leaky aquifers. *Trans. Am. Geophys. Un.* **37**: 702–714.

Hantush, M. S., 1959a. Nonsteady flow to flowing wells in leaky aquifer. *J. Geophys. Res.* **64**: 1043–1052.

Hantush, M. S., 1959b. Analysis of data from pumping wells near a river. *J. Geophys. Res.* **64**: 1921–1932.

Hantush, M. S., 1964a. Hydraulics of wells. In *Advances in Hydroscience*, vol. 1, V. T. Chow (ed.), pp. 281–432. Academic Press, New York and London.

Hantush, M. S., 1964b. Drawdown around wells of variable discharge. *J. Geophys. Res.* **69**: 4221–4235.

Hantush, M. S., 1966. Analysis of data from pumping tests in anisotropic aquifers. *J. Geophys. Res.* **71**: 421–426.

Hantush, M. S., and C. E. Jacob, 1955. Non-steady radial flow in an infinite leaky aquifer. *Am. Geophys. Un. Trans.* **36**: 95–100.

Hantush, M. S., and R. G. Thomas, 1966. A method for analyzing a drawdown test in anisotropic aquifers. *Water Resour. Res.* **2**: 281–285.

Hoffman, G. J., and G. O. Schwab, 1964. Tile spacing prediction based on drain outflow. *Trans. Am. Soc. Agric. Eng.* **7**: 444–447.

Huisman, L., and J. Kemperman, 1951. Bemaling van spanningsgrondwater. *De Ingenieur* **62**: B29–B35.

Klute, A., 1965. Laboratory measurement of hydraulic conductivity of saturated soil. In *Methods of Soil Analysis*, part I, C. A. Black (ed.), Agronomy Monograph No. 9: 210–221. Am. Soc. Agron., Madison, Wis.

Kroszynski, U. I., and G. Dagan, 1975. Well pumping in confined aquifers: the influence of the unsaturated zone. *Water Resour. Res.* **11**: 479–490.

Kruseman, G. P., and N. A. de Ridder, 1970. Analysis and evaluation of pumping test data. *Bull. 11, Internat. Inst. Land Reclam. and Improvement*, Wageningen, The Netherlands, 200 pp.

Lohman, S. W., 1972. Groundwater Hydraulics. *U.S. Geol. Survey Prof. Paper 708*, 70 pp.

Luthin, J. N., and D. Kirkham, 1949. A piezometer method for measuring permeability of soil in situ below a water table. *Soil Sci.* **68**: 349–358.

Maasland, M., 1957. Soil anisotropy and land drainage. In *Drainage of Agricultural Lands*, J. N. Luthin (ed.), Agronomy Monograph No. 7: 216–287. Am. Soc. Agron., Madison, Wis.

Moench, A. F., and T. A. Prickett, 1972. Radial flow in an infinite aquifer undergoing conversion from artesian to water table conditions. *Water Resources Res.* **8**: 494–499.

Neuman, S. P., 1972. Theory of flow in unconfined aquifers considering delayed response of the water table. *Water Resources Res.* **8**: 1031–1045.

Neuman, S. P., 1975. Analysis of pumping test data from anisotropic unconfined aquifers considering delayed gravity response. *Water Resour. Res.* **11**: 329–342.

Neuman, S. P., and P. A. Witherspoon, 1969. Theory of flow in a confined two-aquifer system. *Water Resour. Res.* **5**: 803–816.

Papadopulos, I. S., and H. H. Cooper, Jr., 1967. Drawdown in a well of large diameter. *Water Resour. Res.* **3**: 241–244.

Pinder, G. F., J. D. Bredehoeft, and H. H. Cooper, Jr., 1969. Determination of aquifer diffusivity from aquifer response to fluctuations in river stage. *Water Resour. Res.* **5**: 850–855.

Prickett, T. A., 1965. Type-curve solution to aquifer tests under water-table conditions. *Ground Water* **3**(3): 5–14.

Reeve, R. C., 1957. Measurement of permeability in the laboratory. In "Drainage of Agricultural Lands," J. N. Luthin (ed.), *Agronomy Monograph No. 7*, Am. Soc. Agron., Madison, Wis., pp. 414–419.

Rice, R. C., 1967. Dynamic response of small piezometers. *Trans. Am. Soc. Agr. Eng.* **10**: 80–83.

Stallman, R. W., 1963. Electric analog of three-dimensional flow to wells and its application to unconfined aquifers. *U.S. Geol. Survey Water Supply Paper 1536-H*, pp. 205–242.

Steggewentz, J. H., 1933. "De invloed van de getijbeweging van zeeën en getijrivieren op de stijghoogte van het grondwater." Doctoral dissertation, Techn. University, Delft, The Netherlands.

Sternberg, Y. M., 1967. Transmissibility determination from variable discharge pumping tests. *Ground Water* **5**(4): 27–29.

Talsma, T., 1960. Measurement of soil anisotropy with piezometers. *J. Soil Sci.* **11**: 159–171.

Theis, C. V., 1935. The relation between the lowering of the piezometric surface and the rate and duration of discharge of a well using groundwater storage. *Trans. Am. Geoph. Un.* **16**: 519–524.

Topp, G. C., and M. R. Binns, 1976. Field measurements of hydraulic conductivity with a modified air-entry permeameter. *Can. J. Soil Sci.* **56**: 139–147.

Walton, W. C., 1962. Selected analytical methods for well and aquifer evaluation. *Illinois State Water Survey Bull.* **49**: 81 pp.

Walton, W. C., 1970. "Groundwater Resource Evaluation." McGraw-Hill, New York, 664 pp.

Weeks, E. P., 1969. Determining the ratio of horizontal to vertical permeability by aquifer-test analysis. *Water Resour. Res.* **5**: 196–214.

Youngs, E. G., 1968. Shape factors for Kirkham's piezometer method for determining the hydraulic conductivity of soil in situ for soils overlying an impermeable floor or infinitely permeable stratum. *Soil Sci.* **106**: 235–237.

GROUNDWATER EXPLORATION, WELL CONSTRUCTION, AND PUMPING

6.1 GROUNDWATER EXPLORATION AND WELL-SITE SELECTION

The initial step in successful groundwater development is selection of proper well sites, first ascertaining the presence of suitable aquifers and then finding the best place for the well or wells from a standpoint of quantity, quality, and depth of groundwater, and absence of potential contamination by polluted or other low-quality water. The optimum well location can often be determined without detailed studies if wells already exist in the particular area and the groundwater hydrology is fairly well known. In "new" areas, however, exploration and intensive study may be required, starting with reconnaissance-type surveys to delineate the more promising groundwater areas, and followed by geophysical surveys and test drilling to determine optimum well sites.

6.1.1 Reconnaissance Surveys

Reconnaissance surveys consist of rather extensive studies of the hydrogeology of a given region, using geologic maps, aerial photographs, and ground observations to detect sufficiently permeable strata that by virtue of their relative elevation or depression, geologic history, and hydrology could be water-bearing (Blank and Schroeder, 1973; Davis and DeWiest, 1966). Hydrologic studies may consist of determining water inputs and outputs for areas that could serve as potential recharge sites for the aquifer or aquifers, such as alluvial fans, beach or dune ridges, and other deep, permeable soils and outcrops of permeable, consolidated or unconsolidated strata. Net rates of groundwater accretion can then be evaluated as the difference between inflow (precipitation, stream seepage, etc.) and outflow (runoff, evaporation from soil and vegetation, etc.) of water for these areas. Significant accretion of groundwater may also occur through less permeable surface materials, particularly if they are extensive and rainfall is primarily of low intensity and long duration.

Unconsolidated materials. Aquifers of unconsolidated or nonindurated materials, such as alluvial, glacial, or aeolian deposits, are among the most common sources of groundwater. In alluvial or glacial deposits, buried valleys or old stream beds offer the best groundwater potential. These are ancient stream beds or valleys, primarily consisting of sands and gravels, that have been covered by finer sediments (glacial till, for example). Because of erosion, bedrock tends to be depressed below unconsolidated sediments of ancestoral streams. For these reasons, buried valleys generally are the most productive aquifers in the region. While surface features sometimes are helpful in detecting buried valleys (Saines, 1968), geophysical techniques are more commonly used (see Section 6.1.2).

Alluvial fans, which commonly occur along the periphery of tectonically formed alluvial basins or valleys, consist of more permeable materials than the rest of the basin. Thus, fans are good infiltration areas for the surface runoff entering the basin from the surrounding mountains. However, alluvial fans may not be suitable locations for high-yielding wells because of their proximity to mountain ranges and relatively small depth to bedrock.

Faults often constitute hydraulic barriers to lateral movement of groundwater because permeable strata may be vertically offset or even discontinuous along a fault. Thus, wells should be located upgradient of fault zones. On the other hand, semipervious or leaky faults have also been observed in alluvial deposits (Williams, 1970).

In glaciated regions, sorted and hence permeable deposits such as glacial outwashes, buried stream channels, or pre- or postglacial alluvial deposits offer the best groundwater potential. Old beach ridges, terraces, sand and gravel zones in terminal moraines, and other permeable deposits may also be suitable for groundwater development. The morphological features associated with these landforms can often be identified from aerial photographs. Where glacial features are shallow, geophysical surveys and test drilling must be used to locate underlying aquifers. Eskers, kames, and similar permeable glacial deposits often are situated relatively high and are, therefore, too well drained to contain usable groundwater. Unsorted glacial deposits, such as till, are of low permeability and make poor aquifers unless interspersed with lenses or streaks of sand and gravel that can be tapped by a well.

Of the aeolian deposits, sand dunes in coastal zones and around large inland lakes offer the best potential for groundwater development. The high permeability of dune sand affords ample accretion of groundwater in humid areas. This can create locally elevated water tables that, in coastal zones, depress the underlying salt groundwater from the ocean (see Section 11.2). Loess and other fine-textured aeolian deposits generally are not permeable enough to make good aquifers.

Sedimentary rock. Clean sandstones and cavernous limestones offer the best groundwater potential in sedimentary rock. Solution channels in limestone, created by dissolution of calcium carbonate in acid water (due to dissolved carbon dioxide and organic acids) percolating downward from overlying soils, tend to be concentrated where the flow intensities are greatest. This occurs in the area immediately below the water table and near points where the limestone aquifer discharges to the surface, such as springs, seeps, and streams (Bedinger, 1967). Without solution channels, the groundwater potential of limestone falls in the same group as that of shales and other fine-grained rock. These materials yield groundwater only when fractured by weathering or faulting, and then usually in moderate amounts. Faulted areas often constitute hydraulic barriers to groundwater flow because the permeable strata are offset. The fault zone may also be filled with fine-textured debris. Thus, wells should be located upgradient from faults. On the other hand, major fault and fracture zones, as may be indicated by surface features like reverse drag, could be the best location for wells in moderately permeable sandstones and other sedimentary rock.

Well yields in fractured shale or carbonate rock (including cavernous limestone) vary widely, even over short distances in the same formation. Siddiqui and Parizek (1971) found that wells in the carbonate rocks and shales of Pennsylvania were more productive when located in fracture zones than in nonfracture zones, due to anticline-syncline relations. Of 45 randomly selected wells, 18 were on or near a fracture trace and 27 were not, indicating a 40 percent chance of accidentally hitting a fracture zone in that region. Wells in valley bottoms were generally more productive than those on slopes or uplands. This was also found for the Chalk aquifer in southern England, where transmissivities exceeded 2 500 m^2/day below valleys, but were less than 100 m^2/day below the higher areas (Nutbrown et al., 1975).

Crystalline rocks. Igneous and metamorphic rocks (granites, gneisses, etc.) also yield reasonable (but still moderate) amounts of groundwater only when fractured by faulting or weathering. Davis and Turk (1964) found that most of the interstitial openings in various types of jointed and faulted crystalline rock occurred within a depth of about 30 m. Water yield per meter of well decreased about tenfold if the depth of the well was increased from 30 to 100 m. The economic depth for domestic wells was less than 40 to 80 m and often less than 30 m, and that for larger production wells generally less than 200 m. In another study (Landers and Turk, 1973), optimum well depths were in the range of 20 to 30 m. If adequate well yields could not be obtained after 50 to 60 m of drilling, further drilling was usually futile and it was more economical to drill another well for obtaining additional water. Larger yields are usually obtained from wells in valleys than in hills or uplands, probably because depth of weathering and water movement into the ground are greater for the valley areas (Landers and Turk,

1973; Joiner et al., 1968). For the same reason, crystalline rocks tend to yield more water in humid than in arid areas.

Increased fracturing and weathering of crystalline rock can also be expected in fault and fracture zones. For example, the yield of three wells located along a fault was 5 to 8 times as much as the average well yield in the Archaean bedrock in Sweden (Meier and Petersson, 1951). Fault and fracture zones and pegmatite dikes were the best well locations in the crystalline rocks of the Colorado Rocky Mountains (Florquist, 1973).

Volcanic material. Volcanic rocks differ widely in their suitability for groundwater development. Some recent basalts are extremely permeable and have yielded some of the most productive wells. On the other hand, tuffs and rhyolites, though porous, generally have very low permeabilities. Some porous lavas or basalts are so permeable that they hold water only if it is retained by hydraulic barriers like vertical dikes, faults, mountain ranges, ash beds, dense basalt or lava layers, or other zones of low permeability.

Seeps, springs, and other surface features. Seeps and springs are among the hydrologic surface features that may aid in evaluating the groundwater potential of certain areas. Hillside seeps above outcrops of less permeable strata indicate groundwater in overlying formations. Springs also reveal the presence of groundwater. A few large springs may indicate thick, transmissive aquifers, whereas frequent, small springs tend to indicate thin aquifers of low transmissivity. In arid areas, presence of trees or other deep-rooted vegetation in an otherwise barren landscape may indicate shallow groundwater, for example below ephemeral streams, flood plains, or oases. If the water table in arid areas is close to the surface, evaporation can cause salt accumulation on the surface soil. These salt flats, with or without salt-tolerant vegetation, should not be confused with playas, which are ephemeral lakes formed by surface runoff in low areas with soil of low permeability. Biological indicators of groundwater may even include ants, which are reported to be able to tunnel down as much as 30 m to reach groundwater and are used to detect groundwater in the deserts of Kazahkstan (*Ground Water Newsletter*, 1974).

Well yields. Orders of magnitudes of well yields that can be expected for the various geological materials are:

Sorted or coarse sands and gravels, porous basalts	$1\,000–20\,000$ m^3/day
Cavernous limestones	$500–5\,000$ m^3/day
Sand and gravel mixes with fines, sandstones	$100–2\,000$ m^3/day
Fractured and weathered rock	$10–500$ m^3/day

These values (see also Davis and DeWiest, 1966) should be interpreted as approximate guides. Actual well yields are governed by transmissivity and specific yield or storage coefficient of aquifer, depth of penetration of well into aquifer, drawdown of water level in well, lateral extent of aquifer, and construction and condition of well. For wells in dense rock, where the yield of groundwater is entirely governed by cracks, fissures, solution channels, or other so-called secondary porosity, the productivity of a well depends largely on how carefully the well is located and how lucky the driller is in hitting cracks or other large openings.

6.1.2 Geophysical Methods

Geophysical methods are used to obtain more accurate information about subsurface conditions, such as type and depth of materials (consolidated or unconsolidated), depth of weathered or fractured zone, depth to groundwater, depth to bedrock, and salt content of groundwater. The most common techniques are the electric resistivity and seismic methods, which are discussed below. Gravity surveys have been used to determine depth to bedrock in alluvial deposits (Davis, 1969; Ibrahim and Hinze, 1972; McDonald and Wantland, 1961) and to trace buried channels (Zohdy et al., 1974). Steep gravity gradients may indicate rapid changes in depth to bedrock, like buried fault scarps. Knowledge about the location of such scarps is important in selecting sites for deep wells and in predicting areas that may be prone to surface cracking due to groundwater overdraft and resulting land subsidence (see Section 9.8). Gravity and magnetic surveys may be useful in determining stratigraphy of consolidated rock and in locating major fault and fracture zones. Magnetic surveys have been used to study basalt aquifers and alluvial basins underlain by magnetic bedrock (Zohdy et al., 1974). Temperature surveys may indicate the location of shallow, elongated aquifers. Such aquifers act as heat sinks during warming trends (spring and summer) and as heat sources during cooling trends (fall and winter). This causes temperature anomalies at or near ground surface, which have been measured to detect buried valleys (Cartwright, 1968; Smith, 1974).

Resistivity surveys. Electrical-resistivity surveying is based on evaluating the apparent resistivity R_a of subsurface material by passing a known electric current through the ground and measuring the potential difference between two points. The current is applied with buried tinfoil or metal rods or spikes driven into the ground (Figure 6.1). The distance between the current electrodes is about 10 m to several 100 m, depending on the desired depth of measuring R_a. To avoid polarization, low-frequency alternating current (< 1 Hz) or reversing direct current is applied at potentials of up to about 200 V. Dry soil may have to be wetted around the current electrodes to obtain good electrical contact. The voltage or potential difference is measured with two separate electrodes located symmetrically on a line between the current electrodes (Figure 6.1). The potential electrodes, which consist of metal rods (porous ceramic cups with a saturated copper sulfate solution have also been used), are placed some distance from the current electrodes to avoid areas of rapid voltage change.

The apparent resistivity R_a of underground material is defined as the proportionality factor in the equation relating the measured electrical resistance R of the

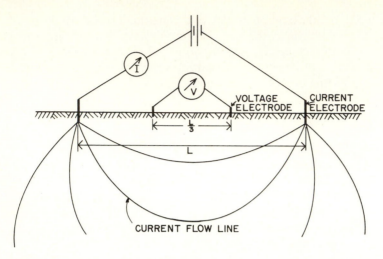

Figure 6.1 Current and voltage electrodes for Wenner configuration in resistivity measurement.

electrical field to the mean path length L_c of the current and the mean cross-sectional area A_c of the electrical field, or

$$R = R_a L_c / A_c \qquad (6.1)$$

Since R is expressed in ohms, the dimension of R_a is $\Omega \cdot m$. R_a increases with increasing porosity of the material, decreasing water content, and decreasing salt content of the water in the formation. Values of R_a range from 1 $\Omega \cdot m$ or less for clay with salty water to 10^8 $\Omega \cdot m$ or more for solid igneous rock and quartz. Weathered rock has a lower R_a value. Sand and gravel aquifers with fresh water have R_a values of 15 to 600 $\Omega \cdot m$. The lower values (15 to 20 $\Omega \cdot m$) are characteristic of aquifers with a relatively high salt content in the water (several hundred to about 1 000 mg/l), as found in the southwestern United States. The higher values (300 to 600 $\Omega \cdot m$) have been observed for coastal aquifers (Maryland) and basaltic aquifers (Idaho). Certain freshwater-bearing sands in California had R_a values of 100 to 250 $\Omega \cdot m$ (Zohdy et al., 1974). R_a values less than 10 $\Omega \cdot m$ are indicative of aquifers with saline or brackish groundwater. R_a is about 50 $\Omega \cdot m$ for fresh water as such, and less than 1 $\Omega \cdot m$ for seawater.

Two configurations are used for the voltage electrodes: the Wenner configuration where the distance between the voltage electrodes is one-third the distance L between the current electrodes (Figure 6.1), and the Schlumberger configuration where the distance between the voltage electrodes is less than $L/5$. For the Wenner configuration, R_a is calculated as

$$R_a = \frac{2}{3} \pi L \frac{V}{I} \qquad (6.2)$$

where V is the potential difference between the voltage electrodes, I is the total current in the electrical field, and L is the distance between the current electrodes.

For the Schlumberger configuration, the equation for R_a is

$$R_a = \pi \frac{(L/2)^2 - (a/2)^2}{a} \frac{V}{I} \tag{6.3}$$

where a is the distance between the voltage electrodes.

With the Wenner configuration, which is widely used in the Western Hemisphere, the depth to which R_a is measured is considered to be about equal to the voltage-electrode spacing. This is true only if the underground material is of fairly uniform resistivity. A layer of low resistivity relatively close to the surface has a dominating effect on the measured value of R_a and reduces the effective depth of the survey. The advantage of the Schlumberger configuration, which is the most widely used in electrical prospecting (Zohdy et al., 1974), is that the voltage electrodes do not have to be moved each time the distance between the current electrodes is increased to measure R_a to greater depth.

A third arrangement is the dipole-dipole array, where the current electrodes and voltage electrodes are arranged in separate pairs or dipoles. The distance between the current electrodes and the distance between the voltage electrodes are much less than the distance between the centers of the dipoles. The electrode pairs can be arranged in various arrays. The method was developed in the late 1940s in Russia, where it has become a common prospecting technique, and it has been used in the 1960s for groundwater investigations in the United States (Zohdy et al., 1974, and references therein).

Resistivity surveys can be carried out laterally or vertically. With lateral surveys, a constant electrode spacing, selected to produce the desired depth of the survey, is used and R_a is measured at different locations to yield a map of isoresistivity lines. Such maps are useful for detecting changes in bedrock or aquifer depth (for example, in tracing buried valleys), vertical discontinuities such as faults and fracture zones, changes in groundwater quality (including travel of contaminated water), and changes in the depth of freshwater-saltwater interfaces (especially in coastal areas).

With vertical surveys, the distance between the electrodes is expanded to increase the depth to which R_a is measured. Plotting the resulting R_a values against the electrode spacing yields a curve which, when properly interpreted, gives information regarding changes in resistivity with depth. To aid in the proper interpretation, theoretically calculated type curves have been prepared showing R_a versus depth for a variety of two-, three-, or four-layered systems with different resistivity values for each layer (Dobrin, 1952; Mooney and Wetzel, 1956; Zohdy et al., 1974). Comparing the field curve of R_a with the type curves then yields information about the thickness and resistivity of the various layers. This is most accurately done for the two-layered case. For multilayered situations, the numerous uncertainties and variables affecting the field data make it difficult to derive a unique interpretation of the lithology. Proper interpretation of field R_a curves is always enhanced if additional information about depth and type of materials—for example, from well logs or seismic surveys—is also available.

The maximum depth range of vertical-resistivity surveys is about 500 m. The depth resolution, however, is restricted to the first layer of low resistivity. To obtain R_a data below such layers, deeper current applications with electrodes in boreholes have been employed (Merkel and Kaminsky, 1972). Vertical-resistivity surveys have been used to determine depths of aquifers, bedrock, fractured or weathered zones in rock, and freshwater-saltwater interfaces. The depth of the water table in unconfined aquifers usually cannot be determined with great accuracy, because the water content in the vadose zone often is too high to yield a significant difference between the resistivity above and below the water table.

Resistivity surveys are relatively inexpensive, and they are among the most widely used techniques in groundwater investigations (Bays, 1950; Bierschenk, 1964; Buhle, 1953; Dudley et al., 1964; Kelley, 1962; McDonald and Wantland, 1961; McGinnis and Kempton, 1961; Page, 1968; Smith, 1974; Spicer, 1952; Swartz, 1937; Volker and Dijkstra, 1955; Zohdy, 1965; and Zohdy et al., 1974). The method is particularly useful when combined with other studies, like seismic surveys, test-well drilling, and aerial photography (Foster and Buhle, 1951; McGinnis and Kempton, 1961). Underground cables, pipelines, transformers on power lines, metal fences, and other conductors in contact with the soil interfere with the measurements and usually give erroneous results within the area of influence.

Seismic surveys. Seismic surveys are based on measuring the velocity of shock or sound waves in the various strata. Since the velocity of sound in underground material increases with increasing bulk density and water content, the results can be interpreted in terms of type, porosity, and water content of the material. The method also enables calculation of the depth of the various strata if their sound velocities are sufficiently different.

The shock or sound waves are generated by setting off a small explosion at a depth of about 1 m or more. The arrival of the shock waves at various distances from this *shot point* is then measured with sound detectors, called *geophones*, placed on the ground surface. The geophones are connected with cables to a central oscillograph or other device for recording the arrival time of the first shock wave after detonation of the charge. Sound velocities range from about 250 m/s in loose, unsaturated material (surface soil) to 5000 m/s or more in dense, crystalline rock. Deeper, unconsolidated materials may have sound velocities of 300 to 1000 m/s when unsaturated and 1500 to 2500 m/s when saturated (Gill et al., 1965; Zehner, 1973). These authors also report sound velocities in bedrock of 3000 to 5500 m/s. Dense, unconsolidated aquifers with fine particles may have sound velocities in excess of 2500 m/s. Sound velocities may vary from 1000 to 1500 m/s in fractured rock, from 2000 to 3000 m/s in hard sandstone, and from 2000 to 5000 m/s in hard limestone.

There are two types of seismic surveys: lateral surveys, called *fan shooting*, and vertical surveys, called *refraction* surveys. With fan shooting, the geophones are arranged on a circle around the shot point, as shown in Figure 6.2, where fan shooting is applied to locate a buried valley. Since the sound velocity in the

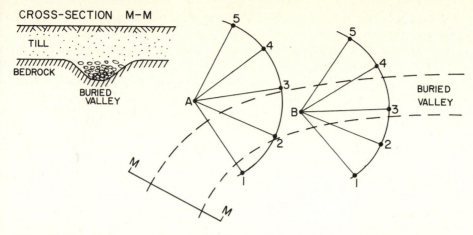

Figure 6.2 Shot point and geophones in fan shooting for determining location of buried valley.

porous sands and gravels of the buried valley will be less than that in the finer, denser overlying sediments, geophones 1 and 2 (Figure 6.2) will record longer arrival times than the other geophones. The next shot point may then be selected above the buried valley (point B), and the geophones are again arranged in a semicircle. This time, geophones 3 and 4 will show the longest arrival times. The process is repeated until the buried valley is sufficiently traced. Fan shooting has been extensively applied in locating salt domes for oil exploration in Texas and Louisiana (Heiland, 1946).

With refraction seismic surveys, the geophones are uniformly spaced on a straight line from the shot point to record the arrival of the first shock waves. These waves may have traveled straight from the shot point to the geophone, or they may have been refracted and reflected in the deeper layers (Figure 6.3).

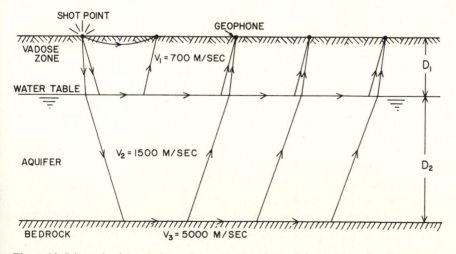

Figure 6.3 Schematic of travel of sound waves in three-layered system with refraction seismic survey.

Refracted and reflected shock waves will reach the more remote geophones sooner than the straight-traveling waves if the velocity of sound in the deeper layers is much greater than that in the surface materials. Plotting the arrival time of the first shock wave at each geophone against distance of geophone from shock point yields a curve which for a layered profile consists of a succession of straight-line sections (Figure 6.4). The first section represents the first or top layer of the profile, the second section the second layer, etc. The sound velocity in each layer is calculated as the reciprocal of the slope of the corresponding straight-line section (Figure 6.4). For examples of actual field curves, reference is made to Gill et al. (1965) and Zohdy (1965).

Several equations have been developed to calculate the depth of the various discontinuities or the thickness of the different layers from plots like Figure 6.4. The equations are based on finding the fastest path of the shock waves between shot point and geophone. Some equations utilize the intercepts of the extended straight-line sections of the curve with the ordinate. Others use the horizontal or "critical" distance of the points where the slope of the curve changes (Dobrin, 1952; Heiland, 1946; Nettleton, 1940; Zohdy et al., 1974). Nettleton presented the following expression for the thickness of a particular layer in a multilayered profile:

$$
D_n = \frac{V_n V_{n+1}}{2\sqrt{V_{n+1}^2 - V_n^2}} \left[t_{i(n+1)} - 2D_1 \frac{\sqrt{V_{n+1}^2 - V_1^2}}{V_1 V_{n+1}} \right.
$$
$$
\left. - 2D_2 \frac{\sqrt{V_{n+1}^2 - V_2^2}}{V_2 V_{n+1}} \cdots - 2D_{n-1} \frac{\sqrt{V_{n+1}^2 - V_{n-1}^2}}{V_{n-1} V_{n+1}} \right] \quad (6.4)
$$

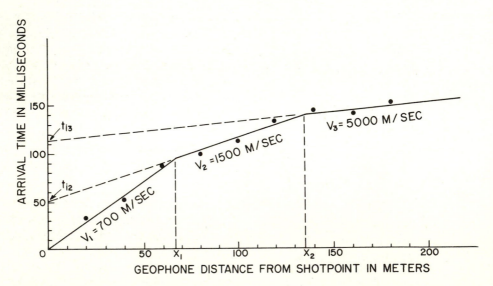

Figure 6.4 Hypothetical relation between arrival time and geophone distance for system in Figure 6.3.

where D_n = thickness of nth layer (for the top layer, $n = 1$)

V_n = sound velocity in nth layer (reciprocal of slope for nth straight section of data curve)

$t_{i(n)}$ = intercept of extended straight line for nth layer with t axis (Figure 6.4)

For the top layer, this expression reduces to

$$D_1 = \frac{V_1 V_2}{2\sqrt{V_2^2 - V_1^2}} t_{i2} \tag{6.5}$$

and for the second layer to

$$D_2 = \frac{V_2 V_3}{2\sqrt{V_3^2 - V_2^2}} \left| t_{i3} - 2D_1 \frac{\sqrt{V_3^2 - V_1^2}}{V_3 V_1} \right| \tag{6.6}$$

An equation commonly used to calculate D_1 from the horizontal distance x_1 of the first break point in the curve (Figure 6.4) is

$$D_1 = \frac{x_1}{2} \sqrt{\frac{V_2 - V_1}{V_2 + V_1}} \tag{6.7}$$

An approximate equation for D_2, presented by Geophysical Specialties Company (1960), is (in modified form)

$$D_2 = \frac{x_2}{2} \sqrt{\frac{V_3 - V_2}{V_3 + V_2}} - \frac{D_1}{6} \tag{6.8}$$

where x_2 is the horizontal distance of the second break point in the data curve.

The depth range of refraction seismic surveys is usually on the order of 100 m, although information to much greater depths has also been obtained. The depth range is restricted to the layer with the highest sound velocity (bedrock, for example), because deflection of sound waves by such a layer makes it impossible to measure sound velocities in underlying material. To get below bedrock, deep shot points in boreholes must be used. In practice, multilayered profiles often do not show enough difference in sound velocities or the layers are too thin to enable use of Eq. (6.4). Wallace (1970) could not obtain accurate information on the depth of bedrock below deep, alluvial fills because the high density of the deeper materials did not offer enough sound-velocity contrast with the bedrock. Zehner (1973) reported errors of 15 to 20 percent in seismically determined depths to bedrock below about 35 m of alluvial deposits. The most common application of seismic methods in groundwater investigations is the determination of the thickness of unconsolidated sediments overlying bedrock, like buried channels (Zohdy et al., 1974).

Seismic surveys require relatively expensive equipment and specialized personnel (Zohdy et al., 1974), especially when dealing with less-than-ideal conditions such as sloping or wedge-shaped formations, linearly increasing sound velocities, and faults or other vertical discontinuities. For refinements in seismic techniques to handle such situations, reference is made to the various textbooks and manuals on geophysics and oil exploration (Dobrin, 1952; Heiland, 1946; Nettleton, 1940; and Zohdy et al., 1974). For shallow exploration (depths not exceeding about 50 m), less expensive, portable seismographs may be used (Hobson and Collett, 1960). In quiet surroundings without noise interference from traffic or aircraft, the shock waves may then be initiated by striking a steel plate on the ground with a sledge hammer (Bianchi and Nightingale, 1975).

6.1.3 Test Drilling

The most accurate information about the geologic profile and the depth and quality of groundwater at a given site is obtained by test drilling. It may even be possible to do some preliminary pumping tests or slug tests on the bore hole to estimate hydraulic properties of the aquifer or aquifers and to calculate potential well yields (see Chapters 4 and 5).

The relative amounts of time and effort that should be spent on geophysical exploration and on test drilling depend on the type of information that is needed and on costs. Where deep groundwater, consolidated materials, or both make test drilling relatively expensive, or where the area is not very accessible to drilling rigs, it may be advantageous to get as much information as possible from geophysical surveys before selecting test-drilling sites. On the other hand, where groundwater is relatively close to the surface and soil materials are unconsolidated, test drilling is relatively inexpensive, and it may be better to minimize geophysical surveys and rely more on test drilling to determine optimum well sites.

Test wells normally are of relatively small diameter and can be drilled at a fraction of the cost of full-sized wells. Considering the cost of complete production wells and the economic advantages of having wells with optimum yields, test drilling usually is economically justified. When a test well indicates a favorable location, it can often be converted into a production well by reaming or redrilling to increase its diameter.

For shallow exploration in unconsolidated materials that are relatively free from gravel and boulders, test wells may be installed by driving or jetting (see Section 6.3.2). For deep wells and hard materials, the cable-tool method is preferable over rotary drilling because it permits more accurate determination of groundwater levels, more accurate sampling of groundwater for quality analysis, and more accurate sampling of the cuttings as brought up with the bailer (see Section 6.2.1). Cutting samples obtained from rotary drilling tend to be a mixture of drilling mud and material that has accumulated from various depths in the bottom of the hole. Sample lag time may also be a problem. On the other hand, rotary drilling is much faster, and the holes do not require casing, so that the test well can be logged with electrical well-logging techniques (see Section 6.2.3).

6.2 WELL LOGGING

Test wells and production wells offer unique opportunities to collect information about the geology and groundwater conditions of a given site by well-logging techniques. When such studies are carried out for a number of wells in a groundwater basin or regional aquifer system, the results can be cross-correlated to yield a complete picture of the groundwater geology of the region. Such information is needed for developing plans for optimum utilization of the groundwater resource. Advanced sampling techniques and geophysical well-logging methods require specialized personnel and equipment. They are, therefore, relatively expensive and not routinely used for low-capacity wells. On the other hand, driller's logs can be maintained relatively simply as the hole is being drilled. Shortcomings notwithstanding, these logs often are the main source of hydrogeologic information.

6.2.1 Driller's Logs

Two types of logs can be kept by the driller as drilling progresses: drilling-time logs and geologic logs. The drilling-time log consists of a record of distance drilled per unit time—for example, meters per hour or minutes per 0.25 m advance. Changes in drilling time with depth indicate discontinuities in the geologic profile (Figure 6.5). The rate of drilling depends not only on the type of material at the bottom of the hole, but also on the equipment (weight of bit, wear on cutting edges, number of blows or rotations per minute, etc.) and on the skill of the operator. Thus, there is no unique relation between type of material and drill rate. Time logs tend to be much more meaningful for rotary drilling than for cable-tool drilling. Highest drilling rates normally are obtained in fine, loose sands with or

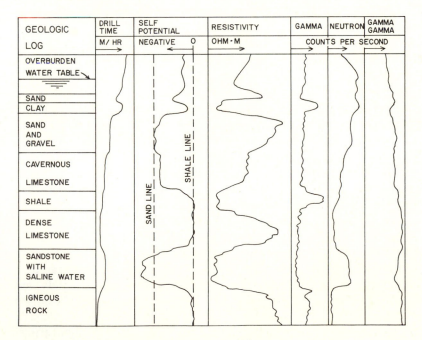

Figure 6.5 Hypothetical driller's and geophysical logs.

without clay, followed by clays, coarse sands, semiconsolidated fine-grained material, gravelly material, boulders, and solid rock.

Geologic logs consist of a record of the cuttings obtained by sampling the contents of the bailer if the cable-tool method is used, or by sampling the drilling-fluid return flow if the rotary method is used (see Section 6.3.2). Drill cuttings often are a mixture of material from the bottom of the hole, drilling mud, and material from higher layers that was still in the hole or that caved in from the wall. Thus, the samples must be carefully analyzed. Often it is better to look for changes in samples than at actual composition. For example, if the bailer initially yielded primarily fine materials, and sand begins to show up, a sand layer may be reached. If gravel chips show up, gravel layers may be reached, etc. Similarly, a reduction in the sand content of the bailer material may indicate that the bottom of a sand bed has been reached. Acid is used to see if rock fragments are of the carbonate type (by effervescence). Experienced drillers with a good knowledge of local subsurface conditions often "know" what kind of material the bit is in from its rate of advance and how it bounces, churns, sounds, or otherwise reacts to the material.

6.2.2 Sampling

The most precise and detailed information about subsurface materials is obtained from actual samples taken at the bottom of the bore hole at selected intervals during drilling. Greenberg et al. (1973), for example, showed a detailed profile of the Mugu aquifer in the Oxnard, California, basin in which 13 layers of different texture (plus three unexamined layers) could be identified in a three-meter section. The textures ranged from silty clay to coarse sand and included clayey silt, silt, silt with fine sand, sandy silt, fine sand, and fine to medium sand.

Numerous techniques have been developed to obtain bore-hole samples with minimum disturbances, and probably for this reason they are called "undisturbed" samples. While it may be possible to obtain samples with minimum disturbance from consolidated rock and from gravel-free clays or fine sands, undisturbed samples of coarse sand and gravel are very difficult to obtain. Coarse sands and gravelly materials require heavy-duty samplers that must be driven down at the risk of damaging or jamming drive shoes or casing ends. The resulting sample often is severely disturbed, and thin layers of fine sand within a predominantly gravel matrix could easily go undetected. This could lead to the selection of well screens with finer slot sizes than could be used if such fine sands were definitely known to be absent. Loose, coarse sands and gravels may have to be solidified in situ by freezing or grouting techniques before they can be brought up with a sampler (Barton, 1974). Air-pressure dewatering of the material prior to sampling may also be effective in preventing loss of the material from the sampler.

The slot sampler often is effective for exploratory sampling of sands and cohesive materials (Barton, 1974; Hvorslev, 1949). This device consists of a slotted tube, closed at the bottom, that is pushed into the bottom of the bore hole. The sampler is then rotated so that the soil material enters the tube through vertical slots in the wall. When the tube is full, it is pulled up and the sample is removed.

More accurate samples are obtained with open-drive and piston-type samplers, of which a number of models and variations are available (Hvorslev, 1949). Open-drive samplers are the simplest. They consist of a thin-walled metal tube, beveled and open at the bottom, that is pushed or driven into the bottom of the bore hole. A ball valve at the top of the sampler, which was open while the sampler was pushed down, closes when the sampler is pulled up to prevent the sample from falling out of the tube. This is not always effective, however, particularly if the material consists of saturated sand or gravel. Some models have a split tube for easy removal of the sample. In others, the sample is collected in a removable metal or transparent-plastic sleeve or liner inside the tube, which also serves as a container for the sample until it is analyzed.

Piston-type samplers are cylindrical tubes closed at the bottom by a piston to prevent entry of unwanted material into the tube while it is lowered. When the sampler has reached the desired depth, the piston is retracted as the sampler is pushed or driven down to take a sample. The retracted piston may then stay in locked position to keep the sample inside the tube when the sampler is pulled up. Common types are the retractable, fixed, and floating piston samplers.

Consolidated rock and strongly cemented materials can be sampled with hollow, diamond-studded core bits operated by rotary-drilling equipment. The samples obtained are usually 10 cm or less in diameter and vary in length from a few decimeters to several meters (Campbell and Lehr, 1974). Sometimes, coring is used as a drilling technique in hard rock. Drill cores may then be 10 to 20 cm in diameter and as much as several meters long, yielding an essentially continuous sample of the rock formations.

6.2.3 Geophysical Logging

Geophysical logging consists of lowering some sensing element into a well or bore hole and recording its output to produce a depth trace, or log, of the parameter in question. Logging techniques, developed since the 1920s by the oil industry, are finding increased application in the water-well industry, including self-potential, resistivity, gamma, neutron, gamma-gamma, temperature, caliper, and well-alignment logging. Self-potential and resistivity logging, collectively called *electric logging*, are among the most commonly used techniques.

Self-potential logging. A log of the self- or spontaneous potential (SP) is obtained by recording the naturally occurring voltage difference between an electrode that is placed in the surface soil near the bore hole and another electrode that is lowered into the hole. The hole must be uncased and is still filled with drilling fluid when the SP log is obtained. Variations in the recorded voltage difference will occur as the hole electrode passes different formations. These variations are due to electrochemical effects between dissimilar layers, different streaming potentials, and other electrokinetic effects associated with movement of water through the various layers. The resulting recorder trace thus serves as a fairly accurate indicator of the depth of discontinuities and types of materials.

The right-side boundary of the SP log is called the *shale line* (Figure 6.5), and it indicates impermeable beds like clays, shale, and rock. The output of the recording potentiometer is so arranged that the shale line represents the zero or reference potential, and so that recorder outputs to the right of the shale line are positive and to the left of the shale line are negative. The left-side boundary of the recorder tracing is called the *sand line* (Figure 6.5), and it represents more permeable strata such as sands, gravels, sandstones, and rock with secondary porosity (fractures, solution channels, etc.). If the permeable layer is relatively thin, the recorder output may not move completely to the sand line (see SP log for sand layer at 20 m in Figure 6.5). Thus, any negative deviation from the shale line may indicate a permeable stratum, particularly if the deviation extends only over a short depth interval. For some strata, such as sand aquifers containing water with a very low salt content, the SP trace may move to the positive or right side of the shale line. The self-potential becomes more negative with increasing salt content of the formation water. Thus, SP logs may also indicate zones of saline water, particularly when used in combination with resistivity logs, which show low resistivity values for such formations (Campbell and Lehr, 1974; Wyllie, 1949 and 1960).

Resistivity logging. To obtain a log of apparent formation resistivities, an alternating current is applied to two electrodes and the potential difference between these electrodes or two other electrodes is measured with a recording potentiometer. Total current is measured with an ammeter. Various electrode arrangements are used, the simplest being the single-point electrode where one current electrode is placed in the surface soil near the well and the other is lowered into the bore hole (Figure 6.6). The hole must be uncased and is usually still filled with drilling mud. The potential difference is measured between the current electrode in the bore

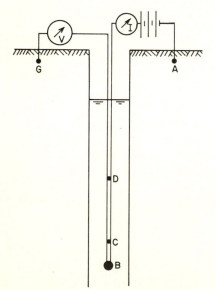

Figure 6.6 Electrode arrangements for resistivity logging of wells. Current electrodes: A and B. Potential electrodes: G and B (single-point device), G and C (short-normal device), or G and D (long-normal device).

hole and a potential electrode placed in the soil near the well (single-point device, Figure 6.6). The resistivity measured with this arrangement mostly applies to a small spherical region around the electrode in the bore hole, and is affected by the resistivity of the drilling mud and that of the local formation. Since the mud resistivity can be considered uniform, any change in recorded resistivity is due to a change in formation material, including changes in the distance that drilling mud has invaded the formation.

In other electrode arrangements, a separate potential electrode is also used in the bore hole (Figure 6.6). This electrode may be located a relatively small distance (about 40 cm) above the current electrode (short-normal arrangement) or a relatively large distance (about 160 cm) above the current electrode (long-normal arrangement). Single-point and short-normal arrangements give better resolution of lithological details, such as thin layers and geologic discontinuities, than do long-normal arrangements. Short-normals record the apparent resistivity of the formation material immediately around the bore hole, which may be invaded by drilling fluid. Long normals have greater lateral penetration and give more accurate values of true formation resistivity, enabling better interpretation of the log for rock type and water quality. For groundwater applications, however, short-normal logs are usually adequate for determining formation resistivity (Campbell and Lehr, 1974). Normal arrangements give poor results in highly resistive rocks (Keys and MacCary, 1971). So-called lateral logs are obtained with both current electrodes in the bore hole at a relatively large distance apart (about 15 m), and with the potential electrodes relatively close together at about 5 m below the lowest current electrode. The large electrode spacing yields resistivity values over a large lateral distance, well beyond the zone that may be invaded by drilling mud. Best results are obtained in relatively thick beds of low to medium resistivity (Keys and MacCary, 1971).

The resistivity of a water-bearing formation primarily depends on the salt content of the water and the porosity of the material (Figure 6.5). Interpretation of resistivity logs usually is most successful when used in conjunction with self-potential logs (Campbell and Lehr, 1974; Wyllie, 1949 and 1960). Resistivity values are highest for dense, solid rock and lowest for clay and shale layers. Medium resistivities in combination with negative self-potentials indicate sand aquifers. Gravelly clays in glaciated regions sometimes exhibit higher resistivities than normally associated with clays. Without prior knowledge of the geology of the area, these layers could easily be interpreted as sand and gravel aquifers (Linck, 1963). If the formation material and porosity are known, the measured resistivity may furnish estimates of the salt content of the formation water. Corrections for well diameter, electrode arrangement, mud resistivity, mud penetration into aquifer, thickness of aquifer, temperature, and other factors may be required, however, to convert measured resistivity into resistivity of formation water (Brown, 1971; Davis and DeWiest, 1966; Keys and MacCary, 1971).

In cased holes, the resistivity method primarily indicates the resistivity of the casing, which is low when the casing is new, but increases with increasing incrustation and corrosion. In old or abandoned wells, resistivity logs can be used to detect depths of casing.

Gamma-ray logging. Logs of the natural gamma-ray emission of the various strata are obtained by lowering a gamma-ray detector into the well and recording its output (counts per second). Since gamma rays pass through metal, the technique can also be used in cased holes. Clays and shales contain much more of the gamma-emitting elements (^{40}K and daughter products of uranium and thorium, for example) than limestones and sands. Granites emit gamma rays at moderate intensities. Some volcanic materials (tuff, for example) may show considerable gamma radiation. Gamma-ray logs, however, are mainly used to distinguish between clay and nonclay materials (Figure 6.5). In this way, gamma logs enhance the interpretation of electrical logs. Some drilling muds and mud additives contain radioactive elements, which can interfere with gamma logging.

Neutron logging. Neutron logs are obtained by lowering a probe with a fast-neutron source (for example, 3 mCi of americium-beryllium) into the bore hole and recording the intensity of the slow neutrons caused by backscatter and attenuation of the fast neutrons by hydrogen in the surrounding formation. The intensity of the slow neutrons, which is measured with a detector in the same probe, can then be related to the water content of the formation material around the probe (Figure 6.5). The method can be used on cased or uncased holes. Correlation factors to relate the intensity of the slow neutrons to the actual water content of the formation must be empirically evaluated for the particular probe-and-well combination (see Section 2.7). Some neutron techniques are based on measuring gamma rays emitted by hydrogen in the formation upon capture of fast neutrons.

Neutron logs yield information about water content and, if the formation is saturated, about porosity of the material around the well. Measured changes in water content may be helpful in locating water tables. When used in conjunction with a falling water table, the difference between the water content above and below the water table enables calculation of specific yield (Meyer, 1962).

Gamma-gamma logging. Gamma-gamma logs are obtained by lowering a probe with a gamma-radiation source (for example, 10 to 35 mCi of ^{60}Co) into a bore hole and measuring the intensity of the backscattered and attenuated gamma rays with a detector in the same probe. This intensity is related to the density of the surrounding material (Figure 6.5), so that bulk density and porosity of formation material around the probe can be detected (see Section 2.7 and Keys and MacCary, 1971).

Neutron and gamma-gamma logs may be used to evaluate the removal of drilling mud from the aquifer around the well. Removal of mud that invaded the aquifer during drilling is necessary for development of the well. If periodic neutron and gamma-gamma logging indicates an increase in water content and porosity around the well, drilling mud and other fines are indeed moving out of the aquifer, as desired for proper development.

Other logging techniques include wall-resistivity logging to detect presence or absence of mud cake, focused-current logging to measure high formation resistivities through a low-resistivity drilling mud in the hole, induction logging for measuring true resistivity by inducing a current to flow in the rocks, caliper logging to measure the diameter of the bore hole (an increased width of the hole may indicate cohesionless material like sand or sorted gravel), temperature logging, acoustic logging to estimate porosity and identify fractures, hole direction logging for plumbness and alignment, and velocity logging. The latter consists of measuring the upward velocity of water in relation to depth in a pumped or flowing well to detect the most productive layer(s). For more information regarding these methods and the other well-logging techniques discussed in this section, including applications and examples of actual logs, reference is made to Campbell and Lehr (1974), Crosley and Anderson (1971), Davis and DeWiest (1966), Guyod (1952, 1966). Johnson, E. E., Inc. (1972), Kelley (1969), Keys (1967, 1968), Keys and MacCary (1971), Linck (1963), Norris (1972), Patten and Bennett (1963), Peterson and Lao (1970), Stratton and Ford (1951), Todd (1959), and Wyllie (1960, 1963).

Interpretation of well logs is most reliable when several techniques are used and the resulting logs are placed side by side (Figure 6.5) to allow cross-checking and successive elimination of alternatives. Above all, intensive well logging and the careful and detailed interpretation of well logs for lithology and groundwater quality are jobs best left to the specialist.

6.3 WELL CONSTRUCTION

6.3.1 Types of Wells

Wells can be classified as to their method of construction and whether they are vertical or horizontal.

Dug wells. The first wells were dug by hand. Prehistoric dug wells have been discovered in archeological excavations and ancient wells can still be found in areas of old civilizations, like the Middle East. Some of these wells have deep rope marks in the stones around the top of the well, mute evidence of many years of use. Dug wells usually are about 1 m in diameter, several meters to several tens of meters deep, and lined with stones or bricks.

Ghanats. The art of manual well construction culminated in the extremely ingenious well-and-tunnel systems called *ghanats* (also known as ganats, kanats, kariz, or foggaras), which probably originated in Iran some 3 000 years ago. From there, the technique spread to other parts of Asia and into Africa and southern Europe. Ghanats are also found in South America (Cressey, 1958). In some parts of the Sahara region, ghanats are still locally known as "Persian works" (Wulff, 1968). The premier country for ghanats is Iran, where presently about 50 000 systems with a total tunnel length of approximately 350 000 km supply roughly 75 percent of the nation's total water needs, including one-third to one-half of the water used for irrigation (Bybordi, 1974).

Ghanats consist of one or more head wells, also called mother wells, which are

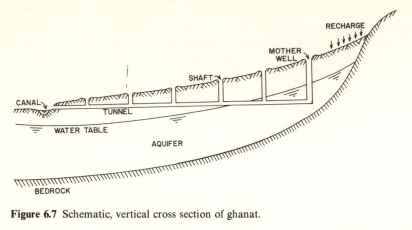

Figure 6.7 Schematic, vertical cross section of ghanat.

usually located in the upper, coarsest sections of alluvial fans along the periphery of desert valleys or basins (Figure 6.7). Runoff from the mountains infiltrates these fans and replenishes the groundwater. The mother wells range in depth from a few meters to 400 m, but are most commonly 30 to 50 m deep (Bybordi, 1974). To get the water to the surface without benefit of pumps, the ghanat diggers or "moghannies" (Bybordi, 1974) constructed a tunnel away from the well at milder slope (usually from less than 1 m to several meters per kilometer) than the slope of the land, so that the tunnel surfaced in the lower part of the valley or basin (Figure 6.7). The tunnels were dug in upstream direction, using oil lamps for proper alignment and detouring around large boulders that proved too much of an obstacle (Wulff, 1968). The tunnels were at least 1.2 m high and 0.8 m wide so that the moghannies could crawl through them. In unstable soils, the tunnels were lined with primitive, oval, baked-clay tiles (Bybordi, 1974). Where the tunnels intersected or were below water tables, they also functioned as drains or infiltration galleries.

As tunnel construction progressed uphill, vertical shafts were dug at intervals of 30 to 100 m to provide ventilation, allow the moghannies to be lowered into and lifted from the tunnel (using a windlass), and enable the removal of excavated material which was dumped around the hole. The resulting spoil heaps have persisted over the centuries, and they lend a characteristic flavor to the Iranian desert landscape (Figure 6.8). The diameter of the vertical shafts is about 0.75 to 1 m. The length of the tunnels varies from a few hundred meters to possibly as much as 70 km, with 4 to 5 km being the most common range. Some ghanats were expanded with lateral tunnels to increase the flow. The flow of water produced by a ghanat ranges from a few hundred cubic meters per day to more than 35 000 m³/day (Bybordi, 1974).

Considering that ghanats were planned and constructed without benefit of modern techniques and tools, the moghannies must be rated among the world's most ingenious geohydrologists and civil engineers. They also worked under extremely unpleasant and hazardous conditions. For more information on ghanats,

Figure 6.8 Top of shafts and spoil banks of ghanat. (*Photograph courtesy of Imperial Iranian Embassy, Washington, D.C.*)

reference is made to Bybordi (1974), Cressy (1958), and Wulff (1968). Because of increasing costs of constructing new ghanats and maintaining old ones, bored wells and diesel-powered pumps are gradually taking their place (Wulff, 1968).

Vertical wells. Almost all wells installed nowadays are vertical, pumped wells. The diameter of the well must be large enough to permit entry of groundwater without undue head losses (see Section 4.5) and to accommodate a pump (except where the groundwater is so shallow that the pump can be placed aboveground). The following relation between expected well yield and optimum well diameter (ID of casing), adapted to the metric system from Johnson (1972), may be used as a guide in selecting well diameters:

Expected well yield, m^3/day	Well diameter, cm
< 500	15
400–1 000	20
800–2 000	25
2 000–3 500	30
3 000–5 000	35
4 500–7 000	40
6 500–10 000	50
8 500–17 000	60

Depths of wells range from a few meters to more than 3 000 m. A common range for production wells is 20 to 500 m. Small, shallow wells for individual residences or farms (domestic wells) may be driven, jetted, or hand-augered down if the underground material is unconsolidated and relatively free from stones. Deeper wells and wells in gravelly or consolidated materials are mechanically drilled, using cable-tool or rotary drilling techniques.

In unconsolidated materials, the bore hole must be lined with pipe called *casing* to prevent the well walls from caving in. Special screen sections or perforated casings are used where groundwater is to enter the well. If the aquifer consists of relatively fine, uniform material, entry of sand into the well may cause excessive wear on pump impellers and bearings, and fill the lower portion of the well. To avoid this situation, an oversized hole is drilled and the annular space between well screen and hole wall is filled with gravel or coarse sand. Such a gravel pack or gravel envelope then effectively prevents sand from moving into the well. Other aquifers may be sufficiently graded to develop their own natural filter zone around the well screen to keep out sand. Wells in consolidated materials do not have to be lined, except where the rock is so heavily fractured that it behaves as unconsolidated material. The top of the well normally is cased, however, because the upper rock layers are often weathered and unconsolidated. Casing the top of the well is also necessary to allow proper sealing to prevent entry of polluted surface water into the well.

Horizontal wells. Horizontal wells are occasionally drilled in mountainsides to tap vertical bedding planes or fracture zones, groundwater trapped behind dikes or other impermeable vertical boundaries, or groundwater above an impermeable, outcropping layer (Figure 6.9). Hillside seeps or springs often are good indicators of groundwater behind dikes or above restricting layers. Horizontal wells are drilled with specially designed rotary equipment and are cased and screened as needed (Summers, 1972). The wells are free-flowing, so that they must be equipped

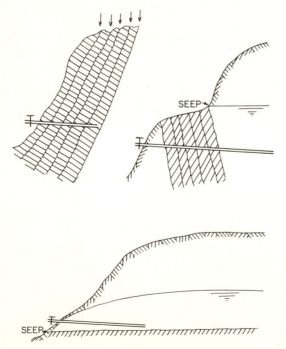

Figure 6.9 Horizontal wells to tap water in vertical bedding planes (top left), behind impermeable dikes (top right), and above impermeable layer (bottom).

with a valve or other flow control device at their outlet. Horizontal wells should be installed with some downward slope into the mountain to avoid possible development of negative pressures at the "bottom" of the wells. For an area in east central Arizona, wells penetrated the mountainsides a distance of 12 to 82 m with an average of 37.4 m. Well yields were mostly in the range of 10 to 50 m^3/day (Welchert and Freeman, 1973). Wells drilled at 45° in a mountainside on the Hawaiian island of Oahu penetrated the mountain about 100 m to tap groundwater behind a lava dike. The yield was about 3 500 m^3/day per well.

In Hawaii, horizontal tunnels are used to tap groundwater that has collected behind impermeable dikes formed by upward lava flows through vertical faults or cracks (Peterson, 1972). One such tunnel is about 520 m long to the dike and extends about 60 m beyond the dike, where it collects groundwater at a rate of approximately 15 000 m^3/day. Elsewhere, horizontal tunnels, shafts, or infiltration galleries have been installed at relatively shallow depths adjacent to or below streams to induce infiltration of water into the stream bottom and to collect "filtered" stream water. These systems were rather popular in the early days of municipal water supply development. Where fresh groundwater is underlain by saline water at a relatively shallow depth, horizontal "skimming" tunnels are more effective in collecting the fresh water than vertical wells, which may yield a mixture of fresh and salt water due to upconing of salt water below the well (see Section 11.2). Such tunnels are used in Hawaii, where they are called *Maui tunnels* (Peterson, 1972). One Maui tunnel on Oahu is about 300 m long and produces 50 000 m^3/day.

Another type of horizontal well is the radial collector well (Figure 4.10), also called *star* well or *Ranney* well (Campbell and Lehr, 1974). These wells consist of poured-in-place, reinforced-concrete caissons with a diameter of at least 4 m, a wall thickness of 0.45 to 0.60 m, and a depth that averages about 20 m but may be as much as 60 m (Spiridonoff, 1964). About 1 m above the bottom slab of the caisson, horizontal collectors are jacked or driven into the aquifer. During driving, the aquifer material ahead of the collector is hydraulically transported to the caisson through a temporary pipe inside the collector. The collector generally consists of 2.4-m sections of perforated pipe with a diameter of 20 to 60 cm, and ranges in length from 30 to 90 m. A typical array consists of 16 collectors extending radially and symmetrically from the caisson. The collectors discharge into the caisson, from where the water is pumped to the surface. The pumphouse is constructed on top of the caisson, often some distance above ground level as protection against floods. Radial collector wells are usually installed next to rivers and sometimes in streambeds. Several such wells may be built in series along a river. Radial collector wells were first developed in the 1930s, and more than 200 of these wells have been constructed in the United States as of 1964 (Spiridonoff, 1964). The median yield of collector wells is in the range of 30 000 to 40 000 m^3/day with some yields as high as 80 000 m^3/day. Radial collector wells produce about 10 to 20 times as much water as infiltration galleries (i.e., 100 to 200 m^3/day per meter length versus 10 m^3/day per meter length of collector) because they can be placed much deeper into the groundwater (Spiridonoff, 1964).

6.3.2 Drilling Techniques

Cable-tool (percussion) and rotary drilling are the two most common methods for drilling vertical wells, while other techniques like jetting, driving, or augering are used only in selected, shallow situations. Over the years, drilling methods have been refined and modified to adapt to special conditions. Experience is an important factor in selecting the best technique and in the efficient execution of the drilling operation. Good equipment, of course, is another requirement for successful well construction. Main aspects of the various techniques are discussed in the following paragraphs, referring to Campbell and Lehr (1974) and Johnson (1972) for further detail. Standards for water-well construction were published by the U.S. Environmental Protection Agency (1976), whereas Campbell and Lehr (1973) developed a guide for planning and engineering rural water systems.

Driving, jetting, and augering. The simplest techniques are driving, jetting, and augering, used singly or in combination to drill wells of relatively small diameter (5 to 20 cm) and shallow depth (less than about 20 m) in soft, unconsolidated materials. Driven wells are constructed by driving a steel pipe with a well point at the lower end into the ground. The well point consists of a perforated or screened section closed at the bottom by a conical, steel drive point. Pipe diameters commonly range from 3 to 10 cm. With manual techniques, using a fencepost driver or similar device, depths up to about 10 m may be reached depending on the soil material. Greater depths can be obtained with heavier weights (about 100 kg) suspended from a tripod or from a cable-tool rig. Holes are started with a shovel or hand-operated soil auger. Some wells are dug entirely with hand augers (for example, bucket or orchard-type augers), which can reach depths of 10 m or more under favorable conditions.

Well points can also be installed by jetting, pumping water down the pipe and out through the well point where the force of the water loosens the surrounding soil material. The water and soil particles then flow back to the surface around the outside of the pipe. Pumping rates are on the order of 50 l/s and pumping pressures about 3 atm (approximately·30 m pressure head). Special, self-jetting well points with a jetting nozzle at the drive point are available. In loose material, the pipe may sink in by its own weight. In denser materials, moving the pipe up and down during jetting may speed up penetration. Where the underground material is too hard to be broken up by jetting alone (for example, dense clays), chisel-type drill bits with nozzles are used to loosen up the material with the combined action of chiseling and jetting. Another technique utilizes a special jetting pipe with cutting teeth at the bottom. Water is pumped down the pipe to loosen soil material, which is removed by the return flow around the outside of the pipe. When dense layers are encountered, the pipe is rotated to break up the material with the cutting teeth. After the desired depth is reached, a pipe with a well point is installed inside the jetting pipe, sand or gravel is placed around the well point if a gravel pack is desired, and the jetting pipe is pulled out.

For shallow aquifers, a number of well points or similar small-diameter wells

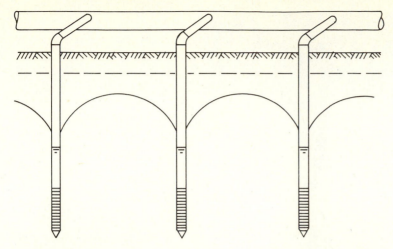

Figure 6.10 Schematic of well points connected by common suction line, static water table (dashed line), and drawdown (solid lines) by the wells.

may be installed relatively close together, for example, 1 to 10 m apart (Figure 6.10). If the suction lift is sufficiently small (see Section 6.4), the wells may be connected to the pump by a common intake manifold. Such well-point systems are commonly used to dewater excavation and construction sites.

Large, machine-operated bucket-type augers can drill relatively shallow wells of large diameter (0.5 to 1 m) in unconsolidated material free from boulders. In sandy materials, concrete or steel casing rings are sunk into the hole by their own weight as soil material is being removed. When the desired depth is reached, the well casing and screen are placed in the hole, the rest of the hole is filled with gravel, and the concrete or steel casing rings are removed. In clays and other cohesive materials, the hole may be sufficiently stable to eliminate the need for temporary casings. Bucket augers cannot remove large rocks and boulders, which must be picked up with an orange-peel bucket, stone tongs, or other special devices. If there are many stones, rotary-bucket drilling is slow and cable-tool or rotary techniques are more appropriate.

Cable-tool technique. With the cable-tool technique (Figure 6.11), also called the *percussion, standard,* or *spudder* method, a vertical hole is drilled by repeatedly dropping a heavy, chisel-type drill bit on the hole bottom. This loosens fine, granular materials, and breaks up gravel, boulders, and consolidated rock. Drilling is periodically interrupted to remove the cuttings with a bailer. The cable-tool method is quite old, having been used in China around 600 B.C. (Campbell and Lehr, 1974).

The length of the drill bit is about 1 m for large-diameter holes (Figure 6.12) and several meters for relatively small diameters. The drill bit is suspended from a cable called the *drill line*, which is strung over a pulley at the top of a near-vertical

Figure 6.11 Cable-tool rig drilling 60-cm-diameter well (notice bailer and stovepipe casing).

mast erected over the hole (Figures 6.11 and 6.12). The lifting and dropping action of the bit is obtained by a spudding beam mounted on the rig at the other side of the vertical mast. The spudding beam is moved up and down by the so-called Pitman arm, which is connected to an engine-driven cam gear. The number of drops of the bit is about 20 to 40 per minute.

The drill line is adjusted so that it is just tight when the drill bit hits the bottom of the hole. Since the drill line is a stranded-wire cable, the resulting stretch will cause the drill line to rotate slightly, which in turn will rotate a swivel located higher up in the drill line. This rotation, which is a few degrees for each stroke, causes the bit to turn and hit a different part of the hole bottom with each stroke, thus producing a round and straight hole.

Well casing is not necessary in consolidated materials, except near the surface to prevent seepage of surface water or shallow groundwater into the hole. Casing must be installed, however, in unconsolidated materials to prevent the hole from caving in. When drilling in clay or other soft material, the casing is usually driven ahead of the hole bottom for about 0.2 to 1 m, after which the plug of clay inside the casing is broken up by the drill and removed as slurry with the bailer. In sands and gravels, the hole is drilled ahead of the casing and the casing is driven down each time drilling has advanced 1 to several meters. Driving is done with the drill string. A heavy clamp is attached to the top of the drill stem, which hits a drive head (Figure 6.12) placed as an anvil on top of the casing when the drill string is dropped. The bottom of the casing is protected by a drive shoe of hard steel. This

Figure 6.12 Close-up of rig in Figure 6.11, showing drill bit suspended above hole. The device on top of the drill bit is the drive head for the casing. The bailer, leaning against the drill rig, is shown on the left.

shoe has a larger diameter than the casing, which reduces friction between casing and hole wall. Eventually, however, friction between casing and hole wall makes it impossible to drive casing any deeper. When this occurs, casing with a smaller diameter is driven into the hole bottom and drilling is continued with a smaller bit. For deep holes, several diameter reductions may be required, yielding a "telescoping" casing. The hole is then started with an oversized diameter so that the bottom portion of the casing will have the desired size. In relatively shallow holes, driving the casing is avoided as much as possible when the target aquifer is reached to minimize disturbance and compaction of aquifer material and resulting reduction in hydraulic conductivity. Instead, the casing is allowed to sink in under its own weight (if possible) by continued drilling and bailing.

The bailer (Figures 6.11 and 6.12) consists of a section of pipe with a valve at the bottom. The valve, which may be of the flat type or the ball-and-tongue type (dart valve), is open when the bailer is lowered into the hole and closed when the bailer is pulled up. A special bailer, called a *sand pump*, is equipped with a plunger which creates a suction when pulled up. This produces more efficient removal of coarse material. The bailer is suspended from a cable called the *sand line*. When the hole is ready for bailing, the drill string is pulled up and the bit set beside the hole. The bailer is lowered, and the hole is bailed several times until the drill cuttings have been removed. As long as the hole has not yet reached groundwater,

water must be added to the hole to produce a slurry. Drilling mud may be needed to prevent rapid loss of water into the formation. Once groundwater is reached, drilling mud usually is no longer necessary.

The drilling rate depends on the type of rock (soft or hard, unconsolidated or consolidated, presence of gravel and boulders, etc.), the weight and diameter of the drill bit, the condition of the cutting edge(s) of the bit, the distance of drop of the bit, the number of drops per minute, the frequency of bailing (accumulation of cuttings in the hole reduces drilling rate, and bailing too frequently causes too much "down" time), and the frequency of driving casing. Experience is an important factor in optimizing the various operations to obtain maximum drilling rate. Drilling rates may be several meters per day in relatively soft, unconsolidated materials with few stones, but as little as 0.5 m/day or less in dense rock or unconsolidated material with many large boulders.

As the bore hole gets deeper, the length and weight of the drill line increase and more time is required to remove and insert tools for drilling, bailing, and driving casing. For these reasons, the depth that can be reached effectively with the cable-tool method is limited. The practical depth limit is between 1 000 and 2 000 m, depending on equipment and field conditions. As a rule, the cable-tool method is not used if the depth of the well will exceed 600 m. The deepest cable-tool-drilled well on record is 3 397 m deep and is located in New York (Campbell and Lehr, 1974).

Over the years, a number of variations have been introduced to adapt the cable-tool method to special conditions. One of these is the California-stove-pipe method, where the drill bit and bailer are combined in a single tool called a *mud scow*. The casing consists of relatively short (about 1 m) sections of thin-walled steel pipe (Figure 6.11) which are welded together as they are lowered into the hole. In another technique, called the *Church* method, bentonite drilling mud is used to provide a mud seal between casing and hole wall to reduce friction during driving and to prevent flow of groundwater between various aquifers along the outside of the casing. This is important where overlying aquifers yield water of lower quality than the target aquifer.

The cable-tool method is extremely versatile and can be used to drill holes in any kind of material. During drilling, accurate samples of the formation material at the bottom of the hole can be obtained free from drilling mud. Also, as casing is driven, new formations are tapped and overlying layers are closed off. Thus, accurate samples of water from the different formations can be obtained, and the equilibrium water level in the well is an accurate measure of the pressure head in the aquifer at the bottom of the hole.

Cable-tool rigs are relatively simple, rugged, and easy to set up, making them practical for use in remote areas. Fuel and water requirements are low. The initial cost of the equipment is much less than for rotary rigs, but this economic advantage may be offset by slower drilling rates. Disadvantages include the depth limitation (usually about 600 m) and the difficulty of pulling casing from deep holes. When casing cannot be pulled, screens and gravel packs cannot be placed, and casing must be perforated in place. This could be objectionable in aquifers with problems of sand moving into the well. For additional detail, reference is made to Campbell and Lehr (1974) and Johnson (1972), including the many publications cited in the former.

Rotary method. Rotary drilling is principally an old method used by the early Egyptians for stone quarrying. The modern development of the technique, however, is largely due to the mining industry (Campbell and Lehr, 1974). Rotary drilling is the standard method for drilling oil wells and is increasingly used for water-well drilling. Drill bits may consist of several conical roller gears (Figure 6.13), flexible carbide fingers, or systematically placed buttons in the bit face. The bit is fastened to the bottom of a hollow drill pipe, which is rotated at a rate of 30 to 60 rpm in the bore hole. Drilling is obtained by the grinding, chipping, and crushing action of the gears, fingers, or buttons as they roll or scrape over the formation material under the weight of the entire drill stem. Drilling fluid consisting of a carefully formulated suspension of bentonite clay in water—plus various additives, if necessary—is pumped down the drill pipe. The fluid leaves the drill pipe through the bit, where it cools and lubricates the abrading surface and picks up drill cuttings. The drill cuttings are transported upward as the fluid returns to ground surface through the annular space between drill pipe and hole wall. The fluid then flows into a pit or settling basin where the cuttings settle out, after which the drilling fluid is pumped back into the drill pipe. The important feature of rotary-drilled holes is that casing is not necessary, because the drilling fluid itself forms a mud lining or filter cake on the hole wall. This lining and fluid pressure

Figure 6.13 Bit of reverse-rotary rig for drilling 60-cm-diameter hole, showing one of three symmetrically located rolling gears.

heads in the bore hole that are maintained higher than those in the formation prevent the hole wall from caving in.

The drilling fluid plays a critical role in rotary drilling. In addition to cooling and lubricating the bit, it removes drill cuttings, forms a skin to prevent the hole wall from caving in, creates higher-pressure heads in the hole than in the formation to prevent breaking of the mud lining and entry of formation water into the hole, and seals the wall to keep drilling fluid inside the hole. These objectives pose different and conflicting demands on the properties of the fluid (density, viscosity, gelling, and filtration characteristics), making it necessary to compromise or to formulate the fluid on the basis of the most critical objective.

A high density of fluid, for example, is needed to ensure that pressure heads inside the well exceed those in the formation and that drill cuttings are kept in suspension while they move upward in the hole (upward flow velocities should be about 0.7 to 1 m/s for adequate removal of cuttings). On the other hand, high-pressure heads resulting from high densities may cause breaks in the mud lining where this lining bridges relatively large pores, cracks, or other openings in the hole wall. Such breaks may then cause loss of drilling fluid (loss of circulation) and invasion of drilling fluid into the aquifer. High-density drilling fluids also can contribute to loss of circulation by hydraulic fracturing of formation material. This fracturing may already occur if the pressure head inside the hole is equal to about 60 percent of the overburden pressure in the rock (see Section 11.10). A high-density drilling fluid also retards settling of cuttings in the mud pit, whereas the high fluid pressures produced on the hole bottom will reduce drilling rates in deep holes. While a high viscosity and high gel strength are desirable for transporting drill cuttings upward in the hole, they will retard settling of the cuttings in the pit. They also increase friction losses in the drill pipe. This could result in decreased circulation rates, which in turn could decrease the drilling rate.

The ability of the mud to form a lining or filter cake on the hole wall is expressed by its filtration properties. Mud with high filtration properties will continue to allow water to move through the mud lining, resulting in a thicker cake as mud continues to accumulate on the inside surface of the liner. The upward-moving drilling fluid, however, will erode the inside surface of the filter cake, thus preventing the inside surface from becoming too thick. When circulation is reduced or stopped, however, the filter cake may become so thick as to eventually touch the drill pipe, which is then sucked against the mud cake by the pressure-head gradient in the cake (pressure heads decrease from the inside surface of the mud cake to the hole wall). The drill pipe will then cease to turn and must be pulled up (which may cause considerable swabbing of the filter cake) and lowered back into the center of the hole in order to free its movement. This phenomenon, known as "stuck pipe," can be avoided by using minimum mud weight to promote a thin mud cake, and a low-filtration-rate drilling fluid to prevent rapid build-up of the mud cake when circulation of drilling fluid is stopped. Friction between the stuck pipe and the mud cake can be reduced by minimizing the sand content of the drilling fluid and adding lubricants or emulsifiers (Campbell and Lehr, 1974). Mud with low filtration properties may form only a thin skin on the hole wall, which then may not be strong enough to prevent caving of cohesionless formation materials.

The basic ingredient of drilling mud is bentonite clay, which has a platelike structure. When mixed with water, the bentonite particles swell as water moves between the plates. Drilling fluids for water wells normally should contain enough bentonite to produce a fluid weight of 1.02 to 1.14 g/cm^3. Native clays are occasionally used as substitutes for bentonite. Attapulgite, another clay mineral, is preferable for use in salt water because of its more fibrous structure. Additives to modify the properties of the drilling fluid include, among others, the following (Campbell and Lehr, 1974, and references therein):

Organic polymers to prevent flocculation of the clay
Pregelatinized starch to reduce filtration properties in salt-saturated muds
Caustic soda for raising the pH
Biocides or modified polysaccharides with fermentation preventatives to avoid
    bacterial degradation of the fluid
Various cellulose preparations to increase viscosity
Guar-gum products to improve suspension and sealing properties
Acrylic polymers to reduce filtration or increase viscosity
Thinners to reduce viscosity and gel development while not affecting fluid density
Barite to increase fluid density
Lime to develop a stiff gel for stabilizing loose sands and gravels in the hole wall
Soda ash to reduce hardness of the water for drilling fluid and to prevent floccula-
    tion of the bentonite
Surfactants for cleaning drill bits
Lubricants

Some of these additives cannot be mixed with bentonite mud.

Many variations in drilling bits are available. The drag bit, consisting of short blades that are cooled and kept clean by jets of drilling fluid ejected from short nozzles, is used for drilling in relatively soft clays and other cohesive materials. The jets of drilling fluid also help loosen the formation material. For harder materials, roller-type bits (Figure 6.13) that chip and crush the rock are used. The cutting teeth on the roller gears are relatively long and widely spaced for soft rock, and short and closely spaced for hard rock.

The rate of drilling depends on numerous factors, including hardness of rock; size, type, and condition of bit; rotation rate of bit; total weight on bit; properties of drilling fluid; fluid pressure on hole bottom; and circulation rate of drilling fluid. Studies on how the rate of drilling is affected by these factors were summarized by Campbell and Lehr (1974). Fast rotation rates produce higher drilling rates but also cause more wear on the bit, requiring more frequent replacement of the bit. The experienced driller optimizes the various factors to obtain the maximum drilling rate and to minimize the risk that something will go wrong. Drilling rates may be on the order of 100 m/day in unconsolidated materials without much gravel or boulders, but on the order of a few meters per day or less in consolidated rock or unconsolidated material with many boulders.

The main complications that may be encountered with rotary drilling are bore-hole instability and circulation loss. Instability of the hole wall, such as caving in, sloughing, heaving, or washout, is normally caused by swelling, deformation, or other reactions between formation material and water that has entered the formation through the mud cake. Damage by bore-hole instability can be minimized by reducing the rate of circulation of the drilling fluid to minimize erosion of the bad section(s). Pressure surges inside the hole should also be avoided.

Circulation loss occurs when pores, cracks, or other openings in the formation are too large to be bridged by the mud cake. Rather than building up an effective seal, drilling fluid will then invade the formation. Apart from loss of drilling fluid, this reduces or even stops circulation; the stopping of circulation in turn can cause stuck pipe, caving in, or flow of formation water into the bore hole. Invasion of drilling fluid into the formation can also damage the aquifer or aquifers. To stop the loss of drilling fluid, granular or fibrous materials are added to the fluid to help bridge and seal openings. When this fails, the well may have to be cased before drilling can be resumed. Another possibility is "blind" drilling, adding enough drilling fluid to keep the bit cooled and lubricated as drilling fluid and drill cuttings continue to "disappear" into the formation.

Rotary-drill rigs (Figure 6.14) are more complicated and more expensive than cable-tool rigs of the same capacity. They take more time to set up, more power to

Figure 6.14 Reverse-rotary rig drilling 60-cm-diameter well.

run, and more trained personnel to operate. Once the rig is set up, however, the faster drilling rate and the absence of the need for bailing and driving casing result in more rapid completion of the well. Rotary drilling thus is most advantageous for deep wells. Since the technique produces uncased holes, electric-well-logging procedures can be applied. When the hole is finished, well casing and screens can be set without difficulty and gravel or sand envelopes can readily be placed (see Section 6.3.3). Complete removal of the mud cake may be difficult when the well is being developed, but this is usually not a serious problem if jetting techniques and dispersants are used (see Section 6.3.5, on well development).

Other rotary-drilling techniques include air and reverse drilling. With air-rotary drilling, the drilling fluid consists of dry air, mist, foam, aerated mud, or other fluid lighter than water. With dry air, the circulation rate must be high enough to produce upward velocities of 10 to 30 m/s in the annular space between drill pipe and hole wall to evacuate the cuttings. One of the reasons for using air or other fluid lighter than water is to increase the drilling rate by reducing the fluid pressure on the hole bottom (Campbell and Lehr, 1974). Air drilling is used primarily in fractured rock. When groundwater starts flowing into the hole at significant rates, foam-type fluids with air-to-liquid ratios on the order to 200 to 1 are used. Stiff foam is preferred for drilling in unconsolidated materials (Campbell and Lehr, 1974).

Reverse-circulation rotary drilling is similar to regular rotary drilling, except that the drilling fluid circulates in the opposite direction, i.e., down through the annular space between drill pipe and hole wall and up through the bit and drill pipe. From there, it is pumped into the settling pit and recirculated by gravity into the bore hole. Because the cross-sectional area of the drill pipe is much smaller than that of the annular space around the drill pipe, upward velocities are much greater than for the hydraulic rotary method. This is especially advantageous when drilling large-diameter holes, as it allows the use of thinner drilling fluids (muddy water instead of drilling mud) and lower circulation rates (about 30 l/s or more) to obtain the same upward-flow velocity. Often, clays and other fines from the formation are sufficient to produce a suitable drilling fluid, eliminating the need for drilling muds.

Stability of the hole wall in reverse-rotary drilling is primarily achieved by the pressure of the drilling fluid inside the hole. For this reason, the fluid level must be maintained at the top of the hole. If the depth to groundwater is small (for example, less than 3 m), it may be necessary to construct a wall around the bore hole so that the fluid level can be maintained above ground surface. The mud cake on the hole wall is thinner than with the regular rotary method. However, there is less erosion of the mud cake because drilling-fluid velocities are lower. Since the mud cake is thinner, water losses into the formation tend to be higher, often requiring makeup water at rates of 2 to 30 l/s. Where hole stability and excessive water losses are a problem, bentonite and other drilling-fluid ingredients may have to be added to build a thicker mud cake.

Reverse-rotary drilling is most suitable for holes of relatively large diameter (0.5 m or more) in unconsolidated materials with few boulders and where the groundwater is relatively deep. Bit rotation rates are on the order of 10 to 40 rpm. Large boulders and gravel are not always readily broken up, especially if they are loose and can turn freely below the bit. Such material must then be removed with an orange-peel bucket, causing interruptions in the drilling. The thinness of the mud cake on the hole wall favors reverse-rotary drilling where artificial gravel envelopes are to be used, because the mud lining is easier to remove during development than the thick mud cakes normally left by conventional rotary drilling.

In extremely hard rock, cable-tool and rotary techniques may fail, in which case core drilling or blasting procedures are used. Blasting is also effective in breaking up large boulders that can hold up drilling progress for hours. For additional detail regarding drilling techniques, reference is made to Campbell and Lehr (1974), who also discussed new developments like turbo and percussion drilling, and to Johnson (1972). Another relatively new technique is the hammer drill. Double-walled casing with a drill bit at the bottom is driven down by a diesel-operated pile hammer. Air is blown down through the annular space in the double casing. The air exits through holes or slots in the drill bit, from where it flows upward through the inside casing, taking drill cuttings with it. After leaving the casing, the air is expanded through an inverted cone where the air escapes at the top and the cuttings accumulate at the bottom. When drilling is completed, the well casing is set (with or without a gravel envelope) and the double-wall casing is removed for use in the next job. The hammer-drill method is faster than the cable-tool method and possibly even the rotary method for putting down relatively shallow and small-diameter holes in gravelly or cobbly materials that are hard to drill. The method cannot be used for drilling in consolidated rock.

6.3.3 Screens, Perforations, and Gravel Envelopes

In cased wells, groundwater enters the well through screens or perforated sections of pipe. The main function of the well screen or perforations is to let water through without undue head loss and risk of encrustation, while keeping sand and other formation material out. Casings and screens normally are not necessary for wells in consolidated materials, where water enters the well through natural pores (interstices), cracks, fissures, solution channels, or other stable openings in the formation.

Types of screens and perforations. Well screens usually consist of sections of "pipe" constructed by winding heavy steel or stainless steel wire around a number of longitudinal rods and welding the contact points (Figure 6.15). This produces a screen with an essentially continuous helical slot. The screens are manufactured with different slot widths to tailor the slot size to the aquifer material. Some well screens are made from round wire, others from triangular wire with the apex pointed inward (Figure 6.15). The triangular wire produces V-shaped slots that may be less subject to blocking by sand and gravel particles, and that may offer less hydraulic resistance than slots formed by round wire.

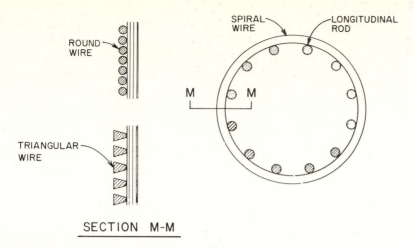

SECTION M-M

Figure 6.15 Cross section of well screen consisting of round (top left) or triangular (bottom left) wire wrapped spirally around longitudinal rods.

Another type of well screen consists of perforated pipe with wire spirally wrapped around the pipe to provide a continuous entry slot. Part of the pipe perforations, however, are then blocked by the wire. To reduce this blockage, longitudinal rods are placed on the outside of the perforated pipe to provide space between the spirally wrapped wire and the exterior pipe surface.

Perforated casings are pipe sections with relatively large slots or other openings in the wall (Figure 6.16). Because the slots are relatively wide (at least a few

Figure 6.16 Prefabricated louver- or shutter-type perforated casing with a diameter of about 50 cm. The casing in the lower right corner has more perforations per meter than the other casing.

millimeters), perforated casings are normally used only in conjunction with gravel envelopes. Perforated casings come prefabricated, like the louver- or shutter-type casing with horizontal slots shown in Figure 6.16, or casings can be perforated on site or in place. On-site perforation is done with a blowtorch prior to installing the casing. The holes are cut vertically, about 10 cm high and 1 cm wide, with four to eight slots per circumference and a vertical slot spacing of a few centimeters to a few decimeters. In-place perforation of casing may be required for cable-tool-drilled wells that are too deep for pulling casing. Vertical slots can then be cut with the so-called Mills knife, which is lowered into the hole and activated to force a cutting blade through the wall of the casing, thereby producing a rough slot of about 5 to 10 cm height and 0.5 to 1 cm width. The device is rotated to make about four to eight slots around the circumference and raised by 5- to 20-cm increments to perforate the desired length of casing. In similar fashion, a louver-type perforator can be lowered in the hole to make horizontal slots in the casing.

Perforated casing without gravel envelope can be used only where the aquifer material is sufficiently coarse and graded to create its own developed zone around the well (see Section 6.3.5). Where these conditions do not exist, continued entry of sand into the well can be expected. This not only causes undue wear on the pump, particularly on impellers and bearings, but it could also lead to collapse of the land surface around the well and resulting damage to the well, pump and power unit, or pumphouse.

Installing screens and perforated sections. Two types of techniques are available for placing well screens or preperforated casings in the bore hole: *pull-back* and *bail-down* methods. The pull-back method is used if the hole is drilled with the cable tool, which leaves the hole completely cased when drilling is finished (Figure 6.17; see also,Johnson, 1972). The well-screen section is lowered into the hole until

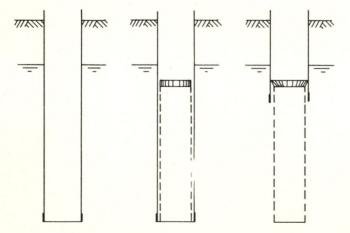

Figure 6.17 Pull-back method for placing well screen, showing cased hole (left), screen placed with lead packer in hole (center), and casing pulled back and packer expanded (right).

it rests on the bottom, after which the casing is pulled up until its lower end is about 30 cm below the top of the screen section. A lead packer ring, placed on top of the screen section, is then flared out by driving a swage block into it, thus creating a seal between the inside of the casing and the screen section (Figure 6.17). The main function of the seal is to keep sand out of the well. It also centers and anchors the top of the screen inside the casing.

The casing is pulled by hydraulic jacks or by bumping upward against a slotted plug screwed into the top of the casing. A third technique consists of placing hooklike devices under the casing bottom which are pulled up by a jarring motion with the drill line. This motion is obtained with a special device in the drill line, called a *jar*. The jar provides a slack of about 20 cm in the line, which causes a momentary delay in pulling action when the drill line is pulled up. If there is too much friction between the casing and the hole wall, as may occur with deep wells in stony, unconsolidated material, the casing cannot be pulled and the bail-down method must be used to set the screen (see next paragraph). The other alternative is in-place perforation of the casing with the Mills knife or similar device.

The pull-back method can also be used in rotary-drilled holes, which are uncased when drilling is completed. To prevent premature caving of the hole, it is often best to set casing all the way to the bottom, then place the screen section and raise the casing as in the pull-back method. If this is not possible or practical, the hole may first be drilled to the desired depth of casing so that permanent casing can be set and grouted into position. Drilling is then resumed until the desired depth of the screen section is reached. The screen section is lowered into place and sealed against the casing by expanding a lead packer at the top of the screen. When the screen section is placed in the hole after setting and grouting the upper casing, but before drilling is resumed, the method is called the bail-down procedure. The screen section is then sunk into the target aquifer by drilling through the screen (including bailing through the screen if the cable-tool method is used).

Perforated casings and screen sections must be located a sufficient distance below the anticipated water level in the well during pumping to avoid free-falling water in the well. This water, also called *cascading water*, causes air entrainment into the well water, which reduces the pumping efficiency.

Gravel envelopes. Gravel envelopes, also called *gravel packs*, should be used in relatively fine textured aquifers (including poorly cemented sandstones) where the effective particle size ($D_{90\%}$, or size of sieve opening retaining 90 percent of the material; see Section 2.2) is less than 0.25 mm and the uniformity coefficient (Section 2.2) is less than 3. Under these conditions, developed filter zones may not be totally effective in preventing sand from moving into the well (Ahrens, 1957). Another advantage of gravel envelopes is that they enable the use of large slot sizes for the screens or perforated sections, thus reducing well losses. Larger slot sizes also reduce the rate of incrustation and are less subject to clogging therefrom (see Section 6.5.1). Finally, gravel packs increase the effective radius of the well.

For these reasons, gravel packs are commonly used for large-capacity industrial and municipal wells.

Gravel envelopes normally are about 15 to 25 cm thick, although properly placed gravel packs as thin as 1.3 cm have also been effective in preventing movement of sand into a well (Hunter Blair, 1970). Where gravel packs are to be used, an oversized hole is drilled to accommodate the desired thickness of the envelope. If the hole is drilled with the cable-tool method, casing extends to the bottom upon completion of the drilling. The next step is to install the well-pipe assembly, consisting of screen (or perforated casing) and solid casing (Figure 6.18), centrally in the hole. The outer casing is then pulled up while simultaneously placing gravel or coarse sand in the annular space between screen and hole wall. Gravel packs normally extend at least 3 m above the screen or perforated section, while the bottom of the outer casing is left at a few decimeters above the top of the gravel pack (Figure 6.18). The rest of the annular space between the inner and outer casing is filled with cement or grout to protect the well water against contamination by surface water, and to anchor the inner casing inside the outer casing. To avoid penetration of the grout into the gravel pack, the gravel should be covered with a layer of sand (sand bridge, Figure 6.18) before grouting. Sometimes, the outer casing is completely removed prior to grouting.

If the hole is rotary-drilled, the well-pipe assembly (casing and screen or perforated casing) is centrally placed in the uncased hole, after which the space between the screen and the hole wall is filled with gravel until the gravel is about

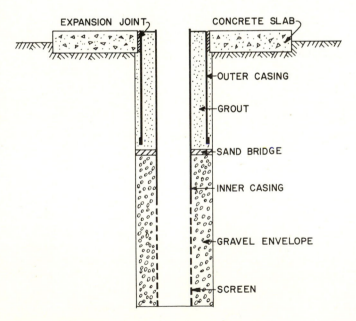

Figure 6.18 Completed, rotary-drilled well with screened section, gravel envelope, sand bridge, grouting between double casing and around outer casing, and concrete slab for pump-and-power unit.

3 m above the screen section. The space between the well casing and hole wall above the gravel pack is then grouted (Figure 6.18).

Gravel packs may consist of uniform or graded material. Uniform gravel can be poured into the bore hole from ground surface, as long as bridging between screen and hole wall is avoided. Graded gravel cannot be dropped into the hole because the differently sized particles will segregate and form bands of predominantly fine and coarse materials in the envelope. This reduces the effectiveness of the pack because aquifer sand could then move through the coarse zones. Particle separation can be minimized by pouring or pumping the gravel through a tremie or conductor pipe, which is kept full of the envelope material and moved around in the well just above the gravel surface to distribute the gravel evenly around the screen.

Selection of screen length and slot and gravel sizes. The length of the screen or perforated section is selected on the basis of the effective open area required to keep the entrance velocity of the groundwater below a certain value. The effective open area is about one-half the total area of screen or slot openings because of blocking by sand and gravel particles. The entrance velocity of the water is selected to avoid excessive well losses and incrustation rates, both of which increase with increasing entrance velocity. Hunter Blair (1970) suggested a maximum value of about 3 cm/s for the entrance velocity, while Walton (1962) proposed an entrance velocity that increases with increasing hydraulic conductivity of the aquifer material, as follows:

Hydraulic conductivity of aquifer, m/day	Optimum entrance velocity through screen or perforation openings, cm/s
< 20	1
20	1.5
40	2
80	3
120	4
160	4.5
200	5
240	5.5
> 240	6

After the entrance velocity has been selected, the total length of screen or perforated casing is calculated from the effective open area per meter length of section and the anticipated yield of the well.

The size of the screen openings, commonly called the *slot size*, is selected on the basis of a sieve analysis of the aquifer material. For relatively fine and uniform materials (uniformity coefficient < 3), the slot size may be taken as the size of the sieve opening that will retain 40 percent of the material $(D_{40\%})$ if the groundwater is noncorrosive, and as $D_{50\%}$ if the groundwater is corrosive (Johnson, 1972). If the aquifer consists of coarse sand and gravel, the slot size may be $D_{30\%}$ to $D_{50\%}$ of the sand fraction. For nonuniform aquifer materials (uniformity coefficient > 6), slot sizes should be equal to about $D_{30\%}$ if the overlying material is stable and to $D_{60\%}$ if the material above the aquifer is unstable and subject to caving in (Ahrens, 1957). If a gravel pack is used, the slot size should be taken as $D_{90\%}$ of the gravel-pack material.

Where the aquifer consists of various layers of different materials, the finest and coarsest layer should be analyzed separately for particle sizes. If $D_{50\%}$ of the coarsest layer is less than 4 times $D_{50\%}$ of the finest layer, a uniform slot or gravel-pack size should be used, based on the finest material. If there is more than a fourfold difference between $D_{50\%}$ of the coarsest and finest layers, slot sizes and gravel packs should be tailored to the individual layers (Ahrens, 1957). If fine material overlies coarse material, the screen for the fine material should extend at least 0.6 m into the underlying coarse material. The slot size for the screen in the coarse material should then not be more than twice the slot size selected for the fine material (Johnson, 1972).

Gravel for artificial envelopes is usually selected so that $D_{50\%}$ of the envelope material is 5 times $D_{50\%}$ of the aquifer material (Smith, 1954). If the aquifer is layered and a uniform pack is used, $D_{50\%}$ of the gravel should be selected on the basis of the finest aquifer material.

For additional information about well screens and perforated casings (including the various types and sizes of casing and screen materials available), reference is made to Campbell and Lehr (1974) and Johnson (1972). More detailed criteria for selecting slot and gravel-pack sizes were presented by Hunter Blair (1970).

6.3.4 Well Cementing

Cementing or grouting of wells, as accomplished by pumping or otherwise placing a cement slurry between casing and hole wall, is done to prevent polluted surface water or low-quality water from other aquifers from seeping along the outside of the casing and contaminating the well or aquifer. Grouting also anchors the casing inside the bore hole, protects the outside of the casing against corrosion, prevents caving in of the land around the well in unstable materials, and enables better development of the productive aquifer section around the well screen because the rest of the well is sealed off. Where double casing is used, the space between the two casings is normally filled with cement (Figure 6.18). Parts of a well can be grouted before drilling is completed—for example, to fill the space between casing and hole wall in rotary-drilled holes prior to resumption of drilling for setting a screen or perforated casing.

The cement slurry normally used for grouting is a mix of about 45 to 55 l of water per 100 kg of cement. Possible additives include bentonite clay, pozzolans, and perlite. Occasionally, clay slurries are used for grouting. Because cement slurries harden rather quickly, they must be mixed and applied in a continuous operation. The slurry is pumped into the hole with a pipe about 5 to 10 cm in diameter. The pipe must extend to the bottom of the section to be grouted, forcing the slurry to flow upward in the bore hole to avoid bridging. Another technique used in conjunction with rotary drilling consists of placing a certain volume of cement slurry in the bore hole prior to setting the casing. The casing, closed at the bottom with a removable or drillable plug and filled with water or drilling mud to increase its weight, is then lowered into the hole, forcing the slurry up between casing and hole wall. When the slurry is hardened, which may take 2 or 3 days, drilling is resumed.

To prevent penetration of slurry into gravel packs or around screens or perforated casings, a *sand bridge* is used. Such a bridge is made by pouring a layer of clean sand of about 0.3 to 0.6 mm particle size down the hole to where the grouting is to start (Figure 6.18). Cement slurries penetrate such sand bridges not more than a few centimeters.

Well casings should extend above ground surface to prevent surface water from running down the hole and contaminating the well water. Often, a concrete slab higher than the surrounding soil is poured around the casing (Figure 6.18) to keep surface water away from the well and to provide a base for the pump and power unit and pumphouse. Direct contact between casing and concrete slab should be avoided because frost heaving, land subsidence, or thermal expansion of the slab can exert large forces which could damage the casing if not separated from the slab by a cylindrical expansion joint (Figure 6.18).

In cold areas, casing often is terminated some distance below ground surface to keep the horizontal discharge pipe and other plumbing below frost level. This often leaves a pit in which surface water can accumulate and seep down to the well water. To avoid this problem, "pitless" adapters are now used which extend the casing above ground surface but allow the well discharge to leave through a horizontal pipe below the frost line.

6.3.5 Development, Stimulation, and Sterilization of Wells

The final phases of well construction are development (including stimulation if well yields should be increased) and sterilization of the well to produce water free from pathogenic organisms.

Well development. Well development is the removal of sand and other fines (including drilling mud) from the aquifer around the well by surging, jetting, intermittent pumping, or other actions which move the fines into the well so that they may be bailed or pumped out. The objective of well development is to create a "developed" or natural filter zone around the well screen or gravel pack that prevents further movement of aquifer particles into the well and that is more permeable than the aquifer itself. The developed zone is coarsest at the screen or envelope surface and grades gradually back to the original aquifer material. The thickness of the developed zone, which also increases the effective radius of the well, may vary from a few centimeters to a few decimeters. With poorly screened wells in relatively uniform, fine-textured aquifers, developed zones may almost never become stable and the wells could pump sand for many years.

The basic principle of development procedures is to create an alternating movement of water from the well into the aquifer and back to break up bridges of fine particles in pores between larger particles. Such bridges may stand up against flow in one direction, but fail when the flow is reversed. The loose particles are then transported into the well.

Oscillating lateral flow in the aquifer adjacent to the well is most effectively obtained by jetting, using a device that shoots out water horizontally at high velocities from two or more symmetrically spaced nozzles. This creates a flow into the aquifer and a local build-up of pressure in the areas opposite the nozzles. The pressure build-up then causes the water to flow back into the well around the areas where it entered. Rotating the jetting device and moving it up and down thus produces a reversing flow in the aquifer around the well. Nozzle diameters normally vary from 0.3 to 0.6 cm, and exit velocities vary from 50 to 100 m/s. Pumping the well while jetting may help move fines into the well. Jetting also is effective in breaking up mud cakes on the walls of rotary-drilled holes, particularly if a dispersant is added to the water.

The most commonly used technique for well development is *surging*, moving a heavy bailer or a special, close-fitting device—like a surge plunger or block—up and down in the well to create flow into and out of the aquifer. Sand and other fines then accumulate on the well bottom, where they are removed with a bailer. The operation is continued until no more sand collects on the bottom. Surging should be started lightly to avoid excessive pressure differences between well and aquifer that could cause collapse of well screens if their openings are plugged by fines. As surging continues, more vigorous action can be employed. Some surge blocks are equipped with a valve allowing water to pass through during the down stroke to avoid excessive pressures below the block. A somewhat similar operation, called *swabbing*, is used to physically wipe off fines and other materials that have accumulated on the inside surfaces of screens or perforated sections.

Surging action can also be created by intermittent pumping, using a turbine pump without a foot valve. When the pump is started, the water level in the well drops and pressure heads in the aquifer around the well decrease. Stopping the pump then causes the water level in the well to rise rapidly as water rushes back through the pump. This produces flow of water into the aquifer. Overpumping to create a larger-than-usual drawdown in the well may also move fine aquifer particles into the well.

Another surging technique frequently used by drillers is *air surging*, injecting air into the well at high rates (about 8 times the expected flow from the well) and high pressures (about 10 atm). Two concentric pipes are placed into the bore hole for this purpose. The bottom of the center pipe or air pipe is initially somewhat above the bottom of the outer pipe, called the *pumping* or *eductor pipe*. Air is pumped into the center pipe to force an air-water mixture up through the annular space between the air pipe and the pumping pipe. When the air-lifted water has become sand-free, the valve on the air compressor is closed to build up pressure in the air tank. The air pipe is lowered so that its bottom is about 0.3 m below that of the pumping pipe. The valve on the air compressor is then opened to send a blast of air down the air pipe. The resulting local pressure increase inside the well will cause a sudden movement of water into the aquifer. The air pipe is pulled up, and water and sand are again air-lifted out of the hole. The pressure reduction due to the decrease in air flow also causes water to move from the aquifer into the well. Air surging is started at the bottom of the screen or perforated section. When the air-lifted water has become sand-free, the pipe assembly is raised about 1 m and the process is repeated until the aquifer around the entire screen or perforated section is developed. A simplified air technique consists of throwing Dry Ice (frozen carbon dioxide) into the well to agitate and churn the water, which may loosen up some of the fines.

Surging is not the most effective way for developing gravel-packed wells because the water tends to slosh back and forth through the gravel rather than through the aquifer material. This is particularly true for rotary-drilled wells with a thick mud cake on the hole wall. Such wells are better developed with jetting, adding polyphosphates or other dispersants to obtain more complete removal of the mud cake.

Well stimulation. In addition to well development, special techniques, collectively called *well stimulation*, are available to increase the productivity of a well. For example, explosives set off in the bore hole can increase the well yield in consolidated materials by enlarging the well diameter, loosening deposits of fine particles on the hole wall, and shattering the rock. Yield increases of 10 to 20 percent have been obtained this way for wells in deep sandstone aquifers (Walton and Csallany, 1962).

Another technique for increasing flow from wells in consolidated aquifers is hydraulic fracturing, which was first developed for oil wells. With this method, liquid is injected into the well at such high pressures that fractures develop in the rock around the screen or perforated section. It was often assumed that the fluid pressure at a given depth in the hole had to be equal to or exceed the overburden pressure in the formation at that depth in order for fracturing to be initiated. It was also assumed that the fractures were horizontal, along bedding planes and other discontinuities. Subsequent theoretical analyses and field data have shown, however, that fracturing can already occur when the fluid pressure has reached a value of 60 percent of the overburden pressure and that the fractures are mostly vertical, extending as radial planes from the bore hole (Hubbert and Willis, 1957; Wolff et al., 1975; see also Section 11.10). The increased productivity of the well is then due to increased vertical communication between the various layers in the aquifer or aquifers, which increases the lateral flow in the more permeable layers. Theoretically, horizontal cracks could develop in tectonically relaxed formations if the fluid pressure exceeds the overburden pressure. However, since vertical cracking could already start at lower fluid pressures, it may not be possible to develop such pressures. To keep the newly created fractures open after the fluid pressure in the hole has been reduced to normal, sand or other "propping" material is mixed with the injection fluid (see Section 11.10).

Wells in carbonate rocks can be stimulated by acid treatment to dissolve particles that could be blocking fractures or solution channels, and to increase secondary porosity. Normally, 15 percent hydrochloric acid is used for this purpose while adding gelatine or other inhibitors to prevent corrosion of the screen and casing. Contact time may range from 1 h to several days, but usually it is about 1 day. After treatment, the well is pumped until the acid is gone, which may take as much as half a day.

Several of the techniques for developing and stimulating wells are also used for rehabilitating old wells whose productivity has declined (see Section 6.5). For additional detail and information regarding well development and stimulation, reference is made to Campbell and Lehr (1974), Johnson (1972), Koenig (1960a, 1960b, 1960c, and 1961), and references therein.

Well sterilization. The last step in well construction is disinfection to kill bacteria and viruses that invariably have entered the well. Drilling equipment, casing, screens, etc., collect microorganisms from contact with the ground, handling by humans, animal droppings, etc. Also, bacteria and viruses enter the hole with soil and other materials that fall in during construction, and with surface water that runs into the hole. Some components, like gravel for gravel packs, may already be disinfected before they are placed in the well. Chlorine or another disinfectant may also be periodically added during drilling to sterilize the hole as it is constructed.

Disinfection normally is done with a chlorine solution of about 50 to 200 mg/l, obtained by adding a solution of sodium hypochlorite or by dissolving calcium hypochlorite or chlorine gas in the water. Contact time should be at least 4 h. Since chlorine and other disinfectants do not kill bacteria or viruses inside solid particles, the well must be thoroughly cleaned to remove all foreign and loose material and suspended solids before disinfection.

The success of chlorine treatment normally is evaluated by determining the number of coliform bacteria in the water after the well has been pumped for a few hours. Coliform bacteria are widespread in the environment. They occur in great numbers in the soil (*Aerobacter aerogenes*, for example) and in the human intestinal tract and feces (*Escherichia coli*). Total coliform tests, which include all coliform organisms, are more useful for assessing the effectiveness of disinfection than fecal coliform tests, which are better indicators of pollution by human waste. The fecal coliform bacteria themselves are harmless, but their presence may indicate the existence in the water of cholera and salmonella (including typhoid and paratyphoid) bacteria, hepatitis virus, and other pathogenic microorganisms found in human waste. For additional detail regarding well sterilization, see Campbell and Lehr (1974) and Johnson (1972).

6.4 PUMP AND POWER UNITS

Free-flowing wells, unfortunately, are rare. In the great majority of wells, the static water level is below ground surface so that the water must be lifted to get it out. Rope-suspended buckets, raised and lowered by hand with or without benefit of a windlass, are still used in many parts of the world, particularly with dug wells. Most wells, of course, are equipped with pumps, normally of the impeller type (centrifugal, mixed-flow, or turbine pumps) and driven by electric motors or internal combustion engines (diesel, propane, gasoline, or natural gas). Power units (particularly combustion engines) are commonly placed aboveground, while the pump is below the water level in the well. The pump is driven via a vertical shaft mounted in the center of the discharge column in the well. Electrically driven pumps are also available as completely submersible, close-coupled pump and power units that are installed below the water level in the well.

6.4.1 Types of Pumps and Performance Characteristics

Different pumps have different performance characteristics. Thus, pumps should be carefully selected so that water is pumped at minimum cost. The main types of pumps are variable-discharge (impeller and jet) and positive-displacement pumps.

Impeller pumps. The working element, or impeller, of centrifugal pumps consists of a hollow, cylindrical unit with an open circumference and spiral vanes mounted semiradially inside (Figure 6.19). The impeller is rotated at speeds of about 1 750 or 3 500 rpm; water enters axially through the center or eye of the impeller and is thrown out radially at the circumference of the impeller. In centrifugal pumps, the water is collected in the volute, which is a spiral-shaped housing around the impeller, and leaves through the discharge end (Figure 6.19). In turbine pumps, water leaves the impeller through stationary diffuser vanes before entering the volute. In mixed-flow pumps, the impeller is shaped so that it also imparts some axial motion to the water [vertical motion if the shaft is vertical (Figure 6.19)]. In the water-well industry, the term *turbine pump* is commonly used for all three types of impeller pumps, although the name correctly applies only to pumps

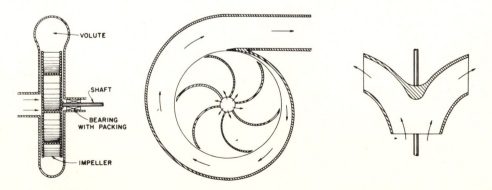

Figure 6.19 Schematic of impeller and volute of centrifugal pump (side and front views, left and center), and of mixed-flow impeller (right).

with diffuser vanes around the impeller. Axial-flow pumps consist of a propeller mounted on a shaft in the center of a pipe. The propeller has about the same diameter as the pipe. When it is rotated the water is moved axially through the pipe.

The total lift of the water, called *total dynamic head* (TDH), is calculated as the sum of the vertical lift from the water level in the well to ground surface, the friction losses in intake strainer and discharge pipe inside the well, and the pressure head in the discharge pipe at the top of the well. Thus, TDH is the vertical distance between the water level in the well and the water level in a piezometer placed in the discharge pipe at the well head (Figure 6.20). To be correct, the velocity head $v^2/2g$ in the discharge pipe should also be included in the TDH.

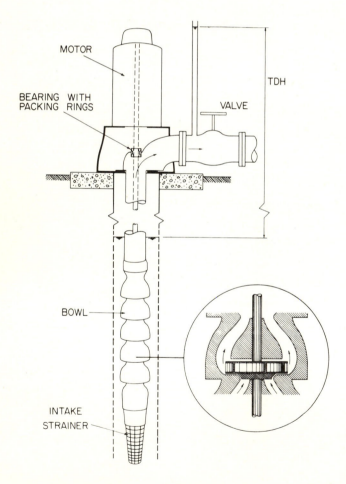

Figure 6.20 Schematic of five-stage turbine pump with suction strainer and aboveground electric motor.

Since the velocity head seldom exceeds 1 m, it can usually be ignored, except where TDH is relatively small. For impeller-type pumps, the pumping rate Q decreases with increasing TDH (Figure 6.21) if the speed (rpm) of the pump is kept constant. For this reason, impeller-type pumps are also called *variable-discharge pumps*. At a given speed, Q varies approximately proportionally with the width and the diameter of the impeller, and TDH increases with the square of the impeller diameter.

The relation between Q and TDH for a given pump speed, called the *pump characteristic*, depends on the type and geometry of the impeller and on the pump design. The TDH value where $Q = 0$ is the shut-off head. It determines the pressure head that will develop at the discharge end of the well when the discharge valve (Figure 6.20) is closed. At this point the pump continues to run unharmed, but heat build-up inside the pump eventually could cause damage. Axial-flow pumps have much lower shut-off heads and TDH values than turbine pumps of similar capacity. For this reason, axial-flow pumps can be used to lift water only from wells with water levels close to the ground surface.

For turbine pumps (including centrifugal and mixed-flow pumps) that are placed inside wells, the diameter of the impeller is limited by the diameter of the screen or casing of the well. This limits the TDH that can be developed by one impeller, making it necessary to use a number of impellers in series if the total required TDH exceeds that of a single impeller. The impellers are stacked on top of each other with the discharge from one impeller being guided into the inlet eye of the impeller above it (Figure 6.20). The resulting multistage pump may contain only a few or as many as 20 or more individual impellers. The impellers and the vanes guiding the flow to the next impeller are mounted inside bowl-shaped units that are, indeed, called *bowls* (Figure 6.20). The flow rate Q from a multistage pump is equal to Q from an individual impeller, but the TDH of the complete unit is equal to the sum of the TDH produced by each impeller.

If the power unit is located aboveground, the impellers are driven by a vertical shaft mounted centrally inside the discharge pipe or column in the well (Figure 6.20). Electrically driven pumps are also manufactured as submersible units with the motor located below the bowls. Such units can be sealed to operate under as much as 150 m of water. Submersible units are more adaptable to bore holes with poor vertical alignment. They also eliminate the need for long drive shafts, which can be important for deep holes, and foundations for power units or pumphouses are not necessary.

Electrically driven turbine pumps run backwards after the motor has been turned off, and water rushes down the discharge column and back into the well through the impellers, which then act as real turbines. This could be undesirable if power is momentarily interrupted because the motor would start against a backward-running pump, which could damage the pump or power unit. To protect the unit against such damage from short power outages, the starter switch should have a built-in time delay of at least several minutes.

The power needed to operate a pump at the desired values of Q and TDH is called the *brake horsepower* (bhp). This is the power input that must be supplied

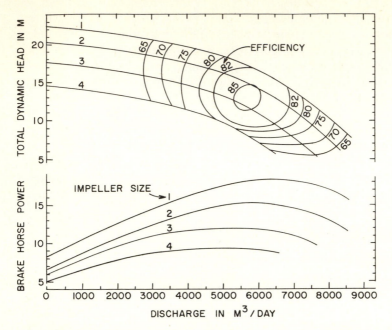

Figure 6.21 Relation between pump discharge, TDH, efficiency, and brake-horsepower requirements (converted to metric units). (*From Bulletin B-180, Pump Division of FMC Corporation, for Peerless Impeller No. 2624332 at 1760 rpm.*)

by the drive shaft. The power output by the pump in terms of lifting water a certain distance is called the *water horsepower* (whp). Since the metric horsepower is 75 kg · m/s[†], whp is calculated as

$$\text{whp} = \frac{Q \times \text{TDH}}{0.075 \times 86\,400} \tag{6.9}$$

where Q = pump discharge in m³/day

TDH = total dynamic head in meters of water

Because of friction, back-slippage of water, and other energy losses inside the pump, whp is always less than bhp. The ratio between whp and bhp is called the *pump efficiency* E_p, or

$$E_p = \frac{\text{whp}}{\text{bhp}} \tag{6.10}$$

The value of E_p normally is expressed as a percentage. For a given impeller and operating speed, E_p is not constant but varies as Q and TDH change. Maximum E_p is obtained for only one particular combination of Q and TDH (Figure 6.21),

---

[†] In English units, the horsepower is 550 ft·lb/s, which converts to 76.04 kg·m/s. Thus the English hp is slightly larger than the metric hp, but the difference is so small that it can be ignored for most practical purposes.

which is a function of the design of the impeller and volute. Pumps must be selected so that their E_p is maximum for the Q-TDH combination at which they are to operate. Good turbine-type pumps often have maximum E_p values in the 80 to 85 percent range. In practice, E_p frequently is much lower (60 percent or less) because of pump wear, clogging and encrustation of the impeller, and pumping at Q and TDH other than the optimum values.

For a given impeller and impeller speed, bhp requirements vary with Q. At the shut-off head, bhp is usually lowest because work is needed only to overcome friction as the impeller turns in a stationary body of water. As Q increases, bhp increases, reaches a maximum, and then decreases as TDH begins to drop more rapidly (Figure 6.21).

Jet pumps. Variable-discharge pumps also include the air-lift and jet pumps. The latter are frequently used for small, domestic-type wells. Water is pumped from aboveground through a downflow pipe into the well (Figure 6.22). At the bottom, the downflow pipe bends upward and the water is passed through a narrow nozzle. The increase in velocity head produces a corresponding decrease in pressure head, which creates a suction that draws water from the well into the pipe through a side opening just above the nozzle. This water is then transported upward with the other water through the upflow pipe. At the top, water can be discharged at the same rate at which well water was drawn in above the nozzle.

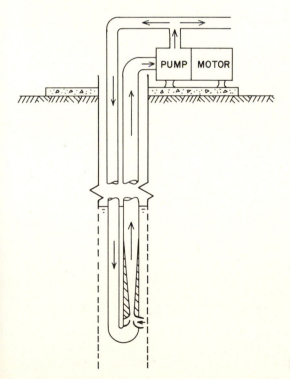

Figure 6.22 Schematic of jet pump.

The rest of the water is recirculated through the downflow pipe. The pump is usually an electrically driven centrifugal pump.

Jet pumps have a low E_p, but this is not a serious disadvantage since they are used mostly for domestic or other purposes with relatively small Q. Advantages of jet pumps include adaptability to small-diameter wells (as little as 5 cm), simplicity and reliability of the jet system (no moving parts underground), and accessibility of pump-and-power unit for maintenance and repair.

Positive-displacement pumps. Piston, gear, and screw pumps are called *positive-displacement* pumps because water is mechanically moved at rates that, for a given pump, are directly related to the speed of the pump (strokes or revolutions per minute) and independent of TDH. They have no shut-off head, and when discharge is blocked, something has to give (relief valves) or damage may result. With piston pumps, water is moved by a reciprocating piston inside a cylinder. The piston may be single-acting or double-acting. Gear pumps or rotary pumps move water in the space between rotating gears or rotors mounted inside a closely fitting housing. Screw pumps consist of a rotating helix inside a pipe, like the Archimedes screw. While screw pumps were traditionally low-lift pumps because of back-slippage of the water, modern screw pumps consisting of a helical-contoured, chrome-plated rotor turning inside a bihelical, contoured rubber stator are capable of pumping water at rates of 45 to 300 m^3/day and total lifts of as much as 370 m (Peerless Hi-Lift pump).

Suction lifts. Ideally, the maximum suction lift, or maximum distance of pump above water level, is equal to the difference between the atmospheric pressure and the vapor pressure of water, both expressed in meters of water. At sea level, this distance is about 10 m for water of moderate temperature (about 20°C). In practice, however, the maximum suction lift is considerably less because of friction losses in the suction strainer and pump intake and because of local pressure reductions inside the pump due to acceleration and separation (water moving away from solid boundaries). These pressure reductions can cause local vaporization of water inside the pump. When the resulting bubbles then move into zones of higher pressure, they collapse. The resulting "implosions" produce noise and can cause pitting, called *cavitation*, of the impeller and other surfaces inside the pump. To avoid cavitation, the theoretical maximum suction lift should be reduced by the net positive suction head (NPSH) specified by the pump manufacturer. Values of NPSH vary from 1 to 10 m or more, the higher values being applicable to turbine-type pumps that operate submerged. The maximum suction lift should also be reduced by friction losses in the intake strainer and suction pipe. Pumps located above the water level should be equipped with foot valves, self-priming devices, or water tanks to facilitate priming the pump after periods of no pumping.

6.4.2 Power and Energy Requirements

The horsepower requirement for the power unit is the brake horsepower (bhp), calculated as whp/E_p [Eq. (6.10)]. The energy input for the power unit is higher than that corresponding to the horsepower output (bhp) because the efficiency E_m of the power unit is less than 100 percent. For electric motors, E_m ranges from 0.8 to 0.95 (80 to 95 percent), with large motors usually being more efficient than small motors. For internal combustion engines, E_m is called the *thermal* efficiency and normally ranges from 0.15 to 0.35. Diesel, natural gas, and high-octane-gas engines have a higher efficiency (0.25 to 0.35) than automobile-type gasoline engines (0.15 to 0.25). Energy input requirements for power units thus are calculated on the basis of bhp/E_m, or whp/$E_p E_m$, with E_p and E_m expressed as fractions. For electrically powered units, the product $E_p E_m$ is called the *wire-to-water efficiency*. Conversely, $E_p E_m$ for pumps driven by internal combustion engines could be called the *fuel-to-water* efficiency.

In SI units, the unit of power is the watt (W), which is $1/g$ (kg·m)/s, or 0.101 97 kg·m/s. A metric horsepower is 75 kg·m/s, or 735.5 W. Lifting 1 m³ of water 1 m in 1 s ideally requires $1\,000/75 = 13.33$ hp, which corresponds to 9 807 W or $9\,807/3\,600 = 2.724$ Wh (watthours). The actual energy requirement for lifting 1 m³ of water a distance of 1 m is $2.724/E_p E_m$ Wh. Thus, an electrically powered unit with $E_p = 0.8$ and $E_m = 0.85$ requires $2.724/0.8 \times 0.85 = 4.006$ Wh for every cubic meter of water lifted 1 m. At this rate, pumping 5 000 m³/day at a TDH of 100 m would require $5\,000 \times 100 \times 4 = 2 \times 10^6$ Wh or 2 000 kWh (kilowatthours). Electricity requirements for other pumping rates, lifts, and wire-to-water efficiencies can be calculated in similar fashion.

Fuel requirements for pumping with internal combustion engines are calculated on the basis of the calorie content or calorific value of the fuel. Since 1 metric hp equals 175.7 cal/s, lifting 1 m³ of water in 1 s corresponds to $175.7/0.075 = 2\,343$ cal/s. This must be divided by the fuel-to-water efficiency $E_p E_m$ to obtain the required energy input for the engine. Taking typical calorific values of the various fuels and selecting values for E_p and E_m that can be obtained with well-designed and properly functioning pump and power units, fuel requirements per cubic meter of water and per meter of lift are calculated in Table 6.1.

The data in Table 6.1, along with the previously calculated electrical energy

Table 6.1 Fuel requirements for lifting 1 m³ of water 1 m, calculated on the basis of calorific value of fuel and fuel-to-water efficiency

Fuel	Calorific value	E_p	E_m	Fuel requirements to lift 1 m³ water 1 m	
				g	cm³
Gasoline	10 000 kcal/kg	0.8	0.25	1.17	1.7
Diesel	10 000 kcal/kg	0.8	0.3	0.975	1.2
Propane	11 000 kcal/kg	0.8	0.25	1.06	1.8†
Natural gas	10 000 kcal/m³	0.8	0.25		1171

† As liquefied petroleum gas (LPG)

requirement of 4 Wh for lifting 1 m³ of water 1 m with a wire-to-water efficiency of 0.68, enable estimation of energy requirements and, hence, of costs of pumping water from wells.

    Example. A well is pumped at 2 500 m³/day against a TDH of 70 m. Assuming $E_p = 0.8$, what are the expected electricity and fuel requirements per day? The product of Q and TDH is 175 000 m·m³/day, which requires 701 kWh, 297 1 of gasoline, 210 1 of diesel fuel, 315 1 of propane (as LPG), or 204 m³ of natural gas. Dividing these values by the well discharge of 2 500 m³/day gives the energy requirement per cubic meter of water.

    These data agree with those based on actual performance tests—such as those presented by Fischbach et al. (1968)—except for the natural gas requirement, which is about 40 percent less than reported by these authors. This may be partly due to the variability in composition of natural gas.

    Cost data based on Table 6.1 or on the electricity requirement of 4 Wh per cubic meter per meter of lift should be interpreted as minimum costs, since in practice E_p and E_m often are less than those used in the calculations. For example, Miles and Longenbaugh (1968) calculated that electrically pumped wells were operating at least 30 percent less efficiently than the maximum possible. In a study of irrigation wells in Kansas (Anonymous, 1969), wire-to-water efficiencies were between 0.36 and 0.70. Fuel-to-water efficiencies were between 0.08 and 0.18 for pumps powered by liquefied petroleum gas, and between 0.12 and 0.19 for diesel-powered units.

6.4.3 Optimum Discharge

The desired pumping rate of a well may be selected on the basis of demand, maximum well yield, or maximum economic benefit. Selection on the basis of demand is done where the required flow is less than the potential yield of the well, as may occur for domestic wells. Pumping at maximum rate, which is obtained when the drawdown of the water level in the well is maximum, is necessary where the demand equals or exceeds the potential well yield. This may happen with new wells if the yield is less than anticipated, or with old wells if the demand has gone up or if the productivity of the well has gone down. Selecting the pumping rate on the basis of maximum economic benefit is possible where the well does not have to meet a particular demand (for example, if various other water sources are also available) and water must be produced as cheaply as possible.

    The economically optimum well discharge is obtained when the difference between pumping costs and economic benefits from the water is greatest (assuming that benefits exceed costs!). Since well losses increase faster than Q (see Section 4.5), the optimum Q is less than the maximum Q. Assuming that the pumping efficiency E_p remains constant and ignoring amortization of well-construction costs and other fixed costs, the cost of pumping water from a given well is directly proportional to the product $Q \times$ TDH. Taking the data of Q and s_{iw} of the example in the section on well losses (Section 4.5), pumping costs were calculated on the basis of $0.05 per 1000 m³ per meter of lift for TDH values of $s_{iw} + 10$ and $s_{iw} + 40$ meters to represent a relatively small and large TDH, respectively. The resulting curves (Figure 6.23) show that cost of pumping increases faster than rate of pumping. In the same graph, benefit lines are plotted with one line representing an income of $5 per 1000 m³ water and the other an income of $10 per 1000 m³ water. The optimum discharge is obtained where the vertical distance between the cost curve and the benefit line is maximum. This occurs where the slope of the cost curve equals that of the benefit line. Nonlinear benefit curves would cause the optimum Q to decrease if the rate of income were to decrease with increasing pumping rates (as may happen with pumping for irrigation), and cause the optimum Q to increase if the rate of (net) income would increase with increased pumping rates. The latter situation may occur with water companies or water districts, which have certain fixed operating expenses. Once enough water is sold to meet those expenses, additional sales then mean greater profit.

Figure 6.23 also shows that the optimum Q increases with decreasing TDH and with increasing economic returns per unit of water. In the same way, it can be shown that the optimum Q increases as the unit cost of pumping decreases. Thus, the optimum Q depends on a number of factors that must be considered for each situation. Johnson (1972) suggests that optimum Q for wells in unconfined aquifers usually is obtained when s_{iw} is about two-thirds the static water depth in the hole. This would mean that only the lower one-third of the well section below

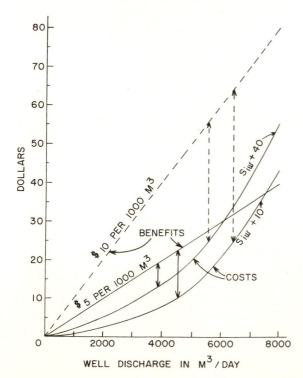

Figure **6.23** Cost curves and benefit lines for determining optimum well discharge (vertical lines indicate maximum difference).

A similar type of corrosion is galvanic corrosion, which is caused by electric potentials generated when two dissimilar metals are in direct contact and immersed in an electrolyte to complete the electric circuit. This will cause electrolysis reactions with corrosion taking place at the anodic (most corrosive) metal and electrolysis products accumulating on the cathodic (least corrosive) metal. The farther the two metals are apart in the electromotive or galvanic series (Table 6.2), the stronger the galvanic corrosion will be. Galvanic corrosion will also occur when there is contact between identical metals in different stages of corrosion—for example, when new pipe is connected to old pipe. Since the old pipe is usually more rusty, which acts as a protective coating, the new pipe will then corrode.

In alloys, galvanic corrosion can take place between crystals of different metals when in contact with an electrolyte. This is called *selective corrosion*, *dezincification*, or *degraphitization* of the metal, in contrast to direct corrosion which destroys the entire metal. Selective corrosion weakens the metal into a porous, spongy condition until it may suddenly disintegrate. Intermediate-sized crystals in alloys will corrode faster than either large or small crystals. Since the crystal size can be controlled by heat treatment, alloys can be produced for maximum corrosion resistance.

Corrosion by bacteria is usually caused by metabolic products that increase corrosion reactions—except for iron bacteria, which consume iron directly as an energy source for synthesizing organic compounds. Sulfate-reducing bacteria use the oxygen of the sulfate ion as hydrogen acceptor in their metabolic processes (like

Table 6.2 Electromotive series of metals

Anodic or	Magnesium
corroded end	Magnesium alloys
	Zinc
	Aluminum 2 S
	Cadmium
	Aluminum 17 ST
	Steel, iron, cast iron
	Chromium-iron (active)
	Ni-Resist
	Stainless steel 18-8 (active)
	Lead, tin, lead-tin solders
	Nickel, Inconel (active)
	Brass, copper
	Bronze, Monel
	Silver solder
	Nickel, Inconel (passive)
	Chromium-iron (passive)
	Stainless steel 18-8 (passive)
Cathodic or	Silver
protected end	Gold, platinum

Source: From the International Nickel Co., Inc., New York.

"breathing" sulfate oxygen) and produce hydrogen sulfide, which increases the acidity of the water and, hence, the corrosion rate. Other, heterotrophic bacteria digest organic matter and produce organic acids. Even if organic substrates are not originally present in the well, autotrophic bacteria may develop which upon death and decay provide the organic energy source for the heterotrophic bacteria.

Corrosion initially leads to enlargement of screen openings or perforations, which could cause entry of sand and other aquifer material into the well. Accumulation of corrosion products could also clog screen openings and aggravate incrustation problems. Ultimately, entire screen or casing sections may disintegrate or collapse under lateral pressures from the aquifer. This could happen in less than 1 year if the groundwater is corrosive and screens and casings have low corrosion resistance.

Corrosion is more severe when the groundwater is acidic (pH < 7) or contains dissolved oxygen, dissolved salts (particularly when in excess of 1 000 mg/l), hydrogen sulfide, carbon dioxide, chloride (when exceeding 300 mg/l), and sulfate. Since all chemical reactions proceed faster at higher temperatures (including biochemical reactions, as long as the optimum temperature is not exceeded), corrosion is more rapid when groundwater temperatures are high. High entrance velocities of the groundwater in the screen openings also increase corrosion rates.

The best protection against corrosion is to provide sufficient screen area to minimize entrance velocities, and to use corrosion-resistant metals or alloys for casings and particularly screens. Johnson (1972) lists metals in the following order of decreasing corrosion resistance:

Monel metal (about 70 percent nickel and 30 percent copper)
Stainless steel (74 percent low-carbon steel, 18 percent chromium, and 8 percent
    nickel)
Everdur metal (96 percent copper, 3 percent silicon, and 1 percent manganese)
Silicon red brass (83 percent copper, 16 percent zinc, and 1 percent silicon)
Yellow brass (67 percent copper and 33 percent zinc)
Low-carbon steel

Techniques used in the oil industry to control corrosion, such as continuous injection of inhibitors or preferential wetting agents, cannot be used for water wells because of their adverse effect on water quality. Use of metal coatings or nonmetallic pipe like asbestos-cement and high-impact plastic casings (Gass, 1977) may increase the useful life of the well. Another method is cathodic protection, consisting of burying a substitute anode at some distance from the well. The material of this anode is more corrosive than that of the well casing and screen. Thus, when stray currents around the well are eliminated by maintaining a direct electric current from the anode to the well, the anode becomes the corroded area and the well the protected area. Cathodic protection tends to protect only exterior surfaces. For this reason, it is commonly used to reduce corrosion of underground pipelines and tanks (Ritchie, 1976).

6.5.3 Abandoned Wells

Wells that are no longer needed or usable should be covered and sealed for reasons of safety and protection of underground water supplies. This can be done by filling the well with cement slurry or other grouting material. Casings and screens may have to be removed or ripped out so that the grout can form an effective seal that prevents surface water from reaching the groundwater and blocks vertical communication between various aquifers.

6.6 COSTS OF WELLS AND PUMPING

Pumping normally accounts for about 60 to 80 percent of the total cost of groundwater production (excepting, of course, free-flowing wells). The remaining 20 to 40 percent consists primarily of amortization of capital investments. Increasing energy costs may change these percentages. General cost figures, as have been collected for well drilling and groundwater pumping in various areas, are useful as guides in planning or comparisons. Accurate cost estimates for actual cases can be obtained only from well drillers and pump suppliers. Operating costs may be estimated from the energy requirements presented in Section 6.4.2. Costs of installing a well depend on the kinds of drilling equipment that are available, local drilling techniques, geologic conditions, site accessibility, availability of water, depth and size of well, labor and fuel costs, cost of casing, screen, gravel envelope, pump and power unit, etc.

Extensive data about costs of well drilling, pumps, and pumping were published by Ackermann (1969) for Illinois conditions. Costs of well drilling were summarized by the following equation

$$\text{Well cost} = aD_d^n \qquad (6.11)$$

where the well cost is in dollars, a and n are constant, and D_d is the depth of the well in meters. Values for a and n for different well diameters (bottom diameter for telescoping wells) and different geologic materials are presented in Table 6.3.

The installed costs of vertical turbine pumps with electric motors aboveground were described by the equation

$$\text{Pump cost} = 7.26Q^{0.453} \times \text{TDH}^{0.642} \qquad (6.12)$$

where Q is the pump capacity in cubic meters per day and TDH the total dynamic head in meters. For turbine pumps with submersible electric motors, the equation was

$$\text{Pump cost} = 4.91Q^{0.541} \times \text{TDH}^{0.658} \qquad (6.13)$$

The constants and exponents in these equations were adapted to the metric system from Ackermann (1969).

Because of inflationary trends, costs as of 1975 were considerably higher than those calculated with Eqs. (6.11), (6.12), and (6.13). Schleicher (1975), for

Table 6.3 Values of a and n in Eq. (6.11) for various geologic materials and well diameters (bottom diameter for telescoping wells)

Type of well and geologic material	Well diameter	a	n
Tubular wells in	15 to 25 cm	1 141	0.299
sand and gravel	30 to 38 cm	1 324	0.373
Gravel-packed wells in	40 to 50 cm	1 104	0.408
sand and gravel	60 to 85 cm	1 205	0.482
	90 to 107 cm	1 779	0.583
Wells in sandstone, limestone	15 cm	3.10	1.413
or dolomite	20 to 30 cm	4.70	1.450
	38 to 60 cm	10.22	1.471
Deep wells in	20 to 30 cm	0.267	1.87
sandstone	38 to 48 cm	7.15	1.429

Source: Adapted to the metric system from Ackermann (1969).

example, listed costs of $60 to $120 per meter for irrigation wells of 20 to 150 m in depth and 30 to 80 cm in diameter in unconsolidated materials in Kansas, Oklahoma, Nevada, Missouri, Montana, and Texas. These costs include casing, development, and in some instances test drilling and gravel packs. The cost of drilling a well of 60 cm diameter and 60 m depth in unconsolidated alluvium with gravel and boulders similar to that shown in Figure 3.18 was about $300 per meter in 1975 (including casing and in-place perforation of the lower 25 m). The well is located in Phoenix, Arizona, and was drilled with the cable-tool rig of Figure 6.11. The cost of drilling with the reverse-rotary method would have been about the same.

Total costs of pumping groundwater (including well amortization, electric

Table 6.4 Total costs of pumping groundwater in the Susquehanna River Basin

Well yield, m³/day	Total cost of water production, dollars per m³	Geologic material
275	0.016	Shale and interbedded sandstone and shale
275	0.013	Metamorphic rock
550	0.011	Shale and interbedded sandstone and shale
550	0.009 8	Metamorphic rock
550	0.008 4	Carbonate rock
2 700	0.005 3	Sandstone
2 700	0.004 0	Carbonate rock
2 700	0.003 2	Glacial sand and gravel
5 500	0.002 4	Glacial sand and gravel

Source: Adapted to metric units from Hollyday and Seaber, 1968.

power, and maintenance) for wells in different geologic materials of the Susquehanna River Basin in Pennsylvania were reported by Hollyday and Seaber (1968). The data (Table 6.4) show a range of $0.002 to $0.02 per cubic meter, with the lowest cost corresponding to the well with the highest yield. Total costs of pumping irrigation water in the alluvial desert valleys of central Arizona were $8.9 to $11.4 per 1 000 m³ (about $0.01 per m³) at TDH values of 110 to 130 m (Nelson and Busch, 1967). Pumping with natural gas was about 20 percent cheaper than with electricity.

PROBLEMS

6.1 Using Eqs. (6.5), (6.6), (6.7), and (6.8), calculate the thickness of the first and second layers in the profile for the seismic data in Figure 6.4.

6.2 Cover the geologic log in Figure 6.5 and try to determine the lithology from the drill time and geophysical logs in the figure. Explain the patterns in these logs.

6.3 Arrange with a local well driller to see cable-tool and rotary drilling in action. Make repeated visits to the sites to observe the various drilling operations. Obtain cost data for well construction and pump and power units.

6.4 A gravel-packed well 60 cm in diameter has been installed to a depth of 300 m in sand and gravel material. The well is to pump 5 000 m³/day at a TDH of 100 m, using a turbine pump with an aboveground electric motor. Calculate the total cost of well and pump with Eqs. (6.11) and (6.12). Assuming annual amortization and maintenance costs of 10 percent of the total capital investment, a wire-to-water efficiency of 60 percent, and an electric rate of $0.03 per kWh (kilowatthour), calculate the total cost of water production per 1 000 m³ water and the percentage of this cost that is due to fixed costs (amortization and maintenance).

REFERENCES

Ackermann, W. C., 1969. Cost of wells and pumps. *Ground Water* 7(1): 35–37.

Ahrens, T. P., 1957. Well design criteria. *Water Well J.* 11 (9, 11), 8 pp.

Anonymous, 1969. Finding ways to save pumping costs. *Johnson Driller's J.* 41(1): 1–4.

Barton, C. M., 1974. Bore hole sampling of saturated uncemented sands and gravels. *Ground Water* 12(3): 170–177.

Bays, C. A., 1950. Prospecting for ground water—geophysical methods. *J. Am. Water Works Assoc.* 42: 947–956.

Bedinger, M. S., 1967. An electrical analog study of the geometry of limestone solution. *Ground Water* 5(1): 24–28.

Bianchi, W. C., and H. I. Nightingale, 1975. Hammer seismic timing as a tool for artificial recharge site selection. *Proc. Soil Sci. Soc. Am.* 39: 747–751.

Bierschenk, W. H., 1964. Geohydrological and geophysical investigations near Izmir, Turkey. *Ground Water* 2(4): 18–24.

Blank, H. R., and M. C. Schroeder, 1973. Geologic classification of aquifers. *Ground Water* 11(2): 3–5.

Brown, D. L., 1971. Techniques for quality-of-water interpretations from calibrated geophysical logs. *Ground Water* 9(4): 25–38.

Buhle, M. B., 1953. Earth resistivity in ground water studies in Illinois. *Min. Eng.* 5: 395–399.

Bybordi, M., 1974. Ghanats: drainage of sloping aquifer. *J. Irrig. Drain. Div., Proc. Am. Soc. Civ. Eng.* 10(IR3): 245–253.

Campbell, M. D., and J. H. Lehr, 1973. *Rural Water Systems—Planning and Engineering Guide.* Commission on Rural Water, Washington, D.C., 150 pp. (Library of Congress Card number 73-77739.)

Campbell, M. D., and J. H. Lehr, 1974. *Water Well Technology.* McGraw-Hill Book Co., New York, 681 pp.

Cartwright, K., 1968. Thermal prospecting for ground water. *Water Resour. Res.* **4**: 395–401.

Cressey, G. B., 1958. Ganats, karez, and foggaras. *Geogr. Review* **58**(1): 27–44.

Crosley, J. W., III, and J. V. Anderson, 1971. Some applications of geophysical well logging to basalt hydrogeology. *Ground Water* **9**(5): 12–20.

Davis, R. W., 1969. Ground water, gravity and rift valleys in Malawi. *Ground Water* **7**(2): 34–37.

Davis, S. N., and R. J. M. DeWiest, 1966. *Hydrogeology.* John Wiley & Sons, Inc., New York, 463 pp.

Davis, S. N., and L. J. Turk, 1964. Optimum depth of wells in crystalline rocks. *Ground Water* **2**(2): 6–11.

Dobrin, M. B., 1952. *Introduction to Geophysical Prospecting.* McGraw-Hill Book Co., New York, 435 pp.

Dudley, W. W., Jr., et al., 1964. Geophysical studies in Nevada relating to hydrogeology. *Univ. of Nevada Tech. Report No. 2*, 46 pp.

Fischbach, P. E., J. J. Sulek, and D. Axthelm, 1968. Your pumping plant may be using too much fuel. *Univ. of Nebraska Extension Service Bulletin EC 68-775.*

Florquist, B. A., 1973. Techniques for locating water wells in fractured crystalline rocks. *Ground Water* **11**(3): 26–28.

Foster, J. W., and M. B. Buhle, 1951. An integrated geophysical investigation of aquifers in glacial drift near Champaign-Urbana, Illinois. *Econ. Geol.* **46**: 367–397.

Gass, T. E., 1977. Installing thermoplastic water well casing. *Water Well J.* **31**(7): 34–35.

Geophysical Specialties Co., 1960. *Instruction Manual Engineering Seismograph MD-1.* Minneapolis. Minn.

Gill, H. E., J. Vecchioli, and W. E. Bonini, 1965. Tracing the continuity of pleistocene aquifers in northern New Jersey by seismic methods. *Ground Water* **3**(4): 33–35.

Greenberg, J. A., J. K. Mitchell, and P. A. Witherspoon, 1973. Coupled salt and water flows in a groundwater basin. *J. Geophys. Res.* **78**: 6341–6353.

Ground Water Newsletter. 1974. Water Information Center, Port Washington, N.Y., N. P. Gillies (ed.), **3**(23), p. 1.

Guyod, H., 1952. *Electrical Well Logging Fundamentals.* Well Instrument Developing Co., Houston, Texas, 164 pp.

Guyod, H., 1966. Interpretation of electric and gamma ray logs in water wells. *Log Anal.*, Jan.–March, 16 pp.

Heiland, C. A., 1946. *Geophysical Exploration.* Prentice-Hall, New York, 1013 pp.

Helweg, O. J., 1975. Determining optimum well discharge. *J. Irrig. Drain. Div., Proc. Am. Soc. Civ. Eng.* **101**(IR3): 201–208.

Hobson, G. D., and L. S. Collett, 1960. Some observations with a hammer refraction seismograph. *Can. Min. Metall. Bull.* **53**: 674–681.

Hollyday, E. F., and P. R. Seaber, 1968. Estimating cost of groundwater withdrawal for river basin planning. *Ground Water* **6**(4): 15–23.

Hubbert, M. K., and E. G. Willis, 1957. Mechanics of hydraulic fracturing. *Trans. Am. Inst. Mech. Eng.* **210**. (Reprinted in "Underground Waste Manag. and Environmental Implications," T. D. Cook (ed.), *Am. Assoc. Petrol. Geol. Memoir* **18**: 239–257, 1972.)

Hunter Blair, A., 1970. Well screens and gravel packs. *Ground Water* **8**(1): 10–21.

Hvorslev, M. J., 1949. *Subsurface Exploration and Sampling of Soils for Engineering Purposes.* Waterways Experiment Station, U.S. Corps of Army Engineers, Vicksburg, Miss., 521 pp.

Ibrahim, A., and W. J. Hinze, 1972. Mapping buried bedrock topography with gravity. *Ground Water* **10**(3): 18–23.

Johnson, E. E., Inc., 1972. *Ground Water and Wells.* Johnson Div., Universal Oil Products Co., St. Paul, Minn., 440 pp.

Joiner, T. J., J. C. Warman, and W. L. Scarborough, 1968. An evaluation of some geophysical methods for water exploration in the Piedmont area. *Ground Water* **6**(1): 19–25.

Kelley, S. F., 1962. Geophysical exploration for water by electrical resistivity. *J. N. Eng. Water Works Assoc.* **76**: 118–189.

Kelley, D. R., 1969. A summary of major geophysical logging methods. *Penn. Geol. Survey Bull. M61,* 88 pp.

Keys, W. S., 1967. Borehole geophysics as applied to ground water. *Mining and Ground Water Geophysics, Geol. Surv. Can., Econ. Geol. Report* **26**: 598–614.

Keys, W. S., 1968. Well logging in ground-water hydrology. *Ground Water* **6**(1): 10–18.

Keys, W. S., and L. M. MacCary, 1971. Application of borehole geophysics to water-resources investigations. Chap. E1 in *Techniques of Water-Resources Investigations of the United States Geological Survey.* U.S. Government Printing Office No. 2401-2053, Washington, D.C., 126 pp.

Koenig, L., 1960a. Survey and analysis of well stimulation performance. *J. Am. Water Works Assoc.* **52**: 333–350.

Koenig, L., 1960b. Economic aspects of water well stimulation. *J. Am. Water Works Assoc.* **52**: 631–637.

Koenig, L., 1960c. Effects of stimulation on well operating costs and its performance on old and new wells. *J. Am. Water Works Assoc.* **52**: 1499–1512.

Koenig, L., 1961. Relation between aquifer permeability and improvement achieved by well stimulation. *J. Am. Water Works Assoc.* **53**: 652–670.

Landers, R. A., and L. J. Turk, 1973. Occurrence and quality of ground water in crystalline rocks of the Llano area, Texas. *Ground Water* **11**(1): 5–10.

Linck, C. J., 1963. Geophysics as an aid to the small water well contractor. *Ground Water* **1**(1): 33–37.

McDonald, H. R., and D. Wantland, 1961. Geophysical procedures in ground-water study. *Trans. Am. Soc. Civ. Eng.* **126**: 122–135.

McGinnis, L. D., and J. P. Kempton, 1961. Integrated seismic, resistivity, and geologic studies of glacial deposits. *Illinois Geol. Survey Circular No. 323,* 23 pp.

Meier, O., and S. G. Petersson, 1951. *Water Supplies in the Archaean Bedrocks of Sweden.* Int. Union of Geodesy and Geophys., Assoc. Sci. Hydrol., Brussels, 252–261.

Merkel, R. H., and J. K. Kaminsky, 1972. Mapping ground water by using electrical resistivity with a buried current source. *Ground Water* **10**(2): 18–25.

Meyer, W. R., 1962. Use of a neutron moisture probe to determine the storage coefficient of an unconfined aquifer. *U.S. Geol. Survey Professional Paper 450-E:* 174–176.

Miles, D. L., and R. L. Longenbaugh, 1968. Evaluation of irrigation pumping plant efficiencies and costs in the High Plains of eastern Colorado. *Colo. State Univ. Exp. St. General Series No. 876.*

Mooney, H. M., and W. W. Wetzel, 1956. "The Potentials about a Point Electrode and Apparent Resistivity Curves for Two-, Three-, and Four-layered Earth." Univ. Minnesota Press, Minneapolis, 146 pp. (plus curves).

Nelson, A. G., and C. D. Busch, 1967. Cost of pumping irrigation water in central Arizona. *Univ. Arizona Tech. Bull. 182,* 44 pp.

Nettleton, L. L., 1940. *Geophysical Prospecting for Oil.* McGraw-Hill Book Co., New York, 444 pp.

Norris, S. E., 1972. The use of gamma logs in determining the character of unconsolidated sediments and well construction features. *Ground Water* **10**(6): 14–21.

Nutbrown, D. A., R. A. Downing, and R. A. Monkhouse, 1975. The use of a digital model in the management of the Chalk aquifer in the South Downs, England. *J. Hydrol.* **27**: 127–142.

Page, L. M., 1968. Use of the electrical resistivity method for investigating geologic and hydrologic conditions in Santa Clara County, California. *Ground Water* **6**(5): 31–40.

Patten, E. P., Jr., and G. O. Bennett, 1963. Application of electrical and radioactive well logging to ground-water hydrology. *U.S. Geol. Survey Water Supply Paper 1544-D,* 60 pp.

Peterson, F. L., 1972. Water development on tropic volcanic islands-type example: Hawaii. *Ground Water* **10**(5): 18–23.

Peterson, F. L., and C. Lao, 1970. Electric well logging of Hawaiian basaltic aquifers. *Ground Water* **8**(2): 11–18.

Ritchie, E. A., 1976. Cathodic protection wells and ground-water pollution. *Ground Water* **14**(3): 146–149.

Saines, M., 1968. Map interpretation and classification of buried valleys. *Ground Water* **6**(4): 32–37.

Schleicher, G., 1975. The well driller's blues. *Irrig. Age* **9**(8): 38–43.

Siddiqui, S. H., and R. R. Parizek, 1971. Hydrogeologic factors influencing well yields in folded and faulted carbonate rocks in Central Pennsylvania. *Water Resourc. Res.* **7**: 1295–1312.

Smith, E. M., 1974. Exploration for a buried valley by resistivity and thermal probe surveys. *Ground Water* **12**: 78–83.

Smith, H. F., 1954. Gravel packing water wells. *Ill. State Water Surv. Circular No. 44.*

Spicer, H. C., 1952. Electrical resistivity studies of subsurface conditions near Antigo, Wisconsin. *U.S. Geol. Survey Circular No. 181*, 19 pp.

Spiridonoff, S. V., 1964. Design and use of radial collector wells. *J. Am. Water Works Assoc.* **56**: 689–698.

Stratton, E. F., and R. D. Ford, 1951. Electric logging. In *Subsurface Geologic Methods*, 2d ed., L. W. LeRoy (ed.). Colorado School of Mines, Golden, pp. 364–392.

Summers, W. K., 1972. Horizontal wells and drains. *Water Well J.* **26**(6): 36–38.

Swartz, J. H., 1937. Resistivity studies of some salt-water boundaries in the Hawaiian Islands. *Trans. Am. Geophys. Union* **18**: 387–393.

Todd, D. K., 1959. *Ground Water Hydrology*. John Wiley & Sons, Inc., New York, 336 pp.

U.S. Environmental Protection Agency, 1976. Manual of water well construction practices. *Environ. Prot. Agency, Office of Water Supply*, Washington, D.C., 156 pp.

Volker, A., and J. Dijkstra, 1955. Détermination des salinités des eaux dans le sous-sol du Zuiderzee par prospection géophysique. *Geophys. Prospect.* **3**: 111–125.

Wallace, D. E., 1970. Some limitations of seismic refraction methods in geohydrological surveys of deep alluvial basins. *Ground Water* **8**(6): 8–13.

Walton, W. C., 1962. Selected analytical methods for well and aquifer evaluation. *Ill. State Water Surv. Bull. No. 49.*

Walton, W. C., and S. Csallany, 1962. Yields of deep sandstone wells in northern Illinois. *Ill. State Water Surv., Report of Investigations No. 43:* 43–47.

Welchert, W. T., and B. N. Freeman, 1973. Horizontal wells. *J. Range Manage.* **26**: 253–256.

Williams, D. E., 1970. Use of alluvial faults in the storage and retention of ground water. *Ground Water* **8**(5): 25–29.

Wolff, R. G., J. D. Bredehoeft, W. S. Keys, and E. Shuter, 1975. Stress determination by hydraulic fracturing in subsurface waste injection. *J. Am. Water Works Assoc.* **67**: 519–523.

Wulff, H. E., 1968. The qanats of Iran. *Sci. Am.* **218**(6): 94–105.

Wyllie, M. R. J., 1949. A quantitative analysis of the electrochemical component of the S.P. curve. *Trans. Am. Inst. Min. Metall. Eng.* **186**: 17–26.

Wyllie, M. R. J., 1960. Log interpretations in sandstone reservoirs. *Geophys.* **25**: 748–778.

Wyllie, M. R. J., 1963. *The Fundamentals of Well Log Interpretation*, Academic Press, New York, 238 pp.

Zehner, H. H., 1973. Seismic refraction investigations in parts of the Ohio River Valley in Kentucky. *Ground Water* **11**(2): 28–37.

Zohdy, A. A. R., 1965. Geoelectrical and seismic refraction investigations near San Jose, California. *Ground Water* **3**(3): 41–48.

Zohdy, A. A. R., G. P. Eaton, and D. R. Mabey, 1974. Application of surface geophysics to ground-water investigations. Chap. D1 in *Techniques of Water-Resources Investigations of the United States Geological Survey*. U.S. Government Printing Office No. 2401-02543, Washington, D.C., 116 pp.

SEVEN

FLOW-SYSTEM ANALYSIS, MODELS, AND UNSATURATED FLOW

Underground flow systems often are more complex than the simple cases of one-dimensional or axisymmetric flow described in Chapters 3 and 4. For example, it may be desirable to predict how groundwater levels are affected by pumping several wells in an aquifer, how drawdown around a well is affected by the proximity of solid boundaries (mountain ranges) or recharging boundaries (streams that are losing water by seepage into the ground), how seepage from losing streams or recharge basins is affected by water-table depth and other subsurface conditions, how groundwater levels will respond to pumping from wells and recharge from streams in entire aquifer systems, how saltwater intrusion can be controlled by groundwater recharge, how flow systems are affected by nonuniformity or anisotropy of the medium, and how flow in vadose zones contributes to underground water movement and to surface-subsurface water relations. Most flow systems are treated as two-dimensional systems, either in a vertical plane (ground water recharge from a losing stream or infiltration basin, seepage below a dam, saltwater intrusion) or in a horizontal plane (flow in aquifers). There are also axisymmetric systems, such as flow to single wells and flow systems in connection with pumping tests, slug tests, piezometer cavities, double-tube installations, or other cylindrical techniques for in-place measurement of T or K. Some flow systems are three-dimensional in nature and should be analyzed as such (multilayered aquifer systems, for example).

The more simple systems often can be analyzed mathematically. Flow in more complex systems, particularly those with irregular boundaries, complex boundary conditions, different soil layers, or other heterogeneities, and those where several flow systems occur simultaneously (pumped wells and recharge, for example), are best solved via models (physical, analog, or numerical models).

7.1 MATHEMATICAL TECHNIQUES

7.1.1 Analytic Solutions

Analytic solutions involve solving the basic differential equation for flow of water through porous media. For two-dimensional, horizontal flow in an aquifer, for example, this equation is derived by expressing inflow and outflow with Darcy's equation for an infinitesimal square element, and expressing storage or release of water in the element as the difference between inflow and outflow (Figure 7.1). Using total head H rather than pressure head h (to readily adapt the resulting equation to systems with vertical flow) and ignoring the minus sign in Darcy's equation for positive flow in the direction of decreasing H, the inflow $q_{x,i}$ for the element in the x direction can be described as

$$q_{x,i} = T_x \left(\frac{\partial H}{\partial x} \right)_i d \qquad (7.1)$$

and the outflow $q_{x,o}$ as

$$q_{x,o} = T_x \left(\frac{\partial H}{\partial x} \right)_o d \qquad (7.2)$$

where $(\partial H/\partial x)_i$ and $(\partial H/\partial x)_o$ represent the hydraulic gradient at the inflow and outflow faces of the element, respectively, T_x is the transmissivity in the x direction, and d is the length of the face (Figure 7.1). Similar expressions can be written

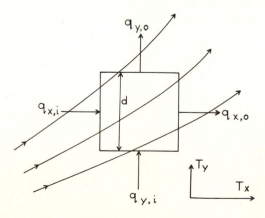

Figure 7.1 Inflow and outflow for infinitesimal square element in system of two-dimensional, horizontal flow in anisotropic aquifer.

for the inflow and outflow in the y direction. The volume rate of water storage or release for the infinitesimal element is calculated as $d^2 S\, \partial H/\partial t$, which is equal to inflow minus outflow, or $q_{x,\,i} + q_{y,\,i} - q_{x,\,o} - q_{y,\,o}$. Thus, the following equation can be written

$$\frac{q_{x,\,i} - q_{x,\,o}}{d^2} + \frac{q_{y,\,i} - q_{y,\,o}}{d^2} = S\frac{\partial H}{\partial t} \tag{7.3}$$

where S is the storage coefficient or specific yield. Substituting Eq. (7.1) and (7.2) for $q_{x,\,i}$ and $q_{x,\,o}$ in the first term of this equation yields

$$T_x \frac{(\partial H/\partial x)_i - (\partial H/\partial x)_o}{d} \tag{7.4}$$

For an infinitesimal value of d, the ratio $[(\partial H/\partial x)_i - (\partial H/\partial x)_o]/d$ is the second derivative of H with respect to x, or $\partial^2 H/\partial x^2$. Similarly, the second term in Eq. (7.3) can be written as $T_x\, \partial^2 H/\partial y^2$, so that Eq. (7.3) becomes

$$T_x \frac{\partial^2 H}{\partial x^2} + T_y \frac{\partial^2 H}{\partial y^2} = S\frac{\partial H}{\partial t} \tag{7.5}$$

This is the basic equation for transient, horizontal two-dimensional flow. Previously, the equivalent of Eq. (7.5) was derived for axisymmetric flow in radial coordinates [Eq. (4.21)]. For steady flow in an isotropic aquifer, $\partial H/\partial t = 0$ and $T_x = T_y$, so that Eq. (7.5) reduces to

$$\frac{\partial^2 H}{\partial x^2} + \frac{\partial^2 H}{\partial y^2} = 0 \tag{7.6}$$

which is the well-known Laplace equation for steady, two-dimensional, potential flow in horizontal or vertical planes. The differential equation for steady, three-dimensional flow in an anisotropic medium can be derived similarly (by writing inflow and outflow for an infinitesimal cube) as

$$K_x \frac{\partial^2 H}{\partial x^2} + K_y \frac{\partial^2 H}{\partial y^2} + K_z \frac{\partial^2 H}{\partial z^2} = 0 \tag{7.7}$$

Analytic solution of underground flow systems is based on finding a function of H in terms of x, y, and z—and t if the flow is transient—that satisfies both the known values of H at boundaries or other points in the flow system and the applicable differential equation [Eq. (7.5), (7.6), or (7.7), for example]. For one-dimensional steady flow, the Laplace equation reduces to $\partial^2 H/\partial x^2 = 0$, which integrates to $\partial H/dx = \text{constant}$—for example, C—and $H = Cx$ plus another constant. This yields a linear variation of H with x, consistent with Darcy's equation.

Once the function of H is found, values of H at different points in the flow system (and at different values of t if the flow is transient) can be calculated, and lines of equal potential, called *equipotentials*, can be determined (see, for example, Figures 7.4, 7.5, and 7.9). In isotropic systems, the direction of flow is normal to the equipotential, so that streamlines showing the flow of water can be drawn as orthogonals to the equipotentials. In steady-state flow systems, streamlines indicate the travel paths of the water molecules, as may be depicted by tracers. In transient systems, a streamline is a curve connecting a series of water particles at a given time so that the velocity vector at each point is tangent to that curve. Since the direction of flow tends to change with time in transient systems, streamlines differ from pathlines, in contrast to steady-state systems where streamlines and pathlines are identical.

Just as equipotentials are characterized by a certain H value, streamlines are numerically indicated by a streamfunction ψ. Because streamlines and equipotentials cross at right angles in isotropic media, ψ and H are related as

$$\frac{\partial H}{\partial x} = \frac{\partial \psi}{\partial y} \tag{7.8}$$

and

$$\frac{\partial H}{\partial y} = -\frac{\partial \psi}{\partial x} \tag{7.9}$$

The orthogonality between H and ψ also means that ψ can be substituted for H in the Laplace equation, which may be advantageous for solving flow systems where the boundaries are mostly streamlines, such as solid boundaries (impermeable material) and symmetry lines. Examples of equipotential boundaries are the wetted perimeter of surface basins or streams (for horizontal models, the slope of the stream may have to be taken into account), the portion of a well below the water level in the well, and the height of the water table or piezometric surface at a great distance from the flow system under consideration. Once the complete network of equipotentials and streamlines has been determined, flow rates can be calculated (see Figure 7.9 for an example of a graphical solution).

Water tables and surfaces of seepage are characterized as the locus of points where $h = 0$ and, hence, H is equal to the elevation head. For groundwater flow systems, the water table (or, rather, the top of the capillary fringe—see Section 7.7) functions as a solid boundary. Evaporation from a water table and storage of water in pore space due to a rising water table are taken into account as upward fluxes across the water table. Infiltration flow down to the groundwater and release of pore water due to a falling water table are represented by downward fluxes across the water table.

The solution of two-dimensional flow systems with the Laplace equation often utilizes conformal mapping procedures, since finding the solution of $H(x, y)$ that satisfies both the known boundary values of H and the Laplace equation is mathematically similar to finding a complex function $W(z)$ where W is the complex variable $H + i\psi$. This technique is applicable to systems that can be represented with simple geometries on an imaginary z plane through conformal mapping, including inverse mapping, Schwarz-Christoffel mapping, and the hodograph technique. Inverse mapping is useful for representing infinite boundaries, while Schwarz-Christoffel mapping reduces complex shapes to simple geometries. The hodograph technique allows solution of steady-state flow systems involving free surfaces, such as water tables or interfaces between two different fluids (fresh and salt water, for example). For additional details and examples of solutions obtained with these techniques, reference is made to DeWiest (1969), Hantush (1964), Harr (1962), Muskat (1946), Polubarinova-Kochina (1962), and the articles cited therein.

7.1.2 Superposition

The basic components of groundwater movement are flow due to sinks (pumped wells, gaining streams, evaporation), sources (natural or artificial recharge, losing streams), and uniform flow (linear, parallel flow). When several such sources, sinks, and/or uniform flows operate simultaneously in a system, the effects of the individual components on the total head H at any point in the system are additive. This principle, known as *superposition*, follows from the linearity of the Laplace equation. Thus, if water is pumped from several wells in one aquifer, the drawdown at each well or any other point in the aquifer is calculated as the sum of the drawdowns at that point caused by each individual well. If groundwater is pumped from an aquifer with a sloping piezometric surface, then drawdowns, calculated in the usual manner, are subtracted from the sloping piezometric surface to obtain the true piezometric surface during pumping. This is illustrated in Figure 7.2, where the dashed lines represent a horizontal, static piezometric surface and the corresponding piezometric surface during pumping. The solid lines represent the original sloping piezometric surface and the corresponding piezometric surface when the well is pumped. At some point downgradient from the pumped well, the piezometric surface is horizontal. This point (point A in Figure 7.2) is called the *stagnation point*, and it separates flow toward the well from continued flow downgradient in the aquifer.

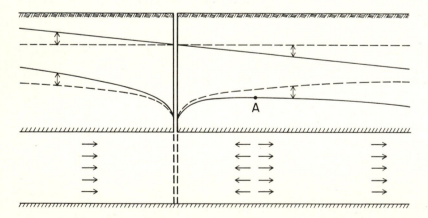

Figure 7.2 Application of superposition principle to determine piezometric surface around pumped well in confined aquifer with sloping piezometric surface (solid lines) from the normal drawdown in a horizontal system (dashed lines).

Superposition is technically correct only for confined aquifers, where T is not affected by the height of the piezometric surface. For unconfined aquifers, T varies directly with H, which introduces an error when superimposing several individual flow systems because T in the composite system will then differ from T in the individual systems. As long as the water-table drawdowns are small compared to the height of the aquifer, however, the error will be small.

7.1.3 Images

A special application of superposition is the method of images which, for example, enables calculation of the piezometric surface around a pumped well near a vertical, solid boundary in the aquifer such as a mountain range or a fault, or near a recharging boundary such as a stream that is in direct hydraulic contact with the aquifer. A vertical, solid boundary acts as a streamline in a horizontal, two-dimensional system. Such a streamline can also be created as a symmetry line, obtained by deleting the solid boundary and placing an imaginary well with the same discharge as the pumped well symmetrically across the location of the boundary, assuming infinite lateral extent of the aquifer. The drawdown around the real well can then be calculated by superposition as the sum of the individual drawdowns due to the real well and the imaginary well (Figure 7.3, top). The resulting flow system (equipotentials and streamlines) in the horizontal plane is shown in Figure 7.4, with the solid boundary as symmetry line between the real and imaginary well (the imaginary system is indicated by dashed lines).

A recharging boundary such as a losing stream is an equipotential. If a pumped well is located near such a boundary, the equipotential can also be created by deleting the stream and assuming an imaginary recharge well with the same flow rate symmetrically across the boundary (Figure 7.3, bottom). The aquifer is again assumed to be of infinite lateral extent, and the drawdowns are calculated as if each well occurred singly in that aquifer (the "drawdowns" due to

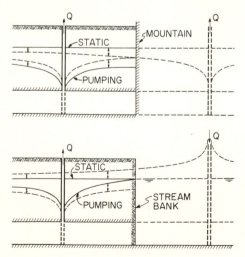

Figure 7.3 Imaginary discharging well to determine drawdown around pumped well near solid, vertical boundary (top), and imaginary recharging well to determine drawdown around well near recharging boundary (bottom).

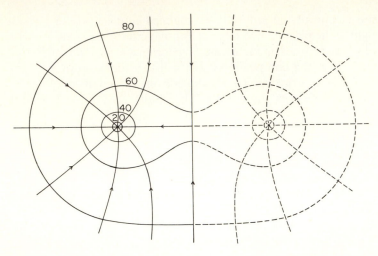

Figure 7.4 Equipotentials and streamlines around well near solid boundary (the numbers on the equipotentials refer to the potential as percentage of the "undisturbed" total head at great distance).

a recharging well are calculated in the same manner as for a discharging well, but they are negative drawdowns or water-level rises; see Section 8.3.3). Adding the individual drawdowns at each point then yields the true drawdowns in the actual flow system between stream and well. An example of such a system is shown in Figure 7.5, with the imaginary part indicated by dashed lines. For additional discussion and examples, reference is made to Ferris et al. (1962).

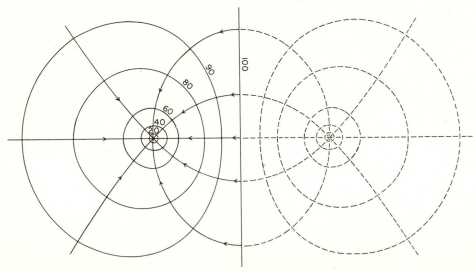

Figure 7.5 Equipotentials and streamlines around well near recharging boundary (the numbers on the equipotentials refer to the potential as percentage of the total-head difference between the recharging boundary and the well).

7.1.4 Numerical Solution

Analytic solution of the basic differential-flow equation (Laplace equation) becomes unmanageable for heterogeneous media and/or complex or irregular boundary conditions. In that case, the differential equation can be solved numerically to find the distribution of H or ψ in the flow system. This is done by placing a network of square meshes over the flow system (Figure 7.6) and calculating H or ψ at each network node by trial and error with the Laplace equation in finite difference form. For this purpose, the gradient $(\partial H/\partial x)_{1-0}$ between nodes 1 and 0, for example, is approximated as $(H_1 - H_0)/d$, where H is the total head at the node and d is the distance between nodes (Figure 7.6). Similarly, $(\partial H/\partial x)_{0-3}$ between nodes 0 and 3 is expressed as $(H_0 - H_3)/d$. The finite-difference expression for $\partial^2 H/\partial x^2$ at node 0 then is $[(\partial H/\partial x)_{1-0} - (\partial H/\partial x)_{0-3}]/d$, or $(H_1 - H_0 - H_0 + H_3)/d^2$. Expressing $\partial^2 H/\partial y^2$ in a similar manner, substituting the results for the second derivatives in Eq. (7.6) and solving for H_0 yields

$$H_0 = (H_1 + H_2 + H_3 + H_4)/4 \tag{7.10}$$

The finite-difference solution of the Laplace equation thus states that H at each node must be equal to the average of H at the four surrounding nodes (provided, of course, that the medium is uniform and the meshes are square). H values at the network nodes are then calculated by trial and error, assigning the proper H values at equipotential boundaries of the flow system and assumed values to the rest of the nodes. Successive corrections are then applied to the assumed values until Eq. (7.10) is satisfied for each node. The differences between H_0 and $(H_1 + H_2 + H_3 + H_4)/4$ that will initially occur are called *residuals*. Special calculation techniques have been developed to most efficiently "relax" the system from its residuals, hence the name *relaxation method* (Southwell, 1946). Once the proper H values have been determined, equipotentials can be constructed and streamlines can be drawn as orthogonals to complete the flow net and calculate flow rates. If the boundaries of the flow system consist primarily of streamlines, time may be saved by solving the flow system in terms of ψ [substituting ψ for H in Eq. (7.10)] and drawing the equipotentials as orthogonals to the streamlines.

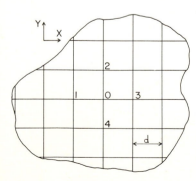

Figure 7.6 Square network for finite-difference solution of Laplace equation.

Manual execution of the relaxation method is tedious and may take hours or days. Time can be saved by starting with a coarse network and using the resulting H values to obtain a better set of assumed H values in a network with one-half the node spacing of the coarse grid. When the proper H values have been determined for the finer network, the H values for the coarse and fine network can then be used to directly calculate a third set of H values that is almost as accurate as would be obtained with a network that is twice as fine as the finer network of the two. This is illustrated for a system of wells equally spaced on a line between a stream as the recharging boundary and a parallel mountain range as the solid boundary (Figure 7.7). The lines through the wells and midway between the wells at right angles to the boundaries are symmetry lines. Thus the flow system will be a repetition of the system in the rectangle $ABCD$. In this system, AB is a source equipotential, the well is a sink equipotential (assuming that the wells are pumped to maintain a constant water level), and BC, CD, and AD are streamlines. Assigning arbitrary values of 100 to the source equipotential AB and 0 to the sink equipotential at the node representing the well, H values within the system were evaluated for a 3×5 grid (Figure 7.8, top) and a 6×10 grid (Figure 7.8, bottom). The H values within the system show the potential at each node as a percentage of the difference between the elevation of the water level in the stream and that at the well. A third solution was obtained with a 12×20 grid. Calling the H values obtained with the 3×5 grid H', those with the 6×10 grid H'', and those with the 12×20 grid H'''', it appeared that H'''' could be estimated from H' and H'' with an average error of less than 2 percent with the equation

$$H'''' = H'' + 2(H'' - H')/3 \qquad (7.11)$$

This equation gave better results than a similar type of equation listed by Liebmann (1950).

Equipotentials at 10 percent increments in the system $ABCD$, obtained by interpolation between the H values in the 12×20 grid, are shown in Figure 7.9. Streamlines can be sketched as orthogonals to the equipotentials, or they can be determined by repeating the relaxation procedure for ψ, assigning an arbitrary

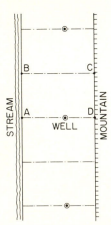

Figure 7.7 Geometry of wells in valley between stream and mountain range.

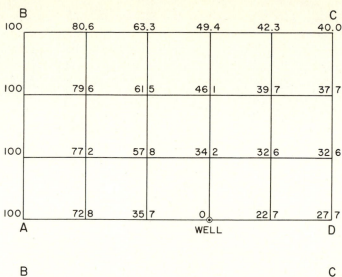

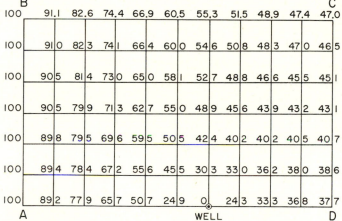

Figure 7.8 Solution of H values in percentage of total potential difference between stream and well for coarse network (top) and fine network (bottom).

value of 100 to the boundary streamline $BCDW$, and a value of 0 to the other boundary streamline from A to the well (Figure 7.9). Using the latter procedure, interpolation between the resulting ψ values at the nodes by 10 percent increments then yielded the streamlines shown in Figure 7.9.

The region between two adjacent streamlines is called a *streamtube*. Each streamtube in Figure 7.9 thus carries 10 percent of the total flow in the area $ABCD$. The flow q in each streamtube can be calculated by applying Darcy's equation to a segment of the tube between two adjacent equipotentials, yielding

$$q = TW \, \Delta H/L \qquad (7.12)$$

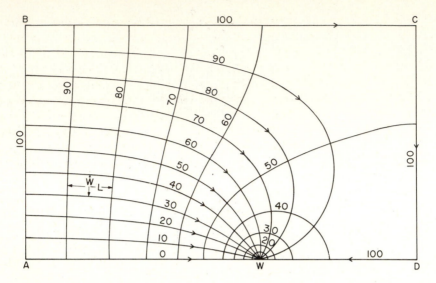

Figure 7.9 Equipotentials and streamlines for flow system in the rectangle *ABCD* of Figure 7.7.

where W and L are the width and length of the segment (Figure 7.9), respectively, ΔH is the total-head difference or equipotential drop across the segment, and T is the transmissivity of the aquifer. Since the streamtubes in Figure 7.9 carry the same flow, the total flow in *ABCD* is $10q$. Because of symmetry, the flow to the well then is $20q$. If the streamlines are sketched as orthogonals to the equipotentials without knowing the ψ values, it is difficult to obtain exactly even distribution of the streamfunction and resulting equal flow in each tube. In that case, Eq. (7.12) should be applied to each streamtube to calculate the flow.

The accuracy of relaxation solutions depends on the density of the network, which typically consists of about 50 to 200 nodes. The technique, however, is quite powerful and can also be applied to heterogeneous media and systems with irregular boundaries or axisymmetric systems. Subdivided networks are used to produce greater node density in areas where the flow is concentrated. Procedures for including heterogeneity and different mesh sizes (including rectangular meshes) are presented in Section 7.3 for numerical solution by resistance-network analogs. Such analogs yield essentially instantaneous relaxation when energized, allowing H values at the nodes to be measured as electrical potentials with a voltmeter. Automatic relaxation can also be obtained through iteration by digital computers. In view of the tremendous time savings obtained with these devices, resistance network analogs and digital computers have superseded the manual relaxation process for obtaining finite-difference solutions of underground flow systems.

Examples of applications of the relaxation method are the solution of problems of flow to underground drains in combination with deep seepage by van Deemter (1950), and the study of surfaces of seepage in wells in unconfined aquifers by Hall (1955).

7.2 SAND TANKS AND HELE-SHAW MODELS

Sand tanks are physical scale models of underground-flow systems, letting water or another liquid flow through sand or some other porous medium. Such "wet" models are suitable for solving flow systems in uniform or layered media with complex boundaries and free surfaces like water tables or interfaces between different liquids. Physical models also are valuable tools for testing the validity of hypotheses or of simplifying assumptions made in mathematical analysis of flow systems. For three-dimensional systems (for example, a radial collector well with unevenly spaced collectors of different lengths), the model may consist of a sand box about 1 or 2 m square. For axisymmetric systems, like flow to a partially penetrating well in an unconfined aquifer, the model may be wedge-shaped to represent a sector of the system (Figure 7.10, top). For two-dimensional flow systems in the vertical plane (seepage from a stream or flow to agricultural drains, for example), the model may be a narrow (about 5 to 20 cm wide) rectangular box to represent a vertical section of the system (Figure 7.10, bottom). Two-dimensional, horizontal flow systems may be modeled in a large box with about 5 to 10 cm of sand on the bottom.

The flow rate in the model can be expressed with Darcy's equation as

$$Q_m = K_m A_m \, \Delta H_m / L_m \qquad (7.13)$$

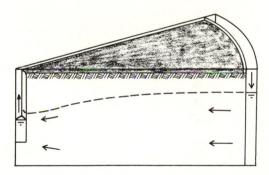

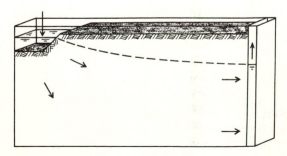

Figure 7.10 Wedge-shaped model for axisymmetric systems (top) and rectangular box for vertical, two-dimensional flow systems (bottom).

where Q_m = flow in model (length^3/time)

K_m = hydraulic conductivity of medium in model

A_m = effective cross-sectional area of flow system

L_m = effective flow length in flow system

ΔH_m = total-head loss in model flow system

The flow Q_p in the real system or prototype can be expressed similarly, changing the subscripts m in Eq. (7.13) to p. Dividing the resulting equation by Eq. (7.13) and solving for Q_p yields

$$Q_p = Q_m \frac{K_p \, A_p \, \Delta H_p \, L_m}{K_m \, A_m \, \Delta H_m \, L_p} \tag{7.14}$$

If the model is a true scale model of the prototype, all dimensions (including the total head H) are reduced by a scale factor S_L defined as L_p/L_m. Substituting this factor for the length ratios and S_L^2 for the area ratio in Eq. (7.14) gives the following equation for calculating the flow in the prototype from the observed flow in the model:

$$Q_p = Q_m \frac{K_p}{K_m} S_L^2 \tag{7.15}$$

If the model represents a layered or otherwise heterogeneous medium, K_p/K_m must be the same for each layer or each point in the system. Anisotropic media can be represented by alternating thin layers of two different materials and calculating K parallel and normal to the layers with Eqs. (3.22) and (3.28) (Bouwer, 1964a), or by changing the vertical scale of the model to transform the medium into an equivalent isotropic medium (see Section 7.6).

If transient-flow systems are modeled, it may be desirable to calculate the prototype time t_p for the water table to rise or fall a certain distance from the observed time t_m in the model. The rate of rise of a water table is calculated as v/f, where v is the vertical upward flow component at the water table and f is the fillable porosity (see Section 2.10). The time it takes for the water table to rise a certain distance L is then calculated as $t = L/(v/f)$. Expressing t_p and t_m in this manner, dividing one expression by the other and solving for t_p then yields

$$t_p = t_m \frac{L_p \, f_p \, v_m}{L_m \, f_m \, v_p} \tag{7.16}$$

However, $L_p/L_m = S_L$ and $v_m/v_p = (K_m \, \Delta H_m/L_m)/(K_p \, \Delta H_p/L_p)$, which reduces to K_m/K_p for a true scale model where $\Delta H/L$ is the same as in the prototype. Thus, Eq. (7.16) can be written as

$$t_p = t_m S_L \frac{f_p}{f_m} \frac{K_m}{K_p} \tag{7.17}$$

This equation also applies to a falling water table. Because of hysteresis, f should be taken as the drainable porosity (specific yield) when the water table is falling, and as the fillable porosity when the water table is rising (see Section 2.10).

In sand models, the height of the capillary fringe may be out of proportion to the height of the flow system below the water table if a free water table is modeled. This yields an unrealistically high contribution of the flow above the water table to the total flow in the model [see Eq. (7.58)]. The height of the capillary fringe in the model can be reduced by using coarse sand or gravel, but this could cause turbulent flow which invalidates Darcy's equation. Turbulent flow, however, can be avoided by using a more viscous liquid (oil or glycerine) in the model.

When a viscous liquid is used in a two-dimensional model (Figure 7.10, bottom), laminar flow can also be created by moving the large walls of the model closer together so that they are about 1 to 10 mm apart, and leaving out the sand or gravel. The resulting model, called the *Hele-Shaw*, *viscous flow*, or *parallel-plate* model, is among the most frequently used "wet" models in studies of groundwater flow. Hele-Shaw models can be vertical similar to Figure 7.10 (bottom), or they can be horizontal to simulate two-dimensional aquifer flow (Figure 7.11). If the aquifer has a certain slope, the Hele-Shaw model can be tilted accordingly. K_m of Hele-Shaw models can be experimentally determined, by, for example, creating horizontal or vertical flow in the model and calculating K_m with Darcy's equation from the measured flow rate, hydraulic gradient, and cross-sectional area between the plates. K_m can also be calculated from laminar-flow theory (see, for example, Fox and McDonald, 1973) as

$$K_m = \frac{d^2 \rho g}{12 \mu} \tag{7.18}$$

where K_m = hydraulic conductivity in cm/s (multiply by 864 to get m/day)
d = distance between parallel plates (cm)
ρ = density of liquid (g/cm^3)
g = acceleration due to gravity (cm/s^2)
μ = absolute viscosity [g/(cm·s)]

For horizontal models, K_m represents T of the aquifer.

If the plate distance d represents the highest K or T values in the system, layers or zones of lower K or T can be simulated by locally reducing the distance between the plates (for example, by cementing plastic strips to the inside of one or both walls). Since K_m is proportional to d^2 [Eq. (7.18)], reducing d by 50 percent

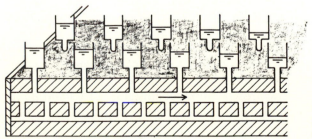

Figure 7.11 Hele-Shaw model with storage for simulating two aquifers separated by an aquitard.

yields a fourfold reduction in K_m, etc. Layers of much lower K, like aquitards, can be simulated by a horizontal strip that completely fills the space between the parallel plates but that has vertical slots or perforations to let liquid through. The size and spacing of the perforations then are selected on the basis of the vertical K of the layer. Anisotropy can be simulated with narrow, parallel, and closely spaced strips cemented to the inside of the wall to locally reduce d. This creates a directional K_m that is greater in the direction of the strips than normal thereto.

Liquid is added to or withdrawn from the model (to simulate sources and sinks, respectively) through fine tubes that extend between the plates. Uniform flow rates like infiltration of rainfall are simulated by an array of tubes. Dyes can be injected to observe streamlines or pathlines.

For vertical Hele-Shaw models, f_m is equal to 1. For horizontal models, storage is obtained by connecting vertical reservoirs to the fluid between the plates (Figure 7.11). The value of f_m is then calculated as the ratio between the surface area of the reservoir and the model area represented by the reservoir. Spatial variations of f_m can be obtained by varying the spacing and/or the diameter of the reservoirs. The model in Figure 7.11 represents two aquifers (the lower aquifer has a lower T) separated by an aquitard.

Physical models of the sand-tank type have been used by Bouwer (1964a), Cahill (1973), Hall (1955), Hansen (1953), Harpaz and Bear (1964), Kimbler (1970), Kraijenhoff van de Leur (1962), Mobasheri and Shahbazi (1969), Peter (1970), Rahman et al. (1969), Rumer and Harleman (1963), and Smith (1967). Studies with Hele-Shaw models were reported by Bear (1960), Bear and Dagan (1964), Çeçen et al. (1969), Collins et al. (1972), Columbus (1966), Dvoracek and Scott (1963), Ibrahim and Brutsaert (1965), Marino (1967), Santing (1958), and Todd (1954). For additional discussions and references, see Guitjens (1974) and Prickett (1975).

7.3 ELECTRICAL AND OTHER ANALOGS

Analogs are devices with similar input-output or cause-and-effect relations as the prototype, but with different physical properties. The most common analogs in the study of subsurface water movement are electrical analogs where current flow through resistors or other conductive media simulates water flow through underground material, and electric potentials correspond to total heads. The analogy is based on the similarity between Ohm's law and Darcy's law. Other analogs are heat-flow analogs, based on the similarity between Fourier's law and Darcy's law, and membrane analogs. The latter utilize stretched membranes and enable determination of water-table shapes by measuring deflection patterns in the membrane in response to local depressions or elevations of the membrane to represent sinks or sources, respectively, in the prototype. Electrical analogs can be divided into analogs with continuous media (electrolytic tanks, conducting sheets) and those with noncontinuous media (resistance network analogs), as discussed in the following sections.

7.3.1 Electrolytic-Tank Analogs

The first electrical analogs utilized an electrolyte (for example, a solution of potassium chloride in water) in a tank as the conducting medium. Variations in T or K in the prototype are simulated by varying the depth of the electrolyte (horizontal systems) or the distance between walls (vertical systems). Layered horizontal systems can be represented by carefully placing electrolytes of different densities in the tank. To avoid polarization, alternating current of about 1000 Hz is used. Solid boundaries are represented as dielectric boundaries (plastic, glass) in the analog. Equipotential boundaries are simulated with strips of brass or other conductors. Voltages are measured with a Wheatstone bridge arrangement (to trace an equipotential, for example) or with a voltmeter of high internal resistance (vacuum-tube or electronic type). Currents are measured with low-internal-resistance ammeters. The conversion from electric units to hydraulic units is the same as discussed for resistance network analogs in the next section.

An interesting feature of tank analogs is the opportunity for almost exactly representing systems of infinite extent or "open" boundaries, utilizing the principle of inversion (Boothroyd et al., 1949). This requires two analogs, one for simulating part of the real system (indicated as I in Figure 7.12) and the other (indicated as II) for simulating the inverse of the rest of the system to infinity. A point outside the real analog—for example, point D in Figure 7.12—is then represented as point D' on the same radial direction in the inverse analog so that $rr' = R^2$ (see Figure 7.12 for meaning of symbols). If $r = \infty, r' = 0$, so that infinity is represented by the center of the inverse analog. Points on the circumference $(r = R)$ of the real analog are represented by corresponding points on the circumference $(r' = R)$ of the inverse analog (points A, B, and C in Figure 7.12). Cartesian coordinates in the real analog become circular in the inverse analog (see Figure 7.12, bottom, which also shows how square elements outside the real analog are represented in the inverse analog). An impermeable layer in the real system must be continued as a circular boundary in the inverse system (for example, an impermeable layer on the horizontal line labeled 4 in the real system must be continued as a solid boundary on the curved line 4 in the inverse system).

The real and inverse analogs must have continuous electrical contact along their boundaries at corresponding points $(A, B, C, \text{etc.})$. For tank analogs, this can readily be achieved by placing a dielectric sheet at half-depth in a cylindrical tank (Figure 7.13). The diameter of the sheet is slightly less than that of the tank, so that the upper and lower electrolyte compartments are electrically connected at their circumferences. The upper level will then be the real system in which the flow system is modeled as such, and the lower level will be the inverse system. The circumference of the real system is then truly an open boundary, accurately representing infinite extent of the analog.

Another type of continuous analog is the conducting-sheet analog. A commercially available conductive paper—Teledeltos paper, for example—has been used to solve various two-dimensional flow systems. Equipotential boundaries are simulated by painting thin lines of a conductive paint (silver paint) or by clamping

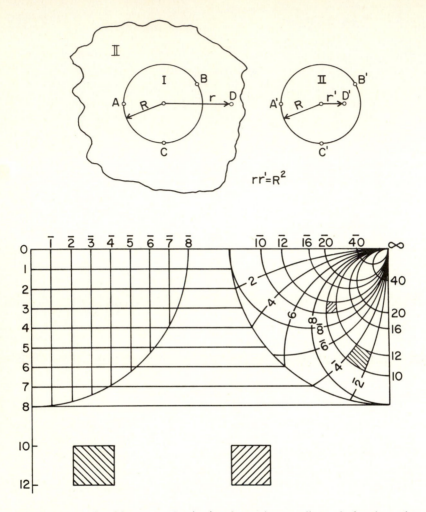

Figure 7.12 Real and inverse analog (top) and cartesian coordinates in fourth quadrant of real and inverse analog (bottom). The hatched areas show how regions outside the real analog are represented in the inverse analog. (*From Bouwer, 1967.*)

or taping copper wire or other conductor to the paper. The paper is cut so that it represents a scale model of the prototype. Edges of the paper are impermeable boundaries. Solid boundaries within the flow system, like a cutoff wall below a dam, are represented by cutting a narrow slot in the paper. Layers or zones of reduced K are simulated by punching small holes in the paper.

Electrolytic-tank and conducting-sheet or paper analogs are relatively easy to

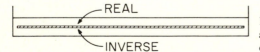

Figure 7.13 Real (top) and inverse (bottom) analog obtained by placing dielectric sheet in cylindrical electrolytic tank.

construct, but their application is limited to relatively simple steady-state flow problems or transient problems that can be handled as a succession of steady states. Before solving a certain flow system, the operator of the analog should first experiment with some simple flow systems (uniform flow or flow to a well, for example) that are governed by simple equations. This will serve as a check on the accuracy of the various analog components and give the operator confidence in the technique. Examples of application of the techniques are the studies by DeBrine (1970), de Josselin de Jong (1962), Mack (1957), Matlock (1966), Reed and Bedinger (1961), Sherwood and Klein (1963), Snell and van Schilfgaarde (1964), Todd and Bear (1961), Youngs (1968), and Zee et al. (1957). For additional discussions and references, see Luthin (1974) and Prickett (1975).

7.3.2 Resistance-Network Analogs

Resistance-network or R analogs are electrical analogs in which the flow medium is represented by a network of resistors (Figures 7.14, 7.15, and 7.16). This makes it possible to readily simulate layered, heterogeneous, or anisotropic media, and to maintain the necessary input and output currents and boundary equipotentials. R analogs can represent steady, two-dimensional systems in vertical or horizontal planes and in axisymmetric systems. Three-dimensional R analogs have been used to study flow in systems of communicating aquifers (Getzen, 1977). R analogs are also relatively simple to construct and operate. Power is supplied by a stable dc source (usually in the 1- to 10-V range) and voltages are measured with a high-internal-resistance voltmeter, preferably of the digital type. Currents are measured with ammeters (Figure 7.16). R analogs are excellent research and teaching tools. They can be used to solve a variety of flow systems; demonstrate basic principles like superposition, images, and flow in anisotropic media (see Section 7.6); and serve as an introduction to the more complex resistance-capacitance networks (RC analogs) and digital-computer solutions.

The principle of R analogs is the same as that for the relaxation method (see Section 7.1), except that the current automatically and essentially instantaneously

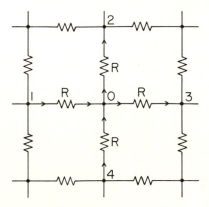

Figure 7.14 Resistors and nodes in R analog.

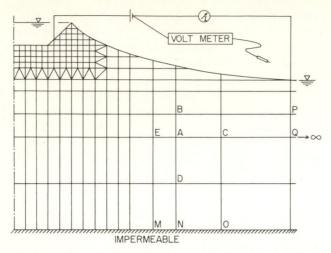

Figure 7.15 Network arrangement with subdivided and rectangular meshes for representing system of seepage from a stream.

relaxes the system from its residuals, which is a great improvement over manual relaxation! For a uniform medium and square meshes, the resistors will all be of the same resistance value [see Eq. (7.20)]. Calling this value $R\ \Omega$ and assuming flow from the lower left to the upper right in Figure 7.14, the current inflow for node 0 can be expressed with Ohm's law as $(V_1 - V_0)/R + (V_4 - V_0)/R$, where V is

Figure 7.16 R analog with variable resistors plugged into banana jacks mounted on masonite board. Also shown are dc power supply, digital voltmeter, and ammeters.

the voltage at the nodes. The outflow for node 0 is similarly expressed as $(V_0 - V_2)/R + (V_0 - V_3)/R$. Assuming steady-state and no storage or release of water, inflow equals outflow. Equating the expressions for inflow and outflow at node 0 and solving for V_0 then yields

$$V_0 = (V_1 + V_2 + V_3 + V_4)/4 \qquad (7.19)$$

which is the electrical equivalent of Eq. (7.10).

Since resistance is the inverse of conductivity, R values are calculated as $1/K$ for two-dimensional, vertical flow systems of unit width, and as $1/T$ for two-dimensional flow in aquifers or other horizontal systems. R values should be in the range of about 100 to 100 000 Ω, so that they are large compared to contact resistances at the nodes and other electrical junctions, and small compared to the internal resistance of the voltmeter. To achieve this range, $1/K$ or $1/T$ is divided by a resistance-scale factor S_R. Thus, R for square meshes is calculated as

$$R = \frac{1}{K S_R} \qquad (7.20)$$

For example, if K for a given flow system is 10 m/day and R is to be 1 000 Ω, S_R is 0.000 1. If the same system contains a restricting layer with $K = 1$ m/day, this layer is then simulated by using R values of 10 000 Ω for the network meshes representing that layer.

To make the most effective use of the available number of network nodes or resistors, it is desirable to use square, subdivided meshes where the flow is concentrated and increasingly rectangular meshes toward the outer reaches of the system (Figure 7.15). To calculate R for rectangular meshes, the horizontal flow q_{AC} through the section represented by the resistor between A and C in Figure 7.15 is expressed with Darcy's equation as

$$q_{AC} = K_{AC} \frac{AB/2 + AD/2}{AC} (H_A - H_C) \qquad (7.21)$$

The electric current I_{AC} through this resistor is given by Ohm's law as

$$I_{AC} = \frac{V_A - V_C}{R_{AC}} \qquad (7.22)$$

To make this equation equivalent to Eq. (7.21), R_{AC} must be equal to $AC/[K_{AC}(AB/2 + AD/2)]$. Incorporating also the resistance-scale factor S_R then yields the following equation for calculating R_{AC}

$$R_{AC} = \frac{1}{K_{AC} S_R} \frac{2AC}{AB + AD} \qquad (7.23)$$

Similarly, the resistance between A and B is calculated as

$$R_{AB} = \frac{1}{K_{AB} S_R} \frac{2AB}{EA + AC} \qquad (7.24)$$

Thus, an R analog model of a flow system may show different R values in the network, depending on network geometry and variations in K. For a network of uniform, rectangular meshes of height d_z and length d_x, Eqs. (7.23) and (7.24) reduce to

$$R_z = \frac{1}{K_z S_R} \frac{d_z}{d_x} \tag{7.25}$$

and

$$R_x = \frac{1}{K_x S_R} \frac{d_x}{d_z} \tag{7.26}$$

for the R values in the vertical and horizontal direction, respectively. For square meshes, $d_x = d_z$ and these equations become identical to Eq. (7.20).

Resistors between nodes on solid boundaries, like nodes N and O in Figure 7.15, are calculated as

$$R_{NO} = \frac{1}{K_{NO} S_R} \frac{2NO}{ND} \tag{7.27}$$

where K_{NO} refers to K between N and O above the solid boundary. This equation, which is derived in the same way as Eq. (7.23), shows that for networks of uniform square or rectangular meshes, resistors representing solid boundaries have twice the resistance of corresponding resistors inside the system. Resistors representing open boundaries, as between P and Q in Figure 7.15, are zero because the distance to the next node (to the right of P and Q in Figure 7.15) is infinite. Thus, open boundaries to simulate infinite extent of the medium are represented by a line of conductors of zero resistance. However, the location of such an open boundary must be selected at sufficient distance from the main flow region, so that its actual position has an insignificant effect on the total flow (Bouwer, 1967).

Sloping boundaries like the water table in Figure 7.15 or irregular boundaries are replaced by the best-fitting step function along the nodes, yielding incomplete meshes for which R can be calculated with Eqs. (7.23) and (7.24). In problems of free-surface flow in the vertical plane (Figure 7.15, for example), the shape of the water table is not known and must be found as part of the solution. For this purpose, the analysis is started with an assumed water table. Voltages at the water-table nodes are then measured, and the water-table shape is adjusted (resistances are changed) until the voltages indicate H values that are the same as the elevation heads of the water-table nodes.

When a portion of the network is subdivided to obtain a greater node density—for example, around the stream in Figure 7.15—diagonal resistors must be employed to obtain a smooth transition between the subdivided and regular network. Values of R in the transition zone, taken from Liebmann (1954), are indicated in Figure 7.17. The R values for small square meshes are, of course, the same as those for large square meshes in the system (if the medium is uniform).

If a pumped well is to be simulated in a two-dimensional R analog model of

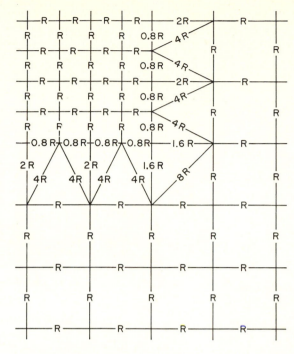

Figure 7.17 Resistance values in transition zone between subdivided and regular network squares. (*From Liebmann, 1954.*)

horizontal flow in an aquifer, current is withdrawn from the node corresponding to the well location in the prototype. Prickett (1967) has shown that a well represented by a single node in a network of square meshes has an effective diameter of $0.208d$, where d is the prototype distance between nodes. Since d may be on the order of several hundred meters or more, this yields unrealistically large well diameters. To obtain the proper r_w value for the well, an additional resistor R_w should be placed between the node representing the well and the electrical contact representing the current sink. R_w is then calculated as

$$R_w = 0.004552R \log (0.208d/r_w) \tag{7.28}$$

where R is the square-mesh resistance value around the well node. This equation was obtained from the R_w equation presented by Prickett (1967) by converting T to metric units and substituting the expressions for the scale factors therein. Thus, if a well 1 m in diameter is to be represented by a node in a square grid with $d = 100$ m and $R = 10\,000\,\Omega$, a resistance of $73.7\,\Omega$ must be placed between the well node and the electrical contact maintaining the proper voltage to simulate the desired water level in the well, or maintaining the proper current to simulate the desired discharge from the well.

Axisymmetric-flow systems for simulating flow to a single well or flow in connection with cylindrical devices for K measurement (see Chapter 5) are represented on R analogs as a sector of the medium (Figure 7.18). The R values are

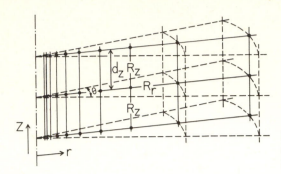

Figure 7.18 Sector of axisymmetric-flow system represented by resistance network (solid lines).

calculated with equations similar to Eqs. (7.23) and (7.24), except that the equation for the radial resistances is based on the Thiem equation (Liebmann, 1950). Since the flow tends to be concentrated toward the central axis, a graded network where the radial node spacing increases with increasing lateral distance from the central axis would allow optimum use of available node space and/or resistors. The R values for such a graded network (Figure 7.18) are then calculated as (Bouwer, 1960)

$$R_z = \frac{1}{KS_R} \frac{d_z(4 - \theta^2)^2}{16r\theta^2}$$ (7.29)

for the vertical resistors, and as

$$R_r = \frac{1}{KS_R} \frac{\ln\left[(2 + \theta)/(2 - \theta)\right]}{\theta d_z}$$ (7.30)

for the horizontal resistors. In these equations, d_z is the vertical node spacing, r is the radial distance, and θ is the angle of the sector in radians (Figure 7.18). The graded network is so designed that the radial node spacing increases linearly with r, or the width of the sector. For this reason, R_r is not affected by r, as shown by Eq. (7.30). The vertical resistances decrease with increasing r, however. The rate of increase in the radial node spacing is determined by θ, which should be selected small if the radial node spacing is not to increase too rapidly.

Known total-head differences in the prototype (for example, between the water level in the stream and the water table at great distance from the stream in Figure 7.15) are simulated by voltage differences in the analog, which are normally in the 1- to 10-V range. The total head H and voltage V are related by the head-scale factor S_H, defined as

$$S_H = \frac{\text{meters in prototype}}{\text{volts in R analog}}$$ (7.31)

Thus, if 1 V in the analog represents an H difference of 10 m in the prototype, $S_H = 10$. The resistance-scale factor S_R is defined as [see Eq. (7.20)]

$$S_R = \frac{1/K}{R}$$ (7.32)

If S_R and S_H were both equal to 1, the current I in amperes would be directly equivalent to Q in cubic meters per day in the prototype, because Darcy's law and Ohm's law are equivalent. In reality, however, the voltages are smaller than the total heads by a factor S_H and the resistances are larger than $1/K$ or $1/T$ by a factor $1/S_R$, so that I is smaller than Q by a factor S_H/S_R. Thus, Q in the actual system is calculated as

$$Q = \frac{S_H}{S_R} I \tag{7.33}$$

where I is the analog current in amperes. For example, if K is 10 m/day and is represented by resistances of $1\,000\ \Omega$ in a square grid, and 1 V in the analog corresponds to 10 m in the prototype, a measured current of 30 mA in the analog would correspond to a flow of $3\,000\ \mathrm{m^3/day}$ in the prototype. If a well in the same model is pumped at a rate of $2\,000\ \mathrm{m^3/day}$, this must be simulated by withdrawing a current of 20 mA from the node representing the well.

R analogs are most versatile when constructed with plug-in-type resistors (Figure 7.16). The supply of resistors should include fixed resistors in various sizes to represent uniform media in square and rectangular node patterns, and variable resistors for boundaries and layers or zones of different K or T. Variable resistors also make it possible to adjust R to H, as may be required for unsaturated flow (see Section 7.7) or for flow in unconfined aquifers where T varies linearly with H.

R analogs are capable of solving only steady-state flow systems. Some transient-flow problems, like the rise of a water table below recharge basins or the fall of a water table in tile-drained land, can be handled as a succession of steady states. This involves measuring electric currents entering or leaving the system across the water table, and calculating the rise or fall of the water table for a given time increment and f value. The "new" water table at the end of the time increment is then simulated on the analog, after which input or output currents are again measured, etc. To analyze transient-flow systems directly, capacitors must be used at water-table nodes to simulate storage or release of water. Such analogs are called *resistance-capacitance* or *RC analogs* (see next section).

R analogs have been used by Bedinger (1967), Bennett et al. (1968), Bouwer (1959a and b, 1962, 1965), Bouwer and Little (1959), Bouwer and Rice (1976), Davies and Herbert (1963), Herbert (1968), Herbert and Rushton (1966), Hunter Blair (1966), Luthin (1952), Stallman (1963a and b), and Vimoke et al. (1962). For additional discussion and references, see Luthin (1974) and Prickett (1975).

7.3.3 Resistance-Capacitance Analogs

RC analogs are typically used to model an entire aquifer or groundwater basin, simulating pumped wells and other depletion (drainage to streams, evaporation) and accretion (natural or artificial recharge, losing streams) to predict future groundwater levels for a number of decades. Different schemes of groundwater pumping and recharge operations can then be simulated to develop criteria for optimum management of the groundwater resource. RC analogs often are about 1.5 × 3 m in size, with nodes about 2 or 3 cm apart, and with fixed resistors to simulate T (including spatial variations in T). Each node is connected to electrical ground via a capacitor (Figure 7.19) to simulate storage.

Assuming square meshes and a uniform T, R is uniform. Applying Kirchoff's law to a certain node (node 0 in Figure 7.19) then yields

$$\frac{1}{R}(V_1 + V_2 + V_3 + V_4 - 4V_0) = C\frac{\partial V_0}{\partial t} \tag{7.34}$$

where C is the capacitance (farads) of the capacitors and V the voltage at the nodes. Finite-difference solution of Eq. (7.5) similar to the derivation of Eq. (7.10) but expressing the difference between inflow and outflow at node 0 as $d^2S\ \partial H_0/\partial t$ yields

$$T(H_1 + H_2 + H_3 + H_4 - 4H_0) = d^2S\frac{\partial H_0}{\partial t} \tag{7.35}$$

where d is the prototype distance between the nodes, S is the specific yield or storage coefficient of the aquifer, and $\partial H_0/\partial t$ is the rate of fall or rise of the water table or piezometric surface at node 0. In addition to the scale factors S_R and S_H [Eq. (7.31) and (7.32)], a time-scale factor S_t is introduced, defined as

$$S_t = \frac{\text{days in prototype}}{\text{seconds in RC analog}} \tag{7.36}$$

Dividing Eq. (7.35) by Eq. (7.34) and expressing $\partial H_0/\partial t$ and $\partial V/\partial t$ in finite-difference form yields

$$TR\frac{H_1 + H_2 + H_3 + H_4}{V_1 + V_2 + V_3 + V_4} = \frac{d^2S}{C}\frac{\Delta H_0}{\Delta V_0}\frac{\Delta t_{\text{analog}}}{\Delta t_{\text{prototype}}} \tag{7.37}$$

According to Eq. (7.20) with T substituted for K, $TR = 1/S_R$. Since $H/V = S_H$

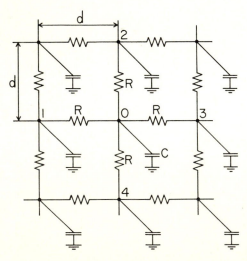

Figure 7.19 Basic arrangement of resistors and capacitors in RC analog.

[Eq. (7.31)], which is constant for a given RC analog, the H/V ratios on the left and right sides of Eq. (7.37) cancel. Also, the ratio $\Delta t_{analog}/\Delta t_{prototype}$ is equal to $1/S_t$ [Eq. (7.36)]. Thus, Eq. (7.37) can be simplified and solved for C as

$$C = d^2 S \frac{S_R}{S_t} \tag{7.38}$$

Since S_R and S_t are constant for a given analog, C is selected in proportion to S and the area represented by each node. Thus, C may vary as S and the node spacing may vary over the aquifer.

Values of S_R and C are selected so that the desired time period for the prototype, which may be 10 to 100 years, is about 0.01 s in the analog. This enables the use of pulse- and wave-form generators (which normally operate at frequencies of about 10 to 100 000 Hz) to simulate pumping schedules and other water input and output relations during the study period. The magnitude of input and output currents and voltages is determined with the same equations as presented for R analogs. The voltage response at selected nodes is observed with an oscilloscope, which produces a voltage curve that shows how the " groundwater " level at each node will vary over the study period.

Representing 10 to 100 prototype years in about 0.01 s analog time places S_t in the range of 3×10^5 to 4×10^6. To obtain an idea of the size of the capacitors used in RC analogs, assume that an aquifer of 10×50 km is to be represented with a network of 20×100 nodes. S of the aquifer is taken as 0.1 and T as 4 000 m^2/day, which is to be simulated by 2 500-Ω resistors. If the capacitors should be selected so that 50 years in the prototype are 0.01 s in the analog, $S_t = 1.825 \times 10^6$. Substituting this value along with $S_R = 1/(4\,000 \times 2\,500) = 10^{-7}$ and $d^2 = 500 \times 500 = 2.5 \times 10^5$ m^2 into Eq. (7.38) yields $C = 2.5 \times 10^{-9}$ F for the capacitors.

If RC analogs are used to model two-dimensional flow systems in the vertical plane with a moving water table (like a transient version of the system in Figure 7.15), capacitors are placed only at the top nodes representing the water table. The capacitors are calculated with Eq. (7.38), where d^2 is then equal to the node spacing because the flow system is of unit width. Such models could be used to study the rise and fall of groundwater mounds below recharge basins, effects of surface streams on groundwater gains or losses and water-table positions, movement of water tables in tile-drained land, etc. An error results from the fact that the network grid and resistors cannot be adjusted to properly represent the water table as it rises or falls. This error can be minimized by selecting the network grid on the basis of the average water-table height in the transient system.

RC analogs have been used extensively to predict groundwater-level trends in aquifers or basins. Examples of these and other applications of RC analogs are the studies of Anderson (1968, 1972), Bredehoeft et al. (1966), Brown (1962), Emery (1966), Hunter Blair (1968), Rushton and Bannister (1970), Skibitzke (1963), Walton (1964), Walton and Prickett (1963), Watkins and Heisel (1970), and White and Hardt (1965). A three-dimensional RC-analog was used by Getzen (1977). For additional discussion and references on RC analogs, see Karplus (1958), Luthin (1974), and Prickett (1975).

7.4 DIGITAL COMPUTERS

7.4.1 Finite-Difference Method

Another technique for relaxation-type solution of the basic differential-flow equation is through iteration by digital computer. A network is again placed over the flow system, and Darcy's equation is applied to develop finite-difference expressions for the flow at each node, equating the difference between inflow and outflow to $SA\ \partial H/\partial t$, where A is the area represented by the node (d^2 for square meshes). Additional sink and source terms can be included in the resulting equation to represent external inputs or outputs of groundwater such as natural or artificial recharge, seepage from losing streams, leakage to or from other aquifers, well flow, drainage to gaining streams, evaporation or water uptake by vegetation, etc. This yields a number of simultaneous equations (one for each node) with the total heads H at each node as the primary unknowns.

Of the several computer methods that are available for solving the simultaneous equations, the backward-difference method is frequently employed (Prickett, 1975; Taylor, 1974). This is an implicit method whereby unknown H values at two or more neighboring nodes are determined simultaneously in terms of known H values at one or more neighboring nodes, in contrast to relaxation methods or explicit techniques where H is determined explicitly in terms of known H at neighboring nodes. Time derivatives are expressed in terms of the difference between known heads at the beginning of the time increment and unknown heads at the end of the increment. The coefficients in the resulting finite-difference equations then form a tridiagonal $N \times N$ matrix if N is the total number of nodes in the system. This matrix is readily solved by computer, using the ADIP (alternating direction implicit procedure). Other techniques, such as the explicit forward-difference method and the intermediate central-difference method, may pose problems of meaningless solutions, instability, or nonlinear finite-difference expressions resulting from nonlinear aquifer situations (Prickett, 1975). The alternating direction explicit procedure (ADEP), while relatively easy to program, poses restrictions regarding the length of the time increments, the degree of aquifer confinement, and the location of wells that can be represented (Tomlinson and Rushton, 1975). Detailed computer programs in Fortran IV and other languages are available in the literature (Bittinger et al., 1967; Pinder, 1970; Prickett, 1975; Taylor, 1974; Trescott et al., 1976; and references therein).

One of the most versatile digital models, presented by Freeze (1972), can handle up to 10 000 nodes in variable grids and is capable of solving two- or three-dimensional, steady or transient, and saturated or unsaturated flow in non-uniform or anisotropic media of varying shapes and with a wide variety of time-variant boundary conditions. The model requires a computer with 750 000 bytes of core storage, and difficult problems may take as much as 1 h execution time on a computer like the IBM 360/91. Simpler problems, however, can be solved with smaller computers, such as a disk minicomputer with 8 000 words (16 bits per word) core storage used by Prickett and Lonnquist (1973). The model by Trescott et al. (1976) will simulate groundwater flow in unconfined and confined aquifers or in a combination of the two. The aquifers may be heterogeneous, anisotropic, and have irregular boundaries. The source term in the equation may include constant recharge, leakage from a confining layer, well discharge, and evapotranspiration.

Digital models are the most versatile technique for solving problems of underground water movement. In addition to the complexities already mentioned, they can handle problems of miscible displacement like saltwater intrusion (Pinder and Cooper, 1970) and of head-dependent conductivities, such as flow in unconfined aquifers with H-dependent transmissivities and unsaturated flow where K is a function of the (negative) pressure head (see Section 7.7). Digital models also yield highly accurate solutions because of their high node density and the ease of accurately representing boundary conditions and inputs and outputs of water. Solution accuracy, however, is usually not a determining factor in choosing the digital-model route, because as a rule uncertainties in input data cause a much greater error in the results than the solution procedure itself (the same is true for manual relaxation and R or RC analogs). An exception may be where the model is to be used to test the validity of certain mathematical hypotheses or to determine how certain factors affect a given flow system.

Disadvantages of computer solutions are that they require availability of the appropriate computer facilities and programming expertise, and that they are nonphysical. The latter limits the experience and "intuition" for flow systems that are invariably gained from working with physical models or passive analogs, and that give insight as to the relative importance of various flow-system parameters and input data. This insight is of great value in developing flow-system models for analysis, because such models, no matter how complex, always are a drastic simplification of the real situation.

Finite-difference solution by computer is a frequently used technique for solving underground flow systems. Examples include the work by Birtles and Wilkinson (1975), Bittinger et al. (1967), Bredehoeft and Pinder (1970), Cooley (1971), Donaldson (1974), Fayers and Sheldon (1962), Freeze (1971), Nutbrown et al. (1975), Pinder and Bredehoeft (1968), Remson et al. (1965), and Taylor and Luthin (1969). For additional discussions and references, see Prickett (1975) and Taylor (1974).

7.4.2 Hybrid Models

Some digital models utilize the "instantaneous" relaxation of a resistance-network analog as a subroutine to reduce the long computer times required for iterative finite-difference solution. The digital computer then provides the proper input data (sources, sinks, media properties, and boundary conditions), which are translated into electrical form by a digital-analog converter and fed into the resistance network through a distributor. The R analog relaxes the system and feeds the node voltages back to the computer, via a multiplexer and analog-digital converter, for further processing. This system, called a *hybrid computer*, is particularly advantageous for solving iteration-intensive problems, such as transient systems that are treated as a succession of steady states and flow in unconfined aquifers with H-dependent T (Karplus, 1967; Moulder and Jenkins, 1969; Vemuri and Dracup, 1967; Vemuri and Karplus, 1969). For additional references, see Prickett (1975)

7.4.3 Finite-Element Method

With the finite-element method, flow systems are solved through an equivalent variational functional, rather than through a finite-difference solution of the differential flow equation. The flow system is considered as a general system of energy dissipation for which the H solution is found as the H distribution that minimizes the rate of energy dissipation. The finite-element method was originally developed (primarily in the 1950s and 1960s) for analysis of frame structures, and it is routinely used in structural, mechanical, and aerospace engineering (Guymon, 1974; Muskat, 1946).

With the finite-element technique, the solution of the differential flow equation, such as Eq. (7.5) with a general source or sink term Q_s added to the left portion, is obtained by finding a solution for H that minimizes an equivalent variational functional of the form (Guymon, 1974; Prickett, 1975; and references therein)

$$F = \iint \left[\frac{K_x}{2}\left(\frac{\partial H}{\partial x}\right)^2 + \frac{K_y}{2}\left(\frac{\partial H}{\partial y}\right)^2 + \left(S\frac{\partial H}{\partial t} - Q_s\right)H \right] dx\, dy \qquad (7.39)$$

To find the solution, the flow domain is divided into a number of subregions or finite elements which, unlike the square or rectangular meshes for the finite-difference method, are triangular and quadrilateral for two-dimensional systems (Figure 7.20) and tetrahedral or parallelepiped for three-dimensional systems. The elements should be as disordered and nonuniform as possible to prevent solutions from going into preferred directions, and smallest where the flow is concentrated.

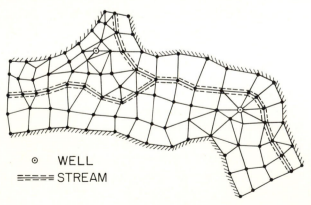

⊙   WELL

===== STREAM

Figure 7.20 Finite elements placed over alluvial valley with tributary stream and two wells.

The irregular form of the elements facilitates representation of irregular boundaries (Figure 7.20). For each solution step, the parameters K_x, K_y, S, and Q_s are kept constant for a given element, but they may vary between elements. To minimize Eq. (7.39) with respect to the potential state H, the differential $\partial F/\partial H$ is evaluated for each node and equated to zero. This produces a number of simultaneous equations which are readily solved by computer (Guymon, 1974; Prickett, 1975; Zienkiewicz, 1971; Zienkiewicz and Cheung, 1967; Zienkiewicz et al., 1966, and references therein).

Whether the finite-element or the finite-difference method is preferable depends on the complexity of the flow system, computer times required, problems of stability and truncation error, and general applicability of computer programs (Guymon, 1974). As long as the relative merits of the two methods have not yet been exhaustively compared, however, the choice between finite-element and finite-difference methods will largely depend on the individual preferences and experience of the person(s) involved.

Applications of the finite-element technique to underground flow systems include the work by Guymon et al. (1970), Javendel and Witherspoon (1968a and b), Neuman and Witherspoon (1970a and b), U.S. Army Corps of Engineers (1970), Witherspoon et al. (1968), and Zienkiewicz et al. (1966). For additional discussion and references, see Guymon (1974) and Prickett (1975).

7.5 INPUT DATA AND MODEL CALIBRATION

If the objective of the model is prediction of groundwater levels or other hydrologic behavior in an actual system, the most difficult part of the procedure is the collection of accurate and adequate input data and simplification of the real situation into a manageable model. For aquifers, all significant inputs and outputs of water, boundary conditions, and the spatial variation of T and S must be known. This may require extensive hydrologic investigations.

Values of T and S can be obtained from pumping tests and other procedures discussed in Chapters 2, 4, and 5. Often, there are not enough wells to adequately characterize the aquifer in this manner, or it may be impractical to perform pumping tests on all wells. Values of T around wells not pump-tested can be inferred by comparing the specific capacities of these wells to those of wells where T is known from pumping tests. This is not an exact procedure, but it is better than guessing.

Where well tests are not available, T can be estimated from a contour map of groundwater levels (water tables or piezometric surfaces). Assuming isotropic conditions, streamlines can be sketched orthogonally to the equipotentials to construct a flow net (the individual segments do not have to be square and may consist of irregular rectangles). If T is known for one segment of a streamtube (from well tests, for example), T of the other segments in the streamtube can then be calculated by applying Darcy's equation to each segment in succession. Assum-

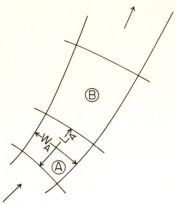

Figure 7.21 Equipotentials and streamlines for determining distribution of T in streamtube.

ing steady, horizontal flow within a segment, the flow in that segment can be calculated as (Figure 7.21)

$$q_A = T_A \, \Delta H_A \, W_A / L_A \qquad (7.40)$$

where q_A = flow in segment A (m³/day)
$\quad T_A$ = transmissivity in segment A (m²/day)
$\quad W_A$ = average width of segment
$\quad L_A$ = average length of segment
$\quad \Delta H_A$ = drop of groundwater level (water table or piezometric surface) across segment

The flow in the next segment B is similarly calculated as

$$q_B = T_B \, \Delta H_B \, W_B / L_B \qquad (7.41)$$

Dividing this equation by Eq. (7.40) and solving for T_B yields

$$T_B = \frac{q_B \, L_B \, W_A \, \Delta H_A}{q_A \, L_A \, W_B \, \Delta H_B} \, T_A \qquad (7.42)$$

which enables calculation of T_B from T_A. Repeating this procedure in succession to the other segments yields the spatial distribution of T for the entire streamtube. Prior knowledge of T for more than one segment enhances the accuracy of the results because T of the other segments can then be calculated in more than one way, allowing averaging of the results. The spatial variation of T in other streamtubes is calculated similarly, yielding T values for the entire aquifer. The ratio q_B/q_A in Eq. (7.42) represents the relative change in flow in the streamtube due to depletion and accretion of groundwater. The various water outputs and inputs must be known to enable evaluation of q_B/q_A, or the ratio can be taken as one if there is reason to believe that q is relatively constant.

The above procedure was used by Hunt and Wilson (1974) for a regional groundwater study in New Zealand. Equations (7.40) and (7.41) are based on replacing curved segments of streamtubes by rectangles of the same average length and width. Nelson (1961) has shown that this produces only a small error in T.

A more rigorous technique for determining the distribution of T or K was developed by Nelson (1962, 1968). First, the distribution of H in the actual system is determined with field measurements, using piezometers. Based on this information, streamlines are evaluated by solving three simultaneous differential equations. If the flow is transient, care should be taken that the H data all refer to the same point in time. The next step is to measure K or T in the field at one point on each streamline. If the flow is two-dimensional in a vertical plane or three-dimensional, point measurement techniques like the piezometer method should be used. If the flow is two-dimensional in an aquifer or other horizontal "plane," T should be evaluated with pumping tests on completely penetrating wells so that T applies to the entire thickness of the aquifer. The resulting K or T data are then used in a computer program to calculate values of K or T for the rest of the medium. For additional discussion and references, see King (1974).

Spatial variation of S can be determined from well tests or with some of the methods presented in Section 2.10. Geologic information may also be helpful to estimate S for parts of the aquifer where wells or pumping-test data are not available.

Once the desired input information has been collected and incorporated into a predictive model (analog or digital), the model should be calibrated against known records of cause-and-effect relations, such as groundwater levels in response to pumping from wells. When the model fails to predict such a record with sufficient accuracy, realistic changes in T or S, or both, and possibly in other input parameters are made until the model correctly duplicates the known response. A correct "hindcasting" capability of the model then enhances the confidence in its forecasts.

7.6 FLOW IN ANISOTROPIC MEDIA

Anisotropic media are readily represented as such in network analogs and digital models. For mathematical solution or for simulation of the system in a sand tank or Hele-Shaw apparatus, it is often more convenient to transform the anisotropic prototype into an equivalent isotropic system. The principles for this transformation can be readily demonstrated, heuristically if not rigorously, with the resistance-network analogy as shown in the following paragraphs. The equation for K at some angle to the major K axis in an anisotropic medium will be derived in similar manner.

Figure 7.22 shows a portion of an R analog that represents a vertical cross section of an aquifer or other porous medium. All horizontal resistors are of a value R_x and all vertical resistors are of a different value R_z. This can mean two things: the analog network consists of square meshes and represents an anisotropic medium (Figure 7.22, top), or the analog network consists of rectangular meshes and represents an isotropic medium (Figure 7.22, bottom). Both systems

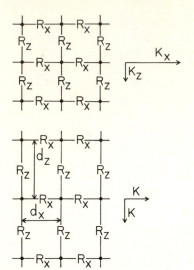

Figure 7.22 R analog with resistances representing square meshes of anisotropic medium (top) or rectangular meshes of equivalent isotropic medium (bottom).

are electrically identical and give, of course, the same solution. Ignoring the resistance-scale factor (assuming $S_R = 1$), the resistors representing the anisotropic system are $R_x = 1/K_x$ and $R_z = 1/K_z$, where K_x and K_z are the horizontal and vertical hydraulic conductivity, respectively. The resistors representing the equivalent isotropic system are $R_x = d_x/(d_z K)$ and $R_z = d_z/(d_x K)$, according to Eqs. (7.25) and (7.26) (see Figure 7.22 for meaning of symbols). Equating the expressions for R_x and R_z yields

$$\frac{1}{K_x} = \frac{d_x}{d_z K} \tag{7.43}$$

and

$$\frac{1}{K_z} = \frac{d_z}{d_x K} \tag{7.44}$$

Solving Eq. (7.43) for d_z/d_x, substituting the resulting expression in Eq. (7.44), and solving for K of the equivalent isotropic system then yields

$$K = \sqrt{K_x K_z} \tag{7.45}$$

which after substitution into Eq. (7.43) and solving for d_z/d_x yields

$$\frac{d_z}{d_x} = \sqrt{\frac{K_x}{K_z}} \tag{7.46}$$

Equations (7.45) and (7.46) show that an anisotropic medium with major conductivity components in the horizontal and vertical directions can be transformed into an equivalent isotropic medium with hydraulic conductivity $\sqrt{K_x K_z}$ by multiplying all vertical dimensions of the medium by $\sqrt{K_x/K_z}$. If there are different layers in the system, each layer is vertically expanded according to its own K_x/K_z ratio (vertically shrunk if this ratio is less than one). Examples of application of these principles are the electrolytic-tank study by Todd and Bear (1961) of seepage from a river into a two-layered anisotropic aquifer, and the extension of the double-tube method for measuring directional K in anisotropic media by Bouwer (1964a).

Streamlines in the equivalent isotropic system are drawn as orthogonals to the equipotentials, or vice versa. When the system is transformed back to the vertical anisotropic system, streamlines and equipotentials no longer intersect at right angles, except where they coincide with the x or z direction. Analytic expressions for the angle between streamlines and equipotentials in anisotropic media have been developed (Wylie, 1951). The simplest way of determining streamlines and equipotentials in an anisotropic medium, however, is to construct streamlines and equipotentials as orthogonals in the equivalent isotropic system and then transforming the flow net back to the anisotropic system. With R analogs, anisotropic systems can be represented as such. Thus, equipotentials can be constructed from voltage measurements at the nodes. Reversing the boundary conditions so that streamlines become equipotentials and vice versa then enables construction of streamlines in like manner. An example of a flow system obtained in this way, showing streamlines and equipotentials in an anisotropic aquifer below a recharge basin, is presented in Figure 7.23 (taken from Bouwer, 1970).

The hydraulic conductivity K_α in an anisotropic medium at some angle to the major anisotropy axis can be calculated as K in a diagonal direction in an R analog model of the medium, using rectangular meshes (Figure 7.24). Assuming $S_R = 1$, the resistances are again calculated as $R_x = d_x/(d_z K_x)$ and $R_z = d_z/(d_x K_z)$. For flow in diagonal direction at angle α with the x direction, the four mesh resistors can be replaced by a single diagonal resistor R_α, which is equal to the shunt resistance $(R_x + R_z)/2$ of the mesh resistors, or

$$R_\alpha = \frac{1}{2}\left(\frac{d_x}{d_z K_x} + \frac{d_z}{d_x K_z}\right)$$
(7.47)

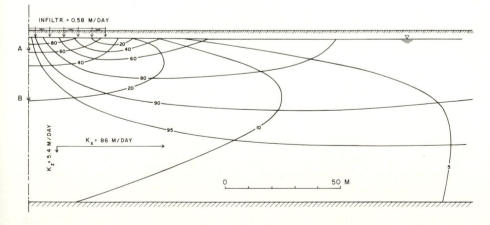

Figure 7.23 Flow system in anisotropic aquifer below recharge basin. (*From Bouwer, 1970.*)

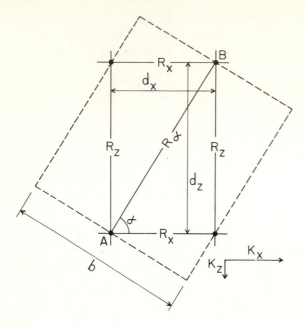

Figure 7.24 Rectangular network mesh of R analog model representing anisotropic medium.

However, R_α can also be considered as representing a slanting rectangular mesh of length AB and width b (Figure 7.24), so that R_α can be calculated with Eq. (7.25) or (7.26) as

$$R_\alpha = \frac{1}{K_\alpha} \frac{AB}{b} \tag{7.48}$$

Since $\cos \alpha = b/2d_z = d_x/AB$, $b = 2d_x d_z/AB$. Substituting this expression for b in Eq. (7.48), equating the resulting expression to Eq. (7.47), and solving for $1/K_\alpha$ yields

$$\frac{1}{K_\alpha} = \frac{d_x^2}{AB^2 K_x} + \frac{d_z^2}{AB^2 K_z} \tag{7.49}$$

which can be written as

$$\frac{1}{K_\alpha} = \frac{\cos^2 \alpha}{K_x} + \frac{\sin^2 \alpha}{K_z} \tag{7.50}$$

This expression for K_α at angle α to the horizontal anisotropy component is the solution of the ellipsoid equation at the end of Section 3.8.

7.7 UNSATURATED FLOW

Unsaturated flow, which is the predominant flow in the vadose zone, is analyzed on the basis of Darcy's equation, with the added complication that K is dependent on the water content θ, which in turn is related to the (negative) pressure head h

(see Section 2.9). In unsaturated material, part of the pore channels are filled with air, which physically obstructs water movement. Water then flows only through the finer pores, which may still be saturated, or in films around the soil particles. The lower the value of θ, the lower the unsaturated hydraulic conductivity will be. Unsaturated flow should theoretically be treated as two-phase flow of water and air. The usual approach, however, is to analyze only the flow of water and consider the air as part of the solid phase. Solution of unsaturated-flow systems thus is a problem of solving the basic differential flow equation for a medium where the unsaturated hydraulic conductivity K_h is a function of the pressure head h, and where θ changes with h. This requires an iterative process, assuming an initial distribution of K_h, solving for H, evaluating h at each point as $H - z$, and determining the K_h values corresponding to these h values according to the relation between K_h and h for the material in question. These K_h values will differ from the initially assumed values. Thus, the procedure is repeated with a new distribution of K_h, etc., until the K_h values corresponding to the values of $H - z$ agree with the K_h values in the system. In transient systems, θ must also be changed as h changes with time. This produces storage or release of pore water, depending on whether h is increasing or decreasing, respectively.

Unsaturated flow has traditionally been the realm of soil physicists. Groundwater hydrologists have only recently become actively interested in the subject. Considering the importance of unsaturated flow in the vadose zone as the link between the groundwater and surface-water environments, unsaturated flow is one of the most neglected areas in groundwater hydrology. Natural and artificial recharge of groundwater, evaporation of groundwater to the atmosphere via plants or directly from the soil, and groundwater contamination by infiltration of polluted water are examples of important issues where unsaturated flow plays a significant role.

7.7.1 Unsaturated Hydraulic Conductivity Relations

Unsaturated hydraulic conductivity is expressed as K_θ or as K_h, depending on whether it is considered in relation to θ or to h. While the relation between K_θ and θ is physically more meaningful, the relation between K_h and h is more useful in flow analysis where K_h is to be adjusted to h as determined from H and z. Relations between K_h and h are governed by pore configurations and must be experimentally determined for each particular material. The most direct method is the long, soil-filled column technique (Figure 7.25) used by Childs (1945) and others (Bouwer and Jackson, 1974, and references therein). Water is applied to the top of the column at a constant rate q that is less than the product of the saturated hydraulic conductivity K and the cross-sectional area A of the column, to produce unsaturated flow in the soil material. The water is allowed to drain freely from the bottom of the column. When equilibrium conditions are established, the bottom of the soil column will be essentially saturated (capillary fringe), but the rest of the soil will be at uniform water content θ and pressure head h (Figure 7.25). A constant h value in the column means a hydraulic gradient of one, so that K_h will

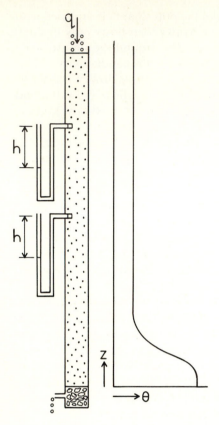

Figure 7.25 Long soil column to determine K_h in relation to h.

be equal to the downward-flow rate q/A of the water in the column. Measuring h with tensiometers in the column (Figure 7.25) will then give one point of the relation between K_h and h. Repeating the test for other q values yields other points, etc., until the $K_h - h$ relation is sufficiently defined. Measurement of θ in the column with, for example, the gamma-ray technique (see Section 2.7), then also yields the $K_\theta - \theta$ relation.

A disadvantage of the long-column technique is the long time required for establishing equilibrium after q is changed, particularly for fine-textured materials. To speed up the procedure, Watson (1967) created a zone of entrapped air in the column by first saturating the material, draining it at the bottom, and then rewetting the soil by ponding water on the surface. When a certain inflow rate q was then maintained at the top, different θ values occurred in the unsaturated zone. Measuring h values at a number of points to determine vertical gradients then enabled calculation of various corresponding values of K_θ and θ for a single q value. The long-column technique has been adapted to field use by Hillel and Gardner (1970) and Bouma et al. (1971) by applying water through artificial surface crusts or impeding layers to create unsaturated, downward flow in the underlying material.

Many other techniques have been developed for determining $K_h - h$ or $K_\theta - \theta$ relations, including advance-of-the-wetting-front, pressure-plate-outflow, instantaneous-profile, and transient-flow methods. There are also procedures for calculating K_h versus h from the relation between θ and h using pore-size distribution models, and computer techniques that develop $K_h - h$ relations by trial and error until they accurately predict a known flow rate in an unsaturated system, such as an infiltration rate. For a review of these methods, reference is made to Bouwer and Jackson (1974).

The relation between K_θ and θ for a given material is usually considered unique and free from hysteresis (Figure 7.26; Topp, 1969; Watson, 1967). Poulo-vassilis and Tzimas (1975), however, reported K_θ differences of as much as 100 percent in equally wet soil materials but with different drying and wetting histories. Sometimes, indirect hysteresis may be observed due to consolidation and swelling of the material as it is dried or wetted, respectively. Since relations between h and θ are very much hysteretic (Figure 7.27), $K_h - h$ relations are likewise affected by hysteresis (Gillham et al., 1976). Figures 7.26 and 7.27 show, for example, that K_h at $h = -60$ cm may vary from 0.01 m/day to 0.36 m/day, depending on whether the h value was reached by draining a wet soil or wetting a dry soil. This is almost a 40-fold difference that could give completely erroneous results if hysteresis were not considered in predicting a certain flow system. The $K_h - h$ curve for wetting often remains below the $K_h - h$ curve for drying even when $h = 0$ or $h > 0$. This means that K after wetting, even at positive pressure heads (like below water tables or surface inundations), will be less than K at saturation. Experience indicates that such "resaturated" K values may be about one-half the K values at complete saturation (Bouwer, 1966).

Experimentally determined relations between K_h and h are generally sigmoid, with coarse-textured materials showing a more abrupt reduction in K_h at higher

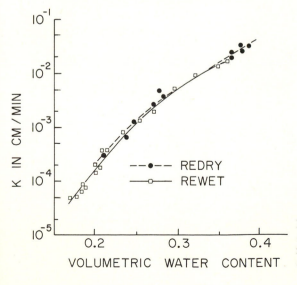

Figure 7.26 Relation between K_θ and θ for Rubicon sandy loam with different wetting and drying histories. (*Redrawn from Topp, 1969.*)

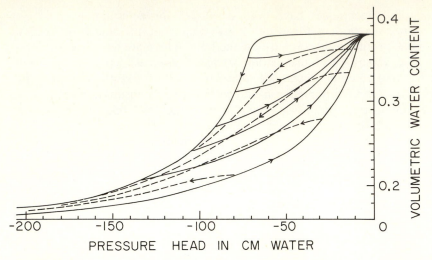

Figure 7.27 Hysteretic relations between h and θ for Rubicon sandy loam. (*From Topp, 1969, as redrawn by Watson, 1974.*)

(less negative) h values than fine-textured materials (Figure 7.28). At low (highly negative) values of h, K_h for clays is greater than for sands, the reverse of the K order at saturation. Figures 7.26 and 7.27 also show that K_h in relatively dry soil materials is insignificantly small compared to K at or near saturation.

Observed $K_h - h$ relations have been expressed in simple, empirical equations to facilitate analytic or numerical solution of unsaturated-flow systems. The simplest approximation is a step function of height K and width h_{cr} with the same

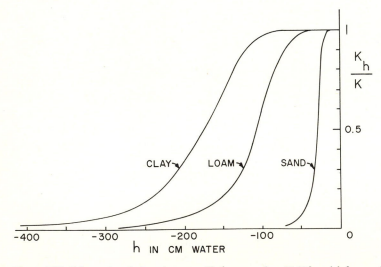

Figure 7.28 Schematic relations between K_h (expressed as K_h/K) and h for sand, loam, and clay.

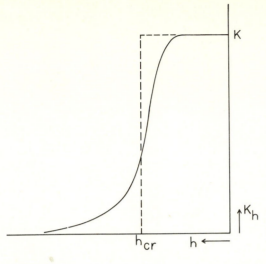

Figure 7.29 Replacement of sigmoid K_h-vs.-h curve by step function of height K and width h_{cr}.

area under the curve as the actual $K_h - h$ curve (Figure 7.29). A detailed discussion of this approximation is given at the end of this section [Eqs. (7.54) to (7.57)]. Another approximation is an equation of the type (Gardner, 1958)

$$K_h = \frac{a}{(-h)^n + b} \qquad (7.51)$$

where a, b, and n are constants that decrease with decreasing particle sizes of the material. For $h = 0$, this equation reduces to $K = a/b$, which is K at saturation. Orders of magnitude of a, b, and n (with h expressed in centimeters and K_h in centimeters per day) are as follows (Bouwer, 1964b):

	a	b	n	$K = a/b$
Medium sands	5×10^9	10^7	5	500 cm/day
Fine sands, sandy loams	5×10^6	10^5	3	50 cm/day
Loams and clays	5×10^3	5×10^3	2	1 cm/day

These data should not be used to calculate actual values of K_h, but they may serve to yield estimates of K/K_h in relation to h. For example, a loamy sand at $h = -60$ cm would have a K_h value that is about 32 percent of K at saturation. For additional models and equations for describing K_h versus h, reference is made to Raats and Gardner (1974).

7.7.2 Flow-System Analysis

In transient unsaturated-flow systems, K_h and θ are continuously changing. For example, infiltration of rain or other water produces an increase in K_h and θ behind a downward-moving wetting front. Stopping the infiltration will then cause K_h and θ to decrease in the wetted zone and to increase in the underlying material. Evaporation at the surface and/or uptake of water by plant roots further reduce K_h and θ in the upper soil layers. The general differential-flow equation for three-dimensional unsaturated flow is developed in the same way as for saturated flow, using an infinitesimal cube with dimensions dx, dy, and dz in a cartesian system and equating the rate of change in the water content to the difference between inflow and outflow rates as expressed with Darcy's equation. In formula

$$\frac{\partial}{\partial x}\left(K_h \frac{\partial h}{\partial x}\right) + \frac{\partial}{\partial y}\left(K_h \frac{\partial h}{\partial y}\right) + \frac{\partial}{\partial z}\left(K_h \frac{\partial h}{\partial z}\right) + \frac{\partial K_h}{\partial z} = \frac{\partial \theta}{\partial t} \qquad (7.52)$$

where the term $\partial K_h / \partial z$ is included to account for gravity flow, since the equation is expressed in terms of h rather than H. Equation (7.52) was derived by Richards (1931). For two-dimensional flow in a vertical plane (stream seepage to a deep groundwater table, infiltration from irrigation furrows, etc.), the second term in Eq. (7.52) is zero. For one-dimensional vertical flow (infiltration of rainfall, evaporation from a water table), the first two terms are zero. For steady systems, $\partial \theta / \partial t$ is zero.

The first solutions of unsaturated-flow systems (primarily one-dimensional) were obtained analytically by treating the flow as a diffusion problem, similar to heat flow. For this purpose, the terms $K_h \partial h/\partial x$, etc., in Eq. (7.52) were replaced by $D_\theta \partial \theta/\partial x$, etc., where D_θ represents the soil-water diffusivity with dimension length^2/time and defined as

$$D_\theta = K_\theta \frac{\partial h}{\partial \theta} \qquad (7.53)$$

The concept of a soil-water diffusivity was introduced by Buckingham (Swartzendruber, 1969, and references therein). Values of D_θ in relation to θ for a given material are obtained by graphically determining $\partial h/\partial \theta$ as the slope of the $h - \theta$ curve at a certain θ and multiplying this slope by K_θ belonging to the same θ. Since D_θ is θ-dependent, the procedure must be repeated for a number of θ values.

At low values of θ (less than 0.1, for example), K_θ is so small that water movement in the vapor phase may no longer be negligible (Swartzendruber, 1969 and references therein). Also, flow of water in films adsorbed on clay particles and other solids may become significant. The value of D_θ then consists of the sum of the diffusivities for flow of water in the liquid phase, the vapor phase, and the adsorbed phase. At low values of K_θ, temperature and osmotic effects may also become significant and the flow may no longer be governed by Darcy's equation (Swartzendruber, 1969).

Since soil-water diffusivity is analogous to thermal diffusivity, the mathematics of heat flow was utilized in obtaining the first solutions for unsaturated-flow problems, including infiltration of water into soil (Philip, 1969 and references therein), evaporation from a water table (Gardner, 1958), and flow systems associated with techniques for measuring D_θ and K_h on soil samples (Bouwer and Jackson, 1974, and references therein).

Solutions of two-dimensional, steady, unsaturated-flow systems have been obtained with R analogs by adjusting resistances to h values obtained from voltage measurements at the nodes until the R values and voltages satisfied the $K_h - h$ relation of the medium. This technique was applied to predict upward, lateral, and downward flow from an underground perforated pipe (Bouwer and Little, 1959) and to assess the significance of flow above the water table in drainage and subirrigation systems (Bouwer, 1959a and b). Transient systems are not readily solved by RC analogs because the relations between h and θ and between K_h and h are difficult to simulate with capacitors and resistors, respectively.

With the development of high-speed, large-memory digital computers, the solution of unsaturated-flow problems in the last decade or so has moved in the direction of digital models that can also take into account hysteresis and other complex situations not amenable to analytic solution with diffusion theory. For the digital approach, Eq. (7.52) is written in finite-difference form and solved with the computer, similar to saturated-flow problems (see Section 7.4). Additional iteration is necessary to satisfy the relations between K_h and h and between h and θ at each node for the material in question (Hornberger et al., 1969; Remson et al., 1971; Rubin, 1968; Watson, 1974; and references therein). The trend in θ (increasing or decreasing) should also be taken into account to include hysteresis.

The computer methods make it possible, at least in principle, to analyze underground-flow systems as one continuum without differentiating between saturated and unsaturated flow. Where the solution yields $h > 0$, K and θ are kept constant at saturated or resaturated values. Where the solution indicates $h < 0$, K_h and θ are adjusted in accordance to the particular K_h-vs.-h and θ-vs.-h relations of the medium. Water tables would then emerge as contours of $h = 0$, which they are, and not as boundaries separating K from impermeability, which they are not but which often is their assigned role in flow-system analysis. Such treatment of underground-flow systems, however, will be warranted only in special situations of conjunctive saturated and unsaturated flow. In most cases, unsaturated flow will be insignificant compared to saturated flow, so that only the latter needs to be considered. Moreover, inclusion of unsaturated flow requires more input data (K_h-vs.-h and K_θ-vs.-θ relations) and more computer time. The outlook thus is still for separate treatment: groundwater movement as saturated flow, and flow in the vadose zone as unsaturated flow, except in special cases.

7.7.3 Simplified Solutions

For certain types of systems, unsaturated flow can be handled in a much simpler way, replacing the actual curve relating K_h to h by a step function (Figure 7.29). The height of this step function is then equal to K at $h = 0$ and the width h_{cr} is such that the area under the step function is the same as that under the actual K_h-vs.-h curve, as will be demonstrated below for the case of steady, one-dimensional lateral flow above a mildly sloping water table (Figure 7.30).

The lateral velocity $v_{x, z}$ at height z above the water table can be expressed with Darcy's equation as

$$v_{x, z} = K_h i \qquad (7.54)$$

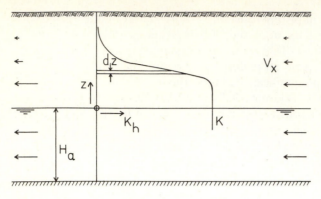

Figure 7.30 Distribution of K_h in vadose zone for steady lateral flow above mildly sloping water table.

where K_h is the unsaturated hydraulic conductivity at height z above the water table and i is the hydraulic gradient or slope of the water table. Since the flow is in the horizontal direction only, h in the vadose zone will be equal to $-z$. Plotting the K_h distribution above the water table thus yields a curve that is identical to the $K_h - h$ relation of the particular soil. The flow rate $q_{x,z}$ in an infinitesimal increment of z is $K_h i\, dz$, which when integrated over the entire vadose zone yields the equation

$$Q_v = i \int_0^{z_w} K_h \, dz \qquad (7.55)$$

where Q_v is the lateral flow in the entire vadose zone and z_w is the height of the vadose zone or depth of the water table. If the vadose zone is replaced by a fictitious saturated capillary fringe of height z_{cr} that produces the same flow as the actual vadose zone, Q_v can be expressed as iKz_{cr}, where K is the saturated K or K at $h = 0$. Equating this expression to Eq. (7.55) and solving for z_{cr} yields

$$z_{cr} = \frac{1}{K} \int_0^{z_w} K_h \, dz \qquad (7.56)$$

Since $h = -z$, this equation can also be written as

$$h_{cr} = \frac{1}{K} \int_0^{h_w} K_h \, dh \qquad (7.57)$$

which shows that, if h_w is taken sufficiently small (negative) so that $K_h \ll K$, the area $h_{cr} K$ of the step function is the same as the area $\int K_h \, dh$ under the actual $K_h - h$ curve.

The width h_{cr} of the step function, called the *critical-pressure head* (Bouwer, 1959a, 1964b), thus is numerically equal to the height of a fictitious, saturated capillary fringe with abrupt reduction of K to zero at its top. This fringe then gives the same steady-state lateral flow above a sloping water table as the entire vadose zone where K_h decreases from K at the water table to essentially zero higher up (Bouwer, 1964b; Mobasheri and Shahbazi, 1969).

The ratio between the lateral flow Q_v above the water table and the lateral flow Q_a below the water table can be expressed as

$$\frac{Q_v}{Q_a} = \frac{-h_{cr}}{H_a} \tag{7.58}$$

where H_a is the height of the aquifer (Figure 7.30). Values of h_{cr} may be -20 cm or more (less negative) for coarse sands, -20 cm to -60 cm for medium and fine sands, and -50 to -200 cm or less (more negative) for structureless loams and clays (Bouwer, 1964b). Thus, for a 50-m-thick unconfined aquifer of medium sand with $h_{cr} = -20$ cm, lateral flow above the water table would only be $0.2/50 = 0.004$ or 0.4 percent of the flow below the water table. On the other hand, lateral flow above the water table would be 30 percent of the flow below the water table in a system of flow toward tile drains in agricultural land consisting of a loam with $h_{cr} = -60$ cm and with an impermeable layer at a depth of 2 m below the water table. The critical-pressure-head concept thus enables inclusion of the flow above the water table in the analysis of lateral groundwater flow by taking the plane of critical-pressure head as the upper boundary of the flow system, instead of the water table which is the plane of zero-pressure head.

The concept of critical-pressure head and associated simplification of the $K_h - h$ curve to a step function is valid not only for lateral flow in the vadose zone, it can also be used to include unsaturated flow in the analysis of systems of seepage from surface water (streams, canals, basins, etc.) to a deep water table or drainage layer (see Section 8.2). In these systems, there is some initial divergence of flow due to the pressure head on the wetted perimeter of the channel or basin and due to lateral unsaturated flow, but further down the flow is predominantly downward (Figure 7.31). If such systems are analyzed with $h = 0$ as the pressure head at the free boundary, unsaturated flow is neglected and the resulting seepage rate will be less than when unsaturated flow is included. However, analyzing the system with h_{cr} as the pressure head at the free boundary accurately accounts for the unsaturated flow. This was demonstrated by a comparison between a computer solution using the actual relation between K_h and h (Reisenauer, 1963) and an R analog solution using the equivalent step function and h_{cr} (Bouwer, 1964b). The larger $-h_{cr}$ is in relation to the width of the water surface in the channel or basin, the more significant the effect of unsaturated flow on total seepage will be.

A third application of the step-function approach is in the analysis of systems of water infiltration into soil. The water enters the soil at the surface as rain or from a ponded condition (sheet runoff, irrigation furrows or basins) and wets the soil while moving downward until it reaches a water table or restricting layer. Infiltration is one of the most researched surface phenomena of the hydrologic cycle, and numerous equations and solutions have been developed (Philip, 1969; see also Section 8.1). A simple model based on a step-function relation between K_h and h, proposed early in this century by Green and Ampt (1911), however, is still one of the most useful techniques for practical application. With this approach,

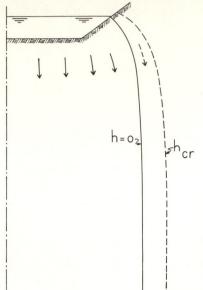

$h = 0$

h_{cr}

Figure 7.31 Boundary of flow system below channel or basin with zero-pressure head and critical-pressure head as boundary condition.

the wetting front is assumed to be abrupt and the water content and K in the wetted zone are assumed to be uniform and constant. Infiltration is thus considered as "piston" flow, which is treated with Darcy's equation to obtain equations relating infiltration rate and accumulated infiltration to time since infiltration began. While Green and Ampt did not specify the physical significance of this pressure head (they called it the capillary coefficient), recent work has indicated that this negative pressure head can be taken as the critical-pressure head h_{cr} (see Section 8.1 for additional discussion and references).

A fourth application of the step-function approach is in calculating the flow through thin surface layers of low K that restrict infiltration or seepage. Such layers, for example, may occur as sediment deposits on the bottom of recharge basins or of open channels. Since K of such layers is much less than K of the underlying material, they restrict the infiltration or seepage rate and cause unsaturated flow in the underlying material. The hydraulic gradient in the deeper material is then theoretically equal to one, and K_h of this material is numerically equal to the seepage or infiltration rate, like in the long column of Figure 7.25. If the K_h-vs.-h relation of the material is taken as a step function, K_h of the underlying material reduces abruptly from K to 0 at h_{cr}. This means that the pressure head in the underlying material will be close to h_{cr} for a wide range in downward flow rates. Thus, h_{cr} can then be taken as the pressure head immediately below the restricting bottom layer when applying Darcy's equation to the flow through this layer for predicting seepage rates, as discussed in Section 8.2.

PROBLEMS

7.1 Derive Eq. (7.7).

7.2 Construct equipotentials in the horizontal plane and sketch streamlines as orthogonals to the equipotentials for the steady-state flow in a confined aquifer with sloping piezometric surface and pumped well, as shown in Figure 7.2. Locate the streamline that separates the flow toward the well from the rest of the flow. Use slope of piezometric surface and drawdown due to pumping from Figure 7.2, or assume your own aquifer conditions and calculate drawdown with the Thiem equation.

7.3 Show imaginary wells for a single well in an alluvial valley between a stream (recharging boundary) and a mountain range (solid boundary), as in Figure 7.7. (*Note:* the imaginary well across one boundary should also be mirrored across the other boundary.) How many imaginary wells are theoretically needed to calculate water-table drawdowns with the superposition principle?

7.4 Using the streamlines and equipotentials in Figure 7.9, calculate the flow to the well if $T = 100$ m^2/day and the water level at the well is 8 m lower than that in the stream.

7.5 Construct a simple R analog as a group project and check the validity of various concepts, such as superposition (Figure 7.2), method of images (Figures 7.4 and 7.5), use of an additional resistor at the node to represent a well of certain diameter [Eq. (7.28)], and evaluating T distribution from streamlines and equipotentials [Eq. (7.42)]. Check the accuracy of the R analog by obtaining solutions for simple two-dimensional flow systems for which exact solutions are available (superimposed point sink and uniform flow, for example).

7.6 Plot curves of K_h versus h (cartesian coordinates) for wetting and drying, using the K_θ-vs.-θ and the extreme h-vs.-θ relations for Rubicon sandy loam shown in Figures 7.26 and 7.27. What is the critical-pressure head of this soil for a falling water table (drying) and a rising water table (wetting)?

7.7 A Rubicon sandy loam is underlain by a restricting layer at a depth of 5 m, which supports perched groundwater. After a long period of rain, during which the entire soil profile was essentially saturated, the perched water table has receded to a depth of 3 m. The perched water drains laterally to a surface stream. What percentage of the total lateral flow occurs above the water table?

7.8 A small industrial well of 0.2 m diameter continuously pumps 500 m^3/day from a confined aquifer with $T = 100$ m^2/day. The equilibrium water level in the well is 6.778 m below the static piezometric surface. Assuming steady state and no well-entry losses, calculate the radius of influence of the well.

A new well is constructed 400 m from the existing well. The new well has a diameter of 0.3 m and will pump continuously at 2000 m^3/day. Assuming that area of influence is proportional to well discharge, calculate the radius of influence of the new well as if it were alone in the aquifer.

Using superposition, calculate the depth of the water level in the old well below the static piezometric surface after the new well has been pumped sufficiently long for steady-state conditions to be reached. How much is the pumping depth in the old well increased?

REFERENCES

Anderson, T. W., 1968. Electrical-analog analysis of ground-water depletion in central Arizona. *U.S. Geol. Survey Water Supply Paper 1860.*

Anderson, T. W., 1972. Electrical-analog analysis of the hydrologic system, Tucson Basin, southeastern Arizona. *U.S. Geol. Survey Water Supply Paper 1939-C.*

Bear, J., 1960. Scales of viscous analogy models for groundwater studies. *J. Hydraul. Div., Am. Soc. Civ. Eng.* **86**(HY2): 11–23.

Bear, J., and G. Dagan, 1964. Moving interface in coastal aquifers. *J. Hydraul. Div., Am. Soc. Civ. Eng.* **90**(HY4): 193–216.

Bedinger, M. S., 1967. An electrical analog study of the geometry of limestone solution. *Ground Water* **5**(1): 24–28.

Bennett, G. D., M. J. Mundorff, and S. A. Hussain, 1968. Electric-analog studies of brine coning beneath fresh-water wells in the Punjab region, West Pakistan. *U.S. Geol. Survey Water Supply Paper 1608-J*.

Birtles, A. B., and W. B. Wilkinson, 1975. Mathematical simulation of groundwater abstraction from confined aquifers for river regulation. *Water Resour. Res.* **11**: 571–580.

Bittinger, M. W., H. R. Duke, and R. A. Longenbaugh, 1967. "*Mathematical Simulations for Better Aquifer Management.*" Proc. Symposium on Artificial Recharge and Management of Aquifers, Haifa, Israel, Int. Assoc. Sci. Hydrol. March 1967, 509–519.

Boothroyd, A. R., E. C. Cherry, and R. Makar, 1949. An electrolytic tank for the measurement of steady-state response, transient response, and allied properties of networks. *Inst. Electr. Eng. (London), Proc.* **96**(part I): 163–177.

Bouma, J., D. Hillel, F. D. Hole, and C. R. Amerman, 1971. Field measurement of unsaturated hydraulic conductivity by infiltration through crusts. *Proc. Soil Sci. Soc. Am.* **35**: 362–364.

Bouwer, H., 1959a. Theoretical aspects of flow above the water table in tile drainage of shallow, homogeneous soils. *Proc. Soil Sci. Soc. Am.* **23**: 260–263.

Bouwer, H., 1959b. Theoretical aspects of unsaturated flow in drainage and subirrigation. *Agric. Eng.* **40**: 395–400.

Bouwer, H., 1960. A study of final infiltration rates from ring infiltrometers and irrigation furrows with a resistance network analog. *Trans. 7th Internat. Congr. of Soil Sci.*, Madison, Wis., vol. I, Comm. VI: 448–456.

Bouwer, H., 1962. Analyzing groundwater mounds by resistance network. *J. Irrig. Drain. Div., Am. Soc. Civ. Eng.* **88**(IR3): 15–36.

Bouwer, H., 1964a. Measuring horizontal and vertical hydraulic conductivity of soil with the double-tube method. *Proc. Soil Sci. Soc. Am.* **28**: 19–23.

Bouwer, H., 1964b. Unsaturated flow in ground-water hydraulics. *J. Hydraul. Div., Am. Soc. Civ. Eng.* **90**(HY5): 121–144.

Bouwer, H., 1965. Theoretical aspects of seepage from open channels. *J. Hydraul. Div., Am. Soc. Civ. Eng.* **91**(HY3): 37–59.

Bouwer, H., 1966. Rapid field measurement of air-entry value and hydraulic conductivity of soil as significant parameters in flow system analysis. *Water Resour. Res.* **2**: 729–738.

Bouwer, H., 1967. Analyzing subsurface flow systems with electric analogs. *Water Resour. Res.* **3**: 897–907.

Bouwer, H., 1970. Groundwater recharge design for renovating waste water. *J. Sanit. Eng. Div., Am. Soc. Civ. Eng.* **96**(SA1): 59–74.

Bouwer, H., and R. D. Jackson, 1974. Determining soil properties. In *Drainage for Agriculture*, J. van Schilfgaarde (ed.), Agronomy Monograph No. 17, Am. Soc. Agron., Madison, Wis. 611–672.

Bouwer, H., and W. C. Little, 1959. A unifying numerical solution for two-dimensional steady flow problems in porous media with an electrical resistance network. *Proc. Soil. Sci. Soc. Am.* **23**: 91–96

Bouwer, H., and R. C. Rice, 1976. A slug test for determining hydraulic conductivity of unconfined aquifers with completely or partially penetrating wells. *Water Resour. Res.* **12**: 423–428.

Bredehoeft, J. D., H. H. Cooper, Jr., and I. S. Papadopulos, 1966. Inertial and storage effects in well-aquifer systems: an analog investigation. *Water Resour. Res.* **2**: 697–707.

Bredehoeft, J. D., and G. F. Pinder, 1970. Digital analysis of areal flow in multiaquifer groundwater systems: a quasi three-dimensional model. *Water Resour. Res.* **6**: 833–888.

Brown, R. H., 1962. Progress in ground water studies with the electric-analog model. *J. Am. Water Works Assoc.* **54**: 943–956.

Cahill, J. M., 1973. Hydraulic sand-model studies of miscible-fluid flow *J. Res. U.S. Geol. Surv.* **1**: 243–250.

Çeçen, K., E. Omay, and A. Siginer, 1969. The study of collector wells by means of viscous flow analogy. *Proc. 13th Congr. Int. Ass. Hydraul. Res.*, Kyoto, Japan, 4, D40-1–D40-10.

Childs, E. C., 1945. The water table, equipotentials, and streamlines in drained land: III. *Soil Sci.* **59**: 405–415.

Collins, M. A., L. W. Gelhar, and J. L. Wilson III, 1972. Hele-Shaw model of Long Island aquifer system. *J. Hydraul. Div., Am. Soc. Civ. Eng.* **98**(HY9): 1701–1714.

Columbus, N., 1966. The design and construction of Hele-Shaw models. *Ground Water* 4(2): 16–22.

Cooley, R. L., 1971. A finite difference method for unsteady flow in variable saturated porous media: application to a single pumping well. *Water Resour. Res.* 7: 1607–1625.

Davies, J., and R. Herbert, 1963. Ground-water flow by electrical analogue. *Water Res. Ass.*, Tech. Pap. No. 32, Medmenham, England.

DeBrine, E. D., 1970. Electrolytic model study for collector wells under river beds. *Water Resour. Res.* 6: 971–978.

De Josselin De Jong, G., 1962. Electric analog models for the solution of geohydrological problems. *Water* 4: 46.

DeWiest, R. J. M., 1969. Fundamental principles of ground-water flow. In *Flow through Porous Media*, R. J. M. DeWiest (ed.), Academic Press, New York, 1–53.

Donaldson, I. G., 1974. Underground waters of the Lower Hutt—a model study. *J. Hydrol. (New Zealand)* 13: 81–97.

Dvoracek, M. J., and V. H. Scott, 1963. Ground-water flow characteristics influenced by recharge pit geometry. *Trans. Am. Soc. Agric. Eng.* 6: 262–268.

Emery, P. A., 1966. Use of analog model to predict stream flow depletion, Big and Little Blue River Basin, Nebraska. *Ground Water* 4(4): 13–19.

Fayers, F. J., and J. W. Sheldon, 1962. The use of a high-speed digital computer in the study of the hydrodynamics of geologic basins. *J. Geophys. Res.* 67: 2421–2431.

Ferris, J. G., D. B. Knowles, R. H. Brown, and R. W. Stallman, 1962. Theory of aquifer tests. *U.S. Geol. Survey Water Supply Paper 1536-E.*

Fox, R. W., and A. T. McDonald, 1973. *Introduction to Fluid Mechanics.* John Wiley & Sons, New York, 630 pp.

Freeze, R. A., 1971. Three-dimensional, transient, saturated-unsaturated flow in a ground-water basin. *Water Resour. Res.* 7: 347–366.

Freeze, R. A., 1972. "A physically-based approach to hydrologic response modeling: phase I: Model development," Completion Rep., Contract No. 14-31-001-3694. *Water Resour. Res.*, Washington, D.C.

Gardner, W. R., 1958. Some steady-state solutions of the unsaturated moisture flow equation with application to evaporation from a water table. *Soil Sci.* 85: 228–233.

Getzen, R. T., 1977. Analog-model analysis of regional three-dimensional flow in the ground-water reservoir of Long Island, New York. U.S. Geol. Survey Prof. Paper No. 982, 49 pp.

Gillham, R. W., A Klute, and D. F. Heermann, 1976. Hydraulic properties of a porous medium: measurement and empirical representation. *Proc. Soil Sci. Soc. Am.* 40: 203–207.

Green, W. H., and G. A. Ampt, 1911. Studies on soil physics. I. The flow of air and water through soils. *J. Agr. Sci.* 4: 1–24.

Guitjens, J. C., 1974. Hydraulic models. In *Drainage for Agriculture*, J. van Schilfgaarde (ed.), *Agronomy Monograph No. 17*, Am. Soc. Agron., Madison, Wis., pp. 537–556.

Guymon, G. L., 1974. Digital computers and drainage problem analysis: Part III—Finite element method. In *Drainage for Agriculture*, J. van Schilfgaarde (ed.), *Agronomy Monograph No. 17*, Am. Soc. Agron., Madison, Wis., pp. 587–607.

Guymon, G. L., V. H. Scott, and L. R. Herrmann, 1970. A general numerical solution of the two-dimensional diffusion-convection equation by the finite element method. *Water Resour. Res.* 6: 1611–1617.

Hall, H. P., 1955. An investigation of steady flow toward a gravity well. *Houille Blanche* 10: 8–35.

Hansen, V. E., 1953. Unconfined groundwater flow to multiple wells. *Trans. Am. Soc. Civ. Eng.* 118: 1098–1130.

Hantush, M. S., 1964. Hydraulics of wells. In *Advances in Hydroscience*, vol. 1, V. T. Chow (ed.), Academic Press, New York, 282–432.

Harpaz, Y., and J. Bear, 1964. Investigations on mixing of waters in underground storage operations. *Int. Assoc. Sci. Hydrol.* 64: 132–153.

Harr, M. E., 1962. *Groundwater and Seepage.* McGraw-Hill Book Co., New York, 315 pp.

Herbert, R., 1968. Time variant ground water flow by resistance network analogues. *J. Hydrol.* 6: 237–264.

Herbert, R., and K. R. Rushton, 1966. Ground-water flow studies by resistance networks. *Geotechnique* **16:** 53–78.

Hillel, D., and W. R. Gardner, 1970. Measurement of unsaturated conductivity and diffusion by infiltration through an impeding layer. *Soil Sci.* **109:** 149–153.

Hornberger, G. M., I. Remson, and A. A. Fungaroli, 1969. Numeric studies of a composite soil moisture ground-water system. *Water Resour. Res.* **5:** 797–802.

Hunt, B. W., and D. D. Wilson, 1974. Graphical evaluation of aquifer transmissivities in Northern Canterbury, New Zealand. *J. Hydrol. (New Zealand)* **13:** 66–80.

Hunter Blair, A., 1966. A ground-water analog investigation of a multi-well pumping scheme at Otterbourne (Southampton Corporation Waterworks). *Water Res. Assoc.*, Tech. Pap. No. 56, Medmenham, England.

Hunter Blair, A., 1968. Simplifications of ground-water data used for an analogue of a coastal aquifer. *Bull. Int. Assoc. Sci. Hydrol.* **XIII**(3): 59–65.

Ibrahim, H. A., and W. Brutsaert, 1965. Inflow hydrographs from large unconfined aquifers. *J. Irrig. Drain. Div., Amer. Soc. Civ. Eng.* **91**(IR2): 21–38.

Javendel, I., and P. A. Witherspoon, 1968a. Application of the finite element method to transient flow in porous media. *J. Soc. Petrol. Eng.* **8:** 241–252.

Javendel, I., and P. A. Witherspoon, 1968b. A method of analyzing transient fluid flow in multilayered aquifers. *Water Resour. Res.* **5:** 856–869.

Karplus, W. J., 1958. *Analog Simulation.* McGraw-Hill Book Co., New York.

Karplus, W. J., 1967. Hybrid computer simulation of groundwater basins. Proc. Nation. Symp. Ground-Water Hydrol., San Francisco, Cal. *Am. Water Resour. Assoc.*, pp. 289–299.

Kimbler, O. K., 1970. Fluid model studies of the storage of freshwater in saline aquifers. *Water Resour. Res.* **6:** 1522–1527.

King, L. G., 1974. Flow through heterogeneous media. In *Drainage for Agriculture*, J. van Schilfgaarde (ed.), *Agronomy Monograph No. 17*, Am. Soc. Agron., Madison, Wis., pp. 271–307.

Kraijenhoff van de Leur, D. A., 1962. Some effects of the unsaturated zone on nonsteady free-surface groundwater flow as studied in a scaled granular model. *J. Geophys. Res.* **67:** 4347–4362.

Liebmann, G., 1950. Solution of partial differential equations with a resistance network analogue. *British Jour. Appl. Physics* **1:** 92–103.

Liebmann, G., 1954. Resistance-network analogues with unequal meshes or subdivided meshes. *Br. J. Appl. Physics* **5:** 362–366.

Luthin, J. N., 1952. An electrical resistance network solving drainage problems. *Soil Sci.* **75:** 259–274.

Luthin, J. N., 1974. Drainage analogues. In *Drainage for Agriculture*, J. van Schilfgaarde (ed.), *Agronomy Monograph No. 17*, Am. Soc. Agron., Madison, Wis., pp. 517–536.

Mack, L. E., 1957. Evaluation of a conducting-paper analog field plotter as an aid in solving ground-water problems," Bull. No. 127, part 2. State Geol. Surv. of Kansas, Lawrence.

Marino, M. A., 1967. Hele-Shaw model study of the growth and decay of groundwater ridges. *J. Geophys. Res.* **72:** 1195–1205.

Matlock, W. G., 1966. Recharge distribution determined by analog model. *Ground Water* **4**(3): 13–16.

Mobasheri, F., and M. Shahbazi, 1969. Steady-state lateral movement of water through the unsaturated zone of an unconfined aquifer. *Ground Water* **7**(6): 28–34.

Moulder, E. A. and C. T. Jenkins, 1969. Analog-digital models of stream-aquifer systems. *Ground Water* **7**(5): 19–24.

Muskat, M., 1946. *The Flow of Homogeneous Fluids Through Porous Media.* J. W. Edwards, Inc., Ann Arbor, Mich., 763 pp.

Nelson, R. W., 1961. In-place measurement of permeability in heterogeneous media. 2. Experimental and computational considerations. *J. Geophys. Res.* **66:** 2469–2478.

Nelson, R. W., 1962. Conditions for determining areal permeability distribution by calculation. *J. Soc. Petrol. Eng.* **2:** 223–224.

Nelson, R. W., 1968. In-place determination of permeability distribution for heterogeneous porous media through analysis of energy dissipation. *J. Soc. Petrol. Eng.* **8:** 33–42

Neuman, S. P., and P. A. Witherspoon, 1970a. Finite element method of analyzing steady seepage with a free surface. *Water Resour. Res.* **6:** 889–897.

Neuman, S. P., and P. A. Witherspoon, 1970b. Variational principles for confined and unconfined flow of ground water. *Water Resour. Res.* **6:** 1376–1382.

Nutbrown, D. A., R. A. Downing, and R. A. Monkhouse, 1975. The use of a digital model in the management of the Chalk aquifer in the South Downs, England. *J. Hydrol.* **27:** 127–142.

Peter, Y., 1970. Model tests for a horizontal well. *Ground Water* **8**(5): 30–34.

Philip, J. R., 1969. Theory of infiltration. In *Advances in Hydroscience*, vol. 5, V. T. Chow (ed.), Academic Press, New York, 215–296.

Pinder, G. F., 1970. A digital model for aquifer evaluation. Chap. C1 in *Techniques for Water-Resources Invest. of the U.S. Geol. Survey*, U.S. Gov't. Printing Office, Washington, D.C., 18 pp.

Pinder, G. F., and J. D. Bredehoeft, 1968. Application of the digital computer for aquifer evaluation. *Water Resour. Res.* **4:** 1069–1093.

Pinder, G. F., and H. H. Cooper, Jr., 1970. A numerical technique for calculating the transient position of the saltwater front. *Water Resour. Res.* **6:** 875–882.

Polubarinova-Kochina, P. Y., 1962. Theory of ground water movement. Translated from the Russian by J. M. R. DeWiest, Princeton University Press, Princeton, N.J., 613 pp.

Poulovassilis, A., and E. Tzimas, 1975. The hysteresis in the relationship between hydraulic conductivity and soil water content. *Soil Sci.* **120:** 327–331.

Prickett, T. A., 1967. Designing pumped well characteristics into electric analog models. *Ground Water* **5**(4): 38–46.

Prickett, T. A., 1975. Modeling techniques for groundwater evaluation. In *Advances in Hydroscience*, vol. 10, V. T. Chow (ed.), Academic Press, New York, 1–143.

Prickett, T. A., and C. G. Lonnquist, 1973. Aquifer simulation model for use on disk supported small computer system. Circ. No. 114, *Ill. State Water Surv.*, Urbana.

Raats, P. A. C., and W. R. Gardner, 1974. Movement of water in the unsaturated zone near a water table. In *Drainage for Agriculture*, J. van Schilfgaarde (ed.), *Agronomy Monograph 17*, Am. Soc. Agron., Madison, Wis., pp. 311–357.

Rahman, M. A., E. T. Smerdon, and E. A. Hiler, 1969. Effect of sediment concentration on well recharge in a fine sand aquifer. *Water Resour. Res.* **5:** 641–646.

Reed, J. E., and M. S. Bedinger, 1961. Projecting the effect of changed stream stages on the water table. *J. Geophys. Res.* **66:** 2423–2427.

Reisenauer, A. E., 1963. Methods for solving problems of multidimensional, partially saturated steady flow in soils. *J. Geophys. Res.* **68:** 5725–5733.

Remson, I., C. A. Appel, and R. A. Webster, 1965. Ground-water models solved by digital computer. *J. Hydraul. Div., Am. Soc. Civ. Eng.* **91**(HY3): 133–147.

Remson, I., G. M. Hornberger, and F. J. Molz, 1971. *Numerical Methods in Subsurface Hydrology.* John Wiley & Sons, New York, 389 pp.

Richards, L. A., 1931. Capillary conduction of liquids through porous mediums. *Physics* **1:** 318–333.

Rubin, J., 1968. Theoretical analysis of two-dimensional transient flow of water in saturated and partially saturated soils. *Proc. Soil Sci. Soc. Am.* **32:** 607–615.

Rumer, R. R., and D. R. F. Harleman, 1963. Intruded salt-water wedge in porous media. *J. Hydraul. Div., Am. Soc. Civ. Eng.* **89**(HY6): 193–220.

Rushton, K. R., and R. G. Bannister, 1970. Aquifer simulation on slow time resistance-capacitance networks. *Ground Water* **8**(4): 15–24.

Santing, G., 1958. A horizontal scale model based on the viscous flow analogy for studying ground-water flow in an aquifer having storage. *Int. Assoc. Sci. Hydrol.* **43:** 105–114.

Sherwood, C. B., and H. Klein, 1963. Use of analog plotter in water-control problems. *Ground Water* **1**(1): 8–15.

Skibitzke, H. E., 1963. The use of analogue computers for studies in ground-water hydrology. *J. Inst. Water Eng.* **17**(3): 216–230.

Smith, W. O., 1967. Infiltration in sands and its relation to groundwater recharge. *Water Resour. Res.* **3:** 539–555.

Snell, A. W., and J. van Schilfgaarde, 1964. Four-well method of measuring hydraulic conductivity in saturated soils. *Trans. Am. Soc. Agr. Eng.* **7:** 83–87, 91.

Southwell, R. V., 1946. *Relaxation Methods in Theoretical Physics.* Oxford University Press, London.

Stallman, R. W., 1963a. Calculation of resistance and error in an electric analog of steady flow through nonhomogeneous aquifers. *U.S. Geol. Survey Water Supply Paper 1544-G.*

Stallman, R. W., 1963b. Electric analog of three-dimensional flow to wells and its application to unconfined aquifers. *U.S. Geol. Survey Water Supply Paper 1536-H.*

Swartzendruber, D., 1969. The flow of water in unsaturated soils. In *Flow through Porous Media*, R. J. M. DeWiest (ed.), Academic Press, New York, pp. 215–292.

Taylor, G. S., 1974. Digital computers and drainage problem analyses. Part II—Finite difference methods. In *Drainage for Agriculture*, J. van Schilfgaarde (ed.), *Agronomy Monograph No. 17*, Am. Soc. Agron., Madison, Wis., pp. 567–586.

Taylor, G. S., and J. N. Luthin, 1969. Computer methods for transient analysis of water-table aquifers. *Water Resour. Res.* **5**: 144–152.

Todd, D. K., 1954. Unsteady flow in porous media by means of a Hele-Shaw viscous fluid model. *Trans. Am. Geophys. Union* **35**: 905–916.

Todd, D. K., and J. Bear, 1961. Seepage through layered anisotropic porous media. *J. Hydraul. Div., Am. Soc. Civ. Eng.* **87**(HY3): 31–57.

Tomlinson, L. M., and K. R. Rushton, 1975. The alternating direction explicit method for analysing groundwater flow. *J. Hydrol.* **27**: 267–274.

Topp, G. C., 1969. Soil-water hysteresis measured in a sandy loam and compared with the hysteretic domain model. *Proc. Soil Sci. Soc. Am.* **33**: 645–651.

Trescott, P. C., G. F. Pinder, and S. P. Larson, 1976. Finite-difference model for aquifer simulation in two dimensions with results of numerical experiments. Chap. C1 in Book 7 of *Automated Data Processing and Computations*, U.S. Geol. Surv., Washington, D.C., 116 pp.

U.S. Army Corps of Engineers, 1970. *Finite Element Solution of Steady State Potential Flow Problems*, Hydrol. Eng. Cent. Generalized Comp. Program 723-G2-L2440. U.S. Army Eng., Sacramento District, Davis, Cal.

van Deemter, J. J., 1950. Theoretical and numerical treatment of flow problems connected with drainage and irrigation (in Dutch). Verslagen Landbouwkundige Onderzoekingen No. 56.7, Government Printing Office, The Hague, Netherlands, 67 pp.

Vemuri, V., and J. A. Dracup, 1967. Analysis of nonlinearities in ground water hydrology: a hybrid computer approach. *Water Resour. Res.* **3**: 1047–1058.

Vemuri, V., and W. J. Karplus, 1969. Identification of nonlinear parameters of ground water basins by hybrid computation. *Water Resour. Res.* **5**: 172–185.

Vimoke, B. S., T. D. Tyra, T. J. Thiel, and G. S. Taylor, 1962. Improvements in construction and use of resistance networks for studying drainage problems. *Proc. Soil Sci. Soc. Am.* **26**: 203–207.

Walton, W. C., 1964. Electric analog computers and hydrogeologic system analysis in Illinois. *Ground Water* **2**(4): 38–48.

Walton, W. C., and T. A. Prickett, 1963. Hydrogeologic electric analog computers. *J. Hydraul. Div., Am. Soc. Civ. Eng.* **89**(HY6): 67–91.

Watkins, F. A., and J. E. Heisel, 1970. Electrical-analog-model study of water resources of the Columbus area, Bartholomew County, Indiana. *U.S. Geol. Surv. Water Supply Paper 1981.*

Watson, K. K., 1967. The measurement of the hydraulic conductivity of unsaturated porous materials utilizing a zone of entrapped air. *Proc. Soil Sci. Soc. Am.* **31**: 716–720.

Watson, K. K., 1974. Some applications of unsaturated flow theory. In *Drainage for Agriculture*, J. van Schilfgaarde (ed.), *Agronomy Monograph No. 17*, Am. Soc. Agron., Madison, Wis., pp. 359–405.

White, N. D., and W. F. Hardt, 1965. Electrical-analog analysis of hydrologic data for San Simon Basin, Cochise and Graham counties, Arizona. *U.S. Geol. Survey Water Supply Paper 1809-R.*

Witherspoon, P. A., I. Javandel, and S. P. Neuman, 1968. Use of finite element method in solving transient flow problems in aquifer systems. *Int. Assoc. Sci. Hydrol. Symp.* **81**: 687–698.

Wylie, C. R., 1951. *Advanced Engineering Mathematics*. McGraw-Hill Book Co., New York.

Youngs, E. G., 1968. Shape factors for Kirkham's piezometer method for determining the hydraulic conductivity of soil in situ for soils overlying an impermeable floor or infinitely permeable stratum. *Soil Sci.* **106**: 235–237.

Zee, C., D. F. Peterson, Jr., and R. O. Bock, 1957. Flow into a well by electric and membrane analogy. *Trans. Am. Soc. Civ. Eng.* **122**: 1088–1112.

Zienkiewicz, O., 1971. *The Finite Element Method in Engineering Science.* McGraw-Hill Book Co., New York.

Zienkiewicz, O., and Y. K. Cheung, 1967. *The Finite Element Method in Structural and Continuum Mechanics.* McGraw-Hill Book Co., New York, 272 pp.

Zienkiewicz, O., P. Mayer, and Y. K. Cheung, 1966. Solutions of anisotropic seepage by finite elements. *J. Eng. Mech. Div., Am. Soc. Civ. Eng.* **92**(EM1): 111–120.

EIGHT

SURFACE-SUBSURFACE WATER RELATIONS

The mechanisms whereby surface water becomes groundwater, and vice versa, are important in groundwater hydrology because they determine water balances and hydrologic safe yields of aquifers and groundwater basins. They are also important in situations of groundwater contamination by polluted surface water (including leachate from garbage disposal sites and other point sources), and of degradation of surface water by discharge of saline or other low-quality groundwater. Surface water becomes groundwater through infiltration of rain and irrigation water, seepage from streams and canals, and artificial recharge with infiltration basins or injection wells. Groundwater returns to the atmosphere by evaporation from soil or vegetation, and to the surface by drainage into streams or other surface water, flow from springs or seeps, discharge of agricultural underground drainage systems, and, of course, flow from pumped or free-flowing wells. Quantitative aspects of these mechanisms are discussed in the following sections.

8.1 INFILTRATION, EVAPOTRANSPIRATION, AND RECHARGE

The maximum rate at which water can move into the soil is called the *infiltration capacity* or *potential infiltration* rate. This is the rate that will occur when the supply of water at the surface is not limiting, as when the soil is covered by ponded rainfall, surface runoff, irrigation water, and streams or other bodies of surface water. If the infiltration is due to rainfall, all the water will infiltrate if the rainfall intensity is less than the potential infiltration rate. However, if the rainfall intensity exceeds the potential infiltration rate, the excess rain cannot move into the soil and will produce surface runoff. The potential infiltration rate is highest at the beginning of an infiltration event, but decreases as infiltration continues and the wetted zone in the soil expands downward. The potential infiltration rate may eventually become constant.

There are two types of equations that express the potential infiltration rate in relation to time and soil hydraulic properties: physical equations and empirical equations.

8.1.1 Physical Infiltration Equations

One of the earliest physical infiltration equations was developed by Green and Ampt (1911), which in light of subsequent research (Bouwer, 1966; Fok, 1975; Mein and Farrell, 1974; Mein and Larson, 1973; Morel-Seytoux and Khanji, 1974; Neuman, 1976) can be written as

$$v_i = K \frac{H_w + L_f - h_{cr}}{L_f} \tag{8.1}$$

where v_i = infiltration rate (length/time)

K = hydraulic conductivity of wetted zone

H_w = depth of water above soil

h_{cr} = critical pressure head of soil for wetting (see Section 7.7.3)

L_f = depth of wetting front (Figure 8.1)

This equation is obtained by applying Darcy's equation to the wetted zone, assuming vertical flow and uniform water content and hydraulic conductivity in the wetted zone. The wetting front is considered as an abrupt interface between wetted and nonwetted material. Thus, the infiltration system is treated as "piston" flow. The term v_i in Eq. (8.1) is the Darcy velocity, which expresses the infiltration rate as the rate of fall dH_w/dt of the water surface above the ground if no water were added to the ponded water. The K value normally is less than K at saturation because entrapped air prevents complete saturation. Thus, K in Eq. (8.1) refers to the "resaturated" K, which may be about one-half of K at

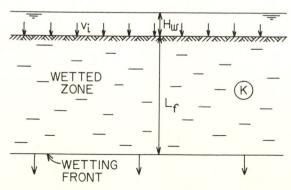

Figure 8.1 Geometry and symbols for piston-flow infiltration system.

saturation (Bouwer, 1966). The water depth H_w may range from zero at incipient ponding (when the potential infiltration rate has decreased so that it has just become equal to the rainfall intensity) to a few millimeters for sheet surface runoff, 5 to 20 cm for flood-type irrigation systems, and several meters or more for streams, canals, or reservoirs. The pressure head at the wetting front as it advances downward is taken as the critical-pressure head (see Section 7.7), which must be selected for the wetting situation. Values of h_{cr} vary from -10 cm or more for coarse material to -100 cm or less for fine soils. Both h_{cr} and the resaturated K can be measured in the field with the air-entry permeameter (Bouwer, 1966). Equation (8.1) shows that as the wet front advances downward (L_f increases), v_i decreases and approaches K when L_f becomes large compared to $H_w - h_{cr}$. Thus, the final infiltration rate of a deep, uniform soil is equal to K.

The rate of advance of the wetting front dL_f/dt is equal to v_i/f, where f is the fillable porosity (difference between volumetric water content of soil before and after wetting). Substituting Eq. (8.1) for v_i in this expression, integrating, and solving for t then yields

$$t = \frac{f}{K}\left[L_f - (H_w - h_{cr})\ln\frac{H_w + L_f - h_{cr}}{H_w - h_{cr}}\right] \qquad (8.2)$$

where t is the time since the start of infiltration. Since the total amount or depth I_t of water that moved into the soil since infiltration began can be expressed as $f L_f$, L_f in Eq. (8.2) can be replaced by I_t/f to obtain the relation between I_t (dimension of length) and t. Thus, if K, h_{cr}, and f are known for a given soil profile, I_t versus t can be calculated with Eq. (8.2). The relation between v_i and t is then determined by taking L_f as I_t/f for various t values and calculating v_i with Eq. (8.1).

Equations (8.1) and (8.2) can also be used to calculate v_i and I_t versus t for soil profiles where f varies with depth. Normally, f decreases with depth because the water content of the soil before wetting tends to increase with depth. The soil profile is then split up into a number of layers of thickness Δz, each with its own f value. The time Δt it takes for the wet front to traverse each layer is calculated with Eq. (8.2), after which the relation between I_t and t is calculated by summing $f\,\Delta z$ and Δt for each layer. The tabular procedure for this calculation can be extended to nonuniform soils that become less permeable or more permeable with depth (Bouwer, 1969a and 1976, respectively). Where the soil becomes more permeable with depth, the upper layers of the profile are completely wetted in the same way as for soils that are uniform or become less permeable with depth. However, when the wetting front in the soil with increasing K reaches a depth where K is equal to v_i, the material below this depth will not be completely wetted and v_i will then remain essentially constant (Whisler et al., 1972; Bouwer, 1976).

The piston-flow assumption of the Green-Ampt equation is a simplification of true infiltration flow where the wetting front is more diffuse and water contents (and thus also hydraulic conductivities) in the wetted zone increase with time, particularly above the wetting front. A solution for the infiltration-flow system based on Eq. (7.52) with a soil-water diffusivity term was obtained by Philip (1969, and references therein). The resulting equation consisted of an infinite series of which the first two terms were the most significant. Thus, the equation could be written as

$$v_i = \tfrac{1}{2}S_i t^{-1/2} + A \tag{8.3}$$

for the infiltration rate and

$$I_t = S_i t^{1/2} + At \tag{8.4}$$

for the accumulated infiltration, where S_i is the sorptivity of the soil (dimension length/time$^{1/2}$) and A is related to K (dimension length/time).

The sorptivity depends on the pore configuration of the soil and is also affected by the initial water content and H_w. Values of S_i for a given soil are best determined experimentally, measuring I_t versus t in the field with cylinder infiltrometers (see next section) or in the laboratory on undisturbed samples. Equation (8.4) shows that for very small t, I_t is dominated by the first term of the equation. Thus, plotting I_t versus $t^{1/2}$ should initially yield a straight line with slope S_i. Talsma (1969) found that such straight lines were indeed obtained if t did not exceed 1 or 2 min. He reported S_i values that ranged from about 0.17 m/day$^{1/2}$ for clay loams to 1.2 m/day$^{1/2}$ for sands.

For large t, I_t is dominated by the second term in Eq. (8.4) as the infiltration rate approaches A [Eq. (8.3)]. Thus, A is related to K of the wetted zone. For long infiltration situations, A may be taken equal to K. For relatively short infiltration events, A may range between $K/3$ and $2K/3$ (Youngs, 1968; Talsma and Parlange, 1972) and can be taken as $K/2$. Equations (8.3) and (8.4) are not readily adapted to nonuniform soils where water content and/or hydraulic conductivity vary with depth.

8.1.2 Empirical Infiltration Equations

In reality, infiltration-flow systems are much more complex than can be expressed with a simple physical equation. Interaction between the infiltrating water and the soil (swelling and flocculation or deflocculation of the clay), rearrangement of soil particles near the surface due to the impact of raindrops or erosion by runoff or other flowing water, accumulation of fines on the surface, bacterial growth on the surface and in the soil, activity of worms and other soil fauna, and the inherent nonuniformity of the soil limit the accuracy with which equations based on Darcy's law or diffusion theory can predict infiltration rates. The other approach, then, is to use empirical equations with "constants" calculated from measured relations between v_i and t or between I_t and t.

The simplest empirical equation is the one by Kostiakov (1932), where I_t is expressed as

$$I_t = Ct^{\alpha} \tag{8.5}$$

The parameters C and α are readily obtained from a plot of the measured values of I_t and t on double logarithmic paper, which should yield a straight line. Differentiating Eq. (8.5) with respect to t gives

$$v_i = C\alpha t^{\alpha - 1} \tag{8.6}$$

Another empirical equation is by Horton (1940), who expressed v_i as

$$v_i = v_\infty + (v_0 - v_\infty)e^{-\beta t} \tag{8.7}$$

where v_0 and v_∞ are the initial and final infiltration rates, respectively, and β is calculated from field measurements as $t^{-1} \ln [(v_0 - v_\infty)/(v_i - v_\infty)]$. A disadvantage of Horton's equation is that theoretically $v_0 = \infty$, so that a realistic value for v_0 is difficult to select. When using infiltration data calculated with one of the physically derived equations, β is not constant and a good fit cannot be obtained. The Horton equation seems to be most suitable for describing infiltration when the water is applied by rain or sprinkling, and then only for relatively short periods.

Field data of v_i, I_t, and t for evaluation of the parameters in empirical infiltration equations must be obtained for the same conditions as will occur for the infiltration system to be predicted with the equations. These conditions include length of infiltration event, type of water application (sprinkling or flooding), depth of flooding, velocity of water above ground (ponded or flowing), soil conditions (surface condition of agricultural fields, vegetation, and soil water-content distribution), and quality of infiltrating water.

Infiltration measurements to predict infiltration for ponded conditions are commonly obtained with *cylinder infiltrometers*. These are cylindrical devices of about 0.2 to 0.5 m in diameter that are pushed or driven 5 to 10 cm in the soil. A constant water level is maintained inside the cylinder, and the amounts of added water are measured to obtain the relation between I_t and t. The main source of error with this technique is lateral divergence of the flow below the cylinder, which may be due to unsaturated flow (Bouwer, 1961; Swartzendruber and Olsen, 1961; Talsma, 1970) or to restricting layers in the soil (Evans et al., 1950). Since this divergence will not occur when infiltration takes place over a large area like a field or a watershed, it will lead to overestimation of I_t. Divergence of flow below the cylinder due to unsaturated flow can be minimized by increasing the diameter of the cylinder. This was more effective than the use of a smaller cylinder placed concentrically inside the regular cylinder and measuring I_t for the inner or "buffered" cylinder only while maintaining equal water levels in both (Bouwer, 1961). Flow divergence due to lateral flow above restricting layers deeper in the profile can be avoided only by using large, almost field-sized areas for the infiltration measurements. Where this is impractical, the relation between I_t and t can better be based on measured K values for the different layers (using the techniques described in Chapter 5), which are then used to calculate infiltration rates with Eqs. (8.1) and (8.2) for layered soil (Bouwer, 1969a and 1976).

Infiltration of water into soil has been the subject of numerous studies, including comparisons between infiltration equations and number of "point" measurements required to obtain the average infiltration rate for entire fields or watersheds. For additional information, reference is made to the publications cited in this section and the articles mentioned therein.

8.1.3 Infiltration of Rainfall

The equations in the previous sections express the infiltration capacity or potential infiltration rate that occurs when water is ponded on the surface (including incipient or sheet runoff when H_w is zero or very small, respectively). If the rainfall intensity v_r is less than the potential infiltration rate v_i of the soil, all rain will infiltrate. If $v_r > v_i$, runoff will be generated at the rate of $v_r - v_i$. Since v_i decreases with time, v_i may initially be greater than a certain constant rainfall intensity v_r. As rain continues, v_i may become equal to v_r (incipient ponding) and then less than v_r (runoff production), as illustrated in Figure 8.2.

Assuming that the depth of the wetting front and the conditions in the wetted zone for infiltration at less than potential rate are the same as for infiltration at potential rate (Mein and Larson, 1973), the time t_p for incipient ponding to occur can be calculated with Eq. (8.1). Calling the accumulated infiltration at incipient ponding I_{tp}, the depth L_f of the wetting front at incipient ponding will be I_{tp}/f. Also, since v_r is constant, t_p is equal to I_{tp}/v_r. Substituting these relations into Eq. (8.1) and solving for t_p then yields

$$t_p = \frac{-h_{cr} f K}{v_r(v_r - K)} \tag{8.8}$$

Infiltration rates for $t > t_p$ are calculated with Eq. (8.1), starting with $L_f = I_{tp}/f$. Since the infiltration rates for $t < t_p$ are less than v_i, the starting point for the $v_i - t$ curve is to the right of the starting point of the rainfall event (Figure 8.2).

The above procedure can be used to calculate how much infiltration and runoff are obtained from a rain with uniform intensity that initially is less than the potential infiltration rate but later exceeds it. The results of this procedure

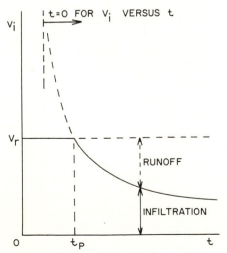

Figure 8.2 Infiltration and runoff for rain of uniform intensity.

compared favorably with a more accurate computer solution of an infiltration event (Mein and Larson, 1973), indicating that the basic assumption presented at the beginning of the previous paragraph is valid.

An extension of this assumption, known as the time-compression approxima-tion (TCA), is the basis for predicting how much infiltration and runoff are ob-tained during a rainfall event with varying intensities, including brief dry or drizzle periods (Reeves and Miller, 1975). According to this approximation, the wetted-zone conditions after a temporary reduction in v_r catch up very rapidly once v_r reaches its original value again, and infiltration proceeds at about the same rate as if the same accumulated infiltration had been obtained without temporary reduc-tion in v_r.

A graphical application of the TCA, developed by Reeves and Miller (1975), is shown in Figure 8.3 for a rainfall event with varying intensities (top) and a given curve of accumulated potential infiltration versus time (bottom). The infiltration curve may have been obtained from cylinder-infiltrometer measurements or from one of the potential-infiltration equations, Eqs. (8.2), (8.4), (8.5), or (8.7). The accumulated-rainfall curve is split into a number of straight-line segments of constant v_r. To determine how much of the rain infiltrated into the soil and how much ran off, each segment is matched against the infiltration curve for the corresponding time portion. When the rainfall segment is steeper than the corre-sponding section of the infiltration curve, the part above the infiltration curve is runoff (shown as dashed vertical lines in Figure 8.3) and the rest is infiltration (solid vertical lines). When the rainfall segment lies below the corresponding part of the infiltration curve, all the rain for that segment infiltrates into the soil. When a rainfall segment ends above the corresponding part of the infiltration curve, the next rainfall segment is plotted from the point of the infiltration curve that is vertically below the end point of the previous rainfall segment (segments *BC* and *DE* in Figure 8.3). When a rainfall segment ends below the corresponding part of the infiltration curve, the next segment is plotted from the point of the infiltration curve that is horizontally to the left of the end point of the previous segment (segments *CD* and *EF*). The total amounts of runoff and infiltration for the entire rainfall event are calculated by summing the runoffs (vertical dashed lines) and infiltrations (vertical solid lines) for each segment of the rainfall curve. These amounts compared reasonably well with computer solutions that more accurately accounted for the effect of varying v_r on infiltration (Reeves and Miller, 1975). The largest errors (about 15 to 20 percent) occurred for uncrusted soils, long durations of initial v_r, and long periods of reduced v_r (drizzle durations).

If dry or drizzle periods are sufficiently long to allow significant drainage of the soil profile, which may take 1 to 2 days for sandy soils and 5 to 10 days for heavy soils, redistribution and reduced water contents in the soil must be taken into account when predicting infiltration for the next rainfall event. If the rains are separated by dry periods, evaporation from the soil and transpiration by plants (collectively called *evapotranspiration*) cause additional reductions in the soil water content, particularly at and near the surface.

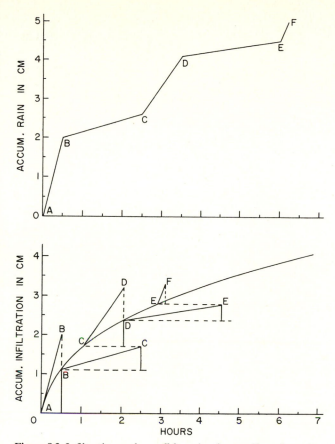

Figure 8.3 Infiltration and runoff for rain of nonuniform intensity.

8.1.4 Profile Drainage

Calculated redistribution of water due to drainage of the soil after infiltration is illustrated in Figure 8.4 for a sand (taken from Watson, 1974; see also Baver et al., 1972). The sand was wetted to a depth of about 35 cm when infiltration was stopped. Then, as drainage started, water contents decreased near the surface and increased below the original wetted zone, causing the wetted zone to move down as a wave that became flatter with increasing time. After about 3 h, the water-content profile began to approach the initial value of about 0.06 prior to wetting, which can be interpreted as the field capacity of the material. The water table was kept at a depth of 180 cm. The increased water content for the 35-cm zone above the water table represents the capillary fringe. The water-content curves in Figure 8.4 were obtained by computer, taking hysteresis effects into account. Using a similar approach, Watson and Lees (1975) developed a hysteretic infiltration-redistribution program to predict rainfall and runoff for a certain rainfall sequence.

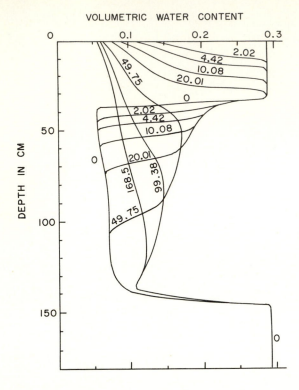

Figure 8.4 Calculated water-content profiles in sand at various times (in minutes on the curves) after cessation of infiltration. (*Redrawn from Watson, 1974.*)

The drainage of a saturated profile above a water table or a permeable drainage layer follows a decay-type curve, which can be described by the equation (Youngs, 1960):

$$\frac{D_t}{D_\infty} = 1 - e^{-d_0 t / D_\infty} \qquad (8.9)$$

where D_t = amount of water (expressed as a depth or volume per unit area) that has drained from the soil profile at time t

D_∞ = amount of water that will ultimately drain from the profile (at $t = \infty$)

d_0 = initial drainage rate after cessation of infiltration (length/time)

t = time since start of drainage

The initial drainage rates can be taken as equal to the average vertical hydraulic conductivity of the wetted zone prior to drainage (K when saturated, K_h when partially saturated). Most of the drainage will take place in the beginning of the drainage period, until the water content has reached field capacity. This is a rather arbitrary point on the drainage curve that is used by soil scientists as indicative of the amount of water that a soil can hold against the force of gravity (Baver et al., 1972). Sometimes, field capacity is taken as the water content at a pressure head of -3.06 m water (-0.3 bar). Field capacities (expressed as volumetric water content) range from about 5 percent for sands to 10 percent for sandy loams, 25 percent for loams, and 40 percent for clays. When a soil has reached field capacity, which may take about a day after complete wetting for sands to 5 days or more for clays, gravity drainage is considerably reduced and evapotranspiration becomes the dominant process for additional water removal, at least from the surface soil or root zone (depending on whether the soil is bare or vegetated).

Knowing evapotranspiration rates and the depth and distribution of water uptake by plant roots (see next section), water-content profiles in the soil after an infiltration event can be calculated, which in turn enables prediction of infiltration rates for the next infiltration event (using, for example, the Green-Ampt equation for nonuniform soil profiles). This procedure can be carried out for selected periods to estimate total infiltration and groundwater accretion. It can also be applied to special situations, such as estimating leaching from garbage disposal sites covered by a layer of soil. In hydrologic models, redistribution and evapotranspiration of soil water between rainfall events are often accounted for in a statistical manner as the number of days between events. While this procedure for describing antecedent soil-water conditions may be sufficiently accurate for predicting peak runoff rates, it may not be sufficiently accurate for predicting infiltration rates, which are more important from a groundwater-recharge standpoint.

8.1.5 Evapotranspiration

The rate at which water is returned from the soil to the atmosphere by evapotranspiration ET (evaporation from soil and transpiration by plants) is controlled by two factors: atmospheric demand and soil-water availability. If soil water at the surface or in the root zone is not limiting, ET is equal to the potential rate as determined by air temperature, wind speed, relative humidity, solar radiation, and other meteorologic conditions. Most evaporation data have been obtained using evaporation pans placed on the ground, like the U.S. Weather Bureau's class A pan, which has a diameter of 1.22 m and a depth of 0.25 m. Because the walls of such pans are exposed to the atmosphere, there is more heat exchange between the water in the pan and the surrounding air than would be true for water in a lake or in the soil. This causes pan evaporation to be greater than lake evaporation (about 25 percent more in Arizona, according to Cooley, 1970) and also greater than the potential ET for soil and vegetation.

The relation between pan evaporation and ET of well-watered crops has been studied for various conditions. A summary of the results by Doorenbos and Pruitt (1974; see also Jensen et al., 1974) shows that ET for well-watered grass ranges from 35 to 85 percent of the evaporation from a class A pan, depending on wind speed, relative humidity, upwind conditions, and pan environment (Table 8.1). Daily pan evaporation data are available from most weather bureaus and agricultural experimental stations. Formulas for calculating pan evaporation from meteorologic data were reviewed and compared by Burman (1976). Annual pan evaporation rates vary from about 1 m for the midwest to 3 m and more for the

Table 8.1 Ratio between evapotranspiration from well-watered grass and evaporation from class A pan

Wind, km/day	Upwind fetch of green crop, m	Case 1: Pan surrounded by short green crop			Upwind fetch of dry fallow, m	Case 2: Pan surrounded by dry-surface ground*		
		Relative humidity percent†				Relative humidity percent†		
		Low 20–40	Med 40–70	High > 70		Low 20–40	Med 40–70	High > 70
Light	0	0.55	0.65	0.75	0	0.7	0.8	0.85
	10	0.65	0.75	0.85	10	0.6	0.7	0.8
< 170 km/day	100	0.7	0.8	0.85	100	0.55	0.65	0.75
	1 000	0.75	0.85	0.85	1 000	0.5	0.6	0.7
Moderate	0	0.5	0.6	0.65	0	0.65	0.75	0.8
	10	0.6	0.7	0.75	10	0.55	0.65	0.7
170–425 km/day	100	0.65	0.75	0.8	100	0.5	0.6	0.65
	1 000	0.7	0.8	0.8	1 000	0.45	0.55	0.6
Strong	0	0.45	0.5	0.6	0	0.6	0.65	0.7
	10	0.55	0.6	0.65	10	0.5	0.55	0.65
425–700 km/day	100	0.6	0.65	0.7	100	0.45	0.5	0.6
	1 000	0.65	0.7	0.75	1 000	0.4	0.45	0.55
Very strong	0	0.4	0.45	0.5	0	0.5	0.6	0.65
	10	0.45	0.55	0.6	10	0.45	0.5	0.55
> 700 km/day	100	0.5	0.6	0.65	100	0.4	0.45	0.5
	1 000	0.55	0.6	0.65	1 000	0.35	0.4	0.45

Source: From Doorenbos and Pruitt, 1974.

* These coefficients apply only to conditions when the surface of soil is dry. Following rains for a day or two in midsummer and for longer periods in fall, winter, and spring, such pans are essentially equivalent to case 1 pans with large fetches. Rains also affect case 1 pans by essentially increasing the 0- and 10-meter fetch to an effective moist surface fetch of 1 000 m or more. One large advantage of case 2 pans is the minimal upkeep involved in such a siting of the pan, i.e., only an effective weed control program is needed for the site.

† Mean of maximum and minimum relative humidities.

southwest of the United States (Kohler et al., 1955). Potential ET rates range from about 2 mm/day for cool, humid climates to 12 mm/day and occasionally even more for hot, dry climates.

Numerous techniques for estimating potential ET have been developed, ranging from empirical equations to physical equations. The latter have been derived from the energy balance of the soil and plant surface, the mass transport of water vapor above the soil and plant surface, or a combination of the two. The success of these methods, which were reviewed by Jensen et al. (1974), depends on how well the equations describe the physical processes and on how accurately the meteorologic input data can be measured.

A well-known combination-method equation is the Penman equation (Penman, 1963; Jensen et al., 1974), which allows calculation of ET for well-watered grass from measurements of net radiation, heat flux in ground, wind speed, vapor pressure, psychrometric constant, and slope of the curve relating saturation vapor pressure to temperature. Jensen and Haise (1963; see also Jensen et al., 1970 and 1974) developed an equation for calculating potential ET from measurements of daily solar radiation and mean air temperature, plus other climatological factors. They used 3 000 measurements of ET obtained over a period of 35 years to fit the data. An empirical equation frequently used in irrigated agriculture in the western U.S. and other parts of the world is the Blaney-Criddle equation (Jensen et al., 1974, and references therein), which expresses potential ET in terms of an experimentally determined crop coefficient, mean monthly air temperature, and mean monthly percentage of annual daytime hours. The crop coefficient varies from 0.5 to 1, depending on type of crop and stage of growth (Jensen et al., 1974).

In a comparative 5-year study carried out at Kimberly, Idaho, for the month of July, the following data were obtained for the average daily potential ET of alfalfa (Jensen et al., 1974):

Actual rate measured with weighing lysimeters	8.1 mm/day
Class A pan (using correction factor of 0.8 from Table 8.1)	7.1 mm/day
Penman equation	7.2 mm/day
Jensen-Haise equation	7.3 mm/day
Blaney-Criddle equation	5.8 mm/day

These data show that the class A pan and the Penman and Jensen-Haise equations all gave about the same estimate of ET, which for this location was approximately 11 percent below the true value. This supports earlier conclusions that evaporation data from well-maintained class A pan installations can yield estimates of ET with an accuracy of about 10 percent—sufficient for most practical purposes (Doorenbos and Pruitt, 1974). Larger errors may occur under conditions of hot, dry weather with strong winds, where pan evaporation may be much higher than crop ET because of advective energy.

Potential ET for vegetated areas is reached only if soil water is not limiting and if plants are actively growing and fully covering the soil. The time for row crops to develop full cover is about 35 days after planting for beans, 65 days for late potatoes, 80 days for early potatoes, 85 days for corn, 110 days for sugar beets, and 140 days for cotton (Jensen et al., 1974). For grain crops and sorghum, full cover is reached when the plants start heading out. When full cover has not yet been attained, ET will be less than potential ET. The ratio between ET at partial cover to that at full cover is called the *crop coefficient*. Values of this coefficient, experimentally determined for irrigated crops in arid regions and for use with the Blaney-Criddle equation, are shown in Table 8.2 (taken from Jensen et al., 1974). For citrus orchards, the crop coefficient is between 0.5 and 0.6.

Table 8.2 Ratio of evapotranspiration at partial cover (top) and at maturity (bottom) to that at full cover for various crops

Crop	Percentage of time from day of planting to day of full cover									
	10	20	30	40	50	60	70	80	90	100
Small grains	0.16	0.18	0.25	0.37	0.51	0.67	0.82	0.94	1.02	1.04
Beans	0.20	0.23	0.30	0.39	0.51	0.63	0.76	0.88	0.98	1.07
Peas	0.20	0.24	0.31	0.40	0.51	0.63	0.75	0.87	0.97	1.05
Potatoes	0.10	0.13	0.20	0.30	0.41	0.53	0.65	0.76	0.85	0.91
Sugar beets	0.10	0.13	0.20	0.30	0.41	0.53	0.65	0.76	0.85	0.91
Corn	0.20	0.23	0.29	0.38	0.49	0.61	0.72	0.82	0.91	0.96
Alfalfa	0.36	0.47	0.58	0.68	0.79	0.90	1.00	1.00	1.00	1.00
Pasture	0.87	0.87	0.87	0.87	0.87	0.87	0.87	0.87	0.87	0.87

	Days after full cover									
	10	20	30	40	50	60	70	80	90	100
Small grains	1.04	0.94	0.74	0.49	0.19	0.10	0.10	0.10	0.10	0.10
Beans	1.02	0.96	0.85	0.73	0.59	0.45	0.31	0.19	0.10	0.10
Peas	0.98	1.02	0.99	0.76	0.20	0.10	0.10	0.10	0.10	0.10
Potatoes	0.90	0.85	0.75	0.60	0.38	0.10	0.10	0.10	0.10	0.10
Sugar beets	0.90	0.90	0.90	0.90	0.90	0.90	0.90	0.90	0.90	0.90
Corn	0.99	0.99	0.93	0.82	0.68	0.54	0.40	0.28	0.20	0.17
Alfalfa	0.75	1.00	1.00	1.00	1.00	1.00	1.00	1.00	1.00	1.00
Pasture	0.87	0.87	0.87	0.87	0.87	0.87	0.87	0.87	0.87	0.87

Source: From Jensen et al., 1974.

A more rigorous approach toward predicting *ET* for crops with incomplete cover was developed by Ritchie (1972), who used the leaf-area index of the crop to separately calculate transpiration by the crop and evaporation from the soil. The soil-water supply to the crop roots was considered unrestricted, but a decrease in evaporation from the soil surface as the soil dried out was taken into account. A similar approach was used by Tanner and Jury (1976), but with different techniques for estimating transpiration and evaporation.

Most of the water used for *ET* is extracted from the upper soil layers because that is where plant roots are concentrated. For irrigated crops, a rule of thumb is that 40 percent of *ET* is derived from the top one-fourth of the root zone, 30 percent from the second one-fourth, 20 percent from the third one-fourth, and 10 percent from the bottom one-fourth of the root zone. Depths of root zones vary from about 0.5 m for shallow rooted crops like certain grasses and vegetables, to 1 to 2 m for most field crops and several meters for small to medium trees (some tree roots may go down 10 m or more). Based on this distribution, the soil-water profile in the root zone can be calculated and used for predicting infiltration rates for a new rainfall or irrigation event.

As the crop reaches maturity and becomes senescent, ET is less than potential ET because the crop no longer is actively growing. Crop coefficients thus decline after full cover is reached, as shown in the lower part of Table 8.2. Exceptions, of course, are forage crops and other plants that continue to grow actively after full cover is reached. The data in Table 8.2, which apply to well-irrigated crops with soil-water conditions not limiting plant growth, can be used as a guide to estimate potential ET for vegetated surfaces with incomplete cover or plants reaching maturity or dormancy.

With continued evapotranspiration, soil-water content declines until it reaches a level where the plant roots can no longer extract the water. The lower limit of soil-water availability to plant roots is the wilting point, which is the water content at a pressure head of -153 m of water (-15 bar). Volumetric water contents at the wilting point range from about 2 percent or less for sands to 30 percent or more for clays. Plants differ in their reaction to decreasing water contents. For some plants, ET remains essentially at potential rate until the wilting point is reached and then suddenly reduces to almost zero. Other plants show a more gradual reduction in ET as the wilting point is approached. The wilting point will be reached first in the upper part of the root zone, from where it can be expected to advance downward as the deeper roots continue to take up water.

Evaporation from bare soil is similar to that from a vegetated surface, except that evaporation sooner is reduced to rates that are less than potential rates, because K_h becomes very small when the surface soil dries out (see Section 8.5). Subsequent evaporation rates are then only a fraction of potential values (Idso et al., 1974; Ritchie, 1972; Tanner and Jury, 1976).

A promising new approach for calculating ET utilizes the temperature of the evaporating surface (soil, plant leaves, or water), as measured with an infrared thermometer. Using this approach, Idso et al. (1975) developed the equation

$$ET = 0.001\,72[S_T - S_R + 1.56(R_A - R_G) + 156] \qquad (8.10)$$

where ET = evaporation in cm/day from a well-watered soil or plant surface
$0.001\,72 = 1/580 =$ conversion from cal/cm^2 per day (langleys/day) to cm/day of water evaporated
S_T = incoming solar radiation (cal/cm^{2}day)
S_R = reflected solar radiation (cal/cm^{2}day)
R_A = incoming long-wave radiation (cal/cm^{2}day)
R_G = outgoing long-wave radiation (cal/cm^{2}day)

The coefficients 1.56 and 156 were empirically determined using measured evaporation from bare soil. The same values have, however, also given good estimates of ET for crops. S_T and S_R can be measured with upright and inverted solarimeters, respectively. For clear days, $S_T - S_R$ can be calculated as $(1 - \alpha)S_T$, where α is the albedo of the evaporating surface. Values of S_T are available from weather bureaus. The albedo or reflectivity coefficient can be measured directly for the surface in question or it can be estimated, using, for example, values of about 0.25 for green vegetation, 0.15 for dark soil, and 0.4 for light-colored soil. R_A is a function of the air temperature and can be calculated with the Idso-Jackson equation (Idso and Jackson, 1969). Values of R_A for different temperatures were tabulated by Brown (1973). To obtain a daily average, R_A is evaluated from the average of the maximum and minimum air temperature on that day. R_G is calculated with the Stefan-Boltzmann equation from the measured temperature of the evaporating surface. Values of R_G for different temperatures were tabulated by Brown (1973). The average of the daily maximum and minimum surface temperature is used to obtain a daily average of R_G. The temperature of the evaporating surface is conveniently measured with an infrared thermometer, which is held 1 or several meters above the evaporating surface. The device can also be mounted in low-flying aircraft, allowing remote sensing of surface temperatures. Sensing from higher altitudes or satellites may also be possible, if corrections for water vapor, particulate matter, and other intervening factors in the atmosphere can be developed.

The advantage of Eq. (8.10) is that, for a number of cases, it will enable calculation of daily ET using only two measurements: the maximum and minimum temperature of the evaporating surface. The other factors can be obtained from weather-bureau data or other sources. The temperature of the evaporating surface integrates a number of factors that influence evaporation, such as wind speed, relative humidity, and advective energy (clothesline effects, etc.). The temperature of the evaporating soil or leaf surface is also affected by the water content of the soil. When soil water is not limiting evaporation or transpiration, the temperature of the evaporating surface will be lower than when soil water is limiting. This may extend the utility of the method to predict ET for situations where evaporation is restricted by limited availability of soil water.

8.1.6 Groundwater Recharge

Assuming that the amount of water that escapes evaporation and transpiration moves down below the root zone as "deep percolation" and eventually joins the groundwater, the rate of groundwater accretion can be calculated as the difference between infiltration and evapotranspiration over a long period. A computer model for estimating seepage to a water table in this manner was developed by King and Lambert (1976), whereas Bultot and Dupriez (1976) developed a computer model of the major hydrologic components of a 1 235-km^2 watershed. This model accurately predicted stream flow at the base of the watershed for an entire year. Other models are not strictly based on physical processes but use statistical parameters to describe rainfall-infiltration-runoff relations (Pattison and McMahon, 1973).

In addition to estimating the infiltration and evapotranspiration components for certain periods with the methods discussed in the previous sections, deep percolation can also be measured directly in the field with, for example, lysimeters. These are round or square tanks of about 1 to 10 m^2 surface area and 1 to 2 m depth that are filled with local soil and placed into the ground so that their surface is on the same level as that of the surrounding land (Fritschen et al., 1973; Hanks and Shawcroft, 1965; Libby and Nixon, 1962; Lourence and Goddard, 1967; Ritchie and Burnett, 1968). The soil inside the lysimeter should duplicate the natural soil profile as much as possible. Monolythic lysimeters are lysimeters where a natural soil block is encased in concrete or other material. The lysimeter is planted to the same crop or vegetation as on the adjacent land. Most lysimeters are of the weighing type, so that infiltration and evapotranspiration are determined as increases and decreases, respectively, of the weight of the entire tank. Weighing mechanisms may range from intricate balance and load-cell installations beneath the tank to resting the tank on water-filled rubber bags or flexible tubing in which the water pressure is measured. A gravel or sand layer is often placed on the bottom of the lysimeter tank so that deep-percolation water can be drained and measured. Such a layer has the disadvantage, however, of creating an artificial boundary of zero-pressure head at the bottom of the soil in the tank which could give erroneous estimates of deep percolation. Better duplication of field conditions is obtained by placing a saturated ceramic plate or layer of fine sand or glass beads at the bottom of the tank and maintaining a negative pressure head in this material that is equal to the pressure head observed with tensiometers at the same depth in the natural soil around the lysimeter. The volume of deep percolation extracted this way is a more accurate measure of the local groundwater recharge rate.

Another field technique for measuring deep-percolation rates utilizes tensiometers, which are placed closely together but with their ceramic cups at different depths (for example, at 20-cm intervals) in the first 1 or 2 m of a representative soil profile. A soil area of several square meters at the tensiometer site is flooded to saturate the profile to at least the deepest tensiometer. When all the water has infiltrated, the area is covered with a plastic sheet or other material to exclude evapotranspiration, so that redistribution of water in the wetted zone is entirely due to downward movement. Downward fluxes can then be calculated from changes in the water content at various depths, as may be measured with the neutron method (see Section 2.7). Combining these fluxes with hydraulic gradients as determined from the tensiometer readings then enables calculation of the unsaturated conductivity K_h and evaluation of the relations between K_h, θ, and h. Once this is done, the plastic sheet is removed so that the tensiometer site returns to normal conditions. Subsequent tensiometer measurements will then enable evaluation of upward and downward fluxes by multiplying vertical hydraulic gradients between two tensiometers by the average K_h value corresponding to the pressure heads indicated by these tensiometers. Downward fluxes between the lowest tensiometers represent deep-percolation rates. Recording tensiometers connected with a rotary switch to a transducer were used by Rice (1975) in an application of this technique. Since the results of this method apply only to a small area, several installations will be required to adequately cover an entire watershed or other large area. The technique is restricted to relatively wet profiles where soil-water pressure heads do not drop below about -8 m.

A third technique for field evaluation of deep percolation is the water-balance approach, where deep percolation is indirectly evaluated as the difference between rainfall and runoff (as determined by stream-flow measurements) plus evapotranspiration.

In irrigated areas, deep percolation from irrigated fields is an important source of groundwater replenishment. This can be an asset if the deep-percolation water is of good quality and the water table remains at least 1 or 2 m below the root zone of the crop, but it is a problem if the water contains too much salt or causes the water table to rise so high that crop yields are reduced. To avoid the latter, groundwater must be removed artificially by horizontal, parallel drains or vertical wells (see Section 8.4) to keep the water table at a depth of at least 1.5 to 2 m. At this depth, the dangers of waterlogging of the root zone and salt accumulation in the soil due to evaporation of groundwater are greatly reduced (see Section 8.5).

The salt content of the deep-percolation water from irrigated land depends on the salt content of the irrigation water and on how much of the irrigation water eventually becomes deep-percolation water. A certain amount of deep percolation is necessary to maintain a salt balance in the root zone (see Section 11.7). This can normally be achieved by applying 10 to 30 percent more irrigation water than needed to meet the evapotranspiration of the crop. Due to inefficient irrigation practices, however, actual amounts of deep percolation are considerably higher, i.e., about 40 to 50 percent of the applied water. These percentages are for flood- or gravity-type systems (sloping borders, furrows, or corrugations). Sprinkler or drip systems are more efficient and produce less deep percolation than gravity systems. According to the salt-balance equation (Section 11.7), however, a reduction in deep percolation due to an increase in irrigation efficiency will produce a higher salt concentration in the deep-percolation water. Irrigation-water requirements to meet the evapotranspiration of the crop(s) range from about 0.5 to 2 m/year, depending on climate and length of growing season. Since most irrigation efficiencies range from 40 to 90 percent, deep percolation from irrigated areas may provide recharge rates of 0.05 to 1 m/year. A realistic average probably is about 0.35 m per year (assuming an application of 1 m/year at an efficiency of 65 percent).

8.2 SEEPAGE FROM SURFACE WATER

If groundwater levels are below water levels in streams, canals, lakes, or reservoirs, water will seep into the ground from these surface waters. Streams that lose water by seepage are called *losing streams*. They have also been called *effluent* streams, but *losing stream* is the preferred term (effluent streams could also refer to the discharge of sewage effluent from treatment plants). The rate of seepage from streams or canals depends on channel geometry, K of bottom material and underlying soil layers, and depth of groundwater table at some distance from the channel. Flow velocity in the channel has no direct effect on seepage (Bouwer et al., 1963), but it could affect seepage indirectly because fine particles and other sediment have more chance to accumulate on the bottom of stagnant or slow-flowing channels than in rapid streams.

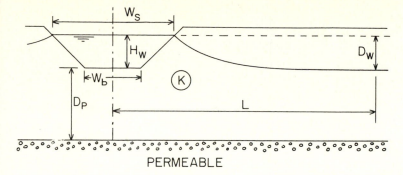

Figure 8.5 Geometry and symbols for channel in soil underlain by permeable material (condition A).

In analyzing systems of seepage flow from surface water (channels as well as impoundments) to groundwater, three basic soil conditions can be distinguished:

A. The soil below the channel is homogeneous and underlain by material of much higher hydraulic conductivity, considered as infinitely permeable in flow-system analyses (Figure 8.5). The water table at some distance from the channel may be above or below the top of the permeable material. If it is below the top of this material, the permeable layer acts as a drainage layer for the seepage water. This situation will be referred to as condition A' (see Figure 8.8).
B. The soil below the channel is homogeneous and underlain by material of much lower hydraulic conductivity, taken as impermeable in the analyses of the flow system (Figure 8.6).
C. The soil below the channel is covered by a relatively thin layer of sediment or other fine material that is of much lower hydraulic conductivity than the underlying soil and restricts the seepage rate (Figure 8.7).

Direct contact between groundwater and surface water exists for condition A if the water table is above the permeable layer, for condition B, and for condition C if the water table is above the channel bottom. In these cases, seepage rates are

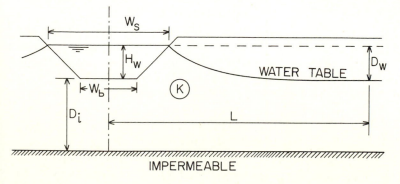

Figure 8.6 Geometry and symbols for channel in soil underlain by impermeable material (condition B).

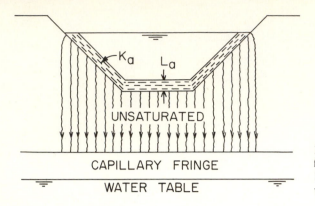

Figure 8.7 Geometry and symbols for channels with a thin layer of low hydraulic conductivity along the wetted perimeter (condition C).

CAPILLARY FRINGE

WATER TABLE

affected by the depth of the groundwater table. If the water table (or, rather, the top of the capillary fringe) is below the top of the drainage layer for condition A, or below the channel bottom for condition C, groundwater and surface water are separated by unsaturated material and seepage rates are not affected by water-table depth.

8.2.1 Steady-State Systems

Steady-state seepage-flow systems for condition A were first analyzed for the situation where the water table is below the top of the permeable material (condition A'). The top of the permeable material is then taken as a line of zero pressure head, which is theoretically only true if the water table coincides with the top of the permeable layer. If the water table (or, rather, the top of the capillary fringe) is lower, the pressure head at the top of the permeable material is governed by the relation between K_h and h of this material and can be approximated as the critical-pressure head (see Section 7.7). However, replacing h_{cr} by zero pressure head will have a negligible effect on the flow system if the depth D_p of the permeable material (Figure 8.5) is large compared to $-h_{cr}$.

Mathematical solutions for the seepage rate under condition A' were obtained by Kozeny for $D_p = \infty$ and by Vedernikov for finite as well as infinite values of D_p (Bouwer, 1969c, and references therein). Equations to calculate seepage for condition A with relatively high groundwater levels (D_w relatively small, Figure 8.5) were developed by Hammad (Bouwer, 1969c, and references therein). Examples of equipotentials and streamlines in seepage-flow systems for conditions A' and A, obtained by R analog, are shown in Figures 8.8 and 8.9 (Bouwer, 1965 and 1969c). The numbers on the equipotentials refer to the percent of the total-head difference in the system. The water table for condition A is taken as a solid boundary, assuming no evaporation or other fluxes across the water table. At great distance from the channel, the water table in Figure 8.9 represents the piezometric surface of the groundwater in the coarse-textured material, at least at static conditions.

The first analyses of seepage for condition B simply consisted of applications of the Dupuit-Forchheimer assumption of horizontal flow (see Section 3.7, Figure

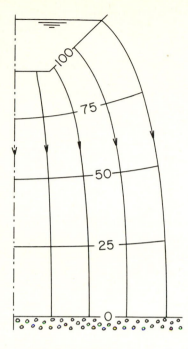

Figure 8.8 Flow system for seepage under condition A' with equipotentials expressed as percentage of total-head loss. *(From Bouwer, 1965 and 1969.)*

3.7), again assuming no flow across the water table (for inclusion of evaporation from the groundwater, see Section 8.6). The resulting solutions give reasonably accurate estimates of seepage rates as long as D_i/W_b is less than about 3 (Bouwer, 1969c). The term D_i is the depth of the impermeable layer below the channel bottom, and W_b is the bottom width of the best fitting trapezoidal cross section with 1 : 1 side slopes of the channel (Figure 8.6). For values of D_i/W_b larger than 3, vertical-flow components below the channel are no longer negligible. These components were taken into account by Dachler, who experimentally determined

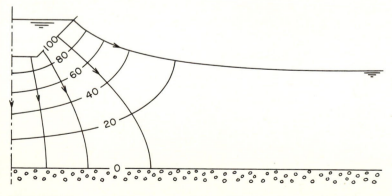

Figure 8.9 Flow system for seepage under condition A with equipotentials expressed as percentage of total-head loss. *(From Bouwer, 1965 and 1969.)*

"form factors" to calculate the head loss in the region near the channel. The head loss in the rest of the flow system was then calculated on the basis of horizontal flow (Bouwer, 1969c, and references therein). Ernst also divided the flow system into two regions, but calculated the head loss in the channel region with radial-flow theory (Bouwer, 1969c, and references therein). Both Dachler's and Ernst's equations yielded seepage rates that agreed well with rates obtained by R analog (Bouwer, 1969c). Figure 8.10 shows an example of a flow system for seepage under condition B with relatively small D_i and, hence, predominantly horizontal flow.

Seepage rates for a wide variety of geometry parameters were obtained with an R analog for conditions A, A', and B (Bouwer, 1965 and 1969c). The seepage rate was expressed as the volume rate of water movement into the soil per unit length of channel and per unit width of the water surface in the channel. This seepage rate I_s has the dimension of a velocity, and it is equal to the rate of fall of the water level in the channel due to seepage as if the channel were ponded. To express the results, I_s was divided by K of the soil around the channel to yield the dimensionless ratio I_s/K. Graphs were prepared to show I_s/K as a function of D_w/W_b (Figures 8.11, 8.12, and 8.13), where D_w is the vertical distance between the water level in the channel and the groundwater table at an arbitrarily taken horizontal distance of $10W_b$ from the channel center. Curves were drawn for different values of D_p/W_b (condition A) and D_i/W_b (condition B) with the curve for $D_p = D_i = \infty$ shared by both conditions. The curves apply to trapezoidal channels with 1 : 1 side slopes, and each graph applies to a certain water depth H_w in the channel (expressed as H_w/W_b). To evaluate I_s/K for channels with other cross sections, the actual cross section is replaced by the best fitting trapezoidal cross section with 1 : 1 side slopes. If H_w/W_b is between or outside the three values covered on the graphs, I_s/K can be determined by inter- or extrapolation, respectively.

The end points of the curves for condition A in the graphs are reached when D_w has become equal to $H_w + D_p$, which is condition A'. Thus, the curve for condition A' is the locus of the end points of the curves for condition A. The abscissa parameter D_w/W_b then represents $(D_p + H_w)/W_b$.

The graphs show that I_s/K for condition A initially increases almost linearly with D_w/W_b, but then at a decreasing rate until condition A' is reached and I_s/K is no longer affected by the water-table position (the top of the permeable material is

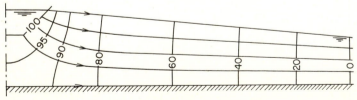

Figure 8.10 Flow system for seepage under condition B with equipotentials expressed as percentage of total-head loss. (*From Bouwer, 1965 and 1969.*)

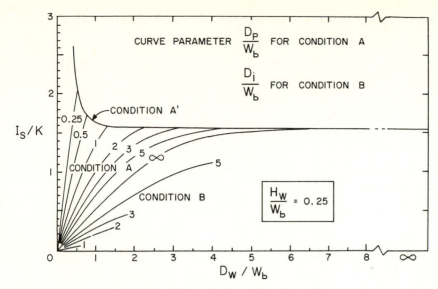

Figure 8.11 I_s/K as a function of D_w/W_b for trapezoidal channels with 1:1 side slopes and $H_w/W_b = 0.25$. (*From Bouwer, 1965.*)

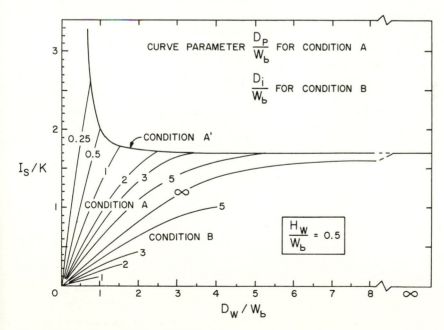

Figure 8.12 As Figure 8.11, but $H_w/W_b = 0.5$.

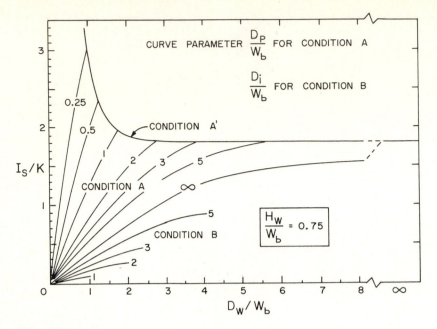

Figure 8.13 As Figure 8.11, but $H_w/W_b = 0.75$.

then taken as zero pressure head, regardless of the depth of the water table in the permeable material). I_s/K for conditions A and A' is highest if D_p is relatively small. For condition B, I_s/K also increases linearly with D_w/W_b only when D_w/W_b is relatively small. For a deep, uniform soil as indicated by $D_p/W_b = D_i/W_b = \infty$ in the graphs, I_s/K increases linearly with D_w/W_b until D_w/W_b is about 2. Then, I_s/K increases at a slower rate and asymptotically approaches the value for condition A' with $D_p = \infty$.

A consequence of the nonlinearity between seepage and water-table depth is that pumping groundwater from a river valley to induce more seepage from the stream is effective only if D_w/W_b is less than about 3 or 4, depending on the water depth in the stream. Values of D_w/W_b in excess of 3 or 4 could readily occur for relatively narrow streams, where W_b is small compared to D_w. When such streams are simulated in analog or digital models, errors in groundwater accretion rates would result if seepage in the models were indiscriminately allowed to vary linearly with groundwater depth.

The permeable and less permeable materials which form the lower boundaries for seepage conditions A and B, respectively, can be taken as infinitely permeable and impermeable if their K values differ from the K value of the overlying soil by a factor of at least 10. If the K difference is less, the seepage-flow system should be treated as a layered system. Solutions for two-layered systems with the channel embedded in the top layer and the second layer underlain by an impermeable boundary were developed by Ernst (Bouwer, 1969c, and references therein). These solutions agreed with those for conditions A and B in Figures 8.11, 8.12, and 8.13 if K of the second layer was at least 10 times larger or smaller, respectively, than K of the top layer.

The layer of clogging material along the channel's wetted perimeter that restricts the seepage rate for condition C (Figure 8.7) may be of natural origin (sedimentation of clay, silt, or fine sand; growth of bacteria or other biological action) or of artificial origin (earth linings or chemical seals for seepage control). The depth of the water table has no effect on seepage for condition C if the groundwater—or, rather, the top of the capillary fringe—is below the channel bottom. In that case, the soil below the clogged zone will be unsaturated and the flow will be vertically downward at unit hydraulic gradient. The unsaturated hydraulic conductivity K_h will then be equal to the flow rate, similar to the flow in the long column of Figure 7.25. If the soil below the clogged layer is relatively coarse-textured, the relation between K_h and h will approach a step function, so that h in this soil will be close to h_{cr} for a wide range in seepage rates (see Section 7.7).

Taking the pressure head below the clogged layer as equal to h_{cr} of the underlying soil, the seepage rate v_i through the clogged layer at the bottom of the channel (Figure 8.7) can be expressed with Darcy's equation as

$$v_i = K_a \frac{H_w + L_a - h_{cr}}{L_a} \tag{8.11}$$

where K_a is the hydraulic conductivity and L_a the thickness of the restricting layer. Since L_a is usually small (sometimes less than 1 mm), it is not very well defined. For this reason, K_a and L_a are usually combined into one parameter L_a/K_a, called the *hydraulic impedance* R_a of the restricting layer. The term R_a has the dimension of time. Since L_a tends to be small compared to $H_w - h_{cr}$, Eq. (8.11) can be written as $v_i = (H_w - h_{cr})/R_a$. Applying this equation to the entire wetted perimeter of a trapezoidal channel and assuming that R_a is uniform and that the flow through the layer on the channel sides is perpendicular to the bank, the following equation for the seepage I_s is obtained:

$$I_s = \frac{1}{W_s R_a} \left[(H_w - h_{cr})W_b + (H_w - 2h_{cr})\frac{H_w}{\sin \alpha} \right] \tag{8.12}$$

where W_s = width of water surface in channel
α = angle between bank and horizontal ($\alpha < 90°$)

If the water table or top of capillary fringe is above the channel bottom, I_s can be calculated with Eq. (8.11), substituting the actual pressure head beneath the clogged layer for h_{cr}. As long as v_i is small compared to K of the material below the clogged layer, the actual pressure head at a given point below the clogged layer can then be taken as the height of the groundwater table next to the channel above that point.

The presence of restricting layers on the bottom of open channels or reservoirs can be detected by visual inspection or by installing piezometers in the center of the channel that reach about 10 to 20 cm into the original soil. If water fails to rise in these piezometers, the pressure head is negative and seepage is controlled by a clogged layer on the bottom. Values of R_a can be determined in situ with falling-head measurements in seepage meters, which are cylindrical devices about 40 cm in diameter that are pushed 5 to 10 cm into the bottom and covered with a leakproof lid to create various pressure heads inside the cylinder. The response of the seepage rate inside the cylinder to changes in pressure head then enables calculation of R_a, or of K if there is no restricting layer on the bottom (Bouwer and Rice, 1963).

8.2.2 Transient Systems

Variations of seepage with time are of interest where water enters a dry channel, like an ephemeral stream or an intermittently used irrigation canal, or where the water level in the channel changes, as in flood flows. When water enters a dry channel, seepage can be calculated with one of the infiltration equations [for example, Eqs. (8.1), (8.2), (8.3), or (8.4)] if the water depth in the channel is small compared to the width. For larger depths, the seepage flow becomes a situation of two-dimensional infiltration (Philip, 1969; Selim and Kirkham, 1973; Zyvoloski et al., 1976).

For the first stages, the system can be treated as one-dimensional infiltration into the bottom (vertically downward) and into the banks (normal to the bank). Subsequent seepage rates can then be estimated by treating the system as a succession of steady states, assuming flat wetting fronts at increasing depths that can be treated as tops of drainage layers for condition A'. Using these procedures, relations between the wetted soil area A_w beneath the channel and the time since infiltration started were determined and plotted in terms of the dimensionless factors A_w/W_s^2 and Kt/fW_s (Bouwer, 1969c). The term W_s is the width of the water surface, f is the fillable pore-space fraction of the soil, and K is the hydraulic conductivity of the soil after wetting (Figures 8.14 and 8.15). The calculations were performed for different water depths H_w, which were considered constant during seepage. Thus, for given values of K, f, and W_s, Kt/fW_s can be calculated for different t and the corresponding values of A_w/W_s^2 can be determined from the graphs. Knowing W_s^2, A_w can be calculated which, after multiplying by f, yields the volume of seepage per unit length of channel at time t. This procedure yields approximate values for seepage after a dry channel is filled. The error, however, is probably less than the error in input data K and f, which are difficult to evaluate with great accuracy for entire channel sections because of soil variations. Once the wetting front has reached the water table, the seepage-flow system will change to that for condition A or B.

Another transient seepage situation occurs after the water surface in a stream undergoes a sudden rise. This problem was analyzed by Collis-George and Smiles (1963) for condition B, assuming a rectangular channel that fully penetrates the aquifer so that the top of the impermeable layer is the channel bottom (D_i in Figure 8.6 equal to zero). Taking static initial conditions (i.e., a horizontal water

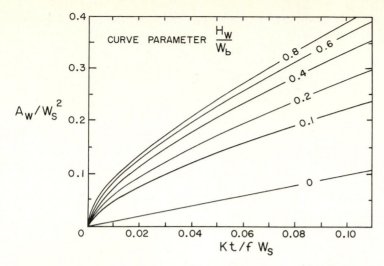

Figure 8.14 Relation between A_w/W_s^2 and Kt/fW_s for trapezoidal channel with $1:1$ side slopes. *(From Bouwer, 1969.)*

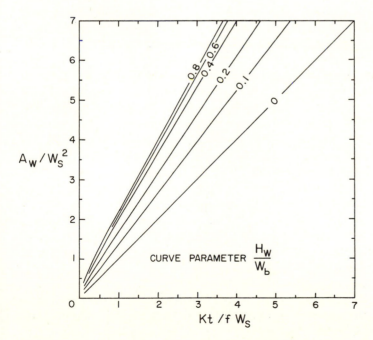

Figure 8.15 As Figure 8.14, but for larger values of Kt/fW_s.

table at the same height as the water level in the channel), the seepage after a sudden rise ΔH of the water level in the channel was calculated as

$$\frac{I_s}{K} = \frac{2\,\Delta H}{W_s}\left(\frac{fH_w}{\pi K t}\right)^{1/2} \tag{8.13}$$

where t is the time after the sudden rise and f is the fillable porosity above the water table. This equation is based on the assumption of horizontal flow and constant transmissivity, which is valid if ΔH is relatively small. Elastic reactions of the aquifer in response to a sudden rise or fall of the water level in a stream or reservoir were taken into account by Streltsova (1975) in predicting transient flow from surface water into an aquifer or vice versa.

The effect of a gradual rise of the water level in a stream on seepage was analyzed by Cooper and Rorabaugh (1963), who developed equations to predict "bank storage" of water during flood flows. Seepage flow should be analyzed as a transient system if the changes in H_w are not small compared to D_w (Figures 8.5 and 8.6). If H_w changes only relatively little, the system can be approximated as a steady-state situation.

Seepage measurement. Direct measurement of seepage can be accomplished with inflow-outflow, ponding, and point-measurement techniques. The inflow-outflow technique consists of measuring the discharge in the stream or channel at two locations which may be several hundred meters apart for small streams and 10 km or more for larger streams. In addition, all other inflows and outflows for the reach (tributaries, waste discharges, diversions, etc.) and changes in water storage due to changing water levels must be measured, so that the seepage loss for the reach can be calculated as the difference between total inflow and outflow. This method is suitable for relatively small streams with essentially constant flow. Difficulties arise where discharge varies too much with time or where discharge is so large in relation to seepage that the difference between inflow and outflow approaches the errors in the individual discharge measurements.

The ponding method consists of measuring the rate of water-level fall (corrected for evaporation if necessary) in a lake or in a section of a canal or stream that is temporarily dammed or ponded. While this technique gives reliable estimates of seepage from lakes or reservoirs or sluggish streams where the water is essentially stagnant anyway, it usually underestimates seepage from fast-flowing channels because sediment accumulates on the bottom of the ponded section where it would not accumulate if water were flowing. The resulting error can be minimized by measuring the water-surface drop immediately after ponding or by taking measurements at different times and extrapolating the seepage rates backward to the time that ponding was started.

Point measurements of seepage can be made with seepage meters, which are essentially cylinder infiltrometers covered with lids. The cylinder is pushed into the bottom or bank of the channel, and seepage is measured as the outflow from the cylinder when the pressure head inside the cylinder is the same as that outside in the stream or channel (Bouwer and Rice, 1963). Where seepage meters are not convenient to use, point measurements of seepage can be obtained by placing salt crystals on the bottom and measuring the rate of advance of the dissolved salt in the bottom material with an electrical-conductivity probe (Bouwer and Rice, 1968).

8.3 ARTIFICIAL RECHARGE

Artificial recharge of groundwater, as can be achieved with basins or other facilities for inducing infiltration of surface water into soil, or with injection wells, is becoming increasingly important in groundwater management and in conjunctive use of surface water and groundwater resources. Artificial recharge is used to reduce, stop, or even reverse declines of groundwater levels. It enables storing of surface water (flood or other surplus water from streams, for example) for future use, which is especially advantageous where surface reservoirs are inadequate or too expensive. Groundwater recharge also is an important technique for protecting fresh groundwater in coastal aquifers against intrusion of salt water from the ocean (see Section 11.2). This is done by creating a groundwater mound parallel to the coast that serves as a fresh-water barrier to inland movement of salt water. Another application of groundwater recharge, using infiltration basins, is to treat or "renovate" sewage effluent or other wastewater. The vadose zone and, to a lesser extent, the aquifers then act as natural filters for chemical and biological purification of the water (see Section 11.4).

8.3.1 Recharge Basins

Systems of groundwater recharge with basins include the "Leaky-Acres Project" in Fresno, California (Figure 8.16), where 10 basins with a total surface area of 47.4 ha are used to replenish the groundwater, which is overdrawn by pumping for municipal water supply (Nightingale and Bianchi, 1973). Water for recharge is obtained from the Kings River and delivered to the basins through an irrigation canal (on the right side of the project in Figure 8.16). The average infiltration rate is about 0.12 m/day when the basins are flooded. However, the basins are dry for about one-fourth of the time, so that the long-term infiltration or hydraulic loading (which includes the time that the basins are dry) is $0.75 \times 0.12 \times 365 = 32.8$ m/year. The total annual recharge volume is 15.6×10^6 m^3 for the entire project. Reasons for interrupting the operation of the recharge basins include high sediment content of the canal water, inadequate flow in the canal, abatement of insect problems (gnats, mosquitoes), and maintenance or repair of basin structures. Other recharge systems frequently consist of infiltration basins located along stream channels or in stream beds to capture runoff. In the Los Angeles, California, area alone, such basins have a combined area of about 1 200 ha (Bulten et al., 1974).

Figure 8.16 Leaky-Acres groundwater recharge project in Fresno, California. *(Photograph courtesy of U.S. Department of Agriculture.)*

Infiltration basins are typically used to recharge unconfined or semiconfined aquifers. The basins should be located so that:

1. The surface soil is sufficiently permeable to yield acceptable infiltration rates (sands to sandy loams are preferred).
2. There are no layers in the vadose zone of such low hydraulic conductivity that they will form perched groundwater mounds that rise into the basins and restrict infiltration to unacceptably low rates.
3. The regional water table is sufficiently deep to keep the groundwater mound below the bottom of the basins, but not so deep that large quantities of water are needed to wet the vadose zone before any recoverable water reaches the water table.
4. The aquifer is unconfined and sufficiently transmissive to allow lateral movement of the recharge water without building up groundwater mounds that rise into the basins.

Selection of the optimum site for recharge basins may require considerable investigation, including soil surveys, geophysical surveys, test drilling, measuring K in the vadose zone and below the water table, and test projects to determine

potential infiltration rates and build-up of groundwater mounds. Where the surface soils consist of loams or clays but the underlying materials are sands or
gravels, it may be advantageous to excavate the basins so that the bottom reaches
into the coarser material. Where groundwater mounds may rise so high that they
will restrict infiltration rates or damage surrounding structures, the infiltration
basins should be long and narrow. Such basins produce lower mounds than round
or square basins of the same surface area [Eq. (8.14), Figure 8.21].

When an infiltration basin is filled, water will infiltrate into the soil and move
downward, as described by Eqs. (8.1) and (8.2), for example. If the wetting front
reaches a layer of reduced hydraulic conductivity, a perching mound will rise and
spread until the water passes through the restricting layer(s) as fast as it arrives
from above (Figure 8.17). Below the restricting layer, the width of the percolation
zone is larger, but the downward velocity is smaller than immediately below the
infiltration basin. When the water finally reaches the actual groundwater, the
downward flux immediately below the water table is less than the arrival rate of
the water, causing the water table to rise and form a groundwater mound. The
recharge water then begins to flow vertically and laterally in the aquifer, causing
the water table adjacent to the infiltration zone to rise also. Examples of streamlines and equipotentials for recharge flow systems with rising mounds and water
tables are shown in Figure 8.18 for a relatively low recharge rate and a shallow

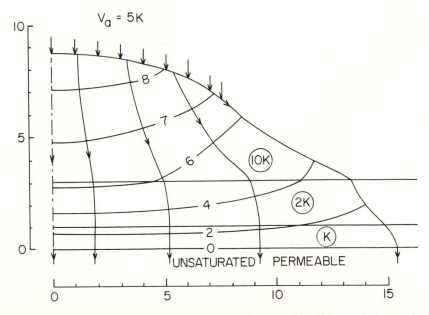

Figure 8.17 Streamlines and equipotentials (in units of total head) in perched groundwater mound
above two restricting layers with K values of 10 and 20 percent of K of the overlying material, as
obtained by R analog. (*From Bouwer, 1962*). The coordinates and total head are expressed in
arbitrary units of length.

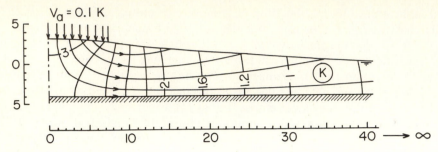

Figure 8.18 Groundwater-recharge flow system in shallow, unconfined aquifer obtained by R analog. *(From Bouwer, 1962).* The coordinates and total-head equipotentials are expressed in arbitrary units of length.

unconfined aquifer, and in Figure 8.19 for a high recharge rate and a deep unconfined aquifer. These systems were obtained with an R analog by considering the rising mound as a succession of steady states (Bouwer, 1962). The rate of rise of groundwater mounds decreases with time, and eventually mounds may reach a pseudoequilibrium position. True equilibrium can be established (at least theoretically) if the water-table height at some distance from the recharge system is kept constant by pumping, discharge of groundwater into a stream or reservoir with constant water level, or by some other mechanism. When infiltration is stopped,

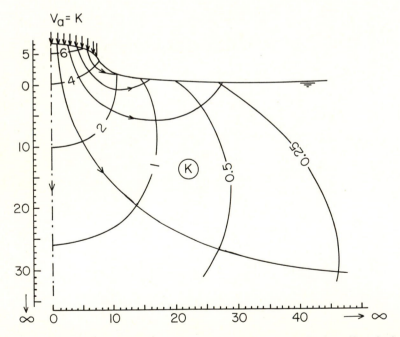

Figure 8.19 Groundwater-recharge flow system in deep, unconfined aquifer, obtained by R analog. *(From Bouwer, 1962.)* The coordinates and total-head equipotentials are expressed in arbitrary units of length.

the groundwater mound will recede and spread until theoretically a horizontal water table again is established.

Equations to predict the rise and fall of groundwater mounds have been developed by Hantush (1967), using horizontal-flow theory. The equation for the rise of the mound in unconfined aquifers below rectangular recharge basins can be written as

$$h_{x,\,y,\,t} - H = \frac{v_a t}{4f} \{F[(W/2 + x)n,\,(L/2 + y)n]$$

$$+ F[(W/2 + x)n,\,(L/2 - y)n]$$
$$+ F[(W/2 - x)n,\,(L/2 + y)n]$$
$$+ F[(W/2 - x)n,\,(L/2 - y)n]\} \qquad (8.14)$$

where $h_{x,\,y,\,t}$ = height of water table above impermeable layer at x, y, and time t (Figure 8.20)

H = original height of water table above impermeable layer

v_a = arrival rate at water table of water from infiltration basin

t = time since start of recharge

f = fillable porosity $(1 > f > 0)$

L = length of recharge basin (in y direction)

W = width of recharge basin (in x direction)

$n = (4tT/f)^{-1/2}$

$F(\alpha, \beta) = \int_0^1 erf(\alpha\tau^{-1/2}) \cdot erf(\beta\tau^{-1/2})\, d\tau$ [which was tabulated by Hantush (Table 8.3)]

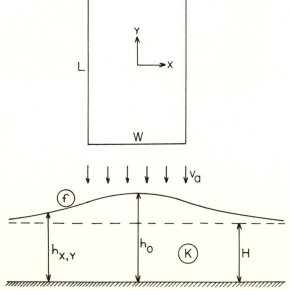

Figure 8.20 Geometry and symbols for rectangular infiltration area (top) and underlying groundwater mound in unconfined aquifer (bottom).

Table 8.3 Values of the function $F(\alpha, \beta)$ in Eq. (8.14) for different values of α and β

α \ β	0.02	0.04	0.06	0.08	0.10	0.14	0.18	0.22	0.26	0.30	0.34	0.38	0.42	0.46	0.50	0.54	0.58	0.62
0.02	0.0041	0.0073	0.0101	0.0125	0.0146	0.0184	0.0216	0.0243	0.0267	0.0288	0.0306	0.0322	0.0337	0.0349	0.0361	0.0371	0.0380	0.0387
0.04	0.0073	0.0135	0.0188	0.0236	0.0278	0.0353	0.0416	0.0470	0.0518	0.0559	0.0596	0.0628	0.0657	0.0683	0.0705	0.0725	0.0743	0.0759
0.06	0.0101	0.0188	0.0266	0.0335	0.0398	0.0509	0.0602	0.0684	0.0754	0.0817	0.0871	0.0920	0.0963	0.1001	0.1035	0.1065	0.1091	0.1115
0.08	0.0125	0.0236	0.0335	0.0425	0.0508	0.0652	0.0776	0.0884	0.0978	0.1060	0.1133	0.1197	0.1254	0.1305	0.1350	0.1389	0.1425	0.1456
0.10	0.0146	0.0278	0.0398	0.0508	0.0608	0.0786	0.0939	0.1072	0.1188	0.1290	0.1381	0.1461	0.1532	0.1595	0.1650	0.1700	0.1744	0.1783
0.14	0.0184	0.0353	0.0509	0.0652	0.0786	0.1025	0.1232	0.1414	0.1573	0.1714	0.1839	0.1949	0.2048	0.2135	0.2212	0.2281	0.2343	0.2397
0.18	0.0216	0.0416	0.0602	0.0776	0.0939	0.1232	0.1490	0.1716	0.1916	0.2094	0.2251	0.2391	0.2515	0.2626	0.2724	0.2812	0.2890	0.2959
0.22	0.0243	0.0470	0.0684	0.0884	0.1072	0.1414	0.1716	0.1984	0.2222	0.2433	0.2621	0.2789	0.2938	0.3071	0.3189	0.3295	0.3389	0.3472
0.26	0.0267	0.0518	0.0754	0.0978	0.1188	0.1573	0.1916	0.2222	0.2494	0.2737	0.2954	0.3147	0.3320	0.3474	0.3612	0.3735	0.3844	0.3941
0.30	0.0288	0.0559	0.0817	0.1060	0.1290	0.1714	0.2094	0.2433	0.2737	0.3009	0.3252	0.3470	0.3665	0.3839	0.3995	0.4134	0.4257	0.4368
0.34	0.0306	0.0596	0.0871	0.1133	0.1381	0.1839	0.2251	0.2621	0.2954	0.3252	0.3520	0.3761	0.3976	0.4169	0.4341	0.4495	0.4633	0.4756
0.38	0.0322	0.0628	0.0920	0.1197	0.1461	0.1949	0.2391	0.2789	0.3147	0.3470	0.3761	0.4022	0.4256	0.4466	0.4654	0.4823	0.4973	0.5108
0.42	0.0337	0.0657	0.0963	0.1254	0.1532	0.2048	0.2515	0.2938	0.3320	0.3665	0.3976	0.4256	0.4508	0.4734	0.4937	0.5119	0.5281	0.5427
0.46	0.0349	0.0683	0.1001	0.1305	0.1595	0.2135	0.2626	0.3071	0.3474	0.3839	0.4169	0.4466	0.4734	0.4975	0.5191	0.5385	0.5559	0.5715
0.50	0.0361	0.0705	0.1035	0.1350	0.1650	0.2212	0.2724	0.3189	0.3612	0.3995	0.4341	0.4654	0.4937	0.5191	0.5420	0.5626	0.5810	0.5975
0.54	0.0371	0.0725	0.1065	0.1389	0.1700	0.2281	0.2812	0.3295	0.3735	0.4134	0.4495	0.4823	0.5119	0.5385	0.5626	0.5842	0.6036	0.6209
0.58	0.0380	0.0743	0.1091	0.1425	0.1744	0.2343	0.2890	0.3389	0.3844	0.4257	0.4633	0.4973	0.5281	0.5559	0.5810	0.6036	0.6238	0.6420
0.62	0.0387	0.0759	0.1115	0.1456	0.1783	0.2397	0.2959	0.3472	0.3941	0.4368	0.4756	0.5108	0.5427	0.5715	0.5975	0.6209	0.6420	0.6609
0.66	0.0394	0.0773	0.1136	0.1484	0.1818	0.2445	0.3020	0.3547	0.4027	0.4466	0.4865	0.5227	0.5556	0.5854	0.6122	0.6364	0.6582	0.6778
0.70	0.0401	0.0785	0.1154	0.1509	0.1849	0.2488	0.3075	0.3612	0.4104	0.4553	0.4962	0.5334	0.5672	0.5977	0.6254	0.6503	0.6728	0.6929
0.74	0.0406	0.0796	0.1171	0.1531	0.1876	0.2526	0.3123	0.3671	0.4172	0.4630	0.5048	0.5429	0.5774	0.6087	0.6371	0.6627	0.6857	0.7064
0.78	0.0411	0.0806	0.1185	0.1550	0.1900	0.2559	0.3166	0.3722	0.4232	0.4699	0.5125	0.5513	0.5865	0.6185	0.6475	0.6736	0.6972	0.7184
0.82	0.0415	0.0814	0.1198	0.1567	0.1921	0.2589	0.3203	0.3768	0.4286	0.4760	0.5192	0.5587	0.5946	0.6272	0.6567	0.6834	0.7074	0.7291
0.86	0.0419	0.0822	0.1209	0.1582	0.1940	0.2615	0.3237	0.3808	0.4333	0.4813	0.5252	0.5653	0.6017	0.6348	0.6648	0.6920	0.7165	0.7386
0.90	0.0422	0.0828	0.1219	0.1595	0.1957	0.2638	0.3266	0.3844	0.4374	0.4860	0.5305	0.5711	0.6080	0.6416	0.6721	0.6996	0.7245	0.7469
0.94	0.0425	0.0834	0.1228	0.1607	0.1971	0.2658	0.3292	0.3875	0.4411	0.4902	0.5351	0.5762	0.6136	0.6476	0.6784	0.7063	0.7316	0.7543
0.98	0.0428	0.0839	0.1236	0.1617	0.1984	0.2676	0.3314	0.3902	0.4442	0.4938	0.5392	0.5807	0.6184	0.6528	0.6840	0.7123	0.7378	0.7608
1.00	0.0429	0.0842	0.1239	0.1622	0.1990	0.2684	0.3324	0.3914	0.4457	0.4955	0.5410	0.5827	0.6206	0.6552	0.6865	0.7150	0.7406	0.7638
1.20	0.0437	0.0858	0.1263	0.1654	0.2030	0.2740	0.3396	0.4001	0.4558	0.5070	0.5540	0.5969	0.6362	0.6719	0.7044	0.7339	0.7605	0.7846
1.40	0.0441	0.0866	0.1275	0.1669	0.2049	0.2767	0.3431	0.4043	0.4608	0.5127	0.5603	0.6039	0.6438	0.6801	0.7132	0.7432	0.7704	0.7949
1.80	0.0444	0.0871	0.1283	0.1680	0.2062	0.2785	0.3454	0.4071	0.4641	0.5165	0.5645	0.6086	0.6489	0.6856	0.7190	0.7494	0.7769	0.8018
2.00	0.0444	0.0871	0.1284	0.1681	0.2064	0.2787	0.3457	0.4075	0.4645	0.5169	0.5651	0.6092	0.6495	0.6863	0.7198	0.7502	0.7778	0.8027
2.20	0.0444	0.0872	0.1284	0.1682	0.2065	0.2788	0.3458	0.4076	0.4646	0.5171	0.5653	0.6094	0.6497	0.6865	0.7200	0.7505	0.7781	0.8030
2.50	0.0444	0.0872	0.1284	0.1682	0.2065	0.2788	0.3458	0.4077	0.4647	0.5172	0.5653	0.6095	0.6498	0.6867	0.7202	0.7506	0.7782	0.8032
3.00	0.0444	0.0872	0.1284	0.1682	0.2065	0.2789	0.3458	0.4077	0.4647	0.5172	0.5654	0.6095	0.6499	0.6867	0.7202	0.7506	0.7782	0.8032

α \ β	3.00	2.50	2.20	2.00	1.80	1.40	1.20	1.00	0.98	0.94	0.90	0.86	0.82	0.78	0.74	0.70	0.66	0.62
0.02	0.0444	0.0444	0.0444	0.0444	0.0444	0.0441	0.0437	0.0429	0.0428	0.0425	0.0422	0.0419	0.0415	0.0411	0.0406	0.0401	0.0394	0.0387
0.04	0.0872	0.0872	0.0872	0.0871	0.0871	0.0866	0.0858	0.0842	0.0839	0.0834	0.0828	0.0822	0.0814	0.0806	0.0796	0.0785	0.0773	0.0759
0.06	0.1284	0.1284	0.1284	0.1284	0.1283	0.1275	0.1263	0.1239	0.1236	0.1228	0.1219	0.1209	0.1198	0.1185	0.1171	0.1154	0.1136	0.1115
0.08	0.1682	0.1682	0.1682	0.1681	0.1680	0.1669	0.1654	0.1622	0.1617	0.1607	0.1595	0.1582	0.1567	0.1550	0.1531	0.1509	0.1484	0.1456
0.10	0.2065	0.2065	0.2065	0.2064	0.2062	0.2049	0.2030	0.1990	0.1984	0.1971	0.1957	0.1940	0.1921	0.1900	0.1876	0.1849	0.1818	0.1783
0.14	0.2789	0.2788	0.2788	0.2787	0.2785	0.2767	0.2740	0.2684	0.2676	0.2658	0.2638	0.2615	0.2589	0.2559	0.2526	0.2488	0.2445	0.2397
0.18	0.3458	0.3458	0.3458	0.3457	0.3454	0.3431	0.3396	0.3324	0.3314	0.3292	0.3266	0.3237	0.3203	0.3166	0.3123	0.3075	0.3020	0.2959
0.22	0.4077	0.4077	0.4076	0.4075	0.4071	0.4043	0.4001	0.3914	0.3902	0.3875	0.3844	0.3808	0.3768	0.3722	0.3671	0.3612	0.3547	0.3472
0.26	0.4647	0.4647	0.4646	0.4645	0.4641	0.4608	0.4558	0.4457	0.4442	0.4411	0.4374	0.4333	0.4286	0.4232	0.4172	0.4104	0.4027	0.3941
0.30	0.5172	0.5172	0.5171	0.5169	0.5165	0.5127	0.5070	0.4955	0.4938	0.4902	0.4860	0.4813	0.4760	0.4699	0.4630	0.4553	0.4466	0.4368
0.34	0.5654	0.5653	0.5653	0.5651	0.5645	0.5603	0.5540	0.5410	0.5392	0.5351	0.5305	0.5252	0.5192	0.5125	0.5048	0.4962	0.4865	0.4756
0.38	0.6095	0.6095	0.6094	0.6092	0.6086	0.6039	0.5969	0.5827	0.5807	0.5762	0.5711	0.5653	0.5587	0.5513	0.5429	0.5334	0.5227	0.5108
0.42	0.6499	0.6498	0.6497	0.6495	0.6489	0.6438	0.6362	0.6206	0.6184	0.6136	0.6080	0.6017	0.5946	0.5865	0.5774	0.5672	0.5556	0.5427
0.46	0.6867	0.6867	0.6865	0.6863	0.6856	0.6801	0.6719	0.6552	0.6528	0.6476	0.6416	0.6348	0.6272	0.6185	0.6087	0.5977	0.5854	0.5715
0.50	0.7202	0.7202	0.7200	0.7198	0.7190	0.7132	0.7044	0.6865	0.6840	0.6784	0.6721	0.6648	0.6567	0.6475	0.6371	0.6254	0.6122	0.5975
0.54	0.7506	0.7506	0.7505	0.7502	0.7494	0.7432	0.7339	0.7150	0.7123	0.7063	0.6996	0.6920	0.6834	0.6736	0.6627	0.6503	0.6364	0.6209
0.58	0.7782	0.7782	0.7781	0.7778	0.7769	0.7704	0.7605	0.7406	0.7378	0.7316	0.7245	0.7165	0.7074	0.6972	0.6857	0.6728	0.6482	0.6420
0.62	0.8032	0.8032	0.8030	0.8027	0.8018	0.7949	0.7846	0.7638	0.7608	0.7543	0.7469	0.7386	0.7291	0.7184	0.7064	0.6929	0.6778	0.6609
0.66	0.8257	0.8257	0.8255	0.8252	0.8243	0.8171	0.8064	0.7846	0.7816	0.7748	0.7671	0.7584	0.7486	0.7375	0.7250	0.7110	0.6953	0.6778
0.70	0.8460	0.8460	0.8458	0.8454	0.8445	0.8370	0.8259	0.8034	0.8002	0.7932	0.7852	0.7762	0.7660	0.7546	0.7417	0.7272	0.7110	0.6929
0.74	0.8642	0.8642	0.8640	0.8636	0.8627	0.8549	0.8434	0.8201	0.8168	0.8096	0.8014	0.7921	0.7816	0.7698	0.7566	0.7417	0.7250	0.7064
0.78	0.8805	0.8805	0.8803	0.8799	0.8789	0.8710	0.8591	0.8351	0.8317	0.8243	0.8159	0.8063	0.7956	0.7834	0.7698	0.7546	0.7375	0.7184
0.82	0.8951	0.8951	0.8949	0.8945	0.8935	0.8853	0.8731	0.8485	0.8450	0.8374	0.8288	0.8190	0.8080	0.7956	0.7816	0.7660	0.7486	0.7291
0.86	0.9081	0.9081	0.9079	0.9075	0.9065	0.8980	0.8855	0.8604	0.8569	0.8491	0.8402	0.8302	0.8190	0.8063	0.7921	0.7762	0.7584	0.7386
0.90	0.9197	0.9197	0.9195	0.9191	0.9180	0.9094	0.8966	0.8710	0.8674	0.8594	0.8504	0.8402	0.8288	0.8159	0.8014	0.7852	0.7671	0.7469
0.94	0.9300	0.9300	0.9298	0.9294	0.9282	0.9195	0.9064	0.8803	0.8767	0.8686	0.8594	0.8491	0.8374	0.8243	0.8096	0.7932	0.7748	0.7543
0.98	0.9391	0.9391	0.9389	0.9384	0.9373	0.9284	0.9151	0.8886	0.8849	0.8767	0.8674	0.8569	0.8450	0.8317	0.8168	0.8002	0.7816	0.7608
1.00	0.9433	0.9432	0.9430	0.9426	0.9414	0.9324	0.9191	0.8924	0.8886	0.8803	0.8710	0.8604	0.8485	0.8351	0.8201	0.8034	0.7846	0.7638
1.20	0.9729	0.9728	0.9726	0.9722	0.9709	0.9614	0.9472	0.9191	0.9151	0.9064	0.8966	0.8855	0.8731	0.8591	0.8434	0.8259	0.8064	0.7846
1.40	0.9878	0.9878	0.9875	0.9871	0.9858	0.9759	0.9614	0.9324	0.9284	0.9195	0.9094	0.8980	0.8853	0.8710	0.8549	0.8370	0.8171	0.7949
1.80	0.9980	0.9979	0.9972	0.9972	0.9959	0.9858	0.9709	0.9414	0.9373	0.9282	0.9180	0.9065	0.8935	0.8789	0.8627	0.8445	0.8243	0.8018
2.00	0.9934	0.9992	0.9990	0.9985	0.9972	0.9871	0.9722	0.9426	0.9384	0.9294	0.9191	0.9075	0.8945	0.8799	0.8636	0.8454	0.8252	0.8027
2.20	0.9998	0.9997	0.9995	0.9990	0.9977	0.9875	0.9726	0.9430	0.9389	0.9298	0.9195	0.9079	0.8949	0.8803	0.8640	0.8458	0.8255	0.8030
2.50	1.0000	1.0000	0.9998	0.9992	0.9979	0.9878	0.9728	0.9432	0.9391	0.9300	0.9197	0.9081	0.8951	0.8805	0.8642	0.8460	0.8257	0.8032
3.00	1.0000	1.0000	0.9998	0.9933	0.9980	0.9878	0.9729	0.9433	0.9391	0.9300	0.9197	0.9081	0.8951	0.8805	0.8642	0.8460	0.8257	0.8032

Source: From Hantush, 1967.

The transmissivity T in the term n should be taken as $K(H + h_{x, y, t})/2$ to allow for the increase in T as $h_{x, y, t}$ increases. Since $h_{x, y, t}$ is not known a priori, T may be taken as KH for first calculation of $h_{x, y, t}$, which is then used to compute T as $K(H + h_{x, y, t})/2$ for the second calculation, etc., until $h_{x, y, t}$ is defined with sufficient accuracy. Equation (8.14) is valid only if $(h_{x, y, t} - H) < 0.5H$.

The terms W and v_a represent the width of the infiltration zone and the downward velocity therein just above the water table. If there are no restricting layers in the vadose zone, W will be essentially equal to the width of the infiltration basin and v_a equal to the infiltration rate. The term t should be taken from the time that the water arrives at the water table, which may be several days after the start of infiltration in the basin. For the center of the mound, x and y are zero, in which case the sum of the F functions between brackets in Eq. (8.14) reduces to $4F[Wn/2, Ln/2]$.

The equation for decay of the groundwater mound after cessation of infiltration is (Hantush, 1967)

$$h_{x, y, t} - H = Z(x, y, t) - Z(x, y, t - t_s) \qquad (8.15)$$

where $h_{x, y, t}$ and H are the same as for Eq. (8.14), t is the time since water began to arrive at the water table, t_s is the time since water ceased to arrive at the water table, and $Z(x, y, t)$ and $Z(x, y, t - t_s)$ represent the right-hand part of Eq. (8.14) with t and $t - t_s$ as time factors. Because of drainage of the wetted zone above the water table, v_a will not immediately reduce to zero when infiltration is stopped [see Eq. (8.9)]. Thus, t_s should be counted from a few days after infiltration was stopped, allowing the vadose zone to reach "field capacity." Also, when calculating mound recession, f should be taken for draining which, since the water content in the vadose zone after draining may be higher than before wetting, tends to be less than the fillable porosity.

Example. An unconfined aquifer with $H = 5.333$ m and $K = 2$ m/day is recharged from a 20×200 m basin with an infiltration rate of 0.2 m/day. If $f = 0.15$, how much will the center of the groundwater mound have risen 6 days after infiltrated water arrived at the water table (assuming straight downward flow in the vadose zone)? Taking $T = 2 \times 5.333 = 10.667$ m^2/day, n is calculated as 0.024 2. Since $x = y = 0$, this yields $F(\alpha, \beta) = F(0.242, 2.42)$, for which the solution is 0.436 (Table 8.3). Substituting 4×0.436 for the term between brackets in Eq. (8.14) then gives $h_{0, 6} - H = 3.5$ m. This agrees reasonably well with the value of 4 m obtained by R analog for a 20-m-wide infinitely long recharge strip (Bouwer, 1962, case C in Figure 4), considering that Hantush's solution begins to lose validity if the water-table rise is more than half the original height of the aquifer. A further refinement could be obtained by calculating $h_{0, 6}$ with the adjusted T value of $2(5.333 + 5.333 + 3.5)/2 = 14.2$ m^2/day, etc.

The second question is: if infiltration is stopped 6 days after it started, what will be the height of the mound above the original water table in the center of the recharge area 6 days and 12 days later? Assuming that arrival of water at the mound ceases at the same time infiltration is stopped, the values of t and $t - t_s$ 6 days after the stop of infiltration are 12 days and 6 days, respectively. For $t = 12$,

Eq. (8.14) yields $h_{0,12} - H = 5.3$ m. Since Eq. (8.14) yielded $h_{0,6} - H = 3.5$ m for $t = 6$ days, the center of the mound 6 days after infiltration was stopped is $5.3 - 3.5 = 1.8$ m above the original water table, according to Eq. (8.15). This is a drop of $3.5 - 1.8 = 1.7$ m. At 12 days after infiltration was stopped, $t = 18$ and $t - t_s = 12$ days. For $t = 18$, Eq. (8.14) yields $h_0 - H = 6.6$ m, and since $h_0 - H$ was 5.3 m for $t = 12$, Eq. (8.15) shows that $h_0 - H = 1.3$ m, or a drop of 0.5 m for the mound center in the second 6 days after infiltration was stopped.

Equation (8.14) applies to rectangular recharge areas. For round areas or very (infinitely) long rectangular areas, the rise and fall of the mound can be predicted with equations especially developed for these geometries (Hantush, 1967), or they can be estimated with Eq. (8.14) by taking the circular area as a square with the same surface area and the infinitely long strip as a long, rectangular area (if $L > 4W$, the rise of the mound already is essentially the same as that for an infinitely long strip; see Figure 8.21).

The Dupuit-Forchheimer assumption of horizontal flow was also used by Glover (1964) to develop equations for mound behavior below recharge areas, and by Marino to develop computer solutions for mound formation below circular areas (Marino, 1975a) and rectangular areas (Marino, 1975b). Marino also developed equations for the growth and decay of groundwater mounds where the recharge area is bounded by a stream or other facility for water-table control on one side (1974a) or on both sides (1974b). Bianchi and Muckel (1970) presented Glover's analyses as a dimensionless graph (Figure 8.21) that enables rapid calculation of the rise of the center of the groundwater mound below square or rectangular recharge areas, including infinitely long strips. Circular areas are treated as squares with the same surface area. The curves in Figure 8.21 show that when $L/W > 4$, the mound rises at almost the same rate as below an infinitely long

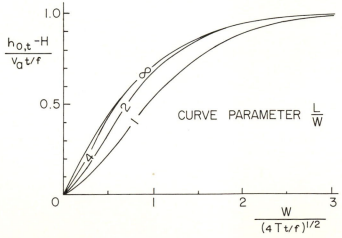

Figure 8.21 Dimensionless graph relating rise of center of groundwater mound to system parameters for square, rectangular, and infinitely long recharge basins. (*From Bianchi and Muckel, 1970.*)

recharge basin. Applying the curve for $L/W = \infty$ to the previous example where $L/W = 10$ yields a mound rise of 3.3 m after 6 days of infiltration, which agrees with the 3.5 m calculated with Eq. (8.14).

The ratio T/f is the aquifer parameter that determines the rise and fall of groundwater mounds, as shown by Eq. (8.14) and Figure 8.21. Conversely, T/f can be calculated for a given situation if the rise or fall of the mound is known. This procedure could be applied to evaluate the suitability of certain aquifers for recharge and to determine the best layout of infiltration basins. The ratio T/f, calculated from the observed rise of the mound below an experimental recharge basin, would then be used to predict the rise and fall of the mound below selected recharge-basin geometries.

While the horizontal-flow assumption has yielded reasonable agreement between observed and calculated groundwater-mound behavior (Bianchi and Haskell, 1975), the equations should be used with caution when H is large compared to W (thick aquifers or narrow basins). In that case, the flow in the aquifer is not evenly distributed over the entire height, as required for the Dupuit-Forchheimer assumption, but concentrated in the upper portion or "active" zone of the aquifer. The deeper portions of the aquifer then contribute very little to the flow and are essentially stagnant or "passive" (see, for example, Figures 8.19 and 7.23). This means that the effective transmissivity T_e of the aquifer for recharge-flow systems is less than T of the entire aquifer. Studies with an R analog have shown that the maximum depth of the active region is about equal to the width of the recharge area for isotropic aquifers (Bouwer, 1962). Thus, the upper limit for the effective transmissivity is approximately KW. If H exceeds W and the actual transmissivity KH is used rather than the effective transmissivity KW, equations based on horizontal flow will underestimate the rate of rise of the mound. Effective transmissivity may also have to be considered when predicting mound behavior below wide recharge basins with T/f calculated from observed mound behavior below narrow basins, because T_e for the flow system below the narrow basin could be less than T_e for the system below the wide basin if the aquifer is sufficiently thick for development of active and passive zones. To obtain more accurate predictions of the rise and fall of groundwater mounds, solution techniques must be used that take vertical-flow components into account. This can be achieved with physical models like the Hele-Shaw apparatus (Marino, 1967), electric analogs (Bouwer, 1962), or digital models (Hunt, 1971; Singh, 1972 and 1976).

8.3.2 Clogging

A major problem in groundwater recharge with basins is the reduction in the infiltration rate caused by accumulation of sediment and other fine material on the bottom and banks of the basins. Such clogging of the soil is undesirable because it increases the land area required for recharge and the percentage of water lost by evaporation. To minimize clogging, the sediment or suspended-solids content of the water must be as low as possible when it is admitted into the infiltration basins. This can be achieved with presedimentation reservoirs or fore-

bays, using a floating or other intake that skims off the surface water for conveyance to the infiltration basins. The detention time in the presedimentation basins
should be large enough to settle the solids, but not so large to enable algae to
multiply to such concentrations that they will clog the bottom of recharge basins.
Other techniques for sediment removal are coagulation and sedimentation
(Rebhun and Hauser, 1967) or slowly filtering the water by overland flow through
dense grass vegetation (Barfield et al., 1975; Wilson, 1967). Clogging of coarse-
textured soils has been observed at sediment concentrations in the recharge water
of as low as 50 mg/l (Behnke, 1969). Rice (1974) reported serious soil clogging
with infiltration of secondary sewage effluent if it contained more than 10 mg/l
suspended (organic) solids.

Clogging is primarily caused by settling of sediment, straining of suspended
material as water moves through the sediment layer and into the soil, and bacterial and other biological action (Behnke, 1969; Ripley and Saleem, 1973). When
water first enters a dry, clean basin, fine solids may actually move a considerable
distance into the bottom soil, especially when this soil is coarse. If the soil
becomes finer farther down, the solids could accumulate on these fine-soil layers,
which can produce serious reductions in downward-flow rates (Berend, 1967).
When sediment begins to accumulate on top of the soil, however, the fine particles
no longer move into the bottom material but are strained out on top of the
sediment layer where they contribute to the hydraulic resistance of the clogging
layer. Clogging thus is primarily a surface phenomenon that rarely extends more
than 10 cm into the soil and often is restricted to the top centimeter or less.
Clogging tends to cause the greatest relative reduction in infiltration rate if the
bottom soil is coarse and the suspended sediment is fine.

Continued clogging eventually reduces infiltration rates to only a fraction
(one-tenth or less, for example) of the original infiltration rate. To maintain
higher average infiltration rates, it is necessary to periodically dry the basin so
that the clogging layer can dry, crack, and curl up, and the organic materials can
decompose. In severe cases of sediment accumulation, removal of the clogging
layer may be necessary. Where recharge is a continuous operation (in contrast to
spurious recharge, as along ephemeral streams), the lengths of infiltration and
drying periods should be selected to produce maximum long-term infiltration or
hydraulic loading. Optimum infiltration and drying schedules are best determined
by local experimentation, measuring declines in infiltration rates during recharge
and recovery of infiltration rates during drying. Sometimes, the length of
infiltration periods is restricted for reasons other than declining infiltration rates,
such as control of mosquitos or other insects breeding in the water.

Mathematical expression of the clogging process generally yields equations
with an exponential decay in infiltration rate (Behnke, 1969). The coefficients in
these equations must be experimentally determined for the sediment concentration of the water, type of sediment, type of bottom material, and infiltration rates
of the particular installation. Berend (1967) developed the empirical equation

$$v_i = v_0 - aMI_t \qquad (8.16)$$

where v_i = infiltration rate at time t (m/day)

$\quad$ v_0 = initial infiltration rate

$\quad$ a = coefficient describing clogging properties of system

$\quad$ M = concentration of suspended matter in water that will be retained on bottom (g/1)

$\quad$ I_t = accumulated infiltration (m)

The time t_r it takes for a certain minimum acceptable infiltration rate v_{min} to be reached is then calculated as

$$t_r = \frac{1}{aM} \ln\left(v_0/v_{min}\right) \tag{8.17}$$

Thus, after t_r days of continuous infiltration, a drying or resting period should be started to allow recovery of the infiltration rate. Values of a must be experimentally determined, measuring v_0, v_i, M, and I_t and calculating a with Eq. (8.16). For flood waters with relatively high sediment content (0.14 to 2 g/1), a varied from 0.1 to 5.5 (Berend, 1967).

8.3.3 Injection Wells

Well injection is practiced for recharge of confined aquifers and of unconfined aquifers if separated from the surface by restricting layers in the vadose zone. It is also used where the groundwater is deep or where the topography or existing land use (urban areas) make basin recharge impractical or too expensive. Deep recharge pits or large-diameter wells or shafts (Signor et al., 1969; McCormick, 1975) can be considered as intermediate between basin recharge and well recharge.

$\quad$ An important application of well injection is the creation of freshwater barriers in coastal aquifers to protect inland, pumped portions of the aquifers against intrusion of salt water from the sea (see Section 11.2). At the West Coast Basin Barrier Project in California, for example, 40×10^6 m^3 of Colorado River water are injected annually through 93 injection wells located near the coast on a 14.5-km-long line between Palos Verdes Hills and the Los Angeles International Airport (Bruington and Seares, 1965; Bulten et al., 1974). Some wells recharge two aquifers from the same hole (Figure 8.22 left), using an inflatable hydraulic packer to separate the well into two recharge compartments that are independently operated. These wells are about 170 m deep. In Orange County, California, freshwater barriers are created by injection through 23 multicasing recharge wells (Bulten et al., 1974). Each well recharges four aquifers with each aquifer being served by a separate casing-and-screen section with its own water supply (Figure 8.22, right). The depths of the aquifers range from 20 to 100 m.

$\quad$ Some wells are dual-purpose wells that are used for recharge to store surplus water underground and for pumping when groundwater is needed. An interesting

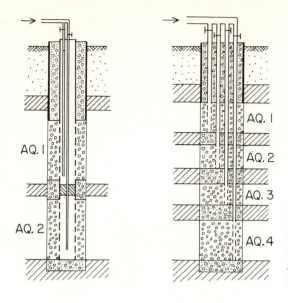

Figure 8.22 Schematic of injection well serving two aquifers with well screens separated by hydraulic packer (left), and of multiaquifer injection well with injection sections separated by concrete plugs (right).

application of this principle is storage of fresh water in saline aquifers. The amount of fresh water that can later be recovered from such storage depends on natural movement of water in the saline aquifer, volume of fresh water stored, and time of storage. Under favorable conditions, 85 to 90 percent of the injected water can be recovered and meet drinking-water standards (Brown and Silvey, 1973; Kimbler et al., 1973). On the other hand, Smith and Hanor (1975) reported recovery percentages of only 24 and 45 percent in an aquifer where the natural rate of groundwater movement was 15 cm/day. At Norfolk, Virginia, reductions of as much as 75 percent in injection rate occurred because the injected fresh water caused dispersion of clay in the saline aquifer. It was anticipated that injection of a polymeric hydroxyl aluminum solution into the aquifer prior to recharge with fresh water would prevent the dispersion of clay, since aluminum ions would be adsorbed by the clay and cause the clay particles to remain flocculated (Brown and Silvey, 1973).

Injection wells are constructed in the same manner as pumped wells (Section 6.3). The cable-tool technique is generally preferred over rotary drilling because the latter may leave a mud cake in the bore hole and mud in the aquifer (Baffa et al., 1965). If rotary drilling is necessary (for example, to enable use of certain casings and screens or gravel envelopes), reverse-rotary drilling is the best technique because it leaves a thinner mud cake or drilling mud may not be needed at all. Careful grouting of the well is required to avoid leakage around the casing and loss of recharge water to nontarget aquifers or seepage to the surface, especially where one well serves more than one aquifer.

The mathematics of injection-well flow systems is the same as that for discharging wells, except that drawdown is "draw-up." The cone of groundwater rise around the well thus is the image of the drawdown cone reflected across the original water table or piezometric surface, and equations describing drawdown for discharging wells (see Chapter 4) are also used to predict build-up cones around recharge wells (Bianchi and Muckel, 1970; Esmaili and Scott, 1968). Mobasheri and Todd (1963) compared recharge- and discharge-flow systems for confined aquifers in a sand-tank model of a well. They found that for laminar flow, total-head loss for recharge was the same as that for discharge at the same flow. If the flow rate was increased to create turbulent flow in the vicinity of the well, the total-head loss for recharge was slightly higher (but not more than 4 percent) than that for discharge at the same flow rate. This difference is not significant for practical purposes. For unconfined aquifers, recharge wells have, of course, no surface of seepage.

The main difference between injection wells and discharge wells is that injection wells are much more sensitive to clogging of the aquifer at the bore hole, because fines accumulate on the hole wall as water enters the aquifer. This clogging is due to suspended solids in the water, to the action of slime-forming and other bacteria in the well, and to the accumulation of incrustation and corrosion products. Additional head losses may also be caused by chemical or physical-chemical reactions between the injected water and the aquifer (precipitation of salts, dispersion of clay) and by air binding. The latter occurs when the injected water has a high dissolved-air content and is colder than the receiving aquifer (Sniegocki, 1963). As the water heats up, air will go out of solution and form air pockets in the aquifer, which reduce the hydraulic conductivity. To avoid air binding, the water should be warmer than the target aquifer, it should have a low dissolved-air content, and it should be piped down into the well to avoid free-falling and splashing water that could become saturated with dissolved air.

To overcome aquifer-entry losses, water can be injected under pressure. This requires careful grouting of the well. Also, injection pressures are limited by the structural integrity of well components and of the aquitards and other restricting underground materials around the well.

Clogging of injection wells is minimized by removing the suspended solids from the water and by chlorinating or otherwise disinfecting the water prior to injection (Baffa et al., 1965). Rebhun and Hauser (1967) reduced the sediment content of runoff water from 200 to 20 mg/l with cationic polyelectrolytes prior to injection. However, the well still sealed up and recharge rates were reduced in several days. Bailing the well effectively removed the clogged layer and restored the recharge capacity of the well. Other well-development techniques such as jetting, air surging, and intermittent pumping (see Section 6.3) have also been effective for restoring recharge rates (Baffa et al., 1965; Bruington and Seares, 1965; McCormick, 1975). Another practice is to alternate injection with pumping of the well—for example, interrupting injection with 1 to several hours' pumping each day, 5 h of pumping each week, or pumping out a certain volume each time a certain volume is injected. The first water pumped after an injection period normally has a high suspended-solids content and may have to be discarded or recycled for injection after sedimentation.

8.4 FLOW OF GROUNDWATER TO THE SURFACE ENVIRONMENT

8.4.1 Wells

Well discharge is the most obvious, human-made return of groundwater to the surface. The total groundwater withdrawal in the entire United States was estimated at 225×10^6 m^3/day in 1965 (Todd, 1970, and references therein) and 260×10^6 m^3/day in 1970 (Murray and Reeves, 1972). Of the latter amount, 170×10^6 m^3/day or 65 percent were used for irrigation (total irrigated area was about 20×10^6 ha in 1970). The relative contribution of groundwater to the total water use in the United States has been rather constant, i.e., 17 percent in 1950 and 18 percent in 1970 (calculated from data by Murray and Reeves, 1972). Rural water supplies obtain more than 95 percent of their water from groundwater. Most of the groundwater pumped for municipal or industrial water supplies eventually is discharged into surface water as sewage effluent or industrial wastewater. Irrigation, however, is primarily a consumptive use of water, since most of the water is returned to the atmosphere via evapotranspiration.

8.4.2 Springs, Seeps, and Gaining Streams

Springs are the most conspicuous forms of natural return of groundwater to the surface. They come in all sizes, from small trickles to large streams. If water oozes out of the soil or rock over a certain area without distinct trickles or rivulets, the discharge is called a *seep*. Springs and seeps occur where downgradient parts of aquifers or other water-carrying materials are exposed to the surface, as in outcrops of aquifers at mountainsides or canyon walls (Figure 1.2) or shallow water tables reaching the surface at the base of long slopes. Springs also form where discontinuities like faults or dikes present hydraulic barriers and force groundwater to flow upward, or where faults cause weak spots in confining layers, allowing water to flow upward and reach the surface if the piezometric surface in the aquifer is sufficiently high. In fractured rock, fissures can fill with rainwater, which then flows through the same fissure system to form springs at lower points. Rock piles and similar debris at the foot of mountains can temporarily store water in periods of rainfall and gradually release the water via springs or seeps along their base. Springs can also occur below surface water, fresh or salt. Large, off-shore springs are said to have supplied sailors in the past with fresh water in the midst of a saltwater environment.

Springs were classified by Meinzer (1923) according to their mean discharge as follows:

Order	Discharge, m^3/day
1st	$> 0.245 \times 10^6$
2d	$0.0245 \times 10^6\text{--}0.245 \times 10^6$
3d	$2\,450\text{--}24\,500$
4th	$545\text{--}2\,450$
5th	$54\text{--}545$
6th	$5.4\text{--}54$
7th	$1.3\text{--}5.4$
8th	< 1.3

The large springs are usually associated with permeable aquifers like cavernous limestone, porous basalt, or sorted gravel. The flow rate from springs depends on the size of the recharge area above them, the rate of rainfall or other water accretion in that area, and T of the aquifer or other water-conducting material. Some springs are rain-fed and flow only during and after periods of rain. Others obtain their water from surface water, like springs below a dam or other impoundment (natural or artificial).

Submerged springs and seeps in streams increase the flow of water in the stream, which is then called a *gaining* stream. Such streams have also been called *influent* streams. General seepage of groundwater into a stream occurs when the water table adjacent to the stream is higher than the water surface in the stream. When the stream runs through a valley or floodplain, the water table normally will not be much higher than the water level in the stream. The rate of groundwater flow into the stream can then be calculated with the same theory presented in Section 8.2 for seepage from losing streams. Thus, steady-state discharge of groundwater into a stream can be determined with Figures 8.11, 8.12, and 8.13, taking D_w as the height of the water table above the water surface in the stream. Discharge of groundwater due to a sudden lowering of the water level in the stream can be calculated with Eq. (8.13) or with the theory developed by Streltsova (1975). If the stream level recedes gradually, like after a flood, the discharge of groundwater (release of bank storage) can be calculated with the equations by Cooper and Rorabaugh (1963). Discharge of groundwater often is an important contribution to stream flow, particularly when there is no surface runoff into the stream and the "base" flow is almost entirely provided by groundwater.

8.4.3 Agricultural Drains

Where groundwater levels are so close to the surface (less than 0.3 to 1 m, for example) that they reduce crop yields and interfere with farming operations, water tables are commonly lowered with underground drains. These drains consist of perforated plastic tubes, clay, or concrete tile 5 to 20 cm in diameter that are installed at depths of 1 to 3 m (Fouss, 1974) to collect groundwater and discharge it into a drainage ditch or other surface water, usually by gravity but sometimes by pumping from collector sumps. Occasionally, groundwater is collected via open ditches instead of by underground drains. Such ditches, however, take land out of production and require considerable cleaning and other maintenance. If the soil is underlain by sand or other permeable material of sufficient transmissivity, water tables in the soil can also be lowered by pumping from a system of wells or well points (Kahn and Kirkham, 1971; Rektorik, 1976). Sometimes, high water tables occur only in a few spots, like seep areas caused by subsurface runoff in sloping fields, or low spots. Such areas are drained by a random system, installing several drain lines in an irregular pattern to intersect as many wet spots as possible. Most drainage systems consist of parallel, equidistant drains to systematically lower the water table in a field. The distance between the drains varies from about 8 to 50 m or more. Optimum stationary water-table depths vary from between 0.6 and 0.9 m for sandy soils to between 1 and 1.5 m for clay soils for a number of crops (Wesseling, 1974, and references therein). Orchards require deeper water tables, whereas pastures can tolerate higher water tables.

The depth and spacing of drains for systematic drainage should be selected so as to obtain the highest monetary return (in terms of increased crop yield) from the drainage investment. Design of drainage systems still relies very much on local experience, even though considerable progress toward rational design has been made (Bouwer, 1974, and references therein). The simplest procedure is based on a steady-state flow system, calculating the drain spacing on the basis of a specified rate of water removal (drainage rate) at a specified maximum height of the water table. Numerous equations have been developed for this purpose (Kirkham et al., 1974, and references therein). The simplest equations are based on the assumption of horizontal flow which, however, is valid only if the impermeable layer is a small distance below the drains. If the impermeable layer is at greater depths, vertical-flow components are no longer negligible and the flow is concentrated in the upper, active zone of the system. As with the recharge-flow systems discussed in Section 8.3.1, the effective transmissivity is then less than the transmissivity of the aquifer at full height. These factors were taken into account by Hooghoudt (1940), who applied radial-flow theory and images to evaluate effective depths of the impermeable layer. This effective depth, which is less than the actual depth of the impermeable layer, was then used in the drain spacing equation, derived on the assumption of horizontal flow as

$$v_a = \frac{4Km(2D_e + m)}{L^2} \tag{8.18}$$

The terms in this equation are:

v_a = drainage rate (rate of water arrival at water table, length/time)

K = hydraulic conductivity of soil

D_e = effective depth of impermeable layer below drain center

m = height of water table above drain centers midway between drains

L = distance between drains (Figure 8.23)

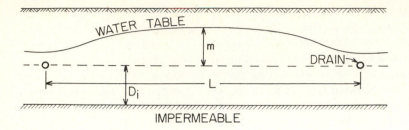

Figure 8.23 Geometry and symbols for tile-drainage system.

Hooghoudt (1940) presented tables showing D_e in relation to the actual depth D_i of the impermeable layer below the drains (Figure 8.23) for different drain spacings L and drain diameters. The effect of the drain diameter on D_e is small, however, and can usually be ignored (Bouwer and van Schilfgaarde, 1963). A graphical representation of D_e versus D_i for different values of L and a drain diameter of 12 cm is shown in Figure 8.24. The curves show that D_e is about equal to D_i if D_i is small, but increasingly less than D_i as D_i increases. If D_i has reached a value of about $0.25L$, D_e has become essentially constant. Thus, as long as the impermeable layer is more than a distance $0.25L$ below the drains, it can be considered as infinitely deep.

The drainage rate v_a is the infiltration rate, rainfall rate, or deep percolation

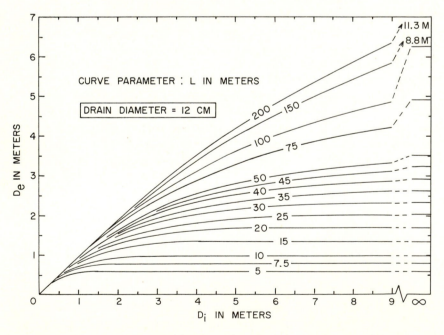

Figure 8.24 Effective depth D_e of impermeable layer below drains as a function of its actual depth D_i for different drain spacings L. (*From tables by Hooghoudt, 1940.*)

rate at the equilibrium position of the water table characterized by m. Values of v_a and m constitute the so-called drainage criterion, which when known enables calculation of L with Eq. (8.18) if K and D_e are known. The drainage criterion in theory can be evaluated from rainfall patterns, capacity of soil to store water, natural drainage rate of soil profile, and relations between crop yield and water-table depth (Bouwer, 1974). However, this is not an easy process, particularly if economic optimization is to be included. Thus, the drainage criterion is often evaluated by simply measuring drain discharges and water-table positions in fields that are known to have "good" drainage. The drainage criterion of $v_a = 7$ mm/day at a maximum water-table height of 0.5 m below field surface that is extensively used in the Netherlands was evaluated in this way. These values were corroborated in subsequent studies where yield reduction was related to frequency of high water-table levels (Bouwer, 1974, and references). The drainage criterion depends on local conditions of climate and soil, and must be evaluated for each distinct physiographic region where drainage is practiced. The Dutch criterion of 7 mm/day at a water-table depth of 0.5 m applies to a rainfall pattern where daily amounts are usually less than 5 mm and rarely more than 30 mm on any one day or 15 mm for several consecutive days. Drainage guides with recommended values for drain spacings or drainage criteria have been prepared for various regions (Bouwer, 1974, and references).

To calculate the drain spacing on the basis of a given drainage criterion, the depth of the drain is selected first. This depth must be large enough to avoid small values of m, but not so large as to require high installation costs. The drains must also be located above an impermeable layer, and preferably in a permeable stratum if present. The value of m is then calculated as the difference between the depth of the drains and the minimum permissible depth of the water table as specified by the drainage criterion. The value of K is determined with the auger-hole method or other appropriate field technique (Chapter 5), and D_i is found by examining the soil profile or taking K measurements at different depths. A value of L is assumed, D_e is evaluated from Figure 8.24, and L is calculated with Eq. (8.18), using the v_a value of the drainage criterion. This calculated L value is used to obtain a second estimate of D_e, which is used to calculate a third L value, etc., until L is determined with sufficient accuracy. If the flow in the capillary fringe must be taken into account, the equivalent fringe height $-h_{cr}$ is added to D_e for calculation of L (Bouwer, 1959). This may be significant where $-h_{cr}$ is not small in relation to D_e.

The steady-state approach is valid in areas with rather prolonged rainfall at moderate intensities. Where rainfall intensities are high or where water is applied with heavy irrigations, the water table may rapidly rise to the top of the root zone and even to field surface. Under those conditions, drainage systems are better designed on the basis of a certain rate of fall of the water table following an infiltration event, to avoid damage to the crop. If the water table rose all the way to field surface, it may be desirable to design the drainage system so that the water table midway between drains drops 30 cm in the first 24 h after cessation of infiltration and 20 cm in the next 24 h (Bouwer, 1974, and references therein).

Many equations have been developed to predict the rate of fall of the water table in relation to L, m, K, D_i, and the drainable pore-space fraction f (van Schilfgaarde, 1974). One of the simplest equations is the integrated Hooghoudt equation, developed by Bouwer and van Schilfgaarde (1963) as

$$\frac{Kt}{f} = \frac{CL^2}{8D_e} \ln \frac{m_0(m_t + 2D_e)}{m_t(m_0 + 2D_e)} \tag{8.19}$$

where C is a coefficient relating the fall midway between the drains to the average fall of the entire water table. If the water table initially is at or near field surface, it will drop faster over the drains than midway between the drains and $C > 1$. Then, when the water table starts falling rather uniformly without change in shape over the entire area between drains, $C = 1$. When the water table approaches the drains, it falls faster midway between the drains than near the drains, in which case $C < 1$. Comparisons with more rigorously derived equations indicate that using an average C value of 0.9 will give good results (van Schilfgaarde, 1974). The other terms in Eq. (8.18) are time t, drainable pore-space fraction f, initial value of m (m_0) and value of m at time t (m_t). The parameters L, K, and D_e are the same as defined for Eq. (8.18).

To design a drainage system on the basis of Eq. (8.18), a certain drain depth is selected and an assumed value is used for L to calculate t for a given water-table drop from m_0 to m_t. If the resulting t is too large, a smaller value is taken for L, or vice versa, etc., until the desired t value is obtained. The desired rate of fall of the water table midway between drains depends on the type and stage of crop growth, weather conditions, antecedent water-table levels, soil fertility, and other factors which are not always readily assessed. Thus, local experience is also important in selecting proper drain spacings for the falling-water-table case.

Steady-state and falling-water-table conditions are special conditions of the true situation, which is a fluctuating water table. Attempts have been made to measure and predict water-table hydrographs for entire crop seasons, and to relate yield reductions to frequencies of high-water-table conditions. The required intensity of a drainage system to limit such frequencies to acceptable values can then be expressed in terms of a simple, steady-state drainage criterion (Bouwer, 1974, and references therein).

The function of drainage systems in irrigated fields in arid climates is to remove deep percolation water in case of insufficient natural drainage, and to keep the water table low enough between irrigations or irrigation seasons to prevent accumulation of salt in the root zone due to evaporation of groundwater. The amount of deep percolation water necessary to maintain a salt balance in the root zone is calculated with the salt-balance equation for an irrigation cycle or season (see Section 11.7). If the water table is not expected to rise too much into the root zone, the drainage rate v_a can simply be calculated as the amount of deep percolation water (expressed as a depth) divided by the irrigation interval or season, respectively. Actual values of v_a for salinity-control range from about 0.5 to 2.5 mm/day (Talsma, 1963).

The equilibrium water-table depth, which determines m for a given drain depth, generally should be larger in irrigated fields in dry climates than in nonirrigated fields in humid climates, to avoid evaporation of groundwater and resulting build-up of salt in the root zone. Recommended equilibrium water-table depths for irrigated land range from 120 cm for coarse-textured soils to at least 190 cm for medium-textured soils (Talsma, 1963). For fine-textured soils, the water table should be kept at least 120 cm below the root zone.

If there is a possibility of temporary high water tables after an irrigation, the drainage system should be designed on the basis of a falling water table or of a certain water-table hydrograph for the entire irrigation season. Bouwer (1969b) used Eq. (8.9) to predict deep percolation rates from root zones in relation to time after irrigation. These rates were then used with Eq. (8.18) to calculate rise and fall of the water table midway between drains for a given drain spacing. Repeating the procedure for other drain spacings enabled selection of the spacing with the best water-table response. Dumm and Winger (1964) developed a procedure for calculating water-table fluctuations in drained irrigated land for an entire season, which when applied to several drain depths and spacings enabled selection of the best design. For additional procedures, see van Schilfgaarde (1974) and references therein.

The removal of deep percolation water from irrigated land by subsurface drains in areas where water tables would otherwise rise and cause problems of waterlogging and salt accumulation is the key to successful, permanent irrigation. Inability to control water tables in this manner may well have forced old civilizations to abandon their fields and perish as a culture.

8.5 EVAPORATION OF GROUNDWATER

Direct evaporation of groundwater occurs when the water table is close to field surface—for example, within a depth of about 0.5 to 2 m (depending on the soil). The unsaturated hydraulic conductivities K_h in the vadose zone may then be sufficiently large to permit upward flow from the water table in response to an evaporative demand at soil surface. This upward flow is in the liquid phase. Actual evaporation of water takes place at or near the surface, as evidenced by salt accumulation (Nakayama et al., 1973). When the topsoil is almost dry, K_h is very small and vapor flow may become significant (see Section 7.7). However, the flow rates will then be very small compared to those at liquid flow.

Evaporation from a bare soil is a complex process that is affected by diurnal patterns in solar radiation, temperature, soil water content, salt concentration of water, and other factors (Idso et al., 1974). The simplest situation is steady-state flow from a constant water table to a bare surface where evaporation takes place. Such systems have been solved analytically by Gardner (1958) for a uniform soil and by Ripple et al. (1972) for a layered soil to calculate the relation between upward-flow rate and water-table depth, knowing the relation between K_h and h of the soil(s). Experimental evaporation data were given by Gardner and Fireman

(1958) and Ripple et al. (1972). A simple way of estimating the relation between upward-flow rate and water-table depth is to split the total drop in pressure head from the water table to the surface into a number of increments, select a certain upward-flow rate, and calculate the depth increment for each pressure-head increment with Darcy's equation, using the $K_h - h$ relation of the soil(s) in question. The calculation is then repeated for other flow rates.

To exemplify this procedure and to illustrate some basic aspects of evaporation of groundwater, the relation between groundwater evaporation and water-table depth will be calculated for Rubicon sandy loam, for which Figure 8.25 shows a logarithmic plot of K_h versus h for wetting and drying. The data points for this plot were taken from Figures 7.26 and 7.27 (see Problem 7.6). Assuming that the water table is slowly receding from an initial position near field surface, the $K_h - h$ curve for drying is the appropriate curve to be used in this calculation. The pressure head h at field surface is selected as -300 cm of water, which is representative of a medium water content. A lower h value—for example, -500 or -1000 cm water—would not significantly affect the water-table depth calculated for a certain evaporation rate, because K_h at these low h values is so low that a very large hydraulic gradient would be necessary to sustain any significant upward flow. This means that h must change very rapidly near the soil surface

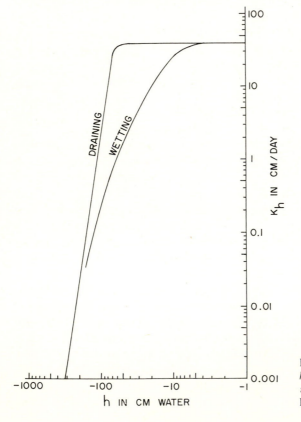

Figure 8.25 Relation between K_h and h for Rubicon sandy loam for draining and wetting, as determined from Figures 7.26 and 7.27.

Table 8.4 Calculation of water-table depth in relation to upward-flow rate for Rubicon sandy loam

Pressure head, cm	K_h, cm/day	$\overline{K}_h$, cm/day	Evaporation rate 0.2 cm/day Δz, cm	0.2 cm/day z, cm	1.2 cm/day Δz, cm	1.2 cm/day z, cm
− 300	0.0014			0		0
		0.0029	0.71		0.12	
− 250	0.0044			0.71		0.12
		0.0117	2.76		0.48	
− 200	0.019			3.47		0.60
		0.0745	13.57		2.92	
− 150	0.13			17.04		3.52
		1.015	41.78		22.91	
− 100	1.9			58.82		26.43
		19.95	49.5		47.16	
− 50	38			108.32		73.59
		38.5	49.7		48.49	
0	39			158.06		122.08

over a distance that is very small compared to the depth of the water table. The decrease in h from zero at the water table to -300 cm at the surface is divided into 50-cm increments (first column, Table 8.4). These increments are sufficient for demonstration purposes, but they should be smaller (5 to 10 cm, for example) in actual calculations to increase the accuracy of the method. The K_h values corresponding to the h values, taken from Figure 8.25 for draining, are listed in the second column of the table. The (arithmetic) average K_h for each increment is shown in column 3. Assuming an evaporative demand or upward-flow rate of 0.2 cm/day, the height increment Δz for each h increment is calculated with Darcy's equation in column 4. Adding the increments then gives the water-table depth (158 cm) corresponding to an upward-flow rate of 0.2 cm/day. This process is repeated for other evaporative demands (the case for 1.2 cm/day is included in Table 8.4), so that a graph of upward-flow rate versus water-table depth can be constructed (Figure 8.26). The maximum flow rate, however, cannot exceed the potential evaporation rate as determined by meteorological conditions. Assuming a potential evaporation of 9 mm/day, Figure 8.26 shows that this rate can be maintained up to a water-table depth of 128 cm. At this point, K_h in the upper part of the vadose zone begins to limit the upward flow and flow rates decrease if the water-table depth increases. When the water-table depth exceeds 200 cm, evaporation of groundwater is very small for the system in Figure 8.26.

The water-table depth where potential evaporation flow can no longer be sustained is less for coarse-textured than for fine-textured soils. On very tight soils, K of the soil at or near saturation may already be so low that it restricts upward flow and causes the soil near the surface to dry, which rapidly reduces evaporation to

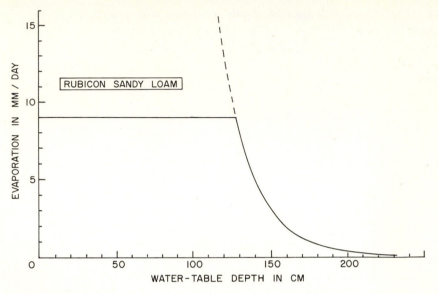

Figure 8.26 Relation between evaporation rate and water-table depth for Rubicon sandy loam calculated with procedure in Table 8.4.

very small values. If the water table is more than a few meters below ground surface or below the bottom of the root zone, evaporation of groundwater may still occur but the water must move upward through the vadose zone in the vapor phase. Deep groundwater "open" to the atmosphere through a vadose zone with no restricting layers thus is not immune to evaporation, but the evaporation rate may be very small. In addition to movement by a vapor-pressure gradient, water vapor may reach the atmosphere because vadose zones tend to "breathe" continuously due to wind effects, changes in surface temperature and barometric pressure, and earth tides. These phenomena may explain why deep vadose zones in desert alluviums are so dry even though they were deposited under water.

8.6 GROUNDWATER USE BY PHREATOPHYTES

When the soil is vegetated, groundwater can be evaporated from greater depths than when the soil is bare, especially if the vegetation is deep-rooted. Water tables within reach of plant or tree roots are valuable in agriculture because they keep crops supplied with water during droughts. The evaporation rates will then remain close to potential evapotranspiration (see Section 8.1.5). Groundwater uptake by nonagricultural plants is a water loss that can be particularly significant in water-short areas. Such plants, collectively called *phreatophytes*, commonly grow along stream channels and in floodplains or stream valleys. Phreatophytes offer scenic value, wildlife habitat, shelter, and windbreaks, but their agricultural or commercial value is low. Examples of common phreatophytes are salt cedar (Tamarix), cottonwood (Populus), mesquite (Prosopis), willow (Salix), rabbit-brush (Chrysothamus), greasewood (Sarcobatus), saltgrass (Distichlis), and bac-charis (Baccharis). Salt cedar is native to the Mediterranean area and western Asia. It was imported into the United States as an ornamental tree in the nineteenth century and has become so well established that presently it is a major phrea-tophyte. An extensive list of phreatophytes containing more than 50 species was prepared by Robinson (1958).

In the western United States alone, phreatophytes are estimated to cover about 6.4×10^6 ha and to use about 3×10^{10} m^3 of water per year (Robinson, 1958), enough to irrigate 3 million ha of land (assuming a water use of 1 m per year) or to supply water for 200 million people (assuming a water use of 400 l per person per day). Some of this water could be saved by removing phreatophytes or replacing them with shallow-rooted, beneficial vegetation like grasses, or by low-ering the water table (Muckel, 1966). Thinning dense stands of trees in floodplains also clears flood channels and, as additional benefit, can reduce flood damage. Phreatophyte removal should be carefully planned so that wildlife and scenic values are preserved.

The amount of groundwater consumed by phreatophytes depends on climate, depth and salinity of groundwater, and on rooting depth, density, and stage of growth or activity of the plants. The data in the previous paragraph indicate an average use of $3/6.4 = 0.47$ m per year. Locally, annual water-use rates may vary from less than 0.3 m to 2.5 m or more. The active rooting depth of a number of phreatophyte trees is on the order of 10 m, although salt-cedar roots were detected as deep as 30 m during construction of the Suez Canal (Renner, 1915). Shrubs and deep-rooted plants may have rooting depths of 1 to 3 m, whereas roots of certain grasses and shallow-rooted plants may go down only 0.5 to 1 m. Most deep-rooted phreatophytes show a significant decrease in the uptake of groundwater when the water table drops below a depth of 2 m (Robinson, 1958). This does not mean that the water use by the plants is also significantly reduced, since roots may continue to absorb water from the vadose zone. Deeper water tables often also produce sparser vegetation, which further reduces the water consumption.

The amount of groundwater used by phreatophytes has been evaluated with various techniques, including growing the plants in large containers or lysimeters with different water table depths and measuring the volumes of water that must be added to keep the water table at constant depth (Horton, 1973, and references therein). Such studies should be carried out with the containers in the same environment as that of the actual phreatophytes to minimize advective energy or clothesline effects which could greatly increase the water use of the plants in the lysimeters. Even then, the results are not always transferable to floodplains with different types of phreatophytes at varying densities and different soil profiles. Field techniques to evaluate groundwater use by phreatophytes include the water-balance approach, measuring all inflow and outflow components for a certain reach of stream and floodplain and determining the water use as the difference between total inflow and total outflow. The disadvantages of this method are that large areas are required to obtain a measurable difference between inflow and outflow and that the errors of the individual components accumulate in the calculated value of the water use. Other field techniques consist of calculating groundwater uptake by phreatophytes from decreases in groundwater flow (measuring gradients and transmissivities in the aquifer), from measurements of the fall of the water table or of decreases in water content of the vadose zone, and from increases in groundwater salinity. Reasonable agreement between the results of these methods has been reported (Gatewood et al., 1950).

The groundwater use by deep-rooted phreatophytes cannot be equated to the water savings that can be obtained by removing such plants, because removal very likely causes water tables to rise in parts of the floodplain to where they could be within reach of shallow-rooted plants or even become subject to direct evaporation from the soil. If stream seepage is the main source of groundwater in a floodplain or alluvial valley, the potential water saving due to phreatophyte removal can be predicted as the difference between stream seepage before and after removal. These seepage rates are then calculated on the basis of the pre- and postremoval relations between evapotranspiration of groundwater and water-table depth (Bouwer, 1975). Examples of these relations are shown in Figure 8.27 for the annual evapotranspiration of groundwater from a dense stand of mature salt cedar in the Gila River bed west of Phoenix, Arizona (data points taken from van Hylckama, 1975), and for the annual evaporation from bare soil (hypothetical curve for sandy soil at same location). At steady-state conditions, stream seepage will be in equilibrium with evapotranspiration in the floodplain. Bouwer (1975)

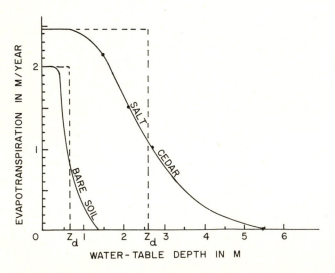

Figure 8.27 Relations between evapotranspiration of groundwater and water-table depth (solid lines) for salt cedar *(data points taken from van Hylckama, 1975)* and bare soil (hypothetical curve for coarse-textured material), and replacement of relations by equivalent step functions (dashed lines).

developed an incremental procedure based on the horizontal-flow assumption to calculate the seepage and groundwater-flow system so that the upward flow (evapotranspiration) from the water table at any point satisfies the relation between evapotranspiration rate and water-table depth for the particular floodplain condition. The results of such a calculation, using the relations in Figure 8.27, are presented in Figure 8.28 for a rectangular channel in an alluvial valley consisting of uniform soil with a K value of 900 m/year (2.466 m/day) and an impermeable layer at a depth of 10 m. The water level in the channel is taken at 0.5 m below the floodplain surface. As can be expected, evapotranspiration of groundwater decreases with distance from the stream and is higher for salt cedar than for bare soil. The water table is lower for the salt-cedar cover than for the bare soil. The lateral flow in the aquifer normal to the stream decreases with distance from the stream and becomes zero when the water table has reached the depth where the rate of groundwater evapotranspiration has become zero (5.5 m for salt cedar and about 1.4 m for bare soil). Calculated seepage rates are higher for salt cedar than for bare soil, and the difference between the seepage rates is the water saving obtainable by removing the salt cedars.

The incremental procedure for calculating water-table shapes and seepage rates for a given relation between evapotranspiration rate ET of groundwater and water-table depth z is rather tedious and can be simplified if seepage is the only quantity that needs to be calculated. The simplified procedure consists of

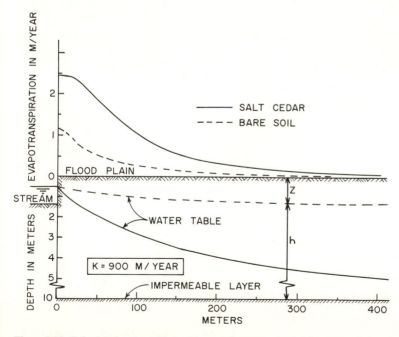

Figure 8.28 Calculated water-table positions and evapotranspiration rates of groundwater if floodplain is covered by salt cedar (solid lines) and when floodplain is in bare soil (dashed lines). *(From Bouwer, 1975.)*

replacing the actual relation between ET and z by a step function with height equal to the potential ET when $z = 0$ and width z_d so that the area under the step function is the same as that under the actual curve. The step functions for the $ET - z$ relations in Figure 8.27 are shown as dashed lines in the figure. Using the parameters of the step function, seepage from the stream can be directly calculated with the following equation, which was derived on the assumption of horizontal flow below the water table (Bouwer, 1975):

$$I_s W_s = 2[ET(h_0^2 - h_d^2)K]^{1/2} \qquad (8.20)$$

where $I_s W_s$ = seepage per unit length of stream (see Section 8.2 for symbols)

ET = potential evapotranspiration rate (height of step function)

h_0 = height of water table above impermeable layer at stream edge

h_d = height of water table above impermeable layer when water-table depth is z_d (width of step function)

K = hydraulic conductivity of soil below water table

The term h_d can also be defined as the depth of the impermeable layer below field surface minus z_d. Seepage rates calculated with Eq. (8.20) were essentially the same as those calculated for the actual relation between ET and z, which indicates the validity of the step-function simplification (Bouwer, 1975).

Orders of magnitude of z_d may be 2 to 3 m for salt cedars and other deep-rooted phreatophytes, 1 m for shallow-rooted plants, and about 0.5 m for bare soil (Bouwer, 1975). Depending on seasonal variations in ET and stream level, Eq. (8.20) can be used to estimate seepage and potential water savings due to phreatophyte control for a year or for shorter periods, such as a season or a month.

PROBLEMS

8.1 For Rubicon sandy loam, $K = 39$ cm/day and h_{cr} for wetting is -17 cm water (see Problem 7.6). Taking $H_w = 0$ and $f = 0.2$, calculate I_t versus t for one day with the Green-Ampt equation [Eq. (8.2)] and plot I_t versus t on cartesian coordinates (take t at 2-min intervals for the first 20 min and at progressively longer intervals for the rest of the 24-h period). On a separate graph, plot I_t versus $\sqrt{t}$ for the I_t values of 0.2, 0.5, and 1 cm, and determine the sorptivity S_i as the slope of the best-fitting straight line. Using this value of S_i, calculate I_t versus t for a period of 1 day with Eq. (8.14) for A values of K and $K/2$. Plot the resulting data on the same $I_t - t$ graph as used for the data calculated with Eq. (8.2) and compare the two curves with the Green-Ampt curve. What A value gives the best fit with the Green-Ampt curve of I_t versus t? Plot I_t and t calculated with Eq. (8.2) on double logarithmic paper, draw the best-fitting straight line for the points between 0.1 and 1 day, and determine C and α for Kostiakov's equation (8.5). Then calculate I_t versus t with Kostiakov's equation, plot the data on the same cartesian graph as used for the other $I_t - t$ relations, and compare the results. Calculate v_i versus t with the Green-Ampt equation (8.1). Select some initial and final values of v_i (for example, 340 and

40 cm/day), and calculate β in Horton's equation (8.7) for different values of t (including small values like 0.01 to 0.1 day and larger values such as 0.1 to 1 day). What is the range of the "constant" β?

8.2 Rain falls with an intensity of 20 cm/day on a clay soil with $K = 5$ cm/day, $f = 0.1$, and $h_{cr} = -50$ cm. How long will it take before runoff starts to develop, and how much water will have infiltrated at that time [use Eq. (8.8)]? How long would it have taken for that amount of water to infiltrate at potential rate [use Eq. (8.2) with $H_w = 0$]?

8.3 Calculate total infiltration and runoff for rainfall event and infiltration curve in Figure 8.3.

8.4 Tensiometers are installed at different depths in Rubicon sandy loam to determine deep-percolation rates. Using the $K_h - h$ relation for wetting in Figure 8.25, calculate the downward flux if the last two tensiometers are at depths of 120 and 150 cm and register pressure heads of -95 and -105 cm of water, respectively (evaluate K_h for average pressure head).

8.5 A soil profile is entirely saturated due to heavy rain. Calculate the distributions of volumetric water content θ and of the fillable pore space f 10 days after the rain if the soil is covered by a crop with a root zone of 1.6 m (assume that θ at saturation is 40 percent, field capacity is 30 percent and reached after 3 days, wilting point is 20 percent, class A pan evaporation is 10 mm/day, humidity is moderate, average wind speed is 250 km/day, and upwind fetch of green plants is 5 km).

8.6 The surface temperature of a well-watered wheat crop on a clear day is found to range from a low of 1.9°C to a high of 16.2°C. Corresponding minimum and maximum air temperatures are 4°C and 17.2°C. The net solar radiation $S_T - S_R$ is 327 langleys/day. Calculate the average evapotranspiration rate for the day with Eq. (8.10) (obtain values of R_G and R_A from Brown, 1973).

8.7 A stream of trapezoidal cross section with 1 : 1 side slopes, 20 m wide at the surface, and 4 m deep in the center, is embedded in a loam with $K = 0.2$ m/day. Very permeable gravel deposits occur at a depth of 6 m below the stream bottom. The groundwater table at a distance of 120 m from the stream center is 8 m lower than the water level in the stream. What is I_s/K (evaluate by interpolation from Figures 8.11, 8.12, and 8.13), and how much water does the stream loose per kilometer of length?

8.8 An ephemeral stream flows for 3 days due to surface runoff from a rainfall period. The stream bottom is 5 m wide, the banks have a slope of 45 percent, the average water depth in the stream during the 3 days of flow is 0.5 m, K of the soil below the bottom is 0.6 m/day, and the fillable pore space of that soil is 0.3. How much water will have seeped into the ground per kilometer of stream length for the 3 days of flow?

8.9 Flood water is used for groundwater recharge with infiltration basins. Assuming $a = 1$, $M = 0.2$ g/l, and $v_0 = 0.6$ m/day, how long should water be maintained in the basins if drying should start when the infiltration rate has dropped to 0.2 m/day? What is the total infiltration for the resulting infiltration period and what is the annual hydraulic loading per hectare of basin if the infiltration periods are rotated with 10-day drying periods?

8.10 For the same system as for Problem 8.8, determine the v_{min} value and length of infiltration periods that, when rotated with 10-day drying periods, give maximum annual hydraulic loading (select values for v_{min} between 0.02 and 0.3 m/day, calculate annual hydraulic loadings, plot against v_{min}, and evaluate v_{min} giving maximum hydraulic loading). What is the maximum annual hydraulic loading rate?

8.11 A tile-drainage system is to be installed to avoid high-water-table conditions in an agricultural field. The drain diameter is 12 cm, and the drains will be installed at a depth of 1.5 m. The soil is a loam with $K = 0.7$ m/day and an impermeable layer at a depth of 4.5 m. Using a drainage criterion of 7 mm/day at a maximum water-table height of 0.5 m below field surface, what is the desired spacing of the drains? Assuming that the drains are 200 m long, what is the discharge of each drain?

8.12 Stream seepage is the main source of groundwater in a narrow valley or floodplain where the soil has a K value of 2.466 m/day and is underlain by impermeable material at a depth of 10 m. The water level in the stream is 0.5 m below soil surface. The floodplain is covered by a dense growth of salt cedars. Using the step functions in Figure 8.27, what is the seepage per unit length of stream due to groundwater uptake by the salt cedars, and what is this seepage after the salt cedars have been removed and the floodplain is in bare soil? How much water can be saved annually per kilometer of stream length by removal of the salt cedars?

REFERENCES

Baffa, J. J., and members of Task Group 2440R, 1965. Experience with injection wells for artificial ground water recharge. *J. Am. Water Works Assoc.* **57**: 629–639.

Barfield, B. G., D. T. Y. Kao, and E. W. Tollner, 1975. Analysis of the sediment filtering action of grassed media. *Research Report No. 90*, Univ. of Kentucky (Lexington), Water Resources Institute, 50 pp.

Baver, L. D., W. H. Gardner, and W. R. Gardner, 1972. *Soil Physics*, 4th ed. John Wiley & Sons, Inc., New York, 498 pp.

Behnke, J. J., 1969. Clogging in surface spreading operations for artificial ground-water recharge. *Water Resour. Res.* **5**: 870–876.

Berend, J. E., 1967. An analytical approach to the clogging effect of suspended matter. *Bull. Int. Assoc. Sci. Hydrol.* **12**(2): 42–55.

Bianchi, W. C., and E. E. Haskell, Jr., 1975. Field observations on transient ground water mounds produced by artificial recharge into an unconfined aquifer. U.S. Dept. of Agriculture, *Agric. Res. Serv. Publ. ARS W-27*, 27 pp.

Bianchi, W. C., and D. C. Muckel, 1970. Ground-water recharge hydrology. U.S. Dept. of Agriculture, *Agr. Research Serv. Publ. ARS 41-161*, 62 pp.

Bouwer, H., 1959. Theoretical aspects of flow above the water table in tile drainage of shallow homogeneous soils. *Proc. Soil Sci. Soc. Am.* **23**: 260–263.

Bouwer, H., 1961. A study of final infiltration rates from cylinder infiltrometers and irrigation furrows with an electrical resistance network. *Trans. 7th Internat. Cong. of Soil Sci.*, Madison, Wis. 1960. Vol. I, Comm. VI, pp. 448–456.

Bouwer, H., 1962. Analyzing ground-water mounds by resistance network. *J. Irrig. Drain. Div., Am. Soc. Civ. Eng.* **88**(IR3): 15–36.

Bouwer, H., 1965. Theoretical aspects of seepage from open channels. *J. Hydraul. Div., Am. Soc. Civ. Eng.* **91**(HY3): 37–59.

Bouwer, H., 1966. Rapid field measurement of air entry value and hydraulic conductivity as significant parameters in flow system analysis. *Water Resour. Res.* **2**: 729–738.

Bouwer, H., 1969a. Infiltration of water into nonuniform soil. *J. Irrig. Drain. Div., Am. Soc. Civ. Eng.* **95**(IR4): 451–462.

Bouwer, H., 1969b. Salt balance, irrigation efficiency, and drainage design. *J. Irrig. Drain. Div., Am. Soc. Civ. Eng.* **95**(IR1): 153–170.

Bouwer, H., 1969c. Theory of seepage from open channels. In *Advances in Hydroscience*, vol. 5, V. T. Chow (ed.), Academic Press, Inc., New York, 121–172.

Bouwer, H., 1970. Groundwater recharge design for renovating waste water. *J. Sanit. Eng. Div., Am. Soc. Civ. Eng.* **96**(SA1): 59–74.

Bouwer, H., 1974. Developing drainage design criteria. In *Drainage For Agriculture*, J. van Schilfgaarde (ed.). Agronomy Monograph No. 17, Am. Soc. Agron., pp. 67–79.

Bouwer, H., 1975. Predicting reduction in water losses from open channels by phreatophyte control. *Water Resour. Res.* **11**: 96–101.

Bouwer, H., 1976. Infiltration into increasingly permeable soils. *J. Irrig. Drain. Div., Am. Soc. Civ. Eng.* **102**(IR1): 127–136.

Bouwer, H., and R. C. Rice, 1963. Seepage meters in recharge and seepage studies. *J. Irrig. Drain. Div., Am. Soc. Civ. Eng.* **89**(IR1): 17–42.

Bouwer, H., and R. C. Rice, 1968. A salt penetration technique for seepage measurement. *J. Irrig. Drain. Div., Am. Soc. Civ. Eng.* **94**(IR4): 481–492.

Bouwer, H., and J. van Schilfgaarde, 1963. Simplified method of predicting fall of water table in drained land. *Trans. Am. Soc. Agric. Eng.* **6**: 288–291.

Bouwer, H. L. E. Myers, and R. C. Rice, 1963. Effect of velocity on seepage and its measurement. *J. Irrig. Drain. Div. Am. Soc. Civ. Eng.* **88**(IR3): 1–14.

Brown, J. M., 1973. Tables and conversions for microclimatology. *General Techn. Rep. NC8*, U.S. Dept. of Agriculture, Forest Service, North Central Forest Exp. Sta., St. Paul, Minn., 32 pp.

Brown, D. L., and W. D. Silvey, 1973. Underground storage and retrieval of fresh water from a brackish-water aquifer. *Proc. 2d Int. Symp. on Underground Waste Management and Artificial Recharge*. J. Braunstein (ed.), Am. Assoc. Pet. Geol., Tulsa, Okla., pp. 379–419.

Bruington, A. E., and F. D. Seares, 1965. Operating a sea water barrier project. *J. Irrig. Drain. Div., Am. Soc. Civ. Eng.* **91**(IR1): 117–140.

Bulten, B., C. Brandes, and J. van Puffelen, 1974. Artificial recharge with wells—report of a trip to the United States of America (in Dutch). *Provincial Water Supply of North Holland*, Netherlands, 112 pp.

Bultot, F., and G. L. Dupriez, 1976. Conceptual hydrological model for an average-sized catchment area. *J. Hydrol.* **29**: 251–272(I) and 273–292(II).

Burman, R. D., 1976. Intercontinental comparison of evaporation estimates. *J. Irrig. Drain. Div., Am. Soc. Civ. Eng.* **102**(IR1): 109–118.

Collis-George, N., and D. E. Smiles, 1963. A study of some aspects of the hydrology of some irrigated soils of western New-South Wales. *J. Soil Res.* **1**: 17–27.

Cooley, K. R., 1970. Evaporation from open water surfaces in Arizona. U.S. Dept. of Agriculture and Univ. of Arizona, Coop. Agric. Ext. Serv. Folder 159, 6 pp.

Cooper, H. H., Jr., and M. I. Rorabaugh, 1963. Ground-water movements and bank storage due to flood stages in surface streams. *U.S. Geol. Survey Water Supply Paper 1536-J*, pp. 343–366.

Doorenbos, J., and W. O. Pruitt, 1974. Guidelines for prediction of crop water requirements. *Food and Agric. Org., Rome, Irrig. and Drain. Paper No. 25*.

Dumm, L. D., and R. J. Winger, 1964. Subsurface drainage system design for irrigated area using transient-flow concept. *Trans. Am. Soc. Agric. Eng.* **7**: 147–151.

Esmaili, H., and V. H. Scott, 1968. Unconfined aquifer characteristics and well flow. *J. Irrig. Drain. Div., Am. Soc. Civ. Eng.* **94**(IR1): 115–136.

Evans, D. D., D. Kirkham, and R. K. Frevert, 1950. Infiltration and permeability in soil overlying an impermeable layer. *Proc. Soil Sci. Soc. Amer.* **15**: 50–54.

Fok, Y. S., 1975. A comparison of the Green-Ampt and Philip two-term infiltration equations. *Trans. Am. Soc. Agric. Eng.* **18**: 1073–1075.

Fouss, J. L., 1974. Drain tube materials and installation. In *Drainage For Agriculture*, J. van Schilfgaarde (ed.), Agronomy Monogr. No. 17, Am. Soc. Agron., pp. 147–177.

Fritschen, L. J., L. Cox, and R. Kinerson, 1973. A 28-meter Douglas-fir in a weighing lysimeter. *Forest Sci.* **19**: 256–261.

Gardner, W. R., 1958. Some steady-state solutions of the unsaturated moisture flow equation with application to evaporation from a water table. *Soil Sci.* **85**: 228–232.

Gardner, W. R., and M. Fireman, 1958. Laboratory studies of evaporation from soil columns in the presence of a water table. *Soil Sci.* **85**: 244–249.

Gatewood, J. S., T. W. Robinson, B. R. Colby, J. D. Hem, and L. C. Halpenny, 1950. Use of water by bottom-land vegetation in lower Safford Valley, Arizona. *U.S. Geol. Survey Water Supply Paper 1103*, 210 pp.

Glover, R. E., 1964. Ground water movement. *U.S. Bureau of Reclamation, Engineering Monograph 31*, 67 pp.

Green, W. H., and G. A. Ampt, 1911. Studies on soil physics. I. The flow of air and water through soils. *J. Agric. Sci.* **4**: 1–24.

Hanks, R. J., and R. W. Shawcroft, 1965. An economical lysimeter for evapotranspiration studies. *Agron. J.* **57**: 634–636.

Hantush, M. S., 1967. Growth and decay of groundwater-mounds in response to uniform percolation. *Water Resour. Res.* **3**: 227–234.

Hooghoudt, S. B., 1940. Bijdragen tot de kennis van eenige natuurkundige grootheden van de grond. *Verslagen Landbouwk. Onderz. No. 46*(14B): 515–707, Government Printing Office, The Hague, Netherlands.

Horton, R. E., 1940. An approach toward a physical interpretation of infiltration capacity. *Proc. Soil Sci. Soc. Am.* **5**: 399–417.

Horton, J. S., 1973. Evapotranspiration and water research as related to riparian and phreatophyte management. Abstract bibliography, *U. S. Dept. of Agriculture Miscell. Publ. 1234*, 192 pp.

Hunt, B. W., 1971. Vertical recharge of unconfined aquifer. *J. Hydraul. Div., Am. Soc. Civ. Eng.* **97**(HY7): 1017–1030.

Idso, S. B., and R. D. Jackson, 1969. Thermal radiation from the atmosphere. *J. Geophys. Res.* **74**: 5397–5403.

Idso, S. B., R. J. Reginato, R. D. Jackson, B. A. Kimball, and F. S. Nakayama, 1974. The three stages of drying of a field soil. *Proc. Soil Sci. Soc. Am.* **38**: 831–837.

Idso, S. B., R. D. Jackson, and R. J. Reginato, 1975. Estimating evaporation: a technique adaptable to remote sensing. *Science* **189**: 991–992.

Jensen, M. E., and H. R. Haise, 1963. Estimating evapotranspiration from solar radiation. *J. Irrig. Drain. Div., Am. Soc. Civ. Eng.* **89**(IR4): 15–41.

Jensen, M. E., D. C. N. Robb, and C. E. Franzoy, 1970. Scheduling irrigations using climate-crop-soil data. *J. Irrig. Drain. Div., Am. Soc. Civ. Eng.* **96**(IR1): 25–28.

Jensen, M. E., et al., 1974. Consumptive use of water and irrigation water requirements. Report by Techn. Comm., *Irrig. Drain. Div., Am. Soc. Civ. Eng.*, 215 pp.

Khan, M. Y., and D. Kirkham, 1971. Spacing of drainage wells in a layered aquifer. *Water Resour. Res.* **7**: 166–183.

Kimbler, O. K., R. G. Kazmann, and W. R. Whitehead, 1973. Saline aquifers—future storage reservoirs for fresh water? *Proc. 2d Int. Symp. on Underground Waste Management and Artificial Recharge.* J. Braunstein (ed.), Am. Assoc. Petrol. Geol., Tulsa, Oklahoma, pp. 192–206.

King, T. G., and J. R. Lambert, 1976. Simulation of deep seepage to a water table. *Trans. Am. Soc. Agric. Eng.* **19**: 50–54.

Kirkham, D., S. Toksöz, and R. R. van der Ploeg, 1974. Steady flow to drains and wells. In *Drainage For Agriculture*, J. van Schilfgaarde (ed.), Agronomy Monograph No. 17, Am. Soc. Agron., pp. 203–244.

Kohler, M. A., T. J. Nordenson, and W. E. Fox, 1955. Evaporation from pans and lakes. *Resource Paper 38*, Weather Bureau, U.S. Dept. of Commerce, Washington, D.C.

Kostiakov, A. N., 1932. On the dynamics of the coefficient of water percolation in soils and on the necessity for studying it from a dynamic point of view for purposes of amelioration. *Trans. 6th Comm. Internat. Soc. Soil Sci., Russian Part A*, pp. 17–21.

Libby, F. J., and P. R. Nixon, 1962. A portable lysimeter adaptable to a wide range of site situations. *Int. Assoc. Sci. Hydrol. Publ. No. 62*, Comm. for Evaporations, pp. 153–158.

Lourence, F. J., and W. B. Goddard, 1967. A water-level measuring system for determining evapotranspiration rates from a floating lysimeter. *J. Appl. Meterol.* **6**: 489–492.

Marino, M. A., 1967. Hele-Shaw model study of the growth and decay of groundwater ridges. *J. Geophys. Res.* **72**: 1195–1205.

Marino, M. A., 1974a. Growth and decay of groundwater mounds induced by percolation. *J. Hydrol.* **22**: 295–301.

Marino, M. A., 1974b. Rise and decline of the water table induced by vertical recharge. *J. Hydrol.* **23**: 289–298.

Marino, M. A., 1975a. Artificial groundwater recharge, I. Circular recharging area. *J. Hydrol.* **25**: 201–208.

Marino, M. A., 1975b. Artificial groundwater recharge, II. Rectangular recharging area. *J. Hydrol.* **26**: 29–37.

McCormick, R. L., 1975. Filter-pack installation and redevelopment techniques for shallow recharge shafts. *Ground Water* **13**(5): 400–405.

Mein, R. G., and D. A. Farrell, 1974. Determination of wetting front suction in the Green-Ampt equation. *Proc. Soil. Sci. Soc. Am.* **38**: 872–876.

Mein, R. G., and C. L. Larson, 1973. Modeling infiltration during a steady rain. *Water Resour. Res.* **9**: 384–394.

Meinzer, O. E., 1923. Outline of ground-water hydrology with definitions. *U.S. Geol. Survey Water Supply Paper 494*, 71 pp.

Mobasheri, F., and D. K. Todd, 1963. "Investigation of the Hydraulics of Flow near Recharge Wells." Contrib. No. 72, Water Resources Center, Univ. of Calif., Berkeley, 32 pp.

Morel-Seytoux, H. J., and J. Khanji, 1974. Derivation of an equation of infiltration. *Water Resour. Res.* **9:** 795–800.

Muckel, D. C., 1966. Phreatophytes-water use and potential water savings. *J. Irrig. Drain. Div., Am. Soc. Civ. Eng.* **92**(IR4): 27–34.

Murray, C. R., and E. B. Reeves, 1972. Estimated use of water in the United States in 1970. *U.S. Geol. Survey Circular 676*, Washington D.C., 37 pp.

Nakayama, F. S., R. D. Jackson, B. A. Kimball, and R. J. Reginato, 1973. Diurnal soil-water evaporation: chloride movement and accumulation near the soil surface. *Proc. Soil Sci. Soc. Am.* **37:** 509–513.

Neuman, S. P., 1976. Wetting front pressure head in the infiltration model of Green and Ampt. *Water Resour. Res.* **12:** 564–566.

Nightingale, H. I., and W. C. Bianchi, 1973. Ground-water recharge for urban use: Leaky-Acres Project. *Ground Water* **11**(6): 36–43.

Pattison, A., and T. A. McMahon, 1973. Rainfall-runoff models using digital computers. *Trans. Civ. Eng. Inst. Eng. Austral. CE15*, pp. 1–4 and 20.

Penman, H. L., 1963. Vegetation and hydrology. *Techn. Comm. No. 53*, Commonwealth Bureau of Soils, Harpenden, England, 125 pp.

Philip, J. R., 1969. Theory of infiltration. In *Advances in Hydroscience*, V. T. Chow (ed.), Academic Press, Inc., New York, pp. 216–296.

Rebhun, M., and V. L. Hauser, 1967. Clarification of turbid water with polyelectrolytes for recharge through wells. *Proc. Symp. on Artificial Recharge and Management of Aquifers*, International Assoc. Scientific Hydrol., pp. 218–288.

Reeves, M., and E. E. Miller, 1975. Estimating infiltration for erratic rainfall. *Water Resour. Res.* **11:** 102–110.

Rektorik, R. J., 1976. Field drainage with manifold well points. *Trans. Am. Soc. Agric. Eng.* **19:** 81–84.

Renner, O., 1915. Wasserversorgung der Pflanze. In *Handworterbuch der Naturwissenschaften*, vol. 10, Jena, Germany, pp. 538–577.

Rice, R. C., 1974. Soil clogging during infiltration of secondary effluent. *J. Water Poll. Contr. Fed.* **46:** 708–716.

Rice, R. C., 1975. Diurnal and seasonal soil water uptake and flux within a bermudagrass root zone. *Proc. Soil Sci. Soc. Am.* **39:** 394–398.

Ripley, D. P., and Z. A. Saleem, 1973. Clogging in simulated glacial aquifers due to artificial recharge. *Water Resour. Res.* **9:** 1047–1057.

Ripple, C. D., J. Rubin, and T. E. A. van Hylckama, 1972. Estimating steady-state evaporation rates from bare soils under conditions of high water table. *U.S. Geol. Survey Water Supply Paper 2019-A*, 39 pp.

Ritchie, J. T., 1972. Model for predicting evaporation from a row crop with incomplete cover. *Water Resour. Res.* **8:** 1204–1213.

Ritchie, J. T., and E. Burnett, 1968. A precision weighing lysimeter for row crop water use studies. *Agron. J.* **60:** 545–549.

Robinson, T. W., 1958. Phreatophytes. *U.S. Geol. Survey Water Supply Paper 1423*, 84 pp.

Selim, H. M., and D. Kirkham, 1973. Unsteady two-dimensional flow of water in unsaturated soil above an impervious barrier. *Proc. Soil Sci. Soc. Am.* **37:** 489–495.

Signor, D. C., V. L. Hauser, and O. R. Jones, 1969. Ground-water recharge through modified shaft. *Trans. Am. Soc. Agric. Eng.* **12:** 486–489.

Singh, R., 1972. Mound geometry under recharge basins. *Calif. State Univ., San Jose, Report No. GK-18526 for National Science Foundation*, 71 pp.

Singh, R., 1976. Prediction of mound geometry under recharge basins. *Water Resour. Res.* **12:** 775–780.

Smith, C. G., Jr., and J. S. Hanor, 1975. Underground storage of treated water: a field test. *Ground Water* **13**(5): 410–417.

Sniegocki, R. T., 1963. Problems in artificial recharge through wells in Grand Prairie Region, Arkansas. *U.S. Geol. Survey Water Supply Paper 1615F*.

Streltsova, T. D., 1975. Unsteady unconfined flow into a surface reservoir. *J. Hydrol.* **27:** 95–110.

Swartzendruber, D., and T. C. Olson, 1961. Model study of the double-ring infiltrometer as affected by depth of wetting and particle size. *Soil Sci.* **92:** 219–225.

Talsma, T., 1963. The control of saline groundwater. *Mededelingen No. 63*, National Agric. Univ., Wageningen, The Netherlands.

Talsma, T., 1969. In situ measurement of sorptivity. *Austral. J. Soil Res.* **7:** 269–276.

Talsma, T., 1970. Some aspects of three-dimensional infiltration. *Austral. J. Soil Res.* **8:** 179–184.

Talsma, T., and J.-Y. Parlange, 1972. One-dimensional vertical infiltration. *Austral. J. Soil Res.* **10:** 143–150.

Tanner, C. B., and W. A. Jury, 1976. Estimating evaporation and transpiration from a row crop during incomplete cover. *Agron. J.* **68:** 239–243.

Todd, D. K., 1970. *The Water Encyclopedia.* Water Information Center, Port Washington, N.Y., 559 pp.

van Hylckama, T. E. A., 1975. Water use by salt cedar in the lower Gila River valley (Arizona). *U.S. Geol. Survey Prof. Paper 491-E.*

van Schilfgaarde, Jan, 1974. Nonsteady flow to drains. In *Drainage For Agriculture*, J. van Schilfgaarde (ed.), Agronomy Monograph No. 17, Am. Soc. Agron., pp. 245–270.

Watson, K. K., 1974. Application of unsaturated flow theory. In *Drainage For Agriculture*, J. Van Schilfgaarde (ed.), Agronomy Monogr. No. 17, Am. Soc. Agron., pp. 359–405.

Watson, K. K., and S. J. Lees, 1975. Simulation of the rainfall-runoff process using a hysteretic infiltration-redistribution model. *Austral. J. Soil Res.* **13:** 133–140.

Wesseling, J., 1974. Crop growth and wet soils. In *Drainage For Agriculture*, J. van Schilfgaarde (ed.), Agronomy Monograph No. 17, Am. Soc. Agron., pp. 7–37.

Whisler, F. D., K. K. Watson, and S. J. Perrens, 1972. Numerical analysis of infiltration into a heterogeneous porous medium. *Proc. Soil Sci. Soc. Am.* **36:** 868–874.

Wilson, L. G., 1967. Sediment removal from flood water. *Trans. Am. Soc. Agric. Eng.* **10:** 35–37.

Youngs, E. G., 1960. The drainage of liquids from porous materials. *J. Geophys. Res.* **65:** 4025–4030.

Youngs, E. G., 1968. An estimation of sorptivity for infiltration studies from moisture moment considerations. *Soil Sci.* **106:** 157–163.

Zyvoloski, G., J. C. Bruch, Jr., and J. M. Sloss, 1976. Solution of equation for two-dimensional infiltration problems. *Soil Sci.* **122:** 65–70.

NINE

SUBSIDENCE AND LATERAL MOVEMENT OF THE LAND SURFACE DUE TO GROUNDWATER PUMPING

An important environmental consequence of groundwater pumping is movement of the land surface. Movements of as much as almost 10 m downward and several meters horizontally have been measured. Downward movement, usually called *subsidence*, can be particularly severe where pumping exceeds the safe yield and water tables or piezometric surfaces are declining. Subsidence is largest where water-level declines are greatest and aquifers or aquitards are thickest and most compressible. Differences in these conditions cause differential subsidence, which is more damaging than uniform subsidence. Horizontal movement of the land surface may also produce cracks and fissures in the earth, which can cause additional damage.

Subsidence and lateral movement of the land surface have damaged buildings, bridges, tunnels, streets, highways, railroads, water and sewer lines, power lines, etc. Well casings have collapsed under the strain of subsiding formations. Differential subsidence has reversed gradients of irrigation and drainage canals. Subsidence also increases the flood hazard of already low areas.

Subsidence caused by long-term overdraft on groundwater can be stopped by reducing pumpage to the safe yield. Small rebounds may occur, particularly if water is injected to increase groundwater levels. Long-term subsidence is, however, essentially irreversible.

9.1 OCCURRENCE OF SUBSIDENCE

Areas with severe subsidence problems include the city of Venice, Italy. This famous city, already low and subject to flooding by high tides, has subsided about

15 cm in the period 1930–1973, mainly due to heavy industrial pumpage of groundwater in the mainland port of Marghera about 7 km away (Gambolati and Freeze, 1973). Gambolati et al. (1974) calculated that Venice will go down another 3 cm if the present rate of groundwater pumping continues, but would rise 2 cm in the next 25 years if all pumping in the area would be stopped now.

Parts of Mexico City have subsided as much as 8 m since intensive groundwater pumping began in 1938 (Poland, 1969). Maximum subsidence of 3 to 4 m has been observed in Tokyo and Osaka, most of which occurred in the period 1928–1943. In the Taipei basin, maximum subsidence was about 1 m (Poland, 1969). Much of London has subsided 6 to 18 cm in the period 1865–1931 due to compaction of thick clay deposits caused by declining piezometric surfaces of the underlying chalk aquifer (Poland and Davis, 1969).

In Baton Rouge, Louisiana, piezometric surfaces have declined about 60 m since 1890 when pumping of groundwater started. This has caused subsidence of 0.3 m in the industrial district where pumping is concentrated. The subsidence thus was 5 cm per 10-m decline of piezometric surface (Davis and Rollo, 1969). For the Houston-Galveston area in Texas, water-level declines were also about 60 m but the maximum subsidence was about 1.5 m, yielding 25 cm subsidence per 10-m water-level decline (Gabrysch, 1969). Later reports indicated that subsidence in the Houston-Baytown had reached as much as 2.7 m and had caused increased flooding of low areas by storms or high tides (Jones and Warren, 1976). Resulting damages and losses in property values in the period 1969–1973 were estimated at 73 million dollars.

Groundwater pumping for irrigation in the San Joaquin Valley, California has caused a total subsidence of as much as 8.5 m and subsidence rates of as much as 0.55 m per year (Lofgren, 1969a; Bull and Miller, 1974; and Bull, 1975). The subsidence was about 60 cm per 10 m water table drop west of Fresno and about 40 cm per 10 m water-table drop 50 km south of Tulare. In the Santa Clara Valley, California, a subsidence of 4 m has been observed (Poland, 1969). Overdraft of groundwater in south central Arizona has caused subsidence of 2.3 m at a water-table decline of 46 m, or 50 cm per 10 m water-table drop (Schumann and Poland, 1969).

Subsidence can also occur where hot water is withdrawn from geothermal fields for electric power development, space heating, and other purposes (see Section 10.8). At Wairakei, New Zealand, for example, an area of more than 65 km^2 is affected by subsidence. The rate is about 0.4 m per year, and the total subsidence since 1956 is almost 4 m at the point of maximum deflection (Axtmann, 1975).

Vertical and lateral movement of the land surface can also result from removal of oil or gas (Yerkes and Castle, 1969). For example, subsidence of more than 9 m and horizontal displacements of as much as 3.7 m have been observed in the area of Wilmington, California, due to pumping of oil (Grant, 1954; Mayuga and Allen, 1969). Damages cost more than $100 million to repair, before massive injection of salt water into the oil reservoirs arrested the subsidence and even caused small rebounds in areas of heaviest injection.

More information on the occurrence of land subsidence is presented by Poland and Davis (1969).

9.2 INTERGRANULAR PRESSURE

The crucial factor in subsidence and lateral movement of the land surface is the intergranular or effective pressure in aquifers and other underground materials. This intergranular pressure is the pressure transmitted through the contact points of the individual particles (gravel, sand, silt, or clay). When the intergranular pressure is increased—for example, by placing a load on the granular material—the individual granules move relative to each other to produce a lower void ratio and, hence, compression of the material.

The intergranular pressure P_i at a given depth is calculated as the difference between the total pressure P_t and the hydraulic pressure P_h at that depth (Terzaghi and Peck, 1948), or

$$P_i = P_t - P_h \qquad (9.1)$$

This equation becomes evident if the vertical forces on an imaginary horizontal plane in an aquifer are considered. The downward force on this plane is equal to the weight of everything above it. There is, however, also an upward force against the bottom of the plane due to hydraulic pressure. The difference between the downward and upward forces is the net load on the plane, which is carried and transmitted by the individual grains at their contact points.

The total pressure at a given depth is calculated as the weight per unit area of all solids and liquids occurring above that point. This is exemplified in Figure 9.1 for an unconfined aquifer with a water-table depth at 10 m, a porosity of 30 percent above as well as below the water table, a density of 2.6 g/cm³ for the solid material in the aquifer and vadose zone, and a volumetric water content of 10 percent above the water table and 30 percent below the water table. The weight of

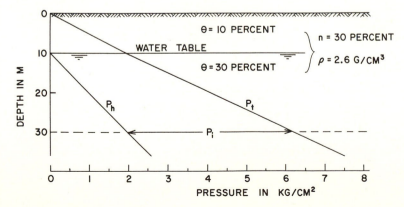

Figure 9.1 Profiles of P_t and P_h in unconfined aquifer.

the solids per cubic centimeter in the vadose zone is $(1 - 0.3)2.6 = 1.82$ g and the weight of the water per cubic centimeter in the vadose zone is $0.1 \times 1 = 0.1$ g. Thus, the unit weight of the material in the vadose zone is 1.92 g/cm^3. Similarly, the unit weight of the material below the water table is calculated as $(1 - 0.3)2.6 + 0.3 \times 1 = 2.12$ g/cm^3. At 10 m depth (the bottom of the vadose zone), the total pressure P_t is $10 \times 100 \times 1.92 = 1920$ g/cm$^2 = 1.92$ kg/cm^2.[†] At a depth of 30 m (arbitrarily selected), P_t is increased by the weight of 20 m of aquifer material, or by $20 \times 100 \times 2.12 = 4\,240$ g/cm$^2 = 4.24$ kg/cm^2. Thus, P_t at 30 m is $1.92 + 4.24 = 6.16$ kg/cm^2. These calculations yield the P_t line shown in Figure 9.1.

The hydraulic pressure is usually considered only below the water table (above the water table, P_h would be negative). Thus, at the water table $P_h = 0$ and at 30 m depth (arbitrarily selected) $P_h = 20 \times 100 \times 1 = 2\,000$ g/cm$^2 = 2$ kg/cm^2, yielding the P_h line in Figure 9.1. The horizontal difference between the P_t and P_h lines is the intergranular pressure P_i. At 30 m depth, for example, $P_i = 4.16$ kg/cm^2.

9.3 INCREASING INTERGRANULAR PRESSURE BY DECLINING WATER TABLE

If the water table in Figure 9.1 were at a depth of 20 m, P_t at 30 m depth would be 5.96 kg/cm^2 and $P_h = 1$ kg/cm^2, yielding $P_i = 4.96$ kg/cm^2. Thus, a water-table drop from 10 to 20 m depth increases the intergranular pressure at 30 m from 4.16 to 4.96 kg/cm^2. An increase in P_i causes a decrease in the void ratio and, hence, subsidence of the land surface (see Section 9.4).

The above increase in P_i of 0.8 kg/cm^2 at 30 m depth is due to the loss of buoyancy of soil particles in the dewatered zone from 10 to 20 m. Since the porosity is 30 percent, the loss in buoyancy is 0.7 g/cm^3, which is the increase in effective weight of the solid material. To this must be added the weight of the 10 percent water remaining in the aquifer material after the water-table drop, or 0.1 g/cm^3. Thus, the material in the dewatered zone has effectively become 0.8 g/cm^3 heavier, which for the 10-m-thick zone amounts to 0.8 kg/cm^2. This is equal to the increase in P_i below the new water-table depth calculated with Eq. (9.1).

Loss of buoyancy thus is the major cause of P_i increases in unconfined aquifers with declining water tables. In confined aquifers with decreasing piezometric surfaces, P_i is increased by a reduction in the upward hydraulic force against the bottom of the upper aquiclude (see Section 9.5). A third process by which P_i can be increased is water movement itself, through a seepage force exerted on the solid particles by the frictional drag of the water as it moves around these particles

---

† This is the metric unit most commonly used for pressure in subsidence calculations. The corresponding SI unit is N/m^2. To convert kg/cm^2 to N/m^2, multiply by $10^4 g$, where g is the acceleration due to gravity in m/s^2. Thus, 20 kg/cm^2 is 1.96×10^6 N/m^2, taking g as 9.8 m/s^2.

(see Section 9.6). Several of these actions may occur simultaneously, having additive effects on P_i. A fourth mechanism whereby P_i can be increased is wetting a dry soil (see Section 9.10).

9.4 CALCULATION OF SUBSIDENCE

To calculate subsidence from an increase in P_i, the relation between P_i and the void ratio must be known for the material in question. These relations are experimentally determined with a consolidometer, placing increasing loads on a saturated sample (preferably undisturbed) and measuring the equilibrium height of the sample after each load increase (Figure 9.2). The sample should be confined at the top and bottom by porous plates to allow drainage of the excess pore water driven out by compression of the sample. The results of such a test are expressed in terms of the void ratio e in the sample and the corresponding value of P_i.

Examples of e-vs.-P_i relations for three materials, taken from Taylor (1958) and Terzaghi and Peck (1948), are shown in Figure 9.3 (the vertical scale for the dense sand is 10 times larger than that for the loose sand and the clay). As P_i increases, e is reduced but at a decreasing rate. The reduction in e is due to a rearrangement of particles in the sample. At very high values of P_i (for example, 100 to 1 500 kg/cm^2), however, fracturing and shattering of individual grains may also occur, which causes further decreases in e and renders sands as compressible as clays and silts (Roberts, 1969). While such high P_i values may happen when oil or gas is removed from great depths, they will usually not be encountered in groundwater situations, except perhaps for very deep aquifers.

When, at a given point on the e-vs.-P_i curve, the load is removed from the sample, e will only slightly increase initially (see curves with left-pointing arrows in Figure 9.3). Then, there is some rebound, but not to the e value at the start of the test. This indicates the inelastic nature of granular materials. The clay exhibited a greater rebound than the loose and dense sand (Figure 9.3). When, after rebound, the load is reapplied, the e-vs.-P_i curve for recompaction differs slightly

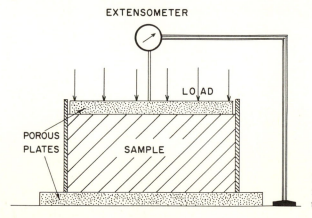

Figure 9.2 Principle of consolidometer for determining relation between e and P_i.

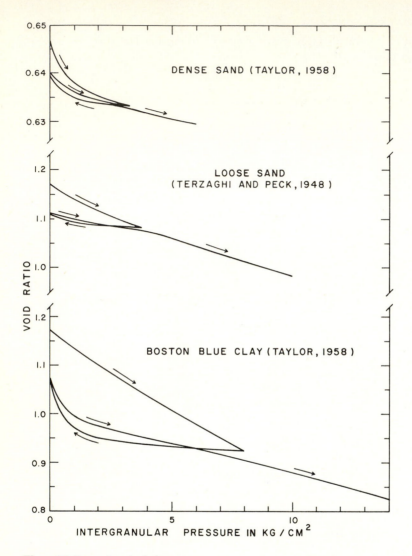

Figure 9.3 Examples of relations between e and P_i for three different soil materials. *(From Terzaghi and Peck, 1948, and Taylor, 1958.)*

from the rebound curve (hysteresis) and then follows the general course of the curve before loading was interrupted. If a core sample is taken at great depth, P_i of the sample after it is brought up will be much less than P_i when it was in place. When such a sample is subjected to a loading test in the laboratory, the e-vs.-P_i curve may then show a discontinuity in the vicinity of the in situ P_i value.

To calculate the relation between the subsidence S_u and a certain change in void ratio in a soil layer of height Z, a unit horizontal area of the layer will be considered so that volumes and heights are numerically the same and interchangeable. All dimensions prior to compression will be indicated with the subscript 1 and those after compression with the subscript 2. Since compression is due to a reduction in the volume of the voids V_v, the subsidence can be expressed as

$$S_u = V_{v1} - V_{v2} \tag{9.2}$$

The height Z_1 (or total volume per unit area) of the layer prior to compression can be expressed as

$$Z_1 = V_{v1} + V_s \tag{9.3}$$

where V_s is the volume of the solids, which of course does not change during compression. Substituting $V_{v1} = e_1 V_s$ into Eq. (9.3) yields

$$Z_1 = e_1 V_s + V_s \tag{9.4}$$

or

$$V_s = \frac{Z_1}{e_1 + 1} \tag{9.5}$$

which after substitution into Eq. (9.3) and solving for V_{v1} yields

$$V_{v1} = Z_1 \frac{e_1}{e_1 + 1} \tag{9.6}$$

The volume of the voids after compression is

$$V_{v2} = e_2 V_s \tag{9.7}$$

Substituting Eq. (9.5) into Eq. (9.7) yields

$$V_{v2} = Z_1 \frac{e_2}{e_1 + 1} \tag{9.8}$$

Substituting Eqs. (9.6) and (9.8) into Eq. (9.2) then gives

$$S_u = Z_1 \frac{e_1 - e_2}{e_1 + 1} \tag{9.9}$$

which expresses the compression of a soil layer in relation to the original height of the layer and the change in void ratio.

Elastic Theory

Two approaches have been used to calculate subsidence. One is based on the elastic theory and the other on the logarithmic theory. With the elastic theory, the subsidence per unit height, or the strain S_u/Z_1, is assumed to vary linearly with the stress increase $P_{i2} - P_{i1}$. The factor of proportionality is called *Young's modulus of elasticity*, symbol E. In formula,

$$\frac{P_{i2} - P_{i1}}{S_u/Z_1} = E \tag{9.10}$$

which yields

$$S_u = (P_{i2} - P_{i1})\frac{Z_1}{E} \tag{9.11}$$

Combining Eqs. (9.9) and (9.11) shows that E is calculated as

$$E = \frac{e_1 + 1}{(e_1 - e_2)/(P_{i1} - P_{i2})} \tag{9.12}$$

This calculation can be done for the e-vs.-P_i curve of the material in question, using small increments in e, or Eq. (9.12) can be put in differential form as

$$E = \frac{e + 1}{de/dP_i} \tag{9.13}$$

which enables calculation of E by determining the slope of the e-vs.-P_i curve at various values of P_i. Applying this procedure to the curves in Figure 9.3 (excluding the rebound and recompression sections) yields the relations between E and P_i shown in Figure 9.4. Sometimes the reciprocal of E is used, which is then called the

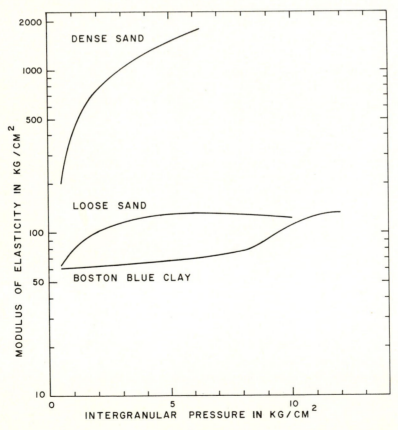

Figure 9.4 Relation between E and P_i calculated from e-vs.-P_i curves in Figure 9.3.

**Table 9.1 Orders of magnitude of E
for different materials**

Material	E, kg/cm^2
Dense gravel and sand	2 000–10 000
Dense sands	500–2 000
Loose sands	100–200
Dense clays and silts	100–1 000
Medium clays and silts	50–100
Loose clays	10–50
Peat	1–5

compressibility index (Gambolati et al., 1974). Lambe and Whitman (1969) defined a constrained modulus of elasticity for use in one-dimensional consolidation theory, which is the reciprocal of the coefficient of volume compressibility used by Terzaghi and Peck (1948).

Figure 9.4 shows that E is not constant, but increases with increasing P_i, particularly for dense sand. Thus, in contrast to true elastic bodies, which have a constant E, the compression of soil and aquifer materials for a given increase in P_i decreases with increasing P_i. In subsidence calculations, therefore, E values must be taken in relation to the P_i values in question. For narrow ranges in P_i, E may not vary too much and may be taken as a constant.

Orders of magnitude of E, taken from Colijn and Potma (1944), Gambolati et al. (1974), and calculated from e-vs.-P_i curves by Taylor (1958) and Terzaghi and Peck (1948) are presented in Table 9.1. These data show that fine-textured materials are much more compressible than sands and gravels. Severe subsidence thus tends to occur only where thick clay deposits are present or where the aquifers are interbedded with silt and clay layers. Materials of organic origin, such as peat and muck, are the most compressible.

Logarithmic Theory

Terzaghi (Terzaghi and Peck, 1948) found that when e was plotted against log P_i, sigmoid curves were obtained that showed flat portions at the low and high values of log P_i but essentially linear sections for the midrange of P_i values (Figure 9.5). The slope of the essentially linear portion of the curve, expressed as

$$\tan \alpha = \frac{e_1 - e_2}{\log P_{i2} - \log P_{i1}} \tag{9.14}$$

is called the compression index C_c of the particular material. Thus, Eq. (9.14) can be written as

$$e_1 - e_2 = C_c \log \frac{P_{i2}}{P_{i1}} \tag{9.15}$$

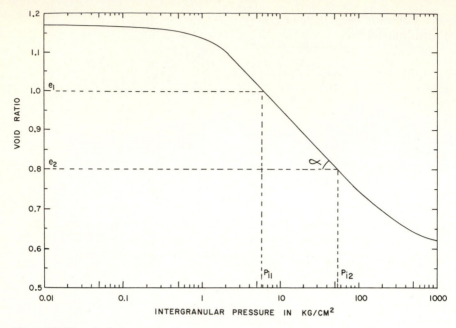

Figure 9.5 Hypothetical example of relation between e and log P_i.

Substituting the right part of this equation for $e_1 - e_2$ in Eq. (9.9) yields the logarithmic compression equation

$$S_u = Z_1 \frac{C_c}{e_1 + 1} \log \frac{P_{i2}}{P_{i1}} \qquad (9.16)$$

Values of C_c, which is dimensionless, generally range between 0.1 and 1 for clays and vary almost linearly with the liquid limit L_q of the clay, according to the equation (Skempton, 1944)

$$C_c = 0.007(L_q - 10 \text{ percent}) \qquad (9.17)$$

The liquid limit is a soil parameter used in soil mechanics to describe the consistency of clays (Terzaghi and Peck, 1948). It is expressed as a gravimetric water content in percent. Equation (9.17) applies to disturbed samples. For clays of low to medium sensitivity (i.e., clays whose consistencies are not greatly affected by disturbing and remolding them for compression tests), field values of C_c may be about 1.3 times those calculated with Eq. (9.17). The value of C_c for the Boston blue clay in Figure 9.3 is 0.40 (Taylor, 1958). For sands, C_c was calculated as 0.1 for dense sand and 0.17 for loose sand from curves presented by Terzaghi and Peck (1948), and as 0.01 for the e-vs.-P_i relation for dense sand shown in Figure 9.3.

Values of e_1 usually range between 0.7 and 1.3, so that the term $e_1 + 1$ in Eq. (9.16) varies from 1.7 to 2.3. This is a variation of only 15 percent from the

Table 9.2 Values of C_u for different materials

Material	C_u
Sand	0.005–0.05
Silt	0.05–0.1
Clay	0.1–0.3
Peat	0.2–0.8

Source: From Colijn and Potma, 1944.

average of 2.0. Since the error in C_c will probably be more than 15 percent, particularly when applying laboratory data to field conditions, allowance for the particular value of e is not warranted. Thus, the term $C_c/(e_1 + 1)$ can be lumped into one compression coefficient C_u (Colijn and Potma, 1944), reducing Eq. (9.16) to

$$S_u = Z_1 C_u \log \frac{P_{i2}}{P_{i1}} \qquad (9.18)$$

Values of C_u, calculated from coefficients listed by Colijn and Potma (1944) for an equation similar to Eq. (9.18), are presented in Table 9.2. The values of C_u for clay and silt are compatible with those for C_c, considering that e may range between 1 and 1.5 for such materials.

Rebound

The equations for calculating subsidence can also be used for calculating rebound if P_i decreases, due, for instance, to rising water tables or piezometric levels. In that case, the values of E, C_c, or C_u must be evaluated from the appropriate rebound portions of e-vs.-P_i relationships (as shown, for example, in Figure 9.3). For the sand curves in Figure 9.3, E for rebound was 2 to 10 times E for compression, depending on P_i. For the Boston blue clay, E for rebound was about 50 percent of E for compression at low values of P_i, but 3 times E for compression at $P_i = 2$ kg/cm^2. Gambolati et al. (1974) found that E of the silt and clay layers below Venice, Italy, was 7 to 10 times larger for rebound than for compression. For the Boston blue clay (Figure 9.4), C_c was 0.08 for rebound and 0.40 for compression (Taylor, 1958).

The relation between compression and rebound parameters for granular materials may also be influenced by the time that the material has been under compression. Whereas complete rebound has been observed after short-term (5 to 30 min) pumping of wells (Davis et al., 1969), rebound of the land surface after prolonged subsidence may be only a fraction of the original subsidence (Gambolati et al., 1974; Mayuga and Allen, 1969).

Example. Figure 9.6 shows an unconfined and a confined aquifer of the same

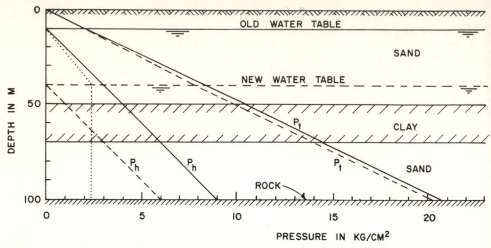

Figure 9.6 Profiles of P_t and P_h before (solid lines) and after (dashed lines) lowering water table in unconfined aquifer. The dotted line indicates the resulting increase in P_i.

material separated by a 20-m-thick clay layer (aquitard). Incompressible bedrock occurs at a depth of 100 m. Overpumping of the upper groundwater causes a decline in the water table from 10 to 40 m depth. Given the following physical properties of the sand and clay, what is the subsidence caused by the water-table decline according to the elastic and logarithmic theories?

	Sand	Clay
Porosity	30 percent	40 percent
Volumetric water content above water table	10 percent	
Volumetric water content below water table	30 percent	40 percent
Density of solid particles	2.6 g/cm^3	2.7 g/cm^3
E	1 000 kg/cm^2	100 kg/cm^2
C_u	0.01	0.2

The weight of the sand above the water table is $0.7 \times 2.6 + 0.1 \times 1 = 1.92$ g/cm^3. The weight of the sand below the water table is similarly calculated as 2.12 g/cm^3. The saturated clay weighs $0.6 \times 2.7 + 0.4 \times 1 = 2.02$ g/cm^3. With the water table at a depth of 10 m, the total pressure P_t at 10 m is $10 \times 100 \times 1.92/1\,000 = 1.92$ kg/cm^2. The weight of the 40 m of underlying saturated sand corresponds to a P_t value of $40 \times 100 \times 2.12/1\,000 = 8.48$ kg/cm^2, yielding a P_t value at 50 m depth (top of clay layer) of $1.92 + 8.48 = 10.4$ kg/cm^2. Similarly, P_t at the bottom of the clay layer is calculated as 14.44 kg/cm^2 and at the bedrock (100 m depth) as 20.8 kg/cm^2. These data yield the P_t line prior to lowering the water table in Figure 9.6. With the water table at 10 m depth, the hydraulic pressure P_h at the bedrock is

Table 9.3 Calculation of compression of underground materials in Figure 9.6 with elastic theory

Layer, m	Z, m	$P_{i2} - P_{i1}$, kg/cm²	E, kg/cm²	S_u, m
10–40	30	1.2	1 000	0.036
40–50	10	2.4	1 000	0.024
50–70	20	2.4	100	0.48
70–100	30	2.4	1 000	0.072
			Total	0.612

$90 \times 100 \times 1/1\,000 = 9$ kg/cm², yielding the P_h line shown in Figure 9.6. The horizontal difference between the P_t and P_h lines is the intergranular pressure P_i.

The P_t and P_h lines after the water table has dropped to a depth of 40 m are calculated in the same manner and shown as dashed lines in Figure 9.6. Comparing P_i before and after lowering the water table shows that P_i is increased by 2.4 kg/cm² in the material below the new water-table position. In the dewatered zone, the P_i increase varies linearly from 0 at 10 m depth to 2.4 kg/cm² at 40 m depth (dotted line in Figure 9.6). In calculating the compression with the elastic theory, the average P_i increase of 1.2 kg/cm² will be used for this zone.

Using Eq. (9.11), the compression of each layer is calculated in Table 9.3. The subsidence of the land surface is equal to the sum of the compressions of each layer, or 0.612 m. This amounts to 0.204 m subsidence per 10 m decline of water table, which is on the same order as has been observed in practice. Of the total subsidence of 0.612 m, 0.48 m or 78 percent is caused by compression of the clay layer.

With the logarithmic theory, the value of P_{i1} for a certain layer will be calculated as the average of P_{i1} at the top and bottom of that layer. The value of P_{i2}

Table 9.4 Calculation of compression of underground materials in Figure 9.6 with logarithmic theory

Depth, m	P_{i1}, kg/cm²	P_{i2}, kg/cm²	Average P_{i1}, kg/cm²	Average P_{i2}, kg/cm²	Z_1, m	C_u	S_u, m
10	1.92	1.92					
			3.6	4.8	30	0.01	0.037
40	5.28	7.68					
			5.84	8.24	10	0.01	0.015
50	6.4	8.8					
			7.42	9.82	20	0.2	0.487
70	8.44	10.84					
			10.12	12.52	30	0.01	0.028
100	11.8	14.2				Total	0.57

will be determined likewise. A refinement may be made by splitting the layer into a number of increments and determining the average P_{i1} and P_{i2} and resulting compression for each increment. Using Eq. (9.18), the compression of the various layers is calculated in Table 9.4. The compression of the clay layer and the total subsidence agree with the results obtained with elastic theory. The same is true for the 10-to-40-m sand layer, but the compression of the deeper sand layer (70 to 100 m) is only about one-third of that calculated with Eq. (9.11) in Table 9.3. This is because the logarithmic theory takes into account that the compression for a certain increase in P_i decreases with increasing P_{i1}.

9.5 INCREASING INTERGRANULAR PRESSURE BY DECLINING PIEZOMETRIC SURFACE

In contrast to unconfined aquifers where a declining water table causes an increase in P_i because of a loss of buoyancy in the dewatered zone, a declining piezometric surface in confined aquifers causes an increase in P_i because of a reduction in the upward hydraulic force against the bottom of the upper confining layer. The increase in P_i for confined aquifers can be calculated in the same manner as for unconfined aquifers, evaluating P_t and P_h before and after the drop in piezometric surface and determining P_i as the difference according to Eq. (9.1).

This is exemplified in Figure 9.7 for a 4-m drop in the piezometric surface. The upper confining layer is taken as a 2-m-thick clay layer which is saturated but not sufficiently permeable to transmit enough water upward to create a water table in the overlying material. Thus, the water table is at the top of the clay layer

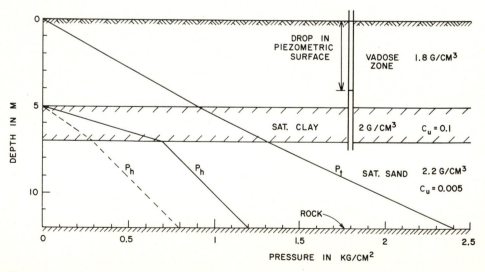

Figure 9.7 Profiles of P_t and P_h before (solid lines) and after (dashed line) lowering piezometric surface in confined aquifer. Unit weights of the materials are indicated in g/cm³.

and the hydraulic pressure in the clay layer increases linearly downward until it reaches the value of the piezometric head at the bottom of the clay layer. This yields the P_h profiles before and after the drop in piezometric surface shown in Figure 9.7. Using the unit weights of materials indicated in Figure 9.7, the P_t profile can also be evaluated. Since the decline in the piezometric surface does not produce a dewatered zone, the P_t profile is not affected by a drop in piezometric surface. The horizontal distance between the P_h and P_t profiles yields P_i, which in the confined aquifer has increased by 0.4 kg/cm^2 due to the 4-m drop in the piezometric level. This increase is equal to the reduction in upward hydraulic pressure against the bottom of the upper confining layer. For the clay layer, the increase in P_i varies linearly from zero at the top of the layer to 0.4 kg/cm^2 at the bottom. The compression of the clay layer and that of the confined aquifer can be calculated with Eqs. (9.11), (9.16), and (9.18) if the proper compression coefficients are known.

The reduction in volume of the compressed layers is the same as the volume of water squeezed out by the increase in P_i. This release of water commonly is the main mechanism whereby groundwater is yielded by confined aquifers upon pumping. Since silts and clays are much more compressible than sands and gravels, most of the water from confined aquifers is released from aquicludes, aquitards, or beds and thin layers of fine-textured material within the aquifer itself. The sands and gravels in confined aquifers then act essentially as conduits, whereas the fine-textured layers act as sources of water.

The compression of aquifers and fine-textured materials can be used to calculate the storage coefficient of the confined aquifer. For example, if the compression of the clay layer in Figure 9.7 is 4.5 cm and that of the underlying sand 0.5 cm, the storage coefficient of the confined aquifer would be $(0.045 + 0.005)/4 = 0.0125$. Aquifers consisting of sands and gravels without layers of fine-textured materials and confined by true aquicludes that do not compress or yield water upon compression will undergo little compression upon lowering the piezometric surface, and, hence, will have low storage coefficients. The compressibility of the water itself is so low (E of water is approximately 20 000 kg/cm^2) that decompression of water contributes very little to storage coefficients.

The volume compression of a layer of saturated porous material undergoing increased loading is equal to the volume of water leaving that material. If the layer is of low hydraulic conductivity and/or is relatively thick, it may take some time before this excess pore water is squeezed out. Consequently, the rate of compression lags beyind the rate of load increase. When this happens, the increased load is not entirely carried by the grains at their contact points, but also by the pore water itself. This causes an increase in the pressure of the pore water (see Section 9.7). When the extra load is applied instantaneously, it will initially be carried entirely by the pore water. As pore water leaves the material and compression takes place, however, P_i increases until eventually there is no excess pore pressure and the entire load increase is reflected in a corresponding increase in P_i. The rate of compression of the material (and, hence, the rate of dissipation of excess pore pressure) varies linearly with K of the material and inversely with the square of the thickness of the layer being compressed (Terzaghi and Peck, 1948; Taylor, 1958; Abbott, 1960; Bull and Poland, 1975). For some thick clay layers, it may take tens of years before excess pore pressure is completely dissipated (Poland and Davis, 1969). Thus, subsidence may lag behind the decline of groundwater levels and could continue long after excessive groundwater pumping is stopped (provided, of course, that groundwater levels do not recover).

9.6 INCREASING INTERGRANULAR PRESSURE BY WATER MOVEMENT

When flow of water through a granular medium is initiated or increased, friction between the moving water and the stationary grains causes a drag which exerts a force on the granular material in the direction of flow. This force, called *seepage force* or *seepage pressure*, increases the intergranular pressure if the flow is downward or horizontal, but decreases the intergranular pressure if the flow is upward. To initiate compression, the seepage force must overcome the resistance of grains to move. This resistance is largely due to intergranular friction. The movement of individual grains due to a change in seepage force and resulting compression take place during acceleration of the flow past grains.† Steady state can be reached again after the compression has taken place. In this section, a simplified treatment will be presented for vertical and horizontal flow, using static conditions as a starting point and steady flow as an end point.

Vertical Flow

The decrease in P_i due to the initiation of upward flow is illustrated in Figure 9.8, which shows profiles of P_t and P_h in a sand with the water table at field surface. The solid lines represent P_t and P_h at static conditions (no flow), and the dashed lines represent P_t and P_h if there is upward flow from below. The assumption is made that water reaching the top of the sand flows off laterally over the surface without build-up of water above the surface, so that the hydraulic pressure at the surface remains essentially zero. Since there is no wetting or dewatering, the P_t line is not affected by the initiation of upward flow. When there is upward flow, P_h at a given depth is greater than where there is no flow, because an upward hydraulic gradient requires additional pressure head. Thus, initiation of upward flow causes the P_h line in Figure 9.8 to swing to the right, which in turn reduces P_i. When the upward hydraulic gradient has increased so much that the P_h line coincides with the P_t line, P_i has become zero. At this point, there is no longer any effective contact between the sand particles, which become essentially suspended in the upward-flowing water. The sand surface then has lost its bearing strength—a condition known as *instability*, *quicksand*, or *boiling sand*.

---

† Donald C. Helm, U.S. Geological Survey, Sacramento, California, personal communication, July, 1975.

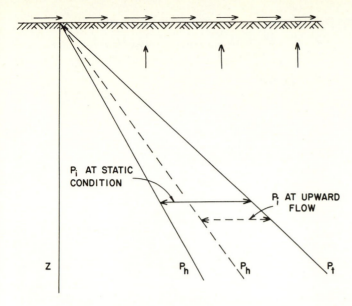

Figure 9.8 Profiles of P_t and P_h at static conditions (solid lines) and after initiation of upward flow (dashed lines) in soil with water table at ground surface.

The upward hydraulic gradient i_q necessary to produce quick conditions can be calculated by expressing P_t at depth Z as

$$P_t = 0.1Z[(1 - n)\rho + \theta \times 1] \text{ kg/cm}^2 \tag{9.19}$$

where Z is in meters, the porosity n is expressed as volume fraction, ρ is the density of the solids in grams per cubic centimeter, and θ is the volumetric water content as volume fraction (multiplied in the equation by the density of water, which is 1). A pressure of 1 kg/cm^2 corresponds to a pressure head of 1 000 cm or 10 m of water. Thus, at quicksand conditions, the pressure head h_z, in meters, at depth Z is equal to $10P_t$, or

$$h_z = Z[(1 - n)\rho + \theta] \tag{9.20}$$

Taking the reference level for the elevation heads at depth Z, the elevation head at Z is zero and that at the surface is Z. The pressure head at the surface is zero, so that the upward hydraulic gradient i_q at quicksand conditions can be expressed as [see Eq. (3.1)]

$$i_q = \frac{Z[(1 - n)\rho + \theta] + 0 - (Z + 0)}{Z}$$

$$= (1 - n)\rho + \theta - 1 \tag{9 21}$$

Typical values for saturated sand are $n = \theta = 0.35$ and $\rho = 2.6$ g/cm^3, for which Eq. (9.21) yields $i_q = 1.04$. Thus, loss of bearing strength, instability, or quicksand conditions can be expected if the upward hydraulic gradient is around 1 (see

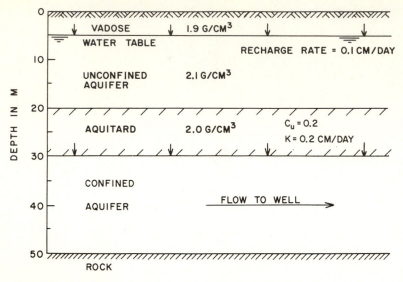

Figure 9.9 Confined aquifer with recharge from overlying unconfined aquifer through aquitard (unit weights and other data to be used in Problem 9.7).

Problem 9.6). This situation could occur below earth dams, dikes, or similar structures that hold water back, or below springs or seep areas.

To remedy quicksand conditions, P_i can be increased by reducing P_h with relief drains or by increasing P_t—for example, by placing a layer of rock or other coarse material on the surface. Such materials must have a high porosity so that water can move over the soil surface without significantly increasing the water depth, leaving the P_h line essentially unaffected.

If a downward flow is initiated, the P_h line in Figure 9.8 will swing to the left, which will cause an increase in P_i and, hence, subsidence, even if there is no drop in the water table. Such a situation could occur where a confined aquifer is hydraulically connected to an unconfined aquifer through an aquitard (Figure 9.9). If water is pumped from the confined aquifer and replenished from the overlying unconfined aquifer via vertical flow through the aquitard, vertical flow is initiated which reduces pore pressures in the aquitard (as well as in the aquifer). This will produce a corresponding increase in P_i and, hence, compression of the material, even if the water table in the unconfined aquifer remains constant. The compression can be estimated by calculating the pressure-head profile in the aquitard with Darcy's equation for the final steady state and evaluating the corresponding P_i profile. Knowing the P_i profile at the initial static condition, the compression of the aquitard can be computed.

Horizontal Flow

A case of lateral compression due to the initiation of horizontal flow is illustrated in Figure 9.10, which shows the pressure heads acting on both sides of a segment

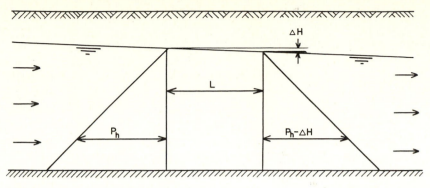

Figure 9.10 Pressure heads on both sides of vertical segment of unconfined aquifer with horizontal flow.

of an aquifer with horizontal flow. The water-table drop across the segment is ΔH. At each depth below the water table, the pressure head on the downstream side of the segment is ΔH less than at the upstream side. Thus, if ΔH is expressed in meters, the decrease in hydraulic pressure at each depth across the segment is 0.1 ΔH kg/cm^2. Assuming that P_t is the same on both sides of the segment, this causes a corresponding increase in P_i across the segment, from which the net horizontal compression of the segment in the direction of flow can be calculated.

Substituting 0.1 ΔH for $P_{i2} - P_{i1}$ in Eq. (9.11) and modifying the symbols for horizontal flow yields

$$S_h = 0.1\ \Delta H\ \frac{L}{E_h} \tag{9.22}$$

where S_h is the horizontal movement in a segment of length L from the time that horizontal flow was initiated until steady state is reached, ΔH is the steady-state water-table drop over distance L, and E_h is the elasticity coefficient in a horizontal direction. All length measurements are in meters, and E_h is in kilograms per square centimeter. S_h can be expressed similarly with the logarithmic theory. Since $\Delta H = iL$, where i is the slope of the water table, Eq. (9.22) can be written as

$$S_h = 0.1\ \frac{iL^2}{E_h} \tag{9.23}$$

Since E increases with P_i and, hence, with depth below the water table, E_h should be selected for the average P_i in the aquifer, or the aquifer should be divided into a number of increments, each with its own P_i and E_h. Deeper layers in an otherwise uniform material will show less horizontal displacement for a given increase in horizontal flow than those closer to the water table, because P_i increases with depth and E_h increases with P_i (Figure 9.4). Where coarse and fine aquifer materials are interbedded, the flow will ultimately be predominantly vertical through the fine material and horizontal through the coarse materials. The various layers may then show different amounts of horizontal movement, which can damage wells.

Horizontal and vertical movement of the land surface has been observed near pumped wells, even with pumping periods of as little as 5 to 30 min (Davis et al., 1969). The movements were small (mostly between 1 and 100 μm) and completely reversible when pumping was stopped. This indicates almost perfect elastic response of the aquifer to short-term pumping. Aquifers consisting of geologically more recent materials were not completely elastic.

For the same reason that aquifer materials are anisotropic with respect to hydraulic conductivity, the elastic properties of aquifers in horizontal directions are likely to differ from those in the vertical direction. Thus E_h should be determined on horizontal samples if used to calculate horizontal movement of the land surface. The same applies to C_c or C_u if the movement is calculated with the logarithmic theory of compression.

Transient, Three-Dimensional Flow

In practice, almost all flow systems are transient and three-dimensional. Thus, problems of vertical and horizontal movement of the land surface due to groundwater flow should be treated accordingly, at least in theory. This was done by Biot (1941, 1955), who developed equations to relate the three-dimensional field of underground fluid flow to the three-dimensional stress field, and also by Verruijt (1969). While theoretically elegant and sound, these approaches have limited practical value because the necessary input information (21 coefficients of anisotropic elasticity, 6 coefficients of anisotropic hydraulic conductivity, and the porosity) is seldom available for field conditions. If the medium is nonuniform, the 28 parameters must be known for each layer in the system. Thus, field situations normally are simplified to manageable systems (using computers if necessary) of one- or two-dimensional (vertical and horizontal) flow (Abbott, 1960; Brown and Burgess, 1973; Gambolati and Freeze, 1973; Gambolati et al., 1974; Gambolati, 1974; Helm, 1975; McNamee and Gibson, 1960).

The methodology for predicting subsidence from aquifer and aquitard parameters can also be used to evaluate such parameters from observed subsidence rates. Riley (1969), for example, showed how equivalent or effective values of hydraulic conductivity, total compressibility, and recoverable compressibility of aquitards can be obtained from records of piezometric levels and deformations in an aquifer system. Use of the resulting data in a finite-difference simulation model gave excellent agreement between observed and calculated compaction of aquitards (Helm, 1975).

9.7 NOORDBERGUM EFFECT

As stated in the last paragraph of Section 9.5, increasing the load on a layer of fine-textured material can produce increases in the pore-water pressure of that material if the drainage of pore water from the layer lags behind the rate of load increase. The increase in pore-water pressure will cause a rise of the water level in a piezometer reaching into that layer. This may seem paradoxical at first if the compression is caused by a falling water table or piezometric surface due to groundwater pumping. This reverse water-level response is known as the " Noordbergum " effect, after the location where it was observed in a well-pumping test (Verruijt, 1969). The effect can readily be expected during the first stages of pumping, particularly if the piezometer extends (unknowingly) into a layer of fine-textured material. In a system with predominantly horizontal flow, reverse water-level response in a piezometer was attributed by Wolff (1970) to distortion of pore space in clay or other fine-textured material resulting from the transfer of horizontal strain of aquifer(s) to less permeable layers via shear. Excess pore-pressure development in systems of three-dimensional flow to a pumped well was analyzed by Gambolati (1974).

9.8 EARTH FISSURES

Groundwater withdrawal can cause cracks and fissures in the earth's surface. Schumann and Poland (1969) reported that fissures in a groundwater basin in Arizona first appeared as narrow cracks, usually less than 2 cm wide and as much as 1.6 km long. The edges of the fissures were initally sharp, and generally there was no evidence of lateral or vertical offset along the cracks, indicating that the fissures were simple tensional breaks. The fissures ran more or less parallel to the surface contours, so that they intercepted rainfall runoff or flow in irrigated fields. The resulting flow and erosion in the fissures then led to widening, deepening, and interconnection of the cracks, ultimately producing fissures as long as 13 km, several meters wide, and several tens of meters deep (Figure 9.11).

The fissures could be formed by tension due to horizontal movement of the land surface toward the area where groundwater withdrawal is concentrated. Another mechanism is differential settlement as may be caused by discontinuities in the groundwater basin. Schumann and Poland (1969) found that the frequency and direction of many fissures conformed to zones of steep gravity gradients adjacent to mountains surrounding the basin. These zones may have reflected buried fault scarps that could yield areas of maximum tensile stress at the surface due to subsidence of the basin (Figure 9.12). If the subsidence is caused by compression of deeper, fine-textured layers, and if the compression increases about linearly from the edge toward the center of the basin, the slab of material above the compressed layers may actually undergo a slight rotational movement around the scarp as the surface of the ground moves from AB to AD (Figure 9.12). Assuming that the pivot point is at the top of the scarp and that $AB = 20$ km, $BD = 4$ m, and $AC = 50$ m, rotation of the slab around C would cause a crack above C that is $4 \times 50/20\,000 = 0.01$ m wide at the surface. This is on the same order as initial crack widths observed in practice. Similar reasoning can be applied to crack formation above local rises in the bedrock floor in other parts of the subsiding basin (Bouwer, 1977). Another possible explanation for crack formation

Figure 9.11 Eroded earthcrack in caliche-cemented alluvium southeast of Chandler, Arizona.

above buried scarps is that there may not be much groundwater in the shallow part of the basin upgradient from the scarp. Pumping groundwater in the center portions of the basin would then cause lateral movement of the land surface to the right of point *A* (Figure 9.12), but not to the left, causing tensile stresses at *A* and possible formation of cracks. For additional theories about crack formation, see Bouwer (1977) and references therein.

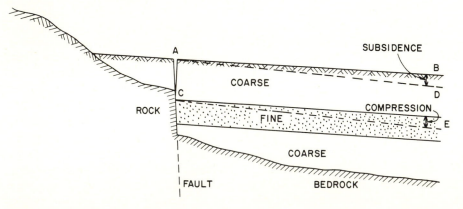

Figure 9.12 Rotation of slab *ABEC* due to linearly increasing compression of fine layer, and resulting fissure above buried scarp.

9.9 SUBSIDENCE OF PEATS AND MUCKS

Organic soils, such as peats and mucks, often make good agricultural soils, but their water table may be too high for farming. When the water table is lowered by draining such soils (see Section 8.4.3), subsidence occurs not only because of compression due to an increase in P_i (organic soils are the most compressible; see Tables 9.1 and 9.2), but also because of shrinking of the organic material as it is dewatered and because of biooxidation of the organic matter by microorganisms as atmospheric oxygen moves deeper into the soil. Drying of organic soils also makes them vulnerable to wind erosion and fire.

These "additional" effects caused subsidence rates of 2 cm/year in Michigan, 3 cm/year in the Everglades of Florida, and 8 cm/year in the Sacramento–San Joaquin delta in California (Stephens and Speir, 1969). The authors also reported a Dutch study, where compression was found to account for 30 percent, oxidation for 20 percent, and shrinkage for 50 percent of the total subsidence of an organic soil after drainage. The average subsidence in the Everglades for the period 1912–1970 was about 2.5 m, of which 67 percent was due to oxidation and 33 percent to compression (Stephens, 1974). Since the oxidation of organic matter is a biologic process, organic soils are oxidized more rapidly in warm than in temperate climates (bacterial processes double in rate with every 10°C rise below their optimum temperature). For organic soils near Minsk, Russia, for example, biooxidation accounted only for about 13 percent of the total subsidence (Stephens, 1974).

Organic soils often are of limited depth. Draining such soils for agriculture eventually leads to their complete disappearance, as continued subsidence requires continual lowering of the water table. Stephens and Speir (1969) predicted that the organic soils in the Everglades will be too shallow for the present agricultural use by the year 2000. The process can be delayed by keeping the water table as high as crops, tillage, and other farming operations will permit.

9.10 HYDROCOMPACTION

A special form of land subsidence, unrelated to groundwater withdrawal, may occur when loose, dry soil is wetted. Wetting a dry soil increases P_i because P_h in a wetted, unsaturated zone is negative. However, the settling of soils most susceptible to this form of subsidence, called *hydrocompaction*, apparently is mainly due to collapse of the soil structure caused by the water itself (Lofgren, 1969b). Most of these soils are loose, dry, and relatively fine deposits of alluvial or aeolian origin, and they occur mostly in dry climates. When dry, these soils can support large loads, but when wetted by infiltrating water they show considerable subsidence (Lofgren, 1969b). Compaction of 25 percent is not unusual. This can result in subsidences of several meters if the soils are relatively deep.

The water that causes hydrocompaction may come from new irrigation canals or irrigated fields, rainfall runoff from roads or rooftops, impoundments of water by dams or levees, leakage from storm drains or sewer lines, etc. Differential subsidence within the wetted areas, or between wetted and nonwetted areas, increases the damage and may cause cracks in the land surface. For new irrigation canals, these cracks tend to run parallel to the canal at distances of 1 to several meters from the water's edge. The damage of hydrocompaction can be minimized by hydraulically or mechanically compacting the loose deposits before constructing the facilities that cause wetting of these soils.

PROBLEMS

9.1 Using the e-vs.-P_i relations in Figure 9.3, plot e versus log P_i for the three soils (ignore the rebound and recompression loops). For the straight-line portions of the e-vs.-log P_i curves, evaluate C_c and, using the average e for that portion, calculate C_u. Compare the results with the values in Table 9.2.

9.2 Using Eq. (9.11), calculate S_u for the dewatered zone from 10 to 40 m in Figure 9.6 by 5-m increments (use average P_i for each increment) and compare with S_u calculated in Table 9.3 for the average increase in P_i in the entire zone.

9.3 As in Problem 9.2, calculate S_u, but with Eq. (9.18) and compare the result with S_u obtained in Table 9.4 for the average P_{i1} and P_{i2} in the entire layer.

9.4 Using Eq. (9.18), calculate S_u for the clay layer from 50 to 70 m depth in Figure 9.6 by using 5-m increments of the layer. Compare the result with S_u calculated for the layer as a whole in Table 9.4.

9.5 Assuming that C_u is 0.1 for the clay layer and 0.005 for the sand in the confined aquifer system of Figure 9.7, calculate the total subsidence caused by the 4-m drop in the piezometric level [the P_i values for Eq. (9.18) are taken as the average of P_i at the top and bottom of each layer before and after the drop in the piezometric surface]. If pore water squeezed from the clay layer and the aquifer is the only source of water release when the aquifer is pumped, what is the storage coefficient of the aquifer? What is the storage coefficient of the aquifer if the upper confining layer is incompressible and water is yielded only by compression of the aquifer itself? What would be the storage coefficient of the aquifer if both the aquifer and the confining layers are incompressible and water is yielded only by the expansion of the water itself due to a decrease in pressure head? The E value for water in the pressure range of 1 to 26 kg/cm² is 19 681.9 kg/cm² at 10°C (round to 20 000 kg/cm² in the calculation). The porosity of the aquifer is 30 percent.

9.6 Using the device of Figure 2.11, tube A is connected via flexible tubing to the bottom of a small reservoir that can be moved up and down. Water is applied to this reservoir to create an upward flow through the sand in the cylinder, allowing the water to spill over the top of the cylinder. The water supply to the reservoir must be sufficiently fast to continuously spill water over the edge of the reservoir. A metal object attached to a string (for retrieval) is placed on top of the sand in the cylinder. The reservoir is slowly raised until the metal object sinks into the sand, indicating that quicksand conditions have been reached. The height of the top of the reservoir is measured, and the upward hydraulic gradient through the sand is calculated. This observed gradient is then compared with the gradient calculated with Eq. (9.21), using known or estimated values of porosity, density, and water content of the sand.

9.7 Using the parameters shown in Figure 9.9, calculate with Eq. (9.18) the compression of the aquitard due to the downward flow (assume a constant water table and negligible head loss in the unconfined aquifer).

9.8 Calculate the lateral movement of the land surface near a pumped well where i is on the order of 0.05 and $L = 100$ m, and in a groundwater basin where i may be around 0.005 and $L = 10$ km. Use $E_h = 10\,000$ kg/cm² in both cases.

9.9 A pumping test is performed on a well in the aquifer system of Figure 9.6 with the water table at a depth of 10 m. The observation well for measuring the drawdown at some distance from the well is inadvertently installed with its bottom at 60 m depth in the center of the clay layer. If the drawdown of the water table at the location of the observation well is 1 m, what would be the maximum rise of the water level in the observation well due to restricted drainage of pore water from the clay layer (assume that the entire increase in P_i is borne by the pore water in the clay and that the water-table drop in the overlying unconfined aquifer has no immediate effect on the pore pressures inside the clay layer)?

9.10 Water seeps into a loose, silty sand that is completely dry to great depth. The sand is covered by a thin layer of fine material, which restricts the infiltration rate and causes unsaturated flow in the underlying silty sand. This produces a wetted zone with a pressure head equal to the water-entry value of the silty sand, which is -100 cm. The initial porosity of the silty sand is 40 percent, the initial water content zero, the water content after wetting 30 percent, and the density of the solid phase is 2.6 g/cm^3. If wetting causes a settlement of 10 percent (1 m per 10 m wetted), what is the effective value of E for the silty sand?

REFERENCES

Abbott, M. B., 1960. One-dimensional consolidation of multi-layered soils. *Geotechnique* **10**: 151–165.

Axtmann, R. C., 1975. Environmental impact of a geothermal power plant. *Science* **187**(4179): 795–803.

Biot, M. A., 1941. General theory of three-dimensional consolidation. *J. Appl. Phys.* **12**(2): 155–164.

Biot, M. A., 1955. Theory of elasticity and consolidation for a porous anisotropic solid. *J. Appl. Phys.* **26**(2): 182–185.

Bouwer, H., 1977. Land subsidence and cracking due to ground-water depletion, *Ground Water* **15**(5): 358–364.

Brown, C. B., and S. J. Burgess, 1973. Steady state ground motions caused by single-well pumping. *Water Resour. Res.* **9**: 1420–1427.

Bull, W. B., 1975. Land subsidence due to groundwater withdrawal in the Los Banos-Kettleman City area, California. Part 2. Subsidence and compaction of deposits. *U.S. Geol. Survey Prof. Paper 437-F*, 90 pp.

Bull, W. B., and R. E. Miller, 1974. Land subsidence due to groundwater withdrawal in the Los Banos-Kettleman City area, California. Part 1. Changes in the hydrologic environment conducive to subsidence. *U.S. Geol. Survey Prof. Paper 437-E*, 71 pp.

Bull, W. B., and J. F. Poland, 1975. Land subsidence due to groundwater withdrawal in the Los Banos-Kettleman City area, California. Part 3. Interrelations of water-level change, change in aquifer-system thickness, and subsidence. *U.S. Geol. Survey Prof. Paper 437-G*, 62 pp.

Colijn, P. J., and J. Potma, 1944. Weg- en Waterbouwkunde. I. Grondmechanica. Kosmos, Amsterdam.

Davis, G. H., and J. R. Rollo, 1969. Land subsidence related to decline of artesian head at Baton Rouge, Lower Mississippi Valley, U.S.A., *Proc. Tokyo Symp. on Land Subsidence*, IASH-UNESCO, pp. 174–184.

Davis, S. N., F. L. Peterson, and A. D. Halderman, 1969. Measurement of small surface displacements induced by fluid flow. *Water Resour. Res.* **5**: 129–138.

Gabrysch, R. K., 1969. Land-surface subsidence in the Houston-Galveston region, Texas. *Proc. Tokyo Symp. on Land Subsidence*, IASH-UNESCO, pp. 43–54.

Gambolati, G., 1974. Second-order theory of flow in three-dimensional deforming media. *Water Resour. Res.* **10**: 1217–1228.

Gambolati, G., and R. A. Freeze, 1973. Mathematical simulation of the subsidence of Venice. 1. Theory. *Water Resour. Res.* **9**: 721–733.

Gambolati, G., P. Gatto, and R. A. Freeze, 1974. Mathematical simulation of the subsidence of Venice. 2. Results. *Water Resour. Res.* **10**: 563–577.

Grant, U. S., 1954. Subsidence of the Wilmington oil field, California. In *Geology of Southern California*, R. H. Johns (ed.), *Calif. Div. Mines Bull. 170*, chap. 10, pp. 19–24.

Helm, D. C., 1975. One-dimensional simulation of aquifer system compaction near Pixley, California. 1. Constant parameters. *Water Resour. Res.* **11**: 465–478.

Jones, L. L., and J. P. Warren, 1976. Land subsidence costs in the Houston-Bayton area of Texas. *J. Am. Water Works Assoc.* **68**: 597–599.

Lambe, T. W., and R. V. Whitman, 1969. *Soil Mechanics.* John Wiley and Sons, Inc., New York, 553 pp.

Lofgren, B. E., 1969a. Field measurement of aquifer-system compaction, San Joaquin Valley, California, U.S.A. *Proc. Tokyo Symp. on Land Subsidence*, IASH-UNESCO, pp. 272–284.

Lofgren, B. E., 1969b. Land subsidence due to the application of water. *Reviews in Engineering Geology II*, Geol. Soc. Am., Boulder, Colo., pp. 271–303.

Mayuga, M. N., and D. R. Allen, 1969. Subsidence in the Wilmington oil field, Long Beach, California, U.S.A. *Proc. Tokyo Symp. on Land Subsidence*, IASH-UNESCO, pp. 66–79.

McNamee, J., and R. E. Gibson, 1960. Displacement functions and linear transforms applied to diffusion through porous elastic media. *Q. J. Mech. Appl. Math.* **13**: 98–111.

Poland, J. F., 1969. Status of present knowledge and needs for additional research on compaction of aquifer systems. *Proc. Tokyo Symp. on Land Subsidence*, IASH-UNESCO, pp. 11–21.

Poland, J. F., and G. H. Davis, 1969. Land subsidence due to the withdrawal of fluids. *Reviews in Engineering Geology II*, Geol. Soc. Am., Boulder, Colo. pp. 187–269.

Riley, F. S., 1969. Analysis of borehole extensometer data from central California. *Proc. Tokyo Symp. on Land Subsidence*, IASH-UNESCO, pp. 423–431.

Roberts, J. E., 1969. Sand compression as a factor in oil field subsidence. *Proc. Tokyo Symp. on Land Subsidence*, IASH-UNESCO, pp. 368–376.

Schumann, H. H., and J. F. Poland, 1969. Land subsidence, earth fissures, and groundwater withdrawal in South-Central Arizona, U.S.A. *Proc. Tokyo Symp. on Land Subsidence*, IASH-UNESCO, pp. 295–302.

Skempton, A. W., 1944. Notes on the compressibility of clays. *Q. J. Geol. Soc. London* **C**: 119–135.

Stephens, J. C., 1974. Subsidence of organic soils in the Florida Everglades—A review and update. *Environments of South Florida, Memoir 2*, Miami Geological Society, pp. 352–361.

Stephens, J. C., and W. H. Speir, 1969. Subsidence of organic soils in the U.S.A. *Proc. Tokyo Symp. on Land Subsidence*, IASH-UNESCO, pp. 523–534.

Taylor, D. W., 1958. *Fundamentals of Soil Mechanics.* John Wiley & Sons, Inc., New York.

Terzaghi, K., and R. B. Peck, 1948. *Soil Mechanics in Engineering Practice.* John Wiley & Sons, Inc., New York.

Verruijt, A., 1969. Elastic storage of aquifers. In *Flow through Porous Media*, R. J. M. DeWiest (ed.), Academic Press, New York, pp. 331–376.

Wolff, R. G., 1970. Relationship between horizontal strain near a well and reverse water level fluctuation. *Water Resour. Res.* **6**: 1721–1728.

Yerkes, R. F., and R. O. Castle, 1969. Surface deformation associated with oil and gas field operations in the United States. *Proc. Tokyo Symp. on Land Subsidence*, IASH-UNESCO, pp. 55–66.

TEN

GROUNDWATER QUALITY

The quality of groundwater, as determined by its chemical and biological constituents, its sediment content, and its temperature, is of great importance in determining the suitability of a particular groundwater for a certain use (public water supply, irrigation, industrial application, cooling, heating, power generation, etc.). The quality of groundwater is the resultant of all processes and reactions that have acted on the water from the moment it condensed in the atmosphere to the time it is discharged by a well or spring. The time spans involved may range from less than 1 day to more than 50 000 years. The chemical composition of groundwater can also be indicative of its origin and history, of the underground materials that the water has been in contact with, and of deep-seated temperatures. In the following sections, natural sources and significance of chemical and biological constituents in groundwater will be examined. Human-made sources are discussed in Chapter 11. Other quality aspects included in this chapter are standards for various uses, dating and tracing of groundwater, and use of hot groundwater for electric power production.

10.1 CHEMICAL CONSTITUENTS IN ATMOSPHERIC PRECIPITATION

Atmospheric precipitation, which is the common source of groundwater, is by no means distilled water when it lands on the earth's surface. Volcanic eruptions; emissions of gases and sublimation of solids from the earth's crust; dust and other windborne solids in the atmosphere (including salt spray from oceans); gaseous and other metabolic products excreted by the earth's biota into the atmosphere; reactions caused by lightning and cosmic rays; and gaseous emissions of industrial and other pollutants are examples of the many ways in which chemicals enter the atmosphere. From there they can return to the earth dissolved in rainfall and snow. Another pathway is dry fallout, which subsequently can be dissolved by rain after it has landed on the earth's surface.

Most of the ammonium and nitrate in atmospheric precipitation is soil-derived (Junge, 1958). Concentrations of these chemicals in precipitation water may vary seasonally and locally and range from 0.01 to 2 mg/l for the United States. Schuman and Burwell (1974) reported average N concentrations of 0.78 mg/l for precipitation at Treynor, Iowa, for the period 1971–1973. About 60 percent was in the ammonium form; the rest was nitrate. Higher nitrogen concentrations were found in rainfall in central Netherlands (Table 10.1). Extensive data on nitrogen and phosphorus contents of rainfall and many literature references were presented by Uttormark et al. (1974).

Average annual chloride concentrations in rainfall varied from 8 mg/l in coastal areas to about 0.1 to 0.2 mg/l in the rest of the United States (Junge and Werby, 1958). Sodium concentrations varied similarly from about 4 to 0.1–0.3 mg/l. The major source of Cl is the ocean, but the Cl/Na ratio in rain is less

Table 10.1 Chemical constituents in mg/l of atmospheric precipitation†

	Lake Ontario, 1970–1971		Menlo Park, 1957–1958	Wageningen, 1973, Aug.–Oct.
	Jan.–Dec.	Apr.–Nov.		
Calcium	4.10	4.00	6.7	
Magnesium	0.65	0.59	2.96	
Potassium	0.50	0.52	0.71	
Sodium	2.01	1.84	3.94	
Bicarbonate (HCO$_3$)			22.5	
Chloride	1.18	0.86	7.28	3
Silica (SiO$_2$)			3.17	
Sulfate (SO$_4$)	8.24	7.63	12.2	9
Ammonium—N	0.594	0.614		2
Nitrate—N	1.11	1.02	0.58	0.6
Nitrite—N			0.008 5	
Total N				3.3
Orthophosphate P	0.029	0.034		0.035
Total P	0.060	0.068		0.072
Cadmium	0.001	0.001		
Copper	0.006	0.006		
Iron	0.031	0.027		
Lead	0.020	0.021		
Nickel	0.004	0.004		
Zinc	0.077	0.078		
Chemical oxygen demand				15

† At Lake Ontario (from Shiomi and Kuntz, 1973); Menlo Park, California (from Whitehead and Feth, 1964); and Wageningen, Netherlands (from van den Berg, 1973).

than in seawater, indicating excess Na, which probably is derived from the soil. The same is true for K (potassium), which varied from about 0.4 mg/l near the coast to 0.1 mg/l in rainwater over the rest of the United States. Calcium concentrations varied from 0.5 to 5 mg/l, with the high values occurring in the southwestern United States where alkali soils and dust storms are most frequent. Sulfate concentrations generally ranged from 1 to 3 mg/l (Junge and Werby, 1958).

An example of more recent analyses is given in Table 10.1 for rain and snow collected around Lake Ontario in the period 1970–1971 (Shiomi and Kuntz, 1973). The data show the influence of industrial and other activities by humans. While the pH of distilled water is almost 7, the pH of rainwater is below 7 because dissolved CO_2 from the atmosphere produces carbonic acid. Air pollution also tends to lower the pH of rainwater. Acid precipitation with pH values in the range of 4.5 to 5.5, for example, occurs over much of the northeastern United States and may be caused by oxidation of SO_2 and NO_2 emitted from industrial areas in the Midwest to H_2SO_4 and HNO_3 in the atmosphere (Cogbill and Likens, 1974).

The importance of dry fallout is illustrated in a study by Whitehead and Feth (1964), who reported mean total-dissolved-solids contents of 1 to 47 mg/l (average 9.7 mg/l) for rainfall as such, but of 3.4 to 324 mg/l (average 52.9 mg/l) for rainfall plus dry fallout (bulk precipitation) at Menlo Park, California, in the period 1957–1958. Mean ion concentrations in the bulk precipitation are listed in Table 10.1. For additional information regarding the chemistry of rainfall, reference is made to Junge (1963).

10.2 SOIL AND PLANT EFFECTS

Minerals are continually mobilized in the soil mantle as clay and other soil particles weather, and plant and animal materials decompose. When it rains, these minerals can be leached out and reach the groundwater. In humid areas, salt concentrations of about 400 to 500 mg/l have been reported in the leachate, as collected by underground drains in agricultural fields (Bower, 1974, and references therein). The dominant ions were Ca, Mg, Na, HCO_3, SO_4, and Cl. In arid areas, there can be a significant concentration effect as plant roots take up water and leave salts behind, and as soil water evaporates from the surface. When these salts are then leached out by rainwater, the leachate that eventually reaches the groundwater will contain considerable amounts of salt, particularly in areas with low rainfall where the leaching rate may be only a few millimeters per year. Leachate from irrigated fields in arid climates may have salt concentrations of several thousand milligrams per liter. Most of these salts are, however, applied with the irrigation water. Concentration of salts by evaporation also occurs where rain or other surface water collects in depressions with relatively impermeable bottom soil (runoff-fed playas, for example). A considerable portion of the water may then have evaporated by the time the remaining water has completely infiltrated. The salt content of surface runoff normally is less than that of leachate or deep-percolation water. Depending on antecedent conditions, particularly evaporation from soil and salt accumulation on the soil surface, runoff water may have a salt concentration of 50 to 500 mg/l.

In addition to increasing the salt concentration, soil and plant effects include ion exchange and other reactions of the water and its constituents with the soil, and uptake of nutrients by plants. Both processes alter the chemical composition of the water as it moves through the root zone. Ion exchange involves primarily the cations, which are adsorbed and exchanged for other cations by the negatively charged clay and organic matter (cation exchange complex; see Section 2.3). The order of replaceability of adsorbed cations depends on the clay mineral and other factors. A common sequence of decreasing replaceability is Li, Na, K, NH_4, Rb, Cs, H, Mg, Ca, Sr, and Ba (Bear, 1964). Thus, adsorbed Li is replaced by Na when present in the soil solution, Na by Ca, etc. Dense layers of clay, such as aquicludes or aquitards, can act as semipermeable membranes that hold back certain ions while passing others (see Sections 3.4, 10.3, and 11.10). Chemical reactions include immobilization of phosphate, which is adsorbed and precipitated in most soils other than pure sands. Also, heavy metals can be fixed in the soil, particularly when the soil contains clay, has a pH above 7, and is aerobic (Page, 1973). In alkaline soils, calcium can precipitate as $CaCO_3$. Acid rainwater, however, can dissolve $CaCO_3$ in the soil. Plant roots will absorb nutrients like N, P, K, S, and certain heavy metals and other trace elements. Thus, the concentrations of these elements will be reduced by the time the rainwater has reached the groundwater. On the other hand, some agricultural fields may receive so much chemical fertilizer that the downward-moving water actually becomes enriched with N, P, and K (see Section 11.7). Carbon dioxide and organic acids produced by the plants and other living material in the soil may lower the pH of the water, which in turn accelerates the rate of weathering of the soil and enhances the mobility of metals.

10.3 AQUIFER EFFECTS

The water that continues to move downward and eventually becomes groundwater further reacts with the soil and rock materials in the vadose zone and the aquifer. These reactions primarily consist of solution of the solid phase in accordance with the solution chemistry of the particular minerals (Hem, 1970). These minerals range from almost insoluble to very soluble (evaporites). The solubility is also affected by temperature and pressure. Other sources of dissolved salts in groundwater are underground waters of marine origin and salty connate waters, which can intrude freshwater aquifers.

Fine-textured formations like clay or shale aquitards can behave like semipermeable membranes, retarding the movement of charged ionic species but allowing relatively unrestricted movement of neutral substances. Also, molecules with high dipole moments like H_2O will be retarded relative to those with low dipole moments, like certain organic compounds. This behavior is caused by the negatively charged surfaces of clay particles, which adsorb cations from the surrounding soil solution. Some anions also get trapped in the adsorbed layers of ions.

When the fine-textured material is very dense, the pores between the clay particles are so small that they are entirely influenced by the adsorbed layers and water and ions no longer move freely through the pores. The clay layer then behaves like a semipermeable membrane, which can produce osmotic-pressure differences, salt sieving or ultrafiltration, and electric-potential differences (Hanshaw, 1972).

Salt sieving and osmotic effects may help explain certain salinity and piezometric properties of water in sedimentary rock that cannot be readily explained by simple-solution chemistry or gravity flow. Various authors, for example, have suggested that the high salt content (sometimes much higher than that of seawater) of certain groundwaters that have not been in contact with evaporites is due to salt sieving as water left the formation through a semipermeable membrane that did not pass the salts (Hanshaw, 1972, and references therein). Also, low pressure heads observed in a confined sandstone overlying shale have been attributed to osmotic cross-formational withdrawal of water through the shale (Hanshaw, 1972, and references).

Meteoric groundwater tends to be of good quality, except where it has been stagnant or otherwise isolated from the hydrologic cycle in mineral-rich or clay-rich aquifers, or has been in contact with salty groundwater or evaporites (Davis and DeWiest, 1966). Thus, igneous and crystalline rocks generally yield groundwater of excellent quality with salt concentrations commonly below 100 mg/l and seldom over 500 mg/l. Depending on the minerals in the rock, the water can be hard, high in silica, high in Mg in comparison to Ca, slightly acid, or slightly alkaline. Volcanic rock also tends to yield water of good quality, usually with a relatively high HCO_3 content. Sedimentary rock generally yields good-quality groundwater. The water from sandstones may be high in Na and HCO_3. Shales may produce slightly acid water high in Fe, SO_4, and F. Limestones yield water that is slightly alkaline and contains Ca and Mg.

Water from alluvial deposits in tectonic basins and valleys may have a relatively high salt content if the basin was closed and/or the water came from deeper aquifers where it was in contact with evaporites or connate water. Water with a lower salt content normally is obtained where there is some outlet for the groundwater and the older groundwater and connate water have been replaced by more recent meteoric water. If there is sedimentary rock in the alluvial fill, the groundwater may contain Ca, Mg, HCO_3, and SO_4. Na and Cl may also be present, depending on how much leaching and concentrating took place in the surface soil. Material of igneous and volcanic origin may contribute SiO_2 to the water. Alluvial river valleys normally yield groundwater with a relatively low salt content, depending on the material. If it contains carbonate rock, the water will contain Ca, Mg, and HCO_3, while gypsum and anhydrite contribute Ca and SO_4. In some valleys, salty water from the bedrock seeps into the aquifers and causes local increases in the salt concentration of the groundwater. Upper aquifers in coastal-plain sediments normally yield good groundwater with a low salt content and with Na, Ca, and HCO_3 as principal ions. Deeper aquifers (from 100 to 300 m) tend to yield water with more Na and HCO_3 and less Ca and SO_4. Below this level, aquifers are generally contaminated with marine or salty connate water.

Sand dunes yield water of good quality, but are vulnerable to seawater intrusion in coastal areas. Groundwater from glacial deposits has a good quality when it is recent and an active part of the hydrologic cycle. If the water has moved long distances through aquifers or if it has been stagnant, like between aquicludes, the salt content may be relatively high. Hard water is yielded where the aquifers contain calcite or dolomite. For additional information on groundwater quality in relation to aquifer characteristics, reference is made to Davis and DeWiest (1966).

Dissolved salts in groundwater primarily consist of Na, Ca, Mg, K, Cl, SO_4, HCO_3, and CO_3. Schemes have been developed to classify groundwater according to the concentrations of these ions, using trilinear diagrams and categories like calcium-sodium-chloride water, sodium-calcium-magnesium-chloride-sulfate water, calcium-bicarbonate water, etc. (Piper, 1953). While such classification may be valuable in describing the general type and geochemical history of a given groundwater, a complete chemical analysis with the results in tabular form is more valuable in determining the suitability of a certain groundwater for a particular use.

In addition to fresh, meteoric groundwater, there is also groundwater of marine origin which is chemically similar to seawater but may contain more Ca. Saline connate water tends to contain more I, B, SiO_2, N, and Ca, but less SO_4 and Mg than seawater, and may be normal to slightly thermal. Metamorphic water may be relatively high in CO_2 and B and relatively low in Cl compared to seawater, moderately high in I, and normal to moderately thermal. Magmatic water may be relatively high in Li, F, SiO_2, B, S, and CO_2, low in I, Br, Ca, Mg, and possibly N, and strongly thermal (Davis and DeWiest, 1966, and references therein).

10.4 GROUNDWATER CONSTITUENTS AND SUITABILITY FOR DRINKING

Detailed, standardized procedures for most analytic techniques for water-quality assessment are presented in *Standard Methods for the Examination of Water and Waste Water* (American Public Health Assoc. et al., 1971). Concentrations of a given element or ion normally are expressed in milligrams per liter (mg/l) or in parts per million (ppm) by weight. For low concentrations, mg/l and ppm are essentially the same, but for high concentrations, mg/l will be more than ppm. Assuming, for example, that a salt solution of 1 l contains 0.9 kg of water and 0.2 kg of salt, the salt concentration would be 200 000 mg/l but 181 818 ppm. Other properties of water, like color, turbidity, pH, and electrical conductivity are expressed in special units. The more important chemical constituents and quality parameters of groundwaters are reviewed in the rest of this section. Most of the discussion on sources of minerals is condensed from the work by Hem (1970), to which reference is made for further detail.

Total Dissolved Solids

The total-dissolved-solids content (TDS) or total salt concentration of ground-water varies from less than 100 to more than 100 000 mg/l. The TDS content often is also expressed in terms of electrical conductance of the water, normally in millimhos per centimeter at 25°C. The relation between conductance and TDS depends on the particular ions in solution. For irrigation water and most other natural waters, 1 millimho customarily is taken as equal to 640 mg/l (Salinity Laboratory, 1954). Groundwater is classified according to its TDS content as (Hem, 1970):

Fresh	< 1 000 mg/l
Moderately saline	3 000–10 000 mg/l
Very saline	10 000–35 000 mg/l
Briny	> 35 000 mg/l

Davis and DeWiest (1966) classified water with a TDS content of 1 000 to 10 000 ppm as brackish, of 10 000 to 100 000 ppm as salty, and of more than 100 000 ppm as brine. By way of comparison, the TDS content of seawater is about 34 000 mg/l and that of a saturated NaCl solution more than 300 000 mg/l. The recommended maximum limit for the TDS content of drinking water is 500 mg/l (see Section 10.5), but water of double or even triple this concentration is used if no other water is available. Additional TDS standards are presented in Section 10.5.

Hardness

Hardness of water relates to its reaction with soap and to the scale and incrustations accumulating in containers or conduits where the water is heated or transported. Since soap is precipitated primarily by Ca and Mg ions, hardness is defined as the sum of the concentrations of these ions expressed as mg/l of $CaCO_3$. This hardness, also called total hardness, calcium-plus-magnesium hardness, or hardness as $CaCO_3$, is calculated by adding the milliequivalents of Ca and Mg per liter and multiplying the sum by 50. Water is classified according to its hardness (mg/l hardness as $CaCO_3$) as (Hem, 1970, and references):

Soft	0–60
Moderately hard	61–120
Hard	121–180
Very hard	> 180

If the hardness as $CaCO_3$ exceeds the alkalinity of the water as $CaCO_3$ (see next parameter), the difference between the two is called *noncarbonate* hardness.

Water for domestic use should not contain more than 80 mg/l total hardness. Waters from aquifers with limestone or gypsum may contain 200 to 300 mg/l hardness or more. Hardness of groundwater can be increased if contaminated by acid leachate from mine spoils, garbage disposal areas, or other sources. Such acid waters can mobilize Ca in underground materials. The disadvantage of hard water is that it precipitates soap, thus increasing soap requirements. In recent years, health aspects are receiving increasing interest because of findings that point to greater incidence of coronary heart disease in areas with soft water than with hard water (Crawford, 1972). The causative factor, however, may be more complex than hardness alone and is probably associated with the effects of hardness on the activity of metals and other constituents of the water, and on their interrelationships. For example, low incidence of heart disease apparently also occurs in areas with very soft water, which led Neri et al. (1975) to postulate that hardness effects on health are due to two competing mechanisms. The very soft waters would not contain enough toxic substances to increase mortality, whereas the hard waters would contain more toxic substances but also enough benign minerals to block and overcome their toxic effects.

Alkalinity, Acidity, and pH

Alkalinity is the capacity of the water to neutralize acid. Since practically all alkalinity of natural water is produced by carbonate and bicarbonate ions, titrated alkalinity is expressed as the equivalent concentration of $CaCO_3$ obtained by adding the equivalents of CO_3^{2-} and HCO_3^- and expressing the sum as mg/1 of $CaCO_3$.

Acidity is the capacity of the water to react with hydroxyl ions. Titrated acidity may be expressed in terms of mg/l of H^+ or as equivalent concentrations of H_2SO_4 or $CaCO_3$. Sources of acidity include HCO_3^- which can react with OH^- to form CO_3^{2-} and H_2O, and undissociated or partly dissociated acids like hydrofluoric acid and bisulfate.

The pH refers to the activity (effective concentration) of hydrogen ions in the water, expressed as the negative logarithm (base 10) of the H^+ activity in moles per liter. At a pH of 7, the H^+ activity is 10^{-7} mol/l and the solution is considered neutral. When the pH is less than 7, the solution behaves as an acid (particularly below pH 4). Above pH, the solution reacts like a base. Most natural waters are within a pH range of 6 to 8.5 (Hem, 1970).

Calcium

Calcium is one of the principal cations in groundwater. Sources of calcium are igneous-rock minerals like silicates, pyroxenes, amphiboles, feldspars, and silicate minerals produced in metamorphism. Since the solubility of these minerals is low, water from igneous or metamorphic rock tends to be low in calcium as well as in TDS. In sedimentary rock, calcium occurs as carbonate (calcite and aragonite), calcium magnesium carbonate (dolomite), calcite (limestone), and calcium sulfate (gypsum and anhydrite). Some calcium fluoride may also be present. Calcium carbonate is one of the main cementing agents for sandstone and other detrital rock. Groundwater from limestone and other calcareous deposits is generally hard and can be expected to be a saturated solution of calcite. Calcium sources in alluvial and other unconsolidated materials consist primarily of the various minerals in the rock and soil fractions in these deposits.

Magnesium

Magnesium in groundwater from igneous rock primarily derives from ferromagnesian minerals like olivine, pyroxenes, amphiboles, and dark-colored micas. For metamorphic and other altered rock, magnesium occurs in minerals like chlorite, montmorillonite, and serpentine. In sedimentary rock, magnesium occurs as magnesite and other carbonates, sometimes mixed with calcium carbonate. Dolomite contains calcium and magnesium in equal amounts. Most groundwaters contain relatively small amounts of Mg, except where they have been in contact with dolomite (amounts of Ca and Mg about the same), or with Mg-rich evaporites that could cause Mg to become the dominant cation in the groundwater.

Sodium

Sodium is primarily derived from feldspars in igneous rock and its weathering products (clay minerals) in other material. Shale and clay layers often yield water with a relatively high sodium content. Other sources of sodium are leachate and deep percolation water from the upper soil layers (including atmospheric precipitation that has been subject to concentration effects), and contamination of groundwater by salty connate water or water of marine origin. Brines and other salty waters which usually occur at great depths contain large amounts of sodium.

Potassium

Potassium is less common than sodium in igneous rock, but more abundant in sedimentary rock as potassium feldspars. These minerals, however, are very insoluble so that potassium levels in groundwater normally are much lower than sodium concentrations.

Strontium

The chemistry of strontium is similar to that of calcium. Groundwater normally contains less than 10 mg/l of strontium, but levels of more than 50 mg/l have been observed.

Iron

Iron is widely distributed in the earth's crust. It occurs in minerals like pyroxenes, amphiboles, biotite, magnetite, and olivine. Pyrite is a common form of iron in sedimentary material, whereas ferric oxides and hydroxides are important iron-bearing minerals. The common form of iron in groundwater is the soluble ferrous ion Fe^{2+}. Concentrations normally are in the 1-to-10-mg/l range. When exposed to the atmosphere, Fe^{2+} is oxidized to the ferric state Fe^{3+}, which is insoluble and precipitates as ferric hydroxide, causing a brown discoloration of the water and the characteristic brown stains in sinks and laundered textiles. Corrosion of well casing and other pipe may also contribute iron to well water. Bacterial activity can increase or decrease iron concentrations in groundwater. The recommended maximum iron concentration for drinking water is 0.3 mg/l, primarily for reasons of taste and to avoid staining of plumbing fixtures and laundered clothes (see Section 10.5).

Manganese

Biotite and horneblende are among the manganese-containing minerals in igneous rock. Manganese oxides and hydroxides are common sources of manganese in other rocks and soils. The divalent ion Mn^{2+} is soluble and present in most groundwaters at concentrations less than those of Fe^{2+}. When exposed to the atmosphere, Mn^{2+} is oxidized to the much-less-soluble hydrated oxides, which form black stains in plumbing fixtures and laundered textiles. For this reason, the maximum concentration of manganese for public water supplies is set at 0.05 mg/l (see Section 10.5).

Aluminum

Aluminum is the third most common element in the earth's outer crust. Groundwater, however, rarely contains more than 0.5 mg/l of Al, except where the pH is below 4 and Al becomes more soluble.

Carbonate and Bicarbonate

Sources of carbonate and bicarbonate include CO_2 from the atmosphere, CO_2 produced by the biota of the soil or by the activity of sulfate reducers and other bacteria in deeper formations, and the various carbonate rocks and minerals. Sodium carbonate can accumulate as evaporite in closed basins, causing high carbonate levels in groundwater that has been in contact with it. Bicarbonate concentrations of more than 200 mg/l are not uncommon in groundwater, and higher concentrations can occur where CO_2 is produced within the aquifer. Carbonate concentrations in groundwater are usually less than 10 mg/l.

Chloride

Primary sources of chloride in groundwater are evaporites, salty connate water, and marine water. Igneous rock materials contribute little chloride. Groundwaters containing significant amounts of chloride also tend to have high amounts of sodium, indicating the possibility of contact with water of marine origin. Leaching of chlorides that have accumulated in upper soil layers may be a significant chloride source in dry climates. The recommended maximum concentration for chloride in drinking water is 250 mg/l, primarily for reasons of taste (see Section 10.5).

Silica

Silicon is the second most abundant element (after oxygen) of the earth's upper crust. Highest silicon concentrations in groundwater are found where the water has been in contact with certain volcanic rocks. Silicon content, expressed in mg/l of silica (SiO_2), is on the order of 20 mg/l for most groundwaters.

Sulfate

Sulfate is formed by oxidation of pyrite and other sulfides widely distributed in igneous and sedimentary rocks. The most important sulfate deposits are found in evaporite sediments (gypsum, anhydrite, sodium sulfate). In arid regions, leaching of sulfate from the upper soil layers may also be significant, causing sulfate to be the principal anion of the underlying groundwater. Sulfate concentrations in drinking water should not exceed 250 mg/l because the water will have a bitter taste and can produce laxative effects at higher levels.

Fluoride

Sources of fluoride in groundwater are minerals like calcium fluoride (fluorite), apatite, certain amphiboles, cryolite (in igneous rocks), and fluorspar (in sedimentary rocks). Groundwaters with more than 1 mg/l of fluoride are often found. Sometimes, F concentrations exceed 10 mg/l and can reach more than 30 mg/l (Cox, 1964). These high concentrations tend to be associated with a high pH. Some F in drinking water is beneficial because it reduces tooth decay. At higher levels, however, mottling of the teeth (fluorosis) occurs. For this reason, maximum F concentrations recommended for drinking water range from 1.4 to 2.4 mg/l, depending on how much water is ingested (see Section 10.5).

Nitrogen

Although some volcanic rocks contain nitrogen, most of the nitrogen in groundwater probably is derived from the biosphere. Molecular nitrogen from the atmosphere can be transformed into organic matter by nitrogen-fixing bacteria (*Rhizobium* species), which live in symbiotic relationship with leguminous plants (clovers, peas, beans, etc.) in nodules on the plant roots. Some non-legumes may also be capable of supporting nitrogen-fixing bacteria on their roots (Brill, 1977; Olsen, 1977). Part of the nitrogen fixed by these bacteria is taken up by the plants themselves. Upon death and decay of the bacteria and plants, the nitrogen originally fixed from the atmosphere is mineralized by soil bacteria into ammonium, which under aerobic conditions is converted into nitrate by nitrifying bacteria. Several free-living microorganisms also have the capability of meta-bolizing atmospheric nitrogen, including the soil bacteria *Azotobacter* and *Clostridium*, and the blue-green algae that normally live in water but also can flourish on soil after rain (Lynn and Cameron, 1973). Death and decay of these organisms add ammonium and nitrate to the soil. Some of the ammonium and nitrate in the soil are absorbed again by plants and microorganisms, and nitrate can be reduced to free nitrogen gas by denitrifying bacteria. The free nitrogen then returns to the atmosphere. Both processes are part of the nitrogen cycle (Lance, 1972; Behnke, 1975).

Ammonium is a cation that can be adsorbed to the clay and organic matter in the soil where it can remain as such under anaerobic conditions, but will be oxidized by nitrifying bacteria to nitrate under aerobic conditions. The nitrate ion is completely mobile and can be leached out of the upper soil layers by infiltrating water. In arid regions, the concentration effect can produce significant nitrate levels in the underlying groundwater. Very high N concentrations may occur below areas where animals tend to concentrate and waste products accumulate and decompose, such as bird colonies, roosting places, and caves with bats. Contamination of groundwater by human activities (agriculture, waste disposal, etc.) is discussed in Chapter 11.

Where the natural N-enrichment processes have been going on for some time, nitrate-nitrogen levels in groundwater may be on the order of 1 to 50 mg/l. Hem (1970) reported NO_3-N concentrations of about 100 mg/l for bat-infested cave areas. Kreitler and Jones (1975) found NO_3-N concentrations of less than 0.2 mg/l to more than 690 mg/l in groundwater of Runnels County in west-central Texas. They concluded that most of this nitrate was derived from natural soil nitrogen under dry-land farming with minimal fertilizer use. This is a good example of how the concentration effect can produce high nitrate levels in water leached from a root zone of essentially unfertilized vegetation where atmospheric nitrogen was the main nitrogen source.

Other forms of nitrogen include nitrite, which is an intermediate product in nitrification and denitrification and occurs in groundwater at much lower concentrations than nitrate. Ammonium-nitrogen levels in groundwater seldom exceed a few milligrams per liter. Organic nitrogen occurs in negligibly small concentrations.

The maximum concentration of nitrate nitrogen for public water supplies is 10 mg/l (45 mg/l when expressed as nitrate). At higher nitrate levels, young infants (generally less than 4 months old) could die from methemoglobinemia or blue-baby disease. Since gastric juices of such infants lack sufficient acidity, nitrate-reducing bacteria can grow in their upper intestinal tracts. When they ingest nitrate, the nitrate can then be reduced to nitrite before the nitrate is completely absorbed in the bloodstream. This results in absorption of nitrite in the blood-stream. The nitrite reacts with the hemoglobin to form methemoglobin, which is ineffective as an oxygen carrier. This produces anoxemia, symptomized by a grey-ish or brownish-blue discoloration of the skin, and can lead to death by asphyxia.

Older infants and adults can tolerate higher nitrate levels in drinking water and food because their stomach pH is too low for nitrate-reducing bacteria. However, nitrate may also play a role in the production of nitrosamines in the stomach, which are known carcinogens (Wolff and Wasserman, 1972). This was considered by Hill et al. (1973) as a possible reason for a higher death rate from gastric cancer in a group of people that had a high nitrate level in their drinking water.

Most cases of methemoglobinemia in Europe and the United States have been associated with waters containing more than 11 mg/l of nitrate nitrogen. In the U.S., 278 cases caused by well water nitrate were reported for the period 1945–1950, resulting in 39 known deaths. The number of reported cases decreased to 40 for the period 1952–1966 and to 10 for the period 1960–1969, with no fatalities (National Research Council, 1972).

Since nitrite is the causative factor in methemoglobinemia, its concentration in drinking water should be low, i.e., less than 1 mg/l as nitrite nitrogen (see Section 10.5). Ammonium nitrogen is undesirable in water for public supplies primarily because of taste and odor. It also reacts with chlorine and increases the amount of chlorine necessary to disinfect the water. For these reasons, the recommended upper limit for ammonium nitrogen in drinking water is 0.5 mg/l (see Section 10.5).

Phosphorus

A common phosphorus-containing mineral is apatite, which has a very low solubility. Thus, phosphorus concentrations in most natural groundwaters are less than 0.1 mg/l.

Boron

The most widely distributed boron-containing mineral is tourmaline. Boric acid and boron fluoride may also be emitted with volcanic gases and return to the earth in rain or as dry fallout. Thus, groundwater from volcanic or geothermal areas may contain relatively high boron concentrations (several milligrams per liter or more). The same is true for water that has been in contact with evaporites. Other groundwaters generally contain less than 0.5 mg/l of boron. Boron is an essential plant food, but it is toxic at higher concentrations. For this reason, irrigation water should contain less than 0.5 to 1 mg/l of boron, depending on crops and soils (see Section 10.5).

Minor and Trace Elements

Many elements are, or can be, present in groundwater, often at concentrations well below 0.1 mg/l, but sometimes much higher, particularly if the water has been in contact with mineralized rock or ore bodies. Several minor elements are of great concern because of their toxic effects, even at low concentrations. Arsenic, for example, should not exceed a concentration of 0.01 to 0.1 mg/l in drinking water (see Section 10.5 for standards on arsenic and other minor elements discussed in this paragraph), but occasionally occurs at concentrations of several milligrams per liter, especially in thermal groundwater. High arsenic concentrations have also been reported for groundwater from mineralized areas, such as the towns of Lukeville and Ajo in southern Arizona (well water in these areas also had a high lead content). Barium can seriously affect the heart, blood vessels, and nerves, and should not exceed a concentration of 1 mg/l in drinking water. Cadmium has been linked to hypertension and accumulates in the liver and kidney. Its maximum permissible concentration in drinking water is 0.01 mg/l. Another toxic element is hexavalent chromium, which should not exceed 0.05 mg/l in drinking water. Chromium concentrations as high as 0.22 mg/l were found in groundwater from an alluvial basin northeast of Phoenix, Arizona (Robertson, 1975). Highest chromium levels were associated with finest-grained sediments. Copper is an essential element in human metabolism, but can cause emesis and liver damage at excessive levels. Drinking water should not have more than 1 mg/l of copper. Lead is toxic at acute and chronic exposures. Lead pipe traditionally has been used in household plumbing. The lead limit for drinking water is set at 0.05 mg/l. Mercury is highly toxic, but its concentration in groundwater usually is below the recommended limit of 0.002 mg/l for drinking water. Selenium produces toxic effects similar to those of arsenic and should not exceed a concentration of 0.01 mg/l. Silver usually occurs at such low concentrations in natural water that it is not necessary to specify an upper limit. Zinc concentrations rarely exceed 0.01 mg/l in groundwater from nonmineralized areas, but values of 345 mg/l have been observed in mine water (Hem, 1970). Zinc is also given off by galvanized pipe. The upper limit for zinc in drinking water is 5 mg/l, primarily for reasons of taste. Excessive metal concentrations often are found in groundwater from mineral-rich areas. Klusman and Edwards (1977), for example, reported that groundwater from the Front Range mineral belt between Leadville and Boulder, Colorado, could be expected to exceed U.S. Public Health Service maximum limits in 14 percent of the samples for Cd, 1 percent for Cu, 51 percent for Fe, 74 percent for Mn, 2 percent for Hg, and 9 percent for Zn.

Radioactive Materials

Most of the natural radioactivity of groundwater is caused by ^{238}U, ^{232}Th, ^{235}U, and, to a lesser extent, by ^{40}K and ^{87}Rb. These nuclides have such long half-lives (4.5×10^9 years for ^{238}U) that they could have been part of the original matter that formed the earth. For this reason, they are called *primordial* nuclides. Radioactive daughter products include ^{226}Ra, which is derived from ^{238}U and which in turn disintegrates into ^{222}Rn. Both products are strongly radioactive. Other radionuclides in groundwater were originally formed in the atmosphere by cosmic-ray bombardment of ^{14}N, ^{16}O, and ^{40}Ar. The best-known products of this activation are ^{3}H and ^{14}C, which have been used to trace and date groundwater (see Section 10.6). A third source is nuclear explosions, where ^{90}Sr and ^{137}Cs are the most hazardous radionuclides in the fallout. Fortunately, however, these ions are readily adsorbed in soils with clay or organic matter. Contamination of groundwater by these nuclides is likely, however, where coarse aquifers crop out, or where fractured or cavernous rock is directly exposed to the atmosphere.

Naturally radioactive groundwaters are found in several areas, especially where there is geothermal activity (Scott and Barker, 1962). Uranium concentrations usually range between 0.0001 and 0.01 mg/l. Radium concentrations of more than 3.3 pCi/l have been found in water from deep aquifers in Iowa (1 curie is the number of disintegrations per second of one gram of radium, or 3.7×10^{10} disintegrations per second). For additional information on radionuclides in groundwater, reference is made to Davis and DeWiest (1966). Standards for drinking water are presented in Section 10.5.

Dissolved Gases

Rainwater contains dissolved gases at concentrations that are in approximate equilibrium with the gas composition of the atmosphere. When rainwater percolates through soil, the dissolved oxygen will be essentially depleted by bacteria and oxidation reactions. Thus, most groundwater is low in dissolved oxygen. Other gases, like CO_2, CH_4, and H_2S, can be generated in the soil or in deeper formations. Some of these gases can give groundwater a bad taste. Noble gases in groundwater have been used to determine the history of the water (see Section 10.6).

Bacteria and Viruses

Natural groundwater from all but very shallow aquifers is considered free from pathogenic bacteria and viruses, and fortunately for the many consumers of marginally treated groundwater (more than 60 million in the United States alone according to Allen and Geldreich, 1975) this is usually true. However, the general absence of pathogens in natural groundwater creates a false sense of security which periodically is brought to the fore when groundwater in aquifers or in distribution systems is contaminated with polluted water and when lack of chlorination or other disinfection causes disease outbreaks. Microbial life in aquifers, however, is entirely possible, and nonpathogenic microorganisms native to the aquifer may show up in well discharges.

Surface soils, of course, contain a myriad of microorganisms, but their numbers decrease rapidly below the root zone. Most studies on subsurface microbiological activity have been carried out on surface soils, and only meager data are available for deeper regions. Physical and chemical properties of deep subsurface environments were examined by NcNabb and Dunlap (1975), who concluded that microbiological activity can and will exist in many subsurface regions. Temperatures favorable for microbial life, for example, extend to a depth of about 2000 m (assuming a normal temperature increase of 3°C per 100 m). Water pressures to this depth are not high enough to deter microbial activity, and many bacteria can live under the high osmotic pressures of saline water. Carbon sources necessary for microbial life are carbonates and other inorganic carbon that is present in most underground materials. Sedimentary deposits also contain organic carbon, mostly as bitumens and other stable, humus-type substances that are only slowly degraded by bacteria. Other essential elements like N, P, S, and trace elements are generally also present in underground materials. Oxidation-reduction levels often are within the tolerance range of bacteria, except perhaps in deep, isolated formations which may be too reduced for microbial growth. Since molecular oxygen is usually absent in deeper regions, anaerobic bacteria will prevail. These bacteria use sulfate, carbon dioxide, nitrate, and simple organic compounds rather than oxygen as electron acceptors in their metabolism.

The study of microbial activity in aquifers and other deep formations is difficult because it is essentially impossible to drill a sterile well. Willis et al. (1975) isolated various bacteria in water pumped from a saline aquifer, but concluded that while the bacteria may be native to the aquifer, the possibility of many of the bacterial genera having been introduced by the drilling of the well cannot be precluded. Spring water could give accurate data for biological conditions in the producing aquifers, but these are usually relatively shallow and not indicative of what may happen in deeper formations. Spring water also is easily polluted, and sampling should be done carefully. The best way to study microbial activity in deep formations is to take core samples during drilling. The outer, contaminated portions of the samples are then removed, leaving the inner, "untouched" portions for bacterial analyses. Despite difficulties of bacterial contamination, the results of such studies often show indigenous bacterial activity, and it is generally accepted that varied and active microfloras exist in deep formations (McNabb and Dunlap, 1975). The bacteria include sulfate reducers, denitrifiers, methane formers, sulfur oxidizers, and hydrocarbon utilizers.

Although bacteria normally grow on solid surfaces, they can break loose and show up in the groundwater pumped from the aquifers. Smith et al. (1976), for example, reported an average of 1 291 denitrifiers per liter in groundwater from 16 wells in the Chalk of London aquifer (the range was 0 to 17 200 per liter). Sulfate-reducing bacteria could be detected in water only from two wells and at concentrations of 14 and 200 per liter. McCabe et al. (1970) found that 60 percent of the water samples from 621 wells had a plate count of aerobic bacteria of 1 000 to 100 000 per liter, 7 percent exceeded a count of 1 000 000 per liter, and 17 percent had a count of zero. While these bacteria normally do not have a direct health significance, they do show that groundwater is not "sterile." Coliform bacteria were found in 9 percent of the samples and fecal coliforms in 2 percent. Presence of coliform bacteria was attributed mostly to contamination of the water resulting from poor construction of the wells.

The origin of bacteria in aquifers is not known. They could have been deposited with the sediments millions of years ago, or they may have migrated recently into the formations with meteoric water or during construction of wells. Bacteria generally do not move large distances in fine-textured soil (less than a few meters, for example), but they can migrate much larger distances in coarse-textured or fractured materials (Romero, 1970; see also Section 11.4). Such materials are vulnerable to bacterial contamination by surface water (especially sewage effluent, leachate from garbage dumps, and other polluted water) and animal feces when exposed to the surface without being protected by a relatively fine textured soil. Pathogenic bacteria, viruses, and other microorganisms not native to the subsurface environment generally do not multiply underground and eventually die (see Section 11.4).

Oxygen Demand and Organic Matter

The biochemical oxygen demand (BOD) is a measure of the biodegradable material in the water. It is determined by incubating a water sample and measuring the decrease in dissolved oxygen as bacteria decompose this material. Since groundwater is at the end of the chain of microbiological processes in surface soils and deeper layers, there is little or no residual biodegradable material and groundwater BODs are essentially zero. Organic matter in peat and muck deposits is relatively stable and will produce low BOD levels in groundwater, even though the total organic carbon content may be relatively high. The chemical oxygen demand (COD, determined by chemical oxidation of a water sample with dichromate or permanganate) is not a useful parameter of organic matter in groundwater either, because reduced inorganic chemicals like ferrous iron and manganese also contribute to the chemical oxygen demand. Common techniques for assessing organic materials in groundwater are the carbon-chloroform and carbon-alcohol extract methods (CCE and CAE, respectively). The most accurate way is to determine the total organic carbon (TOC) content of the water with a special carbon analyzer. Individual organic compounds can be identified with chromatography techniques. Most natural and uncontaminated groundwater can be expected to contain traces of stable organic carbon, which is probably associated with humic-type acids in the soil.

Color

Color of water (expressed in units on the platinum-cobalt scale) is used primarily as an indicator of organic compounds in the water. Light- to dark-brown discolorations have been observed in groundwater exposed to peats or other organic deposits. Most of this color is probably caused by humic- and fulvic-type acids and protein-lignin compounds that comprise the stable organic matter (humus) in the soil. Brownish discoloration can also occur when groundwater with dissolved ferrous iron is exposed to the atmosphere and insoluble ferric hydroxides are formed.

Turbidity

Turbidity of groundwater, usually expressed in terms of reduced light transmission by the water (Jackson turbidity units or JTUs) or as suspended-solids content, is primarily caused by clay, silt, and other fines that enter the well from

Table 10.2 Examples of groundwater with relative high concentrations of one or more chemical constituents, in mg/1

	Thermal		Igneous and crystalline rock							Sedimentary rock						Unconsolidated rock, alluvial fill, glacial outwash								
Sample no.	1	2	3	4	5	6	7	8	9	10	11	12	13	14	15	16	17	118	119	20	21	22	23	24
Aluminum	0.2	0.22		1.3		0.2					28			0.2	0.6					0.1	1.2			
Arsenic	1.5	4.0																						
Bicarbonate (HCO_3)		312	220	296	69	121	133	320	2080	285	0	146	241	241	30	85	202	153	161	101	100	402	412	440
Boron	4.4	48	0.08	4.6					0.4														2.5	
Bromide	1.5	1.5																						
Calcium	0.8	3.6	32	5	17	28	96	88	3	60	424	46	140	35	8.4	277	49	92	32	58		64		126
Carbonate (CO_3)		0	0	10	0				57		0											0	30	
Chloride	405	874	7.9	34	1.1	1.0	25	13	71	12	380	3.5	38	1	1.8	605	246	205	12	39	2.0	30	9.5	8.0
Color (units)			10		5		3	2				5		10	3		2		2		23		1	1
Copper						0.01				0.01										0.01				
Fluoride	25	2.6	0.2	0.8	1	0.1	0.4	0.3	2	0.5	1.8	0.0	0.8	0.9	0.1	0.2	0.1	0.6	0.7	0.0	0.1	0.1	1.7	0.7
Hardness (as $CaCO_3$)	2	9	129	17	49	78	318	250	38	337	1860	132	526	224	27	954	196	386	116	198	15	238	15	490
Iodide	0.3	0.6																						
Iron	0.06	0.52	0.01	0.1	0.33	2.7	1.0	0.02	0.15	0.37	0.88	0.04	0.01	0.39	11		0.00		0.0	0.04	2.9	0.28	0.2	2.3
Lithium	5.2	7																						
Magnesium	0.0	0.0	12	1	1.7	1.9	19	7.3	7.4	31	194	4.2	43	33	1.5	64	18	38	8.8	13	2.0	19	2.1	43
Manganese	0.0	0.00	0.00	0.00	0.00	0.22	0.01	0.00		0.05	9.6				0.32									0.00
Nitrate (NO_3)	1.8	2.7	2.9	0.7	0.0	0.2	0.4	4.6	0.2	0.8	3.1	7.3	4.1	1.2	0.4	35	2.2	83	0.6	0.6	0.6	60	0.6	0.2
pH (units)	9.6	8.9	7.8	8.5	7.1	6.9	7.8	7.5	8.3	7.6	4	7	7.4	8.2	6.3		7.7		7.9	7.0	7.4	7.4	8.7	7.6
Phosphate (PO_4)	1.3	0.24		2.6		0.0					0.0									0.1				
Potassium	24	65	5.2	1.2	-	4.2	1.5	2.8	2.4	4	11	0.8	-	1.3	3.6	-	-	-	-	2.8	1.7	9.5	-	2.1
Selenium																								
Silica (SiO_2)	363	314	49	10	29	31	15	24	16	8.7	98	8.4	13	18	7.9	74	22	23	71	10	12	27	22	20
Sodium	352	660	30	136	7.4*	6.8	18	19	857	12	416	1.5	21*	28	1.5	53*	168*	110*	42*	23	35	114	182*	13
Strontium		0.67								52														
Sulfate (SO_4)	23	108	11	10	6.9	1.4	208	6.7	1.6	111	2420	4	303	88	5.9	113	44	137	54	116	5.6	74	3.5	139
Total diss. solids	1310	2360	257	363	98	137	468	322	2060	440	4190	139	701	329	44	1260	651	764	310	338	101	578	452	571
Zinc																				0.01				

* Sodium concentration includes potassium.

Source: From Hem, 1970; see text for well or spring details.

the aquifer. This kind of turbidity points to poor development of the well or to screen or slot openings that are too large. Water from wells in fractured or cavernous rock exposed to the surface, or from springs which discharge shallow groundwater, can be muddy after rainy periods. Oxidation of dissolved ferrous iron and manganese to insoluble forms also contributes to turbidity, as do rust and other corrosion products from screens and casings, and encrustation products that break loose in the well. A milky appearance of the water discharged by a well may be caused by cascading water inside the well. Bubbles of entrained air are then "homogenized" by the well pump into very fine air bubbles that remain suspended for several minutes in a water sample that is left standing.

Examples

Chemical analyses of groundwaters that have a relatively large amount of one or more elements are presented in Table 10.2 (data taken from Hem, 1970). Sources of the groundwater samples were:

Sample no.	Source
1	Hot spring, Yellowstone National Park, Wyoming, temperature 94°C (high in silica and other elements)
2	Thermal well, 227 m deep. Washoe County, Nevada, bottom temperature 186°C (high in silica, arsenic, and other elements)
3	Well, 232 m deep, Umatilla County, Oregon; water from basalt of Columbia River Group (high in silica)
4	Well, 46 m deep, Lane County, Oregon; water from Fisher Formation (high in metals and other constituents)
5	Well, 122 m deep, Burke County, North Carolina; water from mica schist (high in silica)
6	Well, 51 m deep, Baltimore County, Maryland; water from granitic gneiss (high in aluminum and manganese)
7	Well, 37 m deep, Williamanset, Massachusets; water from Portland Arkose (high in calcium)
8	Well, 30 m deep, Rice County, Arkansas; water from Dakota Sandstone (high in calcium)
9	Well, 152 m deep, Richland County, Montana; water from sandstone and shale (high in sodium)
10	Well, 581 m deep, Waukesha, Wisconsin; water from sandstone (high in metals)
11	Well, 6.7 m deep, Drew County, Arkansas; water from shale, sand, and marl (high in aluminum and manganese, also contained 1.7 pCi/l radium and 0.017 mg/l uranium)
12	Spring, Huntsville, Alabama; water from Tuscumbia Limestone (high in calcium)
13	Well, 257 m deep, Chaves County, New Mexico; water from San Andres Limestone (high in magnesium)
14	Well, 152 m deep, Milwaukee County, Wisconsin; water from Niagara Dolomite (high in magnesium)
15	Well, 64 m deep, Fulton, Mississippi; water from Tuscaloosa Formation (high in iron)
16	Well, Maricopa County, Arizona; water from alluvial fill (high in calcium and nitrate)
17	Well, 152 m deep, Maricopa County, Arizona; water from alluvial fill (high in sodium)
18	Well, 84 m deep, Maricopa County, Arizona; water from alluvial fill (high in nitrate)
19	Average of seven wells, 76–218 m deep, Albuquerque, New Mexico; water from alluvial fill (high in silica)

(continued)

Sample no.	Source
20	Two radial collector wells, 16 m deep, Parkersburg, West Virginia; water from sand and gravel (high in manganese)
21	Well, 399 m deep, Memphis, Tennessee; water from Willcox Formation sand (high in iron)
22	Well, 10 m deep, Lincoln County, Kansas; water from alluvium (high in fluoride)
23	Well, 56 m deep, Raleigh-Durham, North Carolina; water from Coastal Plain sediments (high in sodium)
24	Well, 36 m deep, Columbus, Ohio; water from glacial outwash (high in iron)

10.5 WATER QUALITY STANDARDS FOR DRINKING, IRRIGATION, AND INDUSTRIAL USES

Drinking Water

Quality standards for human drinking water have been developed among others by the Public Health Service (1962) of the U.S. Department of Health, Education, and Welfare, and by the World Health Organization (Cox, 1964, and references therein). In the United States, the 1962 standards of the Public Health Service were enforced only for water used on airplanes, trains, and other interstate carriers subject to federal quarantine inspection. The standards also served as a model for state and local regulations in approving municipal and domestic water supplies. Revision and updating of the 1962 Public Health Service standards were authorized in 1974 with the passage of the Safe Drinking Water Act (Public Law 93-523), which requires the U.S. Environmental Protection Agency to establish national drinking-water standards and, effective June 24, 1977, gives implementation and enforcement responsibility to the states. As a first step, interim standards were developed by the U.S. Environmental Protection Agency (1975). The most complete and up-to-date quality criteria for drinking water were developed by the National Academy of Sciences and the National Academy of Engineering (1972). The voluminous "blue book" produced by these academies also contains water quality standards for other uses, including agricultural and industrial applications. The various standards and criteria for drinking water are summarized in Table 10.3 (see Section 10.4 for a discussion of specific health effects of the various constituents).

Quality criteria for drinking water normally are based on a water intake of 2 l per person per day. Also, the criteria are applicable to relatively unpolluted surface water as the raw water source. It is further assumed that this water receives normal treatment (coagulation, sedimentation, rapid sand filtration, and chlorination or other disinfection) before human consumption. For groundwater, no defined treatment is assumed since groundwater customarily receives no treatment other than chlorination. Thus, water quality recommendations for raw groundwater should be more restrictive than those for raw surface water, where some of the constituents are removed or reduced in concentration by the treatment. While the distinction between surface water and groundwater is not always easily drawn (for example, in the case of shallow groundwater or groundwater collected near a stream or lake), groundwater certainly should meet the quality standards in regard to bacteriological characteristics, toxic substances, and consti-

Table 10.3 Standards and criteria for drinking water in mg/l (see text for sources and references)

Substance or property	Public Health Service, 1962		EPA interim 1975	World Health Org. 1963		Nat. Acad. Sci., Nat. Acad. Eng., 1972
	Desirable max. limit	Absolute max. limit		Max. acceptable	Max. allowable	
Alkyl benzyl sulfonate (ABS, LAS, methylene—blue active substances)	0.5			0.5	1	0.5
Ammonium nitrogen						0.5
Arsenic	0.01	0.05	0.05		0.05	0.1
Barium		1	1		1	1
Cadmium		0.01	0.01		0.01	0.01
Calcium				75	200	
Chloride	250			200	600	250
Chromium (hexavalent)		0.05	0.05		0.05	0.05
Color (Pt-Co units)				5	50	75
Copper	1			1	1.5	1
Cyanide	0.01	0.2			0.2	0.2
Fluoride*	0.6–0.9	0.8–1.7	1.4–2.4			1.4–2.4
Iron (Fe^{2+})	0.3			0.3	1	0.3
Lead		0.05	0.05		0.05	0.05
Magnesium				50	150	
Magnesium and sodium sulfates				500	1000	
Manganese (Mn^{2+})	0.05			0.1	0.5	0.05
Mercury			0.002			0.002
Nitrate nitrogen†	10		10			10
Nitrite nitrogen						1
Organics:						
Carbon chloroform extract	0.2			0.2	0.5	0.3
Carbon alcohol extract						1.5
Pesticides:						
Aldrin						0.001
Chlordane						0.003
DDT						0.05
Dieldrin						0.001
Endrin			0.0002			0.0005
Heptachlor						0.0001
Heptachlor epoxide						0.0001
Lindane			0.004			0.005
Methoxychlor			0.1			1
Toxaphene			0.005			0.005
Organo phosphorus and carbamate insecticides						0.1

(continued)

Table 10.3 (*continued*)

Substance or property	Public Health Service, 1962		EPA interim 1975	World Health Org. 1963		Nat. Acad. Sci., Nat. Acad. Eng., 1972
	Desirable max. limit	Absolute max. limit		Max. accept-able	Max. allow-able	
Herbicides:						
2,4-D			0.1			0.02
2,4,5-TP (Silvex)			0.01			0.03
2,4,5-T						0.002
pH (units)				7–8.5		5–9
Phenolic compounds (as phenol)	0.001			0.001	0.002	0.000 001
Selenium		0.01	0.01		0.01	0.01
Silver		0.05	0.05			
Sulfate	250			200	400	250
Total dissolved solids	500			500	1 500	
Zinc	5			5	15	5

* Maximum fluoride levels are given in relation to annual average daily maximum air temperature, because ingestion of water increases with temperature. The following range is recommended by the National Academy of Sciences and National Academy of Engineering (1972):

    26–32°C   1.4 mg/l F
    22–26°C   1.6
    18–22°C   1.8
    15–18°C   2.0
    12–15°C   2.2
    10–12°C   2.4

† Nitrate-nitrogen limits are also expressed in terms of nitrate (10 mg/l NO_3-N corresponds to 45 mg/l NO_3).

tuents affecting taste. Some substances like mercury, for example, are not listed in the Public Health Service standards because unpolluted surface waters contain such small amounts of these materials that they are no problem. This may not be true, however, for all groundwater.

Bacterial characteristics of water are customarily evaluated by determining the concentration of coliform bacteria. This is a large group of bacteria that are readily identified and counted by membrane-filter or multiple-tube fermentation techniques (see Standard Methods, Am. Public Health Assoc. and others, 1971). Coliform bacteria occur in soils (*Aerobacter aerogenes*, for example) and in the intestinal tract of humans and warm-blooded animals (*Escherichia coli*). The *E. coli* or fecal coliforms are excreted in large numbers with the feces. Raw sewage, for example, contains millions of fecal coliforms per 100 ml. Tests can be run to determine all coliform bacteria in a water sample (yielding the total-coliform concentration), or only the fecal-coliform concentration. The latter is obtained by incubating the membrane filter or fermentation tubes at higher temperatures where the soil coliforms do not multiply. While the total-coliform concentration is a useful indicator of the general bacterial characteristics of a water sample, the fecal-coliform test is more significant as an indicator of fecal pollution of the water. *E. coli* bacteria as a group are not harmful, but certain strains can cause gastroenteritis and, in young children, diarrhea and infection of the urinary tract. The main significance of *E. coli* bacteria, however, is that their presence could indicate the presence of fecal pathogenic bacteria like salmonella, typhoid, cholera, and others that are spread via the contaminated-water route. This route also includes hepatitis and other pathogenic viruses. Documented evidence regarding water-borne disease outbreaks caused by viruses is available only for hepatitis and possibly polio (Sobsey, 1975). Most outbreaks are of the gastroenteritis type, and while sewage contamination is a likely reason, the correct etiology is difficult to determine. Testing water for specific pathogens would be more accurate than the fecal-coliform test for assessing its infectious-disease hazard. This is, however, too time-consuming and expensive for routine use.

Bacterial quality criteria for finished drinking water from public supplies require not more than 1 total-coliform organism per 100 ml as the arithmetic mean of all water samples examined per month, with no more than 4 per 100 ml in more than any one sample if the number of samples is less than 20 per month, or no more than 4 per 100 ml in 5 percent of the samples if the number of samples is more than 20 per month (U.S. Environmental Protection Agency, 1975). These coliform concentrations apply to the membrane-filter technique. Similar values are given for the multiple-tube fermentation technique. The samples should be taken at regular time intervals and at frequencies that vary from one sample per month if the public water supply serves less than 1000 people, to 11 samples per month for 10000 people, 100 samples for 100000 people, 300 samples for 1 million, and 500 samples for more than 4.69 million people (U.S. Environmental Protection Agency, 1975). Raw water that receives the customary treatment before reaching the consumer should not contain more than 2000 fecal coliforms and not more than 20000 total coliforms per 100 ml. Coliform bacteria can be present in water from aquifers that are close to the surface. Water from deep aquifers normally should be coliform-free.

Concern about viruses in drinking water has been rising in the last few years, as better techniques for virus detection became available and viruses have been identified in surface water and groundwater (see Section 11.4). Viruses are more critical than bacteria because they tend to survive longer and one virus unit (identified as one plaque-forming unit in a cell culture) may already set off an infection when ingested. By contrast, ingestion of thousands of pathogenic bacteria may be required before clinical symptoms are produced. For this reason, it has been suggested that the maximum limit of viruses in drinking water be the minimum number of viruses that can be detected with current (1977) technology. This means that drinking water should contain less than 1 virus unit per 400 to 4000 l (Gerba et al., 1975, and references therein). Disinfection by chlorination or ozonation effectively removes all viruses if done adequately and if the water is free from suspended solids that could harbor and protect the viruses against the disinfectant (White, 1975).

Since surface water is much more vulnerable to contamination by fecal matter and human pathogens than groundwater, surface water is almost always disinfected (at least in countries with high hygienic standards), while groundwater is not. It is therefore ironic that most water-borne disease outbreaks in a country like the United States are caused by drinking of nondisinfected groundwater (Craun and McCabe, 1973; Craun et al., 1976). Bacterial contamination of groundwater occurs primarily at the well (due to seepage of polluted surface water into the well) or in the distribution system. Chlorination or other disinfection of all groundwater used for drinking would be a simple way of protecting the public against such disease outbreaks.

The concentration of radionuclides in drinking water should be as low as possible, because any unnecessary exposure to radiation should be avoided. Upper limits recommended by the National Academy of Sciences and National Academy of Engineering (1972) are 0.5 pCi/l for gross alpha activity (keyed to the concentration limit for ^{226}Ra) and 5 pCi/l for gross beta activity (^{90}Sr and isotopes of radio-iodine). Higher radionuclides levels should not constitute grounds for rejection of the water, but investigations should be carried out to determine if the radioactive level of the drinking water can be reduced, and to ascertain that the total exposure of a representative population to radiation (including all pathways) does not exceed the radiation protection guidelines established by the U.S. Federal Radiation Council (National Academy of Sciences and National Academy of Engineering, 1972, and references therein). The Public Health Service standards (1962) call for maximum radioactivity limits of 3 pCi/l for ^{226}Ra and 10 pCi/l for ^{90}Sr.

The pH of finished drinking water should be close to 7, but treatment can cope with a range of 5 to 9. Beyond this range, treatment to adjust to pH 7 becomes less economical. Other quality criteria for surface water, like color, turbidity, taste, and odor, normally are not of significance to groundwater. Most groundwaters are clear, but substantial yellow to brown discoloration can be caused by organic compounds from peats and mucks. This coloring can be removed by filtering through special resins (Gulbrandson et al., 1973). Taste and odor may be caused by iron, hydrogen sulfide, dissolved organics, or excessive concentrations of some of the other chemical constituents shown in Table 10.3.

Irrigation Water

Quality standards for irrigation water are based on (1) the total salt concentration of the water as it affects crop yield through osmotic effects, (2) the concentration of specific ions that may be toxic to plants or that have an unfavorable effect on crop quality, and (3) the concentration of cations that can cause deflocculation of the clay in the soil and resulting damage to soil structure and declines in infiltration rate. The quality requirements of irrigation water vary between crops, types and drainability of soils, and climate. Thus, rigorous universal standards for irrigation water cannot be formulated, and what may be a poor water at one place could be quite acceptable somewhere else. In addition, special farming techniques have been developed to cope with saline irrigation water, including growing salt-tolerant crops, planting furrow-irrigated row crops on the side of the ridges rather than on the top so that plants are away from the zone of salt accumulation, irrigating frequently and periodically applying heavy irrigations to minimize the salt content of the soil water, installing drainage systems where the natural drainage is not adequate to keep the water table sufficiently low, and using surface irrigation techniques (furrows, borders, basins, etc.) rather than sprinkler systems. Sprinkling is avoided to prevent direct contact between the water and the plant leaves, which otherwise could cause leaf burn as salts accumulate on the leaves when the water evaporates after irrigation. For these reasons, irrigation-water standards and classification of irrigation water for various crops and soils have limited significance and must be considered in light of the entire farming situation.

Quality criteria and classification schemes for irrigation water were developed several decades ago by Doneen, Eaton, Scofield, and Wilcox (Wilcox and Durum, 1967, and references therein). Based on these schemes and further experiences, Ayers (1975) developed the set of guidelines shown in Table 10.4. The total dissolved-solids concentrations in this table are expressed in milligrams per liter, which were obtained by multiplying the electrical conductivity in millimhos, as presented in Ayers's original table, by 640. The sodium adsorption ratio (SAR) in Table 10.4 is an adjusted SAR that includes the added effects of precipitation and dissolution of calcium in the soil as related to the concentration of

Table 10.4 Guidelines for interpretation of water quality for irrigation

Problems and quality parameters	No problems	Increasing problems	Severe problems
Salinity effects on crop yield:			
Total dissolved-solids concentration (mg/l)	< 480	480–1 920	> 1 920
Deflocculation of clay and reduction in K and infiltration rate:			
Total dissolved-solids concentration (mg/l)	> 320	<320	< 128
Adjusted sodium adsorption ratio (SAR)	< 6	6–9	> 9
Specific ion toxicity:			
Boron (mg/l)	< 0.5	0.5–2	2–10
Sodium (as adjusted SAR) if water is absorbed by roots only	< 3	3–9	> 9
Sodium (mg/l) if water is also absorbed by leaves	< 69	>69	
Chloride (mg/l) if water is absorbed by roots only	< 142	142–355	> 355
Chloride (mg/l) if water is also absorbed by leaves	< 106	>106	
Quality effects:			
Nitrogen in mg/1 (excess N may delay harvest time and adversely affect yield or quality of sugar beets, grapes, citrus, avocados, apricots, etc.)	< 5	5–30	> 30
Bicarbonate as HCO$_3$ in mg/l (when water is applied with sprinklers, bicarbonate may cause white carbonate deposits on fruits and leaves)	< 90	90–520	> 520

Source: From Ayers, 1975.

$CO_3^{2-} + HCO_3^-$ (see also Section 2.3). The adjusted SAR value is calculated as

$$\text{Adjusted SAR} = \frac{\text{Na}}{\sqrt{\dfrac{\text{Ca} + \text{Mg}}{2}}} [9.4 - p(K_2' - K_c') - p(\text{Ca} + \text{Mg}) - p\text{Alk}] \quad (10.1)$$

where Na, Ca, and Mg refer to the concentrations of these ions in the irrigation water in meq/l and the other terms are shown in Table 10.5 in relation to the

Table 10.5 Values of $p(K_2' - K_c')$, $p(\text{Ca} + \text{Mg})$, and $p\text{Alk}$ for calculation of the adjusted SAR with Eq. (10.1)

Concentration Ca + Mg + Na, meq/l	$p(K_2' - K_c')$	Concentration Ca + Mg, meq/l	$p(\text{Ca} + \text{Mg})$	Concentration $CO_3 + HCO_3$, meq/l	$p\text{Alk}$
0.5	2.11	0.05	4.60	0.05	4.30
0.7	2.12	0.10	4.30	0.10	4.00
0.9	2.13	0.15	4.12	0.15	3.82
1.2	2.14	0.2	4.00	0.20	3.70
1.6	2.15	0.25	3.90	0.25	3.60
1.9	2.16	0.32	3.80	0.31	3.51
2.4	2.17	0.39	3.70	0.40	3.40
2.8	2.18	0.50	3.60	0.50	3.30
3.3	2.19	0.63	3.50	0.63	3.20
3.9	2.20	0.79	3.40	0.79	3.10
4.5	2.21	1.00	3.30	0.99	3.00
5.1	2.22	1.25	3.20	1.25	2.90
5.8	2.23	1.58	3.10	1.57	2.80
6.6	2.24	1.98	3.00	1.98	2.70
7.4	2.25	2.49	2.90	2.49	2.60
8.3	2.26	3.14	2.80	3.13	2.50
9.2	2.27	3.90	2.70	4.0	2.40
11	2.28	4.97	2.60	5.0	2.30
13	2.30	6.30	2.50	6.3	2.20
15	2.32	7.90	2.40	7.9	2.10
18	2.34	10.00	2.30	9.9	2.00
22	2.36	12.50	2.20	12.5	1.90
25	2.38	15.80	2.10	15.7	1.80
29	2.40	19.80	2.00	19.8	1.70
34	2.42				
39	2.44				
45	2.46				
51	2.48				
59	2.50				
67	2.52				
76	2.54				

Source: From Ayers, 1975; National Academy of Sciences and National Academy of Engineering, 1972; and references therein.

concentrations of Na, Ca, and Mg (Ayers, 1975; National Academy of Sciences and National Academy of Engineering, 1972; and references therein). The irrigation water will have a tendency to dissolve lime from the soil through which it moves if the term between brackets in Eq. (10.1) is less than 1, and to precipitate lime if this term is greater than 1.

In addition to the ions listed in Table 10.4, there are many other ions that can be toxic to plants, either by direct uptake or by interfering with the mobility of other ions and reducing their availability to plant roots. Recommendations for maximum concentrations of phytotoxic elements in irrigation water were presented by the National Academy of Sciences and National Academy of Engineering (1972) and are shown in Table 10.6. Plants differ in their tolerance to boron. Relative tolerances of various crops are listed in Table 10.7 (Salinity Laboratory, 1954). Little information is available on the long-term effects of radionuclide accumulation in the soil. For this reason, the permissible maximum radionuclide concentrations in irrigation water were set at the same values as those for drinking water.

Table 10.6 Recommended maximum limits in milligrams per liter for trace elements in irrigation water

	Permanent irrigation of all soils	Up to 20 years irrigation of fine textured neutral to alkaline soils (pH 6 to 8.5)
Aluminum	5	20
Arsenic	0.1	2
Beryllium	0.1	0.5
Boron—sensitive crops*	0.75	2
semitolerant crops	1	
tolerant crops	2	
Cadmium	0.01	0.05
Chromium	0.1	1
Cobalt	0.05	5
Copper	0.2	5
Fluoride	1	15
Iron	5	20
Lead	5	10
Lithium—citrus	0.075	0.075
other crops	2.5	2.5
Manganese	0.2	10
Molybdenum	0.01	0.05†
Nickel	0.2	2
Selenium	0.02	0.02
Vanadium	0.1	1
Zinc	2	10

* See Table 10.7 for boron sensitivity of crops.
† For acid soils only.
   Source: From National Academy of Sciences and National Academy of Engineering, 1972.

Table 10.7 Relative tolerance of plants to boron, listed in decreasing order of tolerance within each group

Tolerant	Semitolerant	Sensitive
Athel (*Tamarix asphylla*)	Sunflower (native)	Pecan
Asparagus	Potato	Black walnut
Palm (Phoenix canariensis)	Acala cotton	Persian (English) walnut
Date palm (*P. dactylifera*)	Pima cotton	Jerusalem artichoke
Sugar beet	Tomato	Navy bean
Mangel	Sweetpea	American elm
Garden beet	Radish	Plum
Alfalfa	Field pea	Pear
Gladiolus	Ragged Robin rose	Apple
Broadbean	Olive	Grape (sultanina and Malaga)
Onion	Barley	Kadota fig
Turnip	Wheat	Persimmon
Cabbage	Corn	Cherry
Lettuce	Milo	Peach
Carrot	Oat	Apricot
	Zinnia	Thornless blackberry
	Pumpkin	Orange
	Bell pepper	Avocado
	Sweet potato	Grapefruit
	Lima bean	Lemon

Source: From Salinity Laboratory, 1954.

Much work has been done on relating crop yield to salinity (TDS content) of soil water, which in turn is related to the salinity of the irrigation water. Because the salt content of soil water varies with soil water content (crop uptake and evaporation of water cause the soil-water content to decrease and the salt concentration of the soil water to increase), the salinity of the soil is commonly expressed as the salt concentration of the saturation extract. This extract is obtained by mixing a soil sample with enough distilled water to produce a paste that glistens in the light (Salinity Laboratory, 1954). The paste is filtered and the salt concentration of the extract is determined, usually by measuring its electrical conductivity and expressing it in millimhos per centimeter at 25°C (symbol EC_e). For most natural waters, one millimho corresponds to a TDS content of about 640 mg/l (Salinity Laboratory, 1954). The salt concentration of the saturation extract normally is about one-half that of the soil water at field capacity, and one-fourth that at the wilting point.

Most studies on the relation between crop yield and soil-water salinity were done by growing plants in pots in which different salinity levels were maintained by frequent application of water with different salt contents. An example of the results of such studies is given in Table 10.8, which shows the salinity levels of the soil water in terms of EC_e of the saturation extract whereby the yield of the plant is reduced about 50 percent. The plants are divided into groups of low, medium, and

Table 10.8 Relative tolerance of crop plants to salinity of soil water, expressed as electrical conductivity EC_e of saturation extract in millimhos

High salt tolerance	Medium salt tolerance	Low salt tolerance
Field crops:		
$EC_e = 16$	$EC_e = 10$	$EC_e = 4$
Barley (grain)	Rye (grain)	Field beans
Sugar beet	Wheat (grain)	
Rape	Oats (grain)	
Cotton	Rice	
$EC_e = 10$	Sorghum (grain)	
	Corn (field)	
	Flax	
	Sunflower	
	Castorbeans	
	$EC_e = 6$	
Forage crops:		
$EC_e = 18$	$EC_e = 12$	$EC_e = 4$
Alkali sacaton	White sweet clover	White Dutch clover
Saltgrass	Yellow sweet clover	Meadow foxtail
Nuttall alkaligrass	Perennial ryegrass	Alsike clover
Bermuda grass	Mountain brome	Red clover
Rhodes grass	Strawberry clover	Ladino clover
Rescue grass	Dallis grass	Burnet
Canada wildrye	Sudan grass	$EC_e = 2$
Western wheatgrass	Hubam clover	
Barley (hay)	Alfalfa (California common)	
Birdsfoot trefoil	Tall fescue	
$EC_e = 12$	Rye (hay)	
	Wheat (hay)	
	Oats (hay)	
	Orchardgrass	
	Blue grama	
	Meadow fescue	
	Reed canary	
	Big trefoil	
	Smooth brome	
	Tall meadow oatgrass	
	Cicer milkvetch	
	Sourclover	
	Sickle milkvetch	
	$EC_e = 4$	
Vegetable crops:		
$EC_e = 12$	$EC_e = 10$	$EC_e = 4$
Garden beets	Tomato	Radish
Kale	Broccoli	Celery
Asparagus	Cabbage	Green beans
Spinach	Bell pepper	$EC_e = 3$
$EC_e = 10$		

(continued)

Table 10.8 (*continued*)

High salt tolerance	Medium salt tolerance	Low salt tolerance
	Cauliflower	
	Lettuce	
	Sweet corn	
	Potatoes (White Rose)	
	Carrot	
	Onion	
	Peas	
	Squash	
	Cucumber	
	$EC_e = 4$	
Fruit crops:		
Date palm	Pomegranate	Pear
	Fig	Apple
	Olive	Orange
	Grape	Grapefruit
	Cantaloupe	Prune
		Plum
		Almond
		Apricot
		Peach
		Strawberry
		Lemon
		Avocado

Source: From Salinity Laboratory, 1954.

high salt tolerance, and they are listed in decreasing order of tolerance in each group. Bernstein (1964) presented similar data, but in graphical form with lines showing yield reductions of 10, 25, and 50 percent. The graph for field crops is shown in Figure 10.1. Data such as in Table 10.8 and Figure 10.1 are useful to indicate relative tolerances of crops to salinity and to select salt-tolerant crops where the TDS content of irrigation water is high. The data also indicate desirable salt concentrations in the deep-percolation water, which is important in determining proper leaching rates for maintaining a salt balance in the root zone at a salinity level that is not damaging to crops (see Section 11.7). Data such as Table 10.8 and Figure 10.1 cannot be used, however, to predict actual yield reductions in the field, because unlike in the pot experiments where a relatively uniform salt content was maintained, the salt content of the water in field root zones increases with depth. Near the surface, soil-water salinities are primarily controlled by the TDS content of the irrigation water. At the bottom of the root zone, soil-water salinity is about equal to that of the deep-percolation water which normally is 3 to 10 times saltier than the irrigation water (Section 11.7). Since plants obtain most of their water from the upper part of the root zone, actual yield reductions in the field tend to be much less than those for comparable uniform salinity levels (van Schilfgaarde et al., 1973).

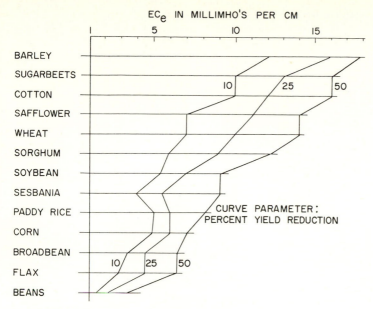

Figure 10.1 Yield reduction of various crops in relation to salt content of soil water, expressed as EC_e of saturation extract. (*Adapted from Bernstein, 1964.*)

Drinking Water for Farm Animals

Maximum concentrations of total salt and specific ions in drinking water for farm animals, as recommended in the report by the National Academy of Sciences and National Academy of Engineering (1974) are shown in Table 10.9. Water with a higher TDS level than the suggested 3 000 mg/l may be suitable for animals other than poultry. Young animals and pregnant or lactating animals are less resistant to high salt levels than old animals. Water with more than 7 000 mg/l TDS becomes increasingly risky for watering farm animals.

Industrial Water

Industrial water use is quite diverse, and water quality requirements vary greatly for different industries and even for different plants within the same industry. Often, water quality at the point of intake is not very significant because the water can be treated to the desired quality at affordable cost. Steam generation requires water that is of sufficient quality to prevent formation of scale or other deposits, corrosion, and foaming or priming (Table 10.10). Also, the silica content should be low to prevent formation of deposits on turbine blades. Cooling water should be noncorrosive and should have sufficiently low concentrations of calcium carbonate, sulfate, and phosphate to be nonscaling (Table 10.10).

Textile industries require water that is low in iron, manganese, other heavy metals, total salt, hardness, color, and turbidity (Table 10.11). Water quality requirements for the pulp and paper industry (Table 10.11) depend on the type of

**Table 10.9 Quality criteria for drinking water
for farm animals**

	Maximum concentration, mg/l
Total dissolved solids	3 000
Aluminum	5
Arsenic	0.2
Boron	5
Cadmium	0.05
Chromium	1
Cobalt	1
Copper	0.5
Fluorine	2
Lead	0.1
Mercury	0.01
NO_3-N plus NO_2-N	100
NO_2-N	10
Selenium	0.05
Vanadium	0.1
Zinc	25

Source: From the National Academy of Sciences
and the National Academy of Engineering, 1972.

products. Chemical industries vary widely as to their water quality requirements.
Often, the requirements are similar to drinking-water standards so that public
water supplies are frequently used. Raw-water quality criteria for the petroleum
industry are summarized in Table 10.11, which also lists the water quality requirements at the point of use for fruit and vegetable processing plants. Soft-drink
industries mostly use water from public supplies. For additional detail and information, reference is made to the "blue book" (National Academy of Sciences and
National Academy of Engineering, 1974).

10.6 DATING AND TRACING OF GROUNDWATER

Chemical constituents of groundwater can often be used to determine the source
of a particular groundwater, its rate of movement, and its natural recharge or
accretion rate. Dating is done primarily by measuring the concentration of certain
isotopes in the groundwater. Tracing of groundwater may utilize natural chemical
constituents peculiar to its source (tritium from nuclear explosions, for example)
or artificial tracers that are added to the groundwater at one point and monitored
for breakthrough at another point. Measurements of groundwater temperatures
have also been used as indicators of groundwater movement. Cartwright (1970),
for example, used area-wide differences of groundwater temperatures at a depth of
152 m to determine areas of vertical flow as indicators of recharge and discharge
of aquifer water.

Table 10.10 Quality requirements of water at point of use for steam generation and cooling in heat exchangers

| Characteristic | Boiler feedwater, quality of water prior to the addition of chemicals used for internal conditioning | | | | Cooling water | | | |
| | Industrial | | | Electric utilities, 102 to 340 atm | Once through | | Makeup for recirculation | |
	Low pressure, 0 to 10 atm	Intermediate pressure, 10 to 48 atm	High pressure, 48 to 102 atm		Fresh	Brackish[a]	Fresh	Brackish[a]
Silica (SiO_2)	30	10	0.7	0.01	50	25	50	25
Aluminum (Al)	5	0.1	0.01	0.01	(b)	(b)	0.1	0.1
Iron (Fe)	1	0.3	0.05	0.01	(b)	(b)	0.5	0.5
Manganese (Mn)	0.3	0.1	0.01	0.01	(b)	(b)	0.5	0.02
Calcium (Ca)	(b)	0.4	0.01	0.01	200	420	50	420
Magnesium (Mg)	(b)	0.25	0.01	0.01	(b)	(b)	(b)	
Ammonium (NH_4)	0.1	0.1	0.1	0.07	(b)	(b)	(b)	(b)
Bicarbonate (HCO_3)	170	120	48	0.5	600	140	24	140
Sulfate (SO_4)	(b)	(b)	(b)	(d)	680	2 700	200	2 700
Chloride (Cl)	(b)	(b)	(b)	(b, d)	600	19 000	500	19 000
Dissolved solids	700	500	200	0.5	1000	35 000	500	35 000
Copper (Cu)	0.5	0.05	0.05	0.01	(b)	(b)	(b)	(b)
Zinc (Zn)	(b)	0.01	0.01	0.01	(b)	(b)	(b)	(b)
Hardness ($CaCO_3$)	350	1.0	0.07	0.07	850	6 250	650	6 250
Alkalinity ($CaCO_3$)	350	100	40	1	500	115	350	115
pH (units)	7.0–10.0	8.2–10.0	8.2–9.0	8.8–9.4	5.0–8.3	6.0–8.3	(b)	(b)

(continued)

371

Table 10.10 (continued)

Boiler feedwater, quality of water prior to the addition of chemicals used for internal conditioning

| Characteristic | Industrial | | | Electric utilities, 102 to 340 atm | Cooling water | | | |
| | Low pressure, 0 to 10 atm | Intermediate pressure, 10 to 48 atm | High pressure, 48 to 102 atm | | Once through | | Makeup for re-circulation | |
					Fresh	Brackish[a]	Fresh	Brackish[a]
Organics:								
Methylene blue active substances	1	1	0.5	0.1	(b)	(b)	1	1
Carbon tetrachloride extract	1	1	0.5	(b, c)	(e)	(e)	1	2
Chemical oxygen demand (COD)	5	5	1.0	1.0	75	75	75	75
Hydrogen sulfide (H_2S)	(b)	(b)	(b)	(b)	—	—	(b)	(b)
Dissolved oxygen (O_2)	2.5	0.007	0.007	0.007	Present	Present	(b)	(b)
Temperature	(b)	(b)	(b)	(b)	(b)	(b)	(b)	(b)
Suspended solids	10	5	0.5	0.05	5000	2 500	100	100

Source: From the National Academy of Sciences and the National Academy of Engineering, 1972.

Note: Unless otherwise indicated, units are mg/l and values that normally should not be exceeded. No one water will have all the maximum values shown.

[a] Brackish water—dissolved solids more than 1 000 mg/l by definition 1963 Census of Manufacturers.

[b] Accepted as received (if meeting other limiting values); has never been a problem at concentrations encountered.

[c] Zero, not detectable by test.

[d] Controlled by treatment for other constituents.

[e] No floating oil.

Table 10.11 Quality requirements of water at point of use for fruit and vegetable processing plants, paper manufacturing plants, and textile plants, and of surface waters that have been used for petroleum industries. The numbers indicate mg/l or special units that normally should not be exceeded

	Fruits and vegetables	Paper	Textile	Petroleum
Acidity (H_2SO_4)	0			
Alkalinity ($CaCO_3$)	250	75–150	50–200	500
Aluminum oxide (Al_2O_3)			8	
Ammonia (NH_3)				40
Bicarbonate (HCO_3)				480
Calcium (Ca)	100			220
Calcium hardness ($CaCO_3$)		0–50		
Carbon dioxide (CO_2)		10		
Chemical oxygen demand (O_2)				1 000
Chloride (Cl)	250	0–200	100	1 600
Chlorine (Cl)	*	0–2		
Color, units	5	5–100	0–5	25
Copper			0.01–5	
Fluoride (F)	1†			1.2
Hardness ($CaCO_3$)	250	100–200	0–50	900
Hydrogen sulfide (H_2S)				20
Iron (Fe)	0.2	0.1–1	0–0.3	15
Magnesium				85
Magnesium hardness ($MgCO_3$)		0–50		
Manganese (Mn)	0.2	0.03–0.5	0.01–0.05	
Nitrate (NO_3)	10†			8
Nitrite (NO_2)	0			
Organics (carbon tetrachloride extractables)	0.2			
Organic growths		0		
pH, units	6.5–8.5			6–9
Silica (SiO_2)	50	20–100	25	85
Sodium and potassium (Na + K)				230
Sulfate (SO_4)	250		100	900
Suspended solids	10	10–100	0–5	5 000
Total dissolved solids	500	200–500	100–200	3 500
Turbidity, units			0.3–5	

Source: Compiled from data by the National Academy of Sciences and National Academy of Engineering, 1972.

* Process water is chlorinated to prescribed levels. Unchlorinated water is used for canning syrups.

† Low values should be used for baby food.

Carbon 14

The isotope ^{14}C is continuously formed in the upper atmosphere by the action of cosmic-ray-produced thermal neutrons on ^{14}N (Pearson and White, 1967, and references therein). It is readily oxidized to $^{14}CO_2$, which mixes with the regular CO_2 reservoir in the atmosphere and is taken up by green plants and algae for photosynthesis of plant tissue, and by microorganisms that can directly assimilate atmospheric CO_2. From there it enters the food chain and ends up in animals and humans. The ^{14}C content of plants, animals, water, and anything else that reacts directly or indirectly with atmospheric ^{14}C will be essentially constant so long as the material is active and in equilibrium with the atmospheric CO_2. When the material is cut off from the atmospheric CO_2 pool, as by death for living material or percolation to aquifers for water, the material is no longer in equilibrium with atmospheric CO_2. After that, its ^{14}C content will gradually decrease because of radioactive decay of ^{14}C and lack of replenishment with atmospheric ^{14}C. The amount of ^{14}C remaining in the material in relation to the original concentration of ^{14}C when it was cut off from the atmosphere then is an indicator of the time elapsed since the cutoff. The half-life of ^{14}C is 5 700 years (Godwin, 1962), so that it can be used to date relatively old groundwater. Ages of 20 000 to 30 000 years are commonly found, and 48 000 years has been reported in one case (Gaspar and Oncescu, 1972, and references therein; see also Section 1.3).

There are several sources of error in ^{14}C dating of groundwater. First, the ^{14}C content of atmospheric CO_2 is not constant because of changes in the sun's activity. One method that has been used to determine past ^{14}C levels in the atmosphere is to measure ^{14}C concentrations of individual rings of trees of known chronology and correcting these concentrations for the time since the ring was formed (Eddy, 1976). This is, of course, difficult if the time period is thousands of years, in which case some average atmospheric ^{14}C concentration must be used. A second source of error is mixing of water of various ages in aquifers and other underground formations. The resulting age may then not be the true age of the water but the apparent or ^{14}C age. A third source of error is dissolution of mineral carbon into the groundwater. In that case, the carbon in groundwater is not entirely derived from the atmosphere at the time of infiltration, but also from limestone and other carbonate rocks in the vadose zone and aquifer. The mineral carbon is much older than the atmospheric carbon in the water and because it contains little or no ^{14}C, it is called "dead" carbon. Without correction for dead carbon, the ^{14}C age of a particular groundwater is much older than its real age. One method to correct for dead carbon is based on the equation (Wigley, 1975)

$$A = C_d A_0 2^{-t/T_h} \qquad (10.2)$$

where A = measured ^{14}C activity in groundwater sample
$\quad A_0 = {}^{14}C$ activity in water at time zero
$\quad\quad t$ = time since water was cut off from atmospheric CO_2 reservoir
$\quad T_h$ = half-life of ^{14}C
$\quad C_d$ = correction factor for dead carbon

The factor C_d normally is taken as the fraction of modern carbon (as opposed to dead carbon) in the sample. This fraction may be evaluated from carbon chemistry, for example by estimating the amount of dead carbon from the solubility of $CaCO_3$ in relation to alkalinity and dissolved CO_2 content of the water (Wigley, 1975, and references therein). Another method for estimating C_d utilizes the stable isotopes ^{13}C and ^{12}C in the sample. Since the ratio of ^{13}C to ^{12}C in plant material differs from that of marine limestone, the ratio of ^{13}C to ^{12}C in the groundwater can be used to indicate how much of the carbon in that water is plant- or atmosphere-derived and how much is dead carbon from the limestone (Pearson and White, 1967; Smith et al., 1976). Both methods for estimating C_d, however, have been known to give erroneous results, and for this reason empirical values of C_d between 0.7 and 0.9 are frequently used (Wigley, 1975).

Tritium

Tritium (symbol ^3H or T) is a hydrogen isotope of atomic mass 3 that is naturally formed in the upper atmosphere by the action of cosmic rays on nitrogen. Tritium is a weak beta emitter with a half-life of 12.26 years. Its concentration is expressed in tritium units (TUs), which is the number of ^3H atoms per 10^{18} atoms of H. The natural ^3H level in rainfall was about 3 to 20 TUs before 1954 (Gaspar and Oncescu, 1972). Thermonuclear explosions then added considerable ^3H to the atmosphere, which resulted in the permanent entry of thermonuclear ^3H in the hydrologic cycle and caused greatly varying TU values in atmospheric precipitation. In the Northern Hemisphere, peak values of ^3H in rain and snow were reached in 1963–1964. These values ranged from 10 000 TU in the extreme northern portions (Vogel et al., 1974) to 4 000 TUs in Chicago (Poland and Stewart, 1975, and references), 1 420 TUs in the Middle East (Dincer et al., 1974), and 500 TU in India (Sukhija and Shah, 1976). After the 1963 ban on atmospheric thermonuclear explosions, ^3H levels in precipitation declined and reached a rather stable level that was on the order of 30 to 40 TUs around 1970.

The difference between the ^3H level in groundwater and rainfall enables determining the apparent age of the groundwater and classifying it as older than 50 years (no detectable ^3H), younger than 50 years but prethermonuclear (low levels of ^3H), and postthermonuclear (TU values exceeding prethermonuclear levels). The peak ^3H levels in atmospheric precipitation of the years 1963 and 1964 also make it possible to identify groundwater that infiltrated in those years, which in turn enables the calculation of natural groundwater recharge rates. Assuming, for example, downward piston-type flow in the vadose zone and an orderly vertical "stacking" of infiltrated water without vertical mixing when it becomes groundwater, the vertical TU profile will show a peak for the water that infiltrated in the 1963–1964 period. Estimating the amount of groundwater and other subsurface water above the depth of the TU peak and dividing this amount by the number of years elapsed since 1963–1964 then yields the average annual recharge rate for that period (Dincer et al., 1974; Münnich, 1968; Sukhija and Shah, 1976; Vogel et al., 1974).

The above technique, which is called the *peak method*, cannot be used if the groundwater has mixed and moved in different directions so that individual water layers can no longer be distinguished. In that case, the total tritium method can be used (Münnich, 1968; Sukhija and Shah, 1976). This method consists of measuring the total amount of ^3H in a certain body of groundwater, including water in the vadose zone, and comparing it to the total amount of ^3H that landed on the surface with atmospheric precipitation during the period in which the groundwater accumulated. The ratio of total ^3H in subsurface water to total ^3H that landed in precipitation then yields the recharge rate as a fraction of the total precipitation.

Kreitler and Jones (1975) applied ^3H dating to high-nitrate groundwater in Runnels County, Texas, and concluded that the water dated from after 1950 when an extensive program of terrace construction was initiated to increase infiltration and water holding capacity of the agricultural lands. Tritium concentrations of groundwater were used by Allison and Hughes (1975) to predict lateral inflow into a South Australian aquifer and vertical recharge by precipitation. The results agreed reasonably well with those obtained from other geohydrological studies. Poland and Stewart (1975) measured ^3H concentrations of groundwater to determine extent and rate of groundwater movement near Fresno, California.

Silicon 32

Another natural product of cosmic radiation is ^{32}Si, which enters the groundwater with precipitation. Its half-life is nearly 500 years, so that it can be used to date groundwater in the age range of 50 to 2 000 years. The use of ^{32}Si for dating, however, is restricted to situations where there has been no exchange between ^{32}Si in the groundwater and silicious minerals in the aquifer (Gaspar and Oncescu, 1972).

Oxygen 18 and Other Tracers

Various isotopes and other chemical constituents of groundwater have been used singly or in combination with ^{14}C and ^3H in a number of geohydrological investigations. For example, Mazor and Verhagen (1976) found that the ^3H content in the water from hot springs (54°C to 100°C) in Rhodesia was much lower than that in adjacent rivers, indicating that the spring water came from deep-seated, igneous sources with little mixing of surface water. Concentrations of noble gases and stable isotopes revealed, however, that the spring water was still of meteoric origin. Moser and Stichler (1975) measured concentrations of ^{18}O (deuterium) and ^3H in spring water, surface water, and precipitation as part of a geohydrological investigation to predict recharge area and recharge rate of an aquifer in Ecuador. They concluded that recharge was too low to justify exploitation of the groundwater. Tenu et al. (1975) determined the age of groundwater in a limestone aquifer in Rumania, as well as the recharge area and the average groundwater velocity, from measurements of ^{18}O, D, ^{13}C, and ^{14}C in the groundwater.

Nitrogen-isotope ratios in recent groundwater of a Texas aquifer were measured by Kreitler and Jones (1975), who concluded that the nitrate was primarily of natural origin.

Concentrations of dissolved helium and argon in groundwater can indicate the age and circulation rate of the water, at least in a qualitative way. This is because these noble gases are produced underground by radioactive decay of uranium, thorium, and potassium. Enrichment of these gases in groundwater may thus indicate relatively low circulation rates (Davis and DeWiest, 1966). Solubility of gases in water decreases with increasing temperature. Thus, dissolved argon concentrations in surface water will be higher in winter than in summer. This principle was used by Sugisaki (1961) to detect zones of summer- and winter-infiltrated water in an aquifer fed by seepage from a river. The age and location of certain groundwater zones then enabled calculation of lateral groundwater velocity.

When water evaporates, isotopes will be enriched in the water remaining behind because the various forms of water [H_2O, HDO, HTO, and $H_2(^{18}O)$] have different vapor pressures (Zimmermann et al., 1966; Dincer, 1968). This phenomenon was used by Dincer et al. (1974) to study downward movement of water in unsaturated sand dunes in the arid Middle East. While the technique was not yet sufficiently developed to make quantitative estimates of recharge rates from isotopic enrichment of water in the vadose zone, the results did establish downward water movement in the dunes as the source of aquifer recharge. Gat and Dansgaard (1972) found that groundwater in mountainous regions of Israel had about the same isotopic composition as the local precipitation water. However, groundwater in the more arid coastal plain and inland valleys was enriched in heavy isotope species relative to precipitation. Changes in isotopic composition (^{18}O and ^3H) and dissolved-salt content of water as it flowed through floodplains, a lake, dune aquifers, and interdune depressions were measured by Roche (1975) in a study of the hydrologic and salt balance of the Lake Chad system. Smith et al. (1976) measured concentrations of ^{18}O and deuterium in groundwater from the Chalk of London Basin to correlate waters of different ^{14}C contents with different recharge mechanisms.

Artificial groundwater tracers that are added to the groundwater at one point and monitored at another point to determine groundwater velocities must be readily detectable at very small concentrations. They must freely move with the groundwater and not be adsorbed, precipitated, or biodegraded in the aquifer. They must be harmless, easy to handle, inexpensive, and not already be present in the groundwater at high background levels. Radioactive tracers that have been added to groundwater to study its movement include ^3H, ^{60}Co, ^{131}I, ^{24}Na, ^{32}P, and ^{82}Br (Gaspar and Oncescu, 1972; Schmotzer et al., 1973). These isotopes are readily detected at small concentrations in the field, except ^3H, which must be determined in the laboratory with liquid scintillation detectors. The use of these tracers, however, is restricted to very low concentrations to avoid health hazards (Schmotzer et al., 1973, and references therein). This can create problems, which can be overcome by using a nonradioactive tracer that can be made radioactive for laboratory detection at low concentrations by postsampling activation. An example of such a tracer is the bromide ion (Schmotzer et al., 1973), which can be made radioactive by radiation with neutrons. Tracer concentrations of as little as 0.02 ppm over background could be detected with this technique. Nontoxic fluorescent dyes capable of being detected at the 0.001-ppm level have been used for tracing groundwater movement in fractured and cavernous rocks (Smart and Smith, 1976).

10.7 GROUNDWATER TEMPERATURE AND GEOTHERMOMETERS

Temperature

Seasonal temperature fluctuations are usually damped out below a depth of 10 to 20 m. Thus, groundwater temperatures in this zone will be essentially constant and the same as local mean annual air temperatures. In the conterminous United States, these temperatures vary from about 4°C in the northern part (10°C in the Northwest) to around 20°C in most of the southern areas (Miller et al., 1962).

The relative constancy of groundwater temperatures can be an important asset for heat-pump temperature control in buildings using groundwater for cooling in the summer and heating in the winter (Gass and Lehr, 1977).

Groundwater temperatures increase with depth because of the hot interior of the earth. This increase normally is 1 to 5°C (average about 2.5°C) per 100 m depth increase (White, 1973). There are also places where the temperature increases much faster with depth. These are called *geothermal anomalies*, and they are mostly associated with youthful mountain-building, tectonic-plate boundaries, and volcanic activity (Figure 10.2). In these areas the earth's mantle is relatively thin, and thermal springs, geysers, and fumaroles abound as surface manifestations of geothermal anomalies. Sometimes, high-temperature groundwater is also found in deep, sedimentary basins because the low thermal conductivity of the sediments has an insulating effect. An example of such groundwater is the water in the " geopressured " deep artesian sediments along the U.S. Gulf Coast (Papadopulos et al., 1975; Dorfman, 1976).

Isotope data and other chemical parameters indicate that geothermal water is primarily meteoric in origin (Fournier et al., 1974, and references therein; Mazor and Verhagen, 1976; Pearl, 1976). Cold water may move from at or near the surface through fault zones or other fissures toward the deeper hot rock and drive the lighter, heated water up to the surface, through other fractures. During this rise, the hot water may mix with cooler water from aquifers or move straight up to form hot-water springs. The various pathways and mixing possibilities are exemplified in Figure 10.3. The cold water moving down and the hot water moving up follow the way of least resistance, which may deviate considerably from the straight down-and-up route. Thus, surface manifestations of geothermal conditions may be kilometers away from the deep-seated geothermal reservoir (Pearl, 1976). If wells are to be drilled to tap the geothermal resource, considerable

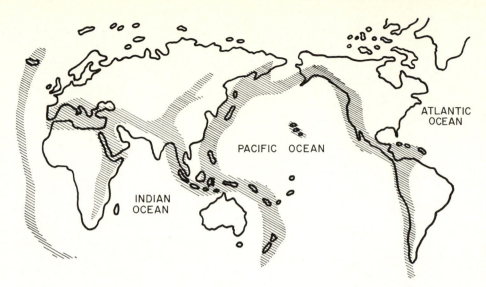

Figure 10.2 Regions of intense geothermal manifestations. *(Redrawn from Koenig, 1973.)*

geophysical investigations and test drilling are normally required to locate the geothermal reservoir.

The temperature of geothermal water from springs or wells varies from a few degrees above mean annual air temperature to that at boiling. The abundance of hot springs, however, decreases rapidly with increasing temperature of the water. Because of high pressures, the temperature of deep-seated geothermal water can be above 100°C, and values of 200 to 300°C are not uncommon. When this hot water rises slowly to the surface, most of its heat can be absorbed by the adjoining rock and the water emerges from the spring below the boiling point.

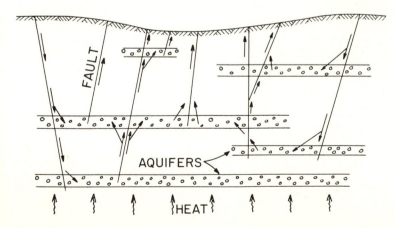

Figure 10.3 Schematic model of hydrothermal systems showing various pathways for meteoric water to seep down and for hot water to move up, with or without mixing with other groundwater. *(Adapted from Fournier and Rowe, 1966.)*

When it rises rapidly, however, heat loss is minimal and the water flashes into steam as the pressure is reduced. Geothermal water from springs or wells has been used for therapeutic purposes (health spas), space heating (homes, greenhouses), industrial applications, fish culture, and, when hot enough to flash into steam, generation of electrical power (Koenig, 1973; see Section 10.8).

Geothermometers

The temperature of deep-seated geothermal reservoirs must be known when determining their exploitability. Power generation, for example, requires reservoir temperatures of at least 180°C, unless the vapor-turbine cycle can be used (see Section 10.8). An idea of the source temperature may be obtained from the chemical composition of the hot water as it flows from springs or wells. Such "geothermometers" are based on the fact that the solubility of certain minerals increases with increasing temperature. Specifically, rock temperatures may be inferred from the SiO_2 concentration of the the hot water (silica geothermometer), from the concentrations of sodium, potassium, and calcium (Na-K-Ca geothermometer), or from both.

The silica geothermometer (Fournier and Rowe, 1966) is based on the effect of temperature on the solubility of quartz in water and on the increase in concentration of dissolved salts as part of the hot water flashes into stream when it rises to the surface. Using steam tables to predict the amount of water that will flash into steam and calculating the resulting concentration of dissolved solids in the liquid phase, Fournier and Rowe (1966) derived a relation between the SiO_2 content of the hot water at the surface and the temperature where the water was last in chemical equilibrium with the rock (Figure 10.4). The curves were calculated on the basis of cooling at constant enthalpy (no heat loss to the surrounding rock) and at constant entropy as part of the water flashed into steam. Thus, if

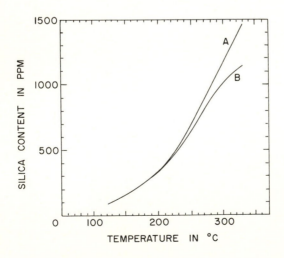

Figure 10.4 Relation between SiO_2 content of boiling water from hot springs or wells and reservoir temperature at source. Curve *A* is for cooling at constant enthalpy and curve *B* for constant entropy. (*Redrawn from Fournier and Rowe, 1966.*)

boiling water is discharged at the surface and its SiO_2 content is 700 mg/l, the underground temperature of the water at the heat source can be expected to be between 250 and 257°C, according to Figure 10.4.

The Na-K-Ca geothermometer (Fournier and Truesdell, 1973) is based on the empirical relation that the molar concentrations of Na, K, and Ca, when plotted as log $(Na/K) + \beta$ log $(Ca^{1/2}/Na)$ versus the reciprocal of the absolute temperature at which the water equilibrated with the rock, yield a straight line (Figure 10.5). The factor β is equal to $1/3$ if the water equilibrated above 100°C, and to $4/3$ if it equilibrated below 100°C. The temperature range of the last water-rock equilibration for which the straight-line relation holds is from 4 to 340°C. To apply the Na-K-Ca geothermometer, the contents of Na, K, and Ca are expressed in molar concentrations, and log $(Ca^{1/2}/Na)$ is calculated. If this number is negative, β is taken as $1/3$ for calculating log $(Na/K) + \beta$ log $(Ca^{1/2}/Na)$. If it is positive, β is taken as $4/3$ in this calculation. Next, the temperature is read from Figure 10.5. If this temperature is higher than 100°C, the value of log $(Na/K) + \beta$ log $(Ca^{1/2}/Na)$ is recalculated using $\beta = 1/3$ and the corresponding temperature is read off Figure 10.5. Otherwise, a β value of $4/3$ is used to determine the temperature. The Na-K-Ca thermometer yielded more accurate estimates of deep-seated water temperatures than the methods based on the Na/K ratio alone (Fournier and Truesdell, 1973, and references therein).

For additional detail regarding the SiO_2 and Na-K-Ca geothermometers, including assumptions, sources of error, and procedures for different flow rates, reference is made to Fournier and Rowe (1966), Fournier and Truesdell (1973), and Fournier et al. (1974). Where the hot water mixes with water from colder aquifers on its way up, the original temperature of the hot water and the fraction of the cold water in the mixture can, under favorable conditions, be estimated

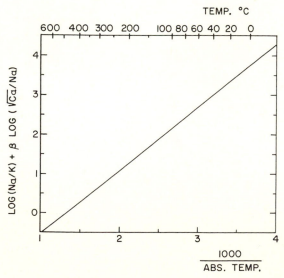

Figure 10.5 Curve for Na-K-Ca geothermometer. (*Adapted from Fournier and Truesdell, 1973.*)

from the temperature and SiO_2 content of the mixture discharged at the surface and the temperature and SiO_2 content of the water in the nonthermal aquifers (Fournier and Truesdell, 1974).

Self-sealing

When hot water rises from its deep-seated position to the surface and cools by loss of heat to the surrounding rock (low upward-flow rates) or by flashing into steam, minerals in solution tend to crystallize, particularly silica, which forms quartz, chalcedony, opal, and β cristobalite (White, 1973). These minerals can effectively seal caprock or other materials overlying the geothermal reservoir. This process is called *self-sealing* of the reservoir. Low-temperature reservoirs (less than 150°C) have little potential for self-sealing because the water does not contain enough silica. As a matter of fact, the hydraulic conductivity of low-temperature convective systems may increase as silica is dissolved by upward-moving hot water (White, 1973).

10.8 GEOTHERMAL POWER GENERATION

The use of geothermal energy for electrical-power generation is receiving increasing interest as additional energy sources must be developed. The amount of heat energy in the earth's crust is so tremendous that it staggers the imagination. Within a depth of 10 km alone (the practical limit of present drilling technology), the total amount of heat stored in all rock is estimated at 3×10^{26} cal (White, 1965), of which about one-fourth is under the continents. This is more than 5000 times the heat energy in world coal resources, more than 6×10^6 times the 1974 world energy use, and more than 6×10^7 times the 1974 world use of electrical power (U.S. National Committee of the World Energy Conference, 1974). In the United States alone, the heat stored within 10 km depth is about 8×10^{24} cal, which is equivalent to 9.3×10^{18} kWh (White and Williams, 1975).

Most of the underground heat, however, is low-grade, and only a fraction of it can be utilized. As is true for other mineral resources, heat can only be "mined" where it is concentrated, easily accessible, and present in sufficient quantity. White (1973) lists the following requirements for geothermal power development to be feasible with currently available technology:

1. Reservoir temperature at least 180°C and preferably more than 200°C
2. Reservoir depth less than 3000 m
3. Presence of natural fluids (water, brines) to transfer heat to the surface
4. Reservoir permeability sufficient to insure sustained delivery of hot fluid to wells at adequate rates
5. Adequate reservoir volume (more than 5 km³) to permit economic exploitation of the geothermal resource

These conditions occur simultaneously in only a few instances, which is the reason for the limited use of geothermal energy. Koenig (1973), for example, estimates that by 1980 the worldwide geothermal generating capacity will be about 2 500 MW. Although this is triple the 1972 capacity, it will still be less than 1 percent of the total generating capacity. Additional technology must be developed before geothermal energy can play a more significant role. Basically, there are three types of geothermal-energy systems: vapor-dominated systems, hot-water-dominated systems, and dry (hot rock) systems, as discussed in the following paragraphs.

Vapor-dominated Systems

In vapor-dominated systems, wells produce essentially dry steam, which after removal of grit and other particulate matter, is sent directly to the turbines for electrical-power generation. Vapor-dominated systems generally have deep groundwater which is heated by underlying rock, and a higher zone where water is in the vapor phase. The higher zone is often capped by material of low permeability, through which the steam can escape only via fractures or boreholes. Also, part of the steam condenses on the margins of the reservoir, where the heat is conducted away by the surrounding formations (White, 1973).

Vapor-dominated systems are the easiest to exploit, but they are also the rarest of geothermal systems. The largest of the three major systems in the world is at The Geysers, about 140 km north of San Francisco, California. The installed capacity of this system in 1975 was about 500 000 kW. By 1980, the capacity is expected to be 1 180 000 kW (Koenig, 1973). The ultimate capacity may be 2 000 000 kW, enough power for the residential needs of 2 000 000 people (assuming 1 kW per person, or 720 kWh per person per month). The geothermal field covers an area of 3 × 12 km. Steam flows upward through fractures in Jurassic-Cretacious graywackes, shales, and basalt and is tapped by wells that average about 1 500 m in depth with the deepest well going down 2 900 m. The total number of wells is 110, of which 85 produce steam (Figure 10.6). The average temperature of the steam is 250°C. The shut-off pressure at the well head varies from 24 to 32 atm (1 atm is 1.033 kg/cm^2), and the operating pressure when delivering steam is kept at 8.2 to 8.8 atm. The yield of steam increases with decreasing operating pressure at the well head. However, the amount of electrical energy generated per kilogram of steam decreases with decreasing steam pressure. Thus, optimization of operating pressure and steam yield is necessary to produce maximum power per well (Budd, 1973).

The average yield for the deeper wells is 70 000 kg of steam per hour per well. Since 9 kg steam are required to generate 1 kWh at the prevailing operating pressures, one well can produce 7 778 kW of electricity (wells tapping shallower reservoirs have a lower capacity). Well yields are gradually declining as reservoir pressures are dropping and more wells are being drilled. Because of heat loss from the conveyance system, steam cannot be transported more than 2 or 3 km. This makes it necessary to build a number of small generating plants throughout the

Figure 10.6 Steam-producing well at The Geysers, California.

Figure 10.7 General aspect of geothermal field and one of the generating plants at The Geysers, California.

area (Figure 10.7) instead of one large, central plant. At The Geysers, power is generated by 11 units, which vary in capacity from 11 000 kW for the oldest unit (installed in 1960) to 106 000 kW for the latest unit (installed in 1975). Most units have a capacity of 53 000 kW.

After the steam has been expanded through the turbines, about 75 percent of it ends up as vapor in the atmosphere, including evaporation from the cooling towers (Bowen, 1973). The remaining 25 percent is condensate and is returned to the geothermal reservoir with injection wells (Figure 10.8). These wells were converted from old production wells whose yields had declined. The reason for reinjection of the condensate is twofold: prolonging the useful life of the geothermal reservoir, and preventing pollution of a nearby stream in which the condensate with its relatively high concentration of ammonia and boron (Table 10.12) would otherwise have to be discharged. The injection wells have a capacity of 3 800 to 5 500 m^3/day per well, and there is one injection well for about every 14 producing wells.

Other major vapor-dominated systems are at Larderello, Italy, and Matsukawa, Japan (Ellis, 1975). The Larderello geothermal reservoir is in Triassic-Jurassic sediments. The average temperature is 200°C (maximum 260°C), the average well depth is 600 m (maximum 1 600 m), and the generating capacity is 406 000 kW. The Matsukawa system is in Quarternary andesites and Miocene

Figure 10.8 Well for reinjecting steam condensate at The Geysers, California.

Table 10.12 Chemical constituents (mg/l) in condensate from geothermal steam at The Geysers

Alkalinity	192
Ammonia nitrogen (NH_3-N)	135
Boron	50.9
Bicarbonate (HCO_3)	234
Calcium	< 1
Carbonate (CO_3)	< 1
Chloride	3.8
Hardness (as $CaCO_3$)	< 1
Iron	< 0.02
Magnesium	< 1
Nitrate	< 0.1
pH (units)	7.8
Potassium	150
Silica (SiO_2)	< 1
Sodium	1.3
Sulfate (SO_4)	105
Total dissolved solids	513

Source: Personal communication, A. J. Chasteen, Union Geothermal Division, Union Oil Company of California, Santa Rosa, California, 21 July 1975.

sandstones. The average temperature is 220°C (maximum 270°C), the average well depth is 1 000 m (maximum 1 500 m), and the generating capacity is 20 000 kW (Ellis, 1975).

Hot-Water-dominated Systems

Liquid-dominated systems are much more abundant than vapor-dominated systems. The geothermal reservoir in liquid-dominated systems yields hot water, which flashes into steam when brought to the surface. The fraction of the water that goes into steam depends on the initial fluid pressure in the reservoir and the separation pressure at the well head. At a separation pressure of 3.4 atm, for example, 33 percent of the hot water (by weight) flashes into steam when the reservoir temperature is 300°C, 11 percent when it is 200°C, while steam may not be formed at all when the reservoir temperature is 150°C because of heat losses when the hot water is brought to the surface (White, 1973). Most operational systems yield 10 to 30 percent steam. The steam-water mixture from the well is passed through a separator, from where the steam goes to the turbine.

Since it takes about 9 kg of dry steam to produce 1 kWh (depending on reservoir temperatures), 30 to 90 kg of hot fluid are required for each kilowatt-hour. If the steam condensate is used for cooling, the total amount of liquid left is 23 to 83 kg per kilowatthour generated (calculated as fluid not flashed into steam

Table 10.13 Examples of liquid-dominated geothermal systems for power generation

Location	Geology	Average and (max.) well depth (m)	Average and (max.) temperature (°C)	Total dissolved solids in water (ppm)	Generating capacity (kW) Installed	Planned
Wairakei, New Zealand	Quaternary rhyolite, andesite	800 (2 300)	230 (260)	4 500	192 000	
Broadlands, New Zealand	Quaternary rhyolite, andesite	1 100 (2 420)	255 (300)	4 000		100 000
Cerro Prieto, Mexico	Sandstone, shales, granite	800 (2 600)	300 (370)	17 000	75 000	75 000
Ahuchapan, El Salvador	Quaternary andesite	1 000 (1 400)	230 (250)	20 000	30 000	50 000
Otake, Japan	Quaternary andesite	500 (1 500)	230 (250)	2 500	11 000	
Pauzhetsk, U.S.S.R.	Quaternary andesite, dacite, rhyolite	(800)	185 (200)	3 000	5 000	7 000
Namafjall, Iceland	Quaternary basalt	1 000 (1 400)	250 (280)	1 000	2 500	

Source: From Ellis, 1975.

from the geothermal reservoir remains on the earth's surface after it is used for power generation. This water must be disposed, for example by reinjection into the geothermal reservoir (prolonging the life of the system) or, as done at the Wairakei system in New Zealand, by discharging it into surface water. In some cases, the water may be of sufficiently good quality for reuse (Bowen, 1973). Most geothermal waters, however, have a high TDS content and high concentrations of SiO_2, Cl, B, Na, K, NH_4, Li, Rb, Cs, As, and others (Tables 10.2 and 10.13; Ellis, 1975). Utilization of geothermal wastewater for heating may cause problems because silica can accumulate in pipes and heat exchangers (Axtmann, 1975).

Details of 15 liquid-dominated systems were presented by Ellis (1975). Some of these are shown in Table 10.13. The largest system (Wairakei) contains about 60 producing wells, which yield approximately 25 percent steam. An important liquid-dominated geothermal field was recently discovered near Milford, Utah.

Stimulation techniques to increase the yield of wells in geothermal reservoirs were discussed by Ramey et al. (1973) and Austin and Leonard (1973).

Dry Systems

Dry (hot-rock) systems are geothermal areas where the heat is stored in dry rock without water or other fluid to transfer the heat to the surface. Dry systems by far contain most of the heat stored in the earth's crust to a depth of 10 km. Ewing (1973), for example, estimates that the energy potential of hot, dry rock is 100 to 1 000 times that of hydrothermal systems—enough to supply U.S. energy needs for

Figure 10.9 Geothermal areas in western United States. *(Redrawn from U.S. Geological Survey as shown by Ewing, 1973.)*

50 000 to 500 000 years at 1972 consumptive rates. The abundance of dry geothermal systems is evident from Figure 10.9, considering that most geothermal areas are of the dry type. A similar map presented by Pearl (1976) shows even greater abundance of geothermal areas.

Use of the energy in dry geothermal systems is dependent upon technology yet to be developed, primarily for creating a permeable zone in the hot rock, transferring the heat by circulating water through this zone, and generating power with the vapor-turbine cycle. The permeable zone may be created by underground explosions, hydraulic fracturing, or other stimulation techniques (Burnham and Stewart, 1973; Sandquist and Whan, 1973; Smith et al., 1973). The transfer of heat could be accomplished by sending surface water down one well, letting it flow through the hot rock, and pumping it up as hot water from another well in the permeable zone. After power generation, the hot water could be recycled through the hot rock.

Since the amount of geothermal heat available at temperatures of 100 to 180°C may be 100 times as much as the easily available heat above 180°C (White, 1973), special turbines that can be driven by steam with a temperature less than 180°C will undoubtedly play a key role in future utilization of geothermal energy for power generation. A promising development is the vapor-turbine cycle, where the turbine is not directly driven by the steam, but by isobutane or another working fluid in a closed loop (Anderson, 1973). The geothermal fluid is then used to heat the isobutane in a heat exchanger.

Economic and Environmental Aspects

The cost of generating geothermal power compares favorably with that of generating power with fossil-fuel or nuclear energy (Koenig, 1973), at least for vapor-dominated systems and liquid-dominated systems. The cost of generating power with the energy from dry, hot rock cannot yet be determined, because it depends on the cost (and success) of new technology for transferring the heat to the surface and using the vapor-turbine cycle.

Environmental impacts unique to geothermal power generation as compared to other power generation primarily result from gaseous emissions into the atmosphere, discharge of low-quality water to the surface, and possibly increased seismic activity and land subsidence. The impact of developing a geothermal field on land use should be minimal because wells and pipes do not occupy much space. With proper design and use of vegetation, much of the facilities for collecting and transporting steam can actually be hidden from the public eye.

Gaseous emissions are produced by the power plant and its cooling system, and by direct venting of steam from wells whose full yield is not needed. Such venting is preferred over a shut-down of the well because the latter often decreases the productivity of the well. Since geothermal fluid is by no means pure water, the steam is not all water vapor. At The Geysers, for example, the steam consists of 99.5 percent water (Bowen, 1973). Of the noncondensable gases, 80 percent is CO_2 and 4.5 percent is H_2S. The rest is primarily CH_4, H_2, N_2, and NH_3. The hydrogen sulfide presents the most serious environmental problem, particularly for large systems. At The Geysers, efforts are being made to reduce the H_2S content of the steam. Radioactivity of the gases is at or near natural background levels (Bowen, 1973, and references therein). At Wairakei, New Zealand, the steam from the boreholes contains 0.004 percent H_2S (by volume), 0.000 5 percent CH_4, 0.000 8 percent NH_3, 0.06 percent CO_2, and various other gases (Axtmann, 1975). While H_2S is the most noticeable, it is not causing great public dismay. The large amounts of CO_2 in the steam could be a valuable resource when used for manufacturing carbonic acid, Dry Ice, or methyl alcohol (Axtmann, 1975).

The geothermal waste fluid left after power generation presents no environmental problems at The Geysers because it is reinjected into the geothermal reservoir. In other cases, the waste fluid can be a serious threat to surface water, particularly in liquid-dominated systems where large volumes of waste fluid are produced. For the vapor-turbine cycle, the entire volume of geothermal fluid produced by the well becomes waste water if not returned to the geothermal reservoir.

Reinjection of geothermal wastewater seems to be the best solution because it puts the water back to its source, thus enhancing the capacity and life of the geothermal resource and reducing the chances for subsidence and seismic activity.

Reinjection usually presents no problem in vapor-dominated systems. Liquid-dominated systems normally have permeable geothermal reservoirs, so that injection pressures do not need to be high and the chances for hydraulic fracturing and increased seismicity are low. If the wastewater has a high silica content, however, silica could precipitate due to the lower temperature of the injected fluid. This could lead to eventual reduction of the reservoir's permeability. Also, injection of waste fluid may interfere with the operation of the geothermal reservoir (Pearl, 1976, and references therein). Thus, careful geohydrologic investigations should be made to determine if reinjection is feasible.

At Wairakei, the waste fluids (TDS concentration of 4 400 mg/l) are discharged into a stream. The average stream flow is about 70 times the average waste flow (127 m^3/s versus 1.8 m^3/s), but the disposal still produces significant increases in the concentration of various chemicals; for example, B is increased by 0.27 mg/l, As by 0.039 mg/l, and Hg by 1.5×10^{-6} mg/l (Axtmann, 1975). At low stream flows, As concentrations could exceed the limit for drinking water. In addition, silica in the waste fluid precipitates in the discharge canals, requiring periodic removal with pneumatic equipment. The concentration of H_2S in the waste fluid is high enough to be potentially dangerous to fish in the receiving stream (Axtmann, 1975). Waste brines from the Cerro Prieto system in Mexico are discharged into large evaporation basins. Thermal pollution of surface water by cooling water can be more serious for geothermal than for fossil-fuel or nuclear power plants, because geothermal plants produce more waste heat per kilowatt-hour than other plants (Axtmann, 1975, and references therein).

Seismicity and land subsidence normally are of no concern in vapor-dominated systems because the reservoir usually consists of consolidated rock with enough integrity to withstand pressure reductions in the reservoir as steam is withdrawn. They could be a problem, however, in liquid-dominated systems where the formations may be unconsolidated and large amounts of fluid are withdrawn and not reinjected. One of the few places for which subsidence has been reported is the Wairakei system, where subsidence rates at the point of maximum deflection are about 0.40 m/year and the total subsidence since operation of the field began is 4 m (Axtmann, 1975). The total affected area is about 65 km^2. Subsidence was still going on after the pressures in the geothermal reservoir had stabilized, indicating delayed reactions or possibly other mechanisms (Axtmann, 1975, and references therein). Increases in seismic activity due to geothermal power development usually are not of great concern because most geothermal areas are in regions that are already seismically active.

To put the environmental impacts of geothermal power plants in proper perspective, they should be compared with the total environmental impact of other systems. For fossil-fuel plants, this includes environmental effects of mining and transporting coal or producing and transporting fuel oil or natural gas. For nuclear plants, it includes mining and enrichment of uranium and disposal of radioactive wastes. When these factors are taken into consideration, geothermal plants may well have a lower total environmental impact. The cleanest power plants, of course, are hydroelectric plants.

PROBLEMS

10.1 Study the references by Davis and DeWiest (1966) and Hem (1970) and obtain additional information on groundwater quality and radionuclides in groundwater.

10.2 Determine the suitability of the waters listed in Table 10.2 for drinking, irrigation, and electrical-power plants.

10.3 Calculate the thermal gradient in the earth's upper crust for the thermal waters in the first two columns of Table 10.2 and compare with the normal temperature gradient.

REFERENCES

Allen, M. J., and E. E. Geldreich, 1975. Bacteriological criteria for groundwater quality. *Ground Water* **13**(1): 45–51.

Allison, G. B., and M. W. Hughes, 1975. The use of environmental tritium to estimate recharge to a South-Australian aquifer. *J. Hydrol.* **26**: 245–254.

Am. Publ. Health Assoc., Am. Water Works Assoc., and Water Poll. Contr. Fed., 1971. *Standard Methods for the Examination of Water and Waste Water*, 13th ed. Am. Public Health Assoc., Washington, D.C., 874 pp.

Anderson, J. H., 1973. The vapor-turbine cycle for geothermal power generation. In *Geothermal Energy*, P. Kruger and C. Otte (eds.), Stanford University Press, Stanford, Cal., pp. 163–176.

Austin, C. F., and G. W. Leonard, 1973. Chemical explosive stimulation of geothermal wells. In *Geothermal Energy*, P. Kruger and C. Otte (eds.), Stanford University Press, Stanford, Cal., pp. 269–292.

Axtmann, R. C., 1975. Environmental impact of a geothermal power plant. *Science* **187**: 795–803.

Ayers, R. S., 1975. Quality of water for irrigation. *Proc. Irrig. Drain. Div., Specialty Conf., Am. Soc. Civ. Eng.* August 13–15, Logan, Utah, pp. 24–56.

Bear, F. E., Chemistry of the Soil, 1964. Monogr. Am. Chem. Soc., Reinhold Publ. Corp., New York, 515 pp.

Behnke, J., 1975. A summary of the biogeochemistry of nitrogen compounds in ground water. *J. Hydrol.* **27**: 155–167.

Bernstein, L., 1964. Salt tolerance of plants. *U.S. Dept. of Agriculture Information Bulletin No. 283*, Washington, D.C., 24 pp.

Bowen, R. G., 1973. Environmental impact of geothermal development. In *Geothermal Energy*, P. Kruger and C. Otte (eds.), Stanford University Press, Stanford, Cal., pp. 197–215.

Bower, C. A., 1974. Salinity of drainage waters. In *Drainage For Agriculture*, J. van Schilfgaarde (ed.), Agronomy Monograph No. 17, Am. Soc. Agron., pp. 471–487.

Brill, W. J., 1977. Biological nitrogen fixation. *Scient. American* **236** (3): 68–81.

Budd, C. F., Jr., 1973. Steam production at The Geysers geothermal field. In *Geothermal Energy*, P. Kruger and C. Otte (eds.), Stanford University Press, Stanford, Cal., pp. 129–144.

Burnham, J. B., and D. R. Stewart, 1973. Recovery of geothermal energy from hot, dry rock with nuclear explosives. In *Geothermal Energy*, P. Kruger and C. Otte (eds.), Stanford University Press, Stanford, Cal., pp. 223–230.

Cartwright, K., 1970. Groundwater discharge in the Illinois Basin as suggested by temperature anomalies. *Water Resour. Res.* **6**: 912–918.

Cogbill, C. V., and G. E. Likens, 1974. Acid precipitation in northeastern United States. *Water Resour. Res.* **10**: 1133–1137.

Cox, C. R., 1964. Operation and control of water treatment processes. *Monograph Series No. 49*, World Health Organization, Geneva, Switzerland, 390 pp.

Craun, G. F., and L. J. McCabe, 1973. Review of the causes of waterborne-disease outbreaks. *J. Am. Water Works Assoc.* **65**: 74–84.

Craun, G. F., L. J. McCabe, and J. M. Hughes, 1976. Waterborne disease outbreaks in the U.S. *J. Am. Water Works Assoc.* **68**: 420–424.

Crawford, M. D., 1972. Hardness of drinking water and cardiovascular disease. *Proc. Nutr. Soc.* **31:** 347–353.

Davis, S. N., and R. J. M. DeWiest, 1966. *Hydrogeology.* John Wiley & Sons, Inc., New York, 463 pp.

Dincer, T., 1968. The use of oxygen-18 and deuterium concentrations in the water balance of lakes. *Water Resour. Res.* **4:** 1289–1306.

Dincer, T., A. Al-Mugrim, and U. Zimmermann, 1974. Study of the infiltration and recharge through the sand dunes in arid zones with special reference to the stable isotopes and thermonuclear tritium, *J. Hydrol.* **23:** 79–109.

Dorfman, M. H., 1976. Water required to develop geothermal energy. *J. Am. Water Works Assoc.* **68:** 370–375.

Eddy, J. A., 1976. The Maunder minimum. *Science* **192:** 1189–1202.

Ellis, A. J., 1975. Geothermal systems and power development. *Am. Sci.* **63:** 510–521.

Ewing, A. H., 1973. Stimulation of geothermal systems. In *Geothermal Energy,* P. Kruger and C. Otte (eds.), Stanford University Press, Stanford, Cal., pp. 217–222.

Fournier, R. O., and J. J. Rowe, 1966. Estimation of underground temperatures from the silica content of water from hot springs and wet-steam wells. *Am. J. Sci.* **264:** 685–697.

Fournier, R. O., and A. H. Truesdell, 1973. An empirical Na-K-Ca geothermometer for natural waters. *Geochim. Cosmochim. Acta* **37:** 1255–1275.

Fournier, R. O., and A. H. Truesdell, 1974. Geochemical indicators of subsurface temperature—Part 2, Estimation of temperature and fraction of hot water mixed with cold water. *J. Research U.S. Geol. Survey* **2**(3): 263–270.

Fournier, R. O., D. E. White, and A. H. Truesdell, 1974. Geochemical indicators of subsurface temperature—Part 1, Basic assumptions. *J. Res. U.S. Geol. Survey* **2**(3): 259–262.

Gaspar, E., and M. Oncescu, 1972. *Radioactive Tracers in Hydrology.* Elsevier Publ. Comp., Amsterdam, 342 pp.

Gass, T. E., and J. H. Lehr, 1977. Ground water energy and the ground water heat pump. *Water Well J.* **31**(4): 42–47.

Gat, J. R., and W. Dansgaard, 1972. Stable isotope survey of the fresh water occurrences in Israel and the northern Jordan Rift Valley. *J. Hydrol.* **16:** 177–212.

Gerba, C. P., C. Wallis, and J. L. Melnick, 1975. Viruses in water: the problem, some solutions. *Env. Sci. Technol.* **9:** 1122–1126.

Godwin, H., 1962. Half-life of radiocarbon. *Nature* **195:** 984.

Gulbrandsen, R., H. Klosterman, C. M. Janeck, and R. L. Witz, 1973. Removal of organic coloring from groundwaters with macroreticular resins. *Trans. Am. Soc. Agric. Eng.* **16:** 1085–1087.

Hanshaw, B. B., 1972. Natural-membrane phenomena and subsurface waste emplacement. In *Underground Waste Management and Environmental Implications,* T. D. Cook (ed.), *Am. Assoc. Pet. Geol. Memoir* **18:** 308–317.

Hem, J. D., 1970. Study and interpretation of the chemical characteristics of natural water, Second Ed. *U.S. Geol. Survey Water Supply Paper 1473,* 363 pp.

Hill, M. J., G. Hawksworth, and G. Tattersall, 1973. Bacteria, nitrosamines, and cancer of the stomach. *Br. J. Cancer* **28:** 562–567.

Ingerson, E., and F. J. Pearson, Jr., 1964. Estimation of age and rate of motion of groundwater by the C-14 method. In *Recent Researches in the Fields of Atmosphere, Hydrosphere, and Nuclear Geochemistry,* Sugawara Festival Volume, Maruzen Co., Tokyo, pp. 263—283.

Junge, C. E., 1958. The distribution of ammonia and nitrate in rain water over the United States. *Eos Trans. Am. Geoph. Union* **39:** 241–248.

Junge, C. E., 1963. *Air Chemistry and Radioactivity.* Academic Press, New York, 382 pp.

Junge, C. E., and R. T. Werby, 1958. The concentration of chloride, sodium, potassium, calcium, and sulfate in rainwater over the United States. *J. Meteorol.* **15:** 417–425.

Klusman, R. W., and K. W. Edwards, 1977. Toxic metals in ground water of the Front Range, Colorado. *Ground Water* **15**(2): 160–169.

Koenig, J. B., 1973. Worldwide status of geothermal resources development. In *Geothermal Energy,* P. Kruger and C. Otte (eds.), Stanford University Press, Stanford, Cal., pp. 15–58.

Kreitler, C. W., and D. C. Jones, 1975. Natural soil nitrate: the cause of the nitrate contamination in Runnels County, Texas. *Ground Water* **13**(1): 53–61.

Lance, J. C., 1972. Nitrogen removal by soil mechanisms. *J. Water Poll. Contr. Fed.* **44**: 1352–1361.

Lynn, R. I., and R. E. Cameron, 1973. The role of algae in crust formation and nitrogen cycling in desert soils. *Research Memor. RM 73–40*, Utah State Univ., Logan, 26 pp.

Mazor, E., and B. T. Verhagen, 1976. Hot springs of Rhodesia: their noble gases, isotopic, and chemical composition. *J. Hydrol.* **28**: 29–43.

McCabe, L. J., J. M. Symons, R. D. Lee, and G. G. Robeck, 1970. Survey of community water supply systems. *J. Am. Water Works Assoc.* **62**: 670–687.

McNabb, J. F., and W. J. Dunlap, 1975. Subsurface biological activity in relation to ground-water pollution. *Ground Water* **13**(1): 33–44.

Miller, D. W., J. J. Geraghty, and R. S. Collins, 1962. *Water Atlas of the United States.* Water Information Center, Inc., Port Washington, N.Y. (40 plates and text.)

Moser, H., and W. Stichler, 1975. Use of environmental isotope methods as a reconnaissance tool in groundwater exploration near San Antonio de Pichincha, Ecuador. *Water Resour. Res.* **11**: 501–505.

Münnich, K. O., 1968. Infiltration and deep percolation. In: *Nuclear Techniques in Hydrology*, IAEA, Vienna, Chapter V-3a, pp. 191–197.

National Academy of Sciences and National Academy of Engineering, 1972. *Water Quality Criteria 1972.* Report prepared by Committee of Water Quality Criteria at request of U.S. Environmental Protection Agency, Washington, D.C., 594 pp.

National Research Council, 1972. Accumulation of nitrate. *Nat. Acad. of Sci.*, Washington, D.C., 106 pp.

Neri, L. C., D. Hewitt, G. B. Schreiber, T. W. Anderson, J. S. Mandel, and A. Zdrojewski, 1975. Health aspects of hard and soft waters. *J. Am. Water Works Assoc.* **67**: 403–409.

Olsen, S., 1977. Nodulated non-legumes. *Crops and Soils* **29**(6): 15–16.

Page, A. L., 1973. Fate and effects of trace elements in sewage sludge when applied to agricultural lands. *U.S. Environmental Protection Agency*, Cincinnati, Ohio. (Literature review study for Ultimate Disposal Research Program.)

Papadopulos, S. S., R. H. Wallace, J. B. Wasselman, and R. E. Taylor, 1975. Assessment of onshore geopressured-geothermal resources in the northern Gulf of Mexico basin. In *Assessment of Geothermal Resources of the United States.* U.S. Geol. Survey Circ. No. 726, pp. 125–146.

Pearl, R. H., 1976. Hydrological problems associated with developing geothermal energy systems. *Ground Water* **14**(3): 128–137.

Pearson, F. J., Jr., and D. E. White, 1967. Carbon 14 ages and flow rates of water in Carrizo Sand, Atascosa County, Texas. *Water Resour. Res.* **3**: 251–261.

Piper, A. M., 1953. A graphic procedure in the geochemical interpretation of water analyses. *U.S. Geol. Survey Ground Water Note 12.*

Poland, J. F., and G. L. Stewart, 1975. New tritium data on movement of groundwater in western Fresno County, California. *Water Resour. Res.* **11**: 716–724.

Public Health Service, 1962. *Drinking Water Standards*, U.S. Dept. of Health, Education and Welfare, Washington, D.C., 61 pp.

Ramey, H. J., Jr., P. Kruger, and R. Raghavan, 1973. Explosive stimulation of hydrothermal reservoirs. In *Geothermal Energy*, P. Kruger and C. Otte (eds.), Stanford University Press, Stanford, Cal., pp. 231–250.

Robertson, F. N., 1975. Hexavalent chromium in the ground water in Paradise Valley, Arizona. *Ground Water* **13**(6): 516–527.

Roche, M. A., 1975. Geochemistry and natural ionic and isotopic tracing: two complementary ways to study the natural salinity regime of the hydrologic system of Lake Chad. *J. Hydrol.* **26**: 153–171.

Romero, J. C., 1970. The movement of bacteria and viruses through porous media. *Ground Water* **8**(2): 37–49.

Salinity Laboratory, 1954. Diagnosis and improvement of saline and alkaline soil. *U.S. Dept. of Agriculture Handbook No. 60*, 160 pp.

Sandquist, G. M., and G. A. Whan, 1973. Environmental aspects of nuclear stimulation. In *Geothermal Energy*, P. Kruger and C. Otte (eds.), Stanford University Press, Stanford, Cal., pp. 293–313.

Schmotzer, J. K., W. A. Jester, and R. R. Parizek, 1973. Groundwater tracing with post sampling activitation analysis. *J. Hydrol.* **20:** 217–236.

Schuman, G. E., and R. E. Burwell, 1974. Precipitation nitrogen contribution relative to surface runoff discharges. *J. Environ. Qual.* **3:** 366–369.

Scott, R. C., and F. B. Barker, 1962. Data on uranium and radium in ground water in the United States, 1954–1957. *U.S. Geol. Survey Prof. Paper 426*, 115 pp.

Shiomi, M. T., and K. W. Kuntz, 1973. Great Lakes precipitation chemistry: Part 1. Lake Ontario basin. *Proc. 16th Conf. Great Lakes Res., 1973. Int. Assoc. Great Lakes Research*, pp. 581–602.

Smart, P. L., and D. I. Smith, 1976. Water tracing in tropical regions, the use of fluorometric techniques in Jamaica. *J. Hydrol.* **30:** 179–195.

Smith, M., R. Potter, D. Brown, and R. L. Aamodt, 1973. Induction and growth of fractures in hot rock. In *Geothermal Energy*, P. Kruger and C. Otte (eds.), Stanford University Press, Stanford, Cal., pp. 251–268.

Smith, D. B., R. A. Downing, R. A. Monkhouse, R. L. Otlet, and F. J. Pearson, 1976. The age of groundwater in the Chalk of London Basin. *Water Resour. Res.* **12:** 392–404.

Sobsey, M. D., 1975. Enteric viruses and drinking water supplies. *J. Am. Water Works Assoc.* **67:** 414–418.

Sugisaki, R., 1961. Measurement of effective flow velocity of ground water by means of dissolved gases. *Am. J. Sci.* **259:** 144–153.

Sukhija, B. S., and C. R. Shah, 1976. Conformity of groundwater recharge rate by tritium method and mathematical modelling. *J. Hydrol.* **30:** 167–178.

Tenu, A., P. Noto, G. Cortecci, and S. Nuti, 1975. Environmental isotopic study of the Barremian-Jurassic aquifer in South Dobrogea (Roumania). *J. Hydrol.* **26:** 185–198.

U.S. Environmental Protection Agency, 1975. National interim primary drinking water regulations, Part 141. *Federal Register* **40**(248), Dec. 24, 1975, pp. 59566–59588.

U.S. National Committee of the World Energy Conference, 1974. *Proc. World Energy Conference, Survey of Energy Resources*. New York, 460 pp.

Uttormark, P. D., J. D. Chapin, and K. M. Green, 1974. Estimating nutrient loadings of lakes from non-point sources. *U.S. Environmental Protection Agency Report 660/3-74-020*, Washington, D.C., 99 pp.

van den Berg, C., 1973. Annual report 1973. *Institute of Soil and Water Management*, Wageningen, Netherlands, 111 pp.

van Schilfgaarde, J., L. Bernstein, J. D. Rhoades, and S. L. Rawlins, 1973. Irrigation management for salt control. *Proc. Irrig. Drain. Div., Specialty Conf., Am. Soc. Civ. Eng.*, April 22–24, Fort Collins, Colo., pp. 647–672.

Vogel, J. C., L. Thilo, and M. van Dijken, 1974. Determination of groundwater recharge with tritium. *J. Hydrol.* **23:** 131–140.

White, D. E., 1965. Geothermal energy. *U.S. Geol. Survey Circ. No. 519*, 17 pp.

White, D. E., 1973. Characteristics of geothermal resources. In *Geothermal Energy*, P. Kruger and C. Otte (eds.), Stanford University Press, Stanford, Cal., pp. 69–94.

White, G. C., 1975. Disinfection: the last line of defense for potable water. *J. Am. Water Works Assoc.* **67:** 410–413.

White, D. E., and D. L. Williams, 1975. Assessment of geothermal resources in the United States. *U.S. Geol. Survey Circ. No. 726*, 155 pp.

Whitehead, H. C., and J. H. Feth, 1964. Chemical composition of rain, dry fallout, and bulk precipitation at Menlo Park, California, 1957–1959. *J. Geophys. Res.* **69:** 3319–3333.

Wigley, T. M. L., 1975. Carbon 14 dating of groundwater from closed and open systems. *Water Resour. Res.* **11:** 324–328.

Wilcox, L. V., and W. H. Durum, 1967. Quality of irrigation water. In *Irrigation of Agricultural Land*, R. M. Hagan, H. R. Haise, and T. W. Edminster (eds.), Agronomy Monograph No. 11, Am. Soc. Agron., Madison, Wis., pp. 104–122.

Willis, C. J., G. H. Elkan, E. Horvath, and K. R. Dail, 1975. Bacterial flora of saline aquifers. *Ground Water* **13**(5): 406–409.

Wolff, I. A., and A. E. Wasserman, 1972. Nitrates, nitrites, and nitrosamines. *Science* **177**: 15–19.

Zimmermann, U., D. Ehhalt, and K. O. Münnich, 1966. Soil water movement and evapotranspiration: Changes in the isotopic composition of water. *Proc. Symp. Isotopes in Hydrology, Int. A.E.C.*, Vienna, pp. 567–585.

ELEVEN

GROUNDWATER CONTAMINATION

The intensive use of natural resources and the large production of wastes in modern society often pose a threat to groundwater quality and already have resulted in many incidents of groundwater contamination. Degradation of groundwater quality can take place over large areas from plane or diffuse sources like deep percolation from intensively farmed fields, or it can be caused by point sources such as septic tanks, garbage disposal sites, cemeteries, mine spoils, and oil spills or other accidental entry of pollutants into the underground environment. A third possibility is contamination by line sources of poor-quality water, like seepage from polluted streams or intrusion of salt water from oceans.

Because groundwater tends to move very slowly, many years may elapse after the start of pollution before affected water shows up in a well. For the same reason, many years may be required to rehabilitate contaminated aquifers after the source of pollution has been eliminated. This long delay can force abandonment of wells and may require costly development of alternate water supplies. Prevention of contamination thus is the best way for protecting groundwater quality. Slow movement of groundwater, however, is a favorable factor where the contaminants are biodegradable, radioactive, or consist of bacteria and viruses that decompose, decay, or die, respectively, with time. In those cases, long underground detention times may result in essentially complete removal of the undesired substances.

Where contamination of groundwater is caused by downward movement of polluted water from the surface or from the upper vadose zone (deep percolation from agricultural fields, leachate from garbage disposal areas or septic tanks, etc.), the concentration of pollutants will be highest at the top of the groundwater. Thus, groundwater samples for assessing the severity of contamination should be taken as close to the water table as possible. Brines or other solutions with a significantly higher density than the groundwater will move to the bottom of the aquifer (Pettyjohn, 1976). After the polluted water has reached the groundwater, it tends to move laterally in the direction of groundwater flow. As the water moves away from the pollution source, the concentration of contaminants decreases due to dispersion and other attenuation effects. These processes are discussed in the next section. Major pollution sources and case histories of groundwater contamination are reviewed in the rest of the chapter.

11.1 DISPERSION AND OTHER ATTENUATION

When poor-quality water moves into a freshwater aquifer, the concentration of contaminants decreases with increasing distance of flow because of hydraulic dispersion. Longitudinal dispersion occurs in the direction of flow and is caused by different macroscopic velocities as some parts of the invading fluid move through wider and less tortuous pores than other parts. For example, poor-quality water entering at point B in Figure 11.1 will advance faster than that entering at point A. The different rates of advance produce the typical S-shaped breakthrough curves of the invading fluid at some distance from the source (Figure 11.2). Transverse dispersion occurs normal to the direction of flow and results from the repeated splitting and deflection of the flow by the solid particles in the aquifer. Transverse dispersion is effective only at the edges of a contamination source (point C in Figure 11.1). Where the invading fluid enters over a broad front (seawater intrusion, for example), the effects of transverse dispersion within the contaminated zone cancel each other, and only longitudinal dispersion needs to be considered. With point sources of pollution, however, both longitudinal and transverse dispersion are effective, yielding the typical contamination "plumes" downgradient from the source (Figure 11.3).

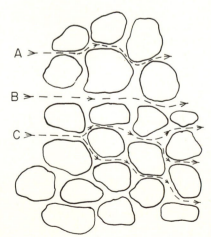

A >

B >

C >

Figure 11.1 Schematic of pathlines causing longitudinal dispersion (A and B) and transverse dispersion (C) of a foreign fluid invading a saturated porous medium.

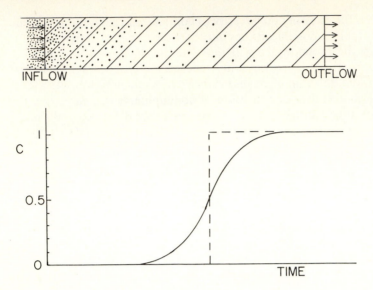

Figure 11.2 Schematic of relative concentration of salt solution invading a freshwater sand column (top), and sigmoid and piston-flow (dashed line) breakthrough curves of salt solution at outflow end (bottom).

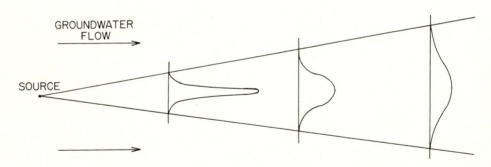

Figure 11.3 Schematic of contamination plume downgradient from point pollution source with Gaussian concentration distributions across plume.

The general differential equation describing hydrodynamic dispersion in homogeneous, isotropic media (see, for example, Bachmat and Bear, 1964, and Scheidegger, 1961) can be reduced to the following equation for two-dimensional dispersion of a fluid invading a porous medium from a line source normal to steady, one-dimensional flow in the medium (Bruch and Street, 1967; Hoopes and Harleman, 1967)

$$D_L \frac{\partial^2 C}{\partial x^2} + D_T \frac{\partial^2 C}{\partial y^2} - v_x \frac{\partial C}{\partial x} = \frac{\partial C}{\partial t} \qquad (11.1)$$

where C = relative concentration of invading fluid in the original fluid of the medium $(0 < C < 1)$

D_L = longitudinal dispersion coefficient

D_T = transverse dispersion coefficient

v_x = fluid velocity in x direction

x = coordinate in direction of flow

y = coordinate normal to flow

t = time since start of invasion

The fluid velocity often is taken as the mean pore velocity calculated from the time of travel of a tracer between two points (Oakes and Edworthy, 1976). The dimension of dispersion coefficients is length^2/time. An analytic solution of Eq. (11.1) was obtained by Bruch and Street (1967). If the invading fluid enters the porous medium over a broad front, there is no lateral dispersion and Eq. (11.1) reduces to

$$D_L \frac{\partial^2 C}{\partial x^2} - v_x \frac{\partial C}{\partial x} = \frac{\partial C}{\partial t} \tag{11.2}$$

This equation was solved analytically by Shamir and Harleman (1967a) to predict longitudinal dispersion in downward flow through a horizontally layered medium.

Dispersion coefficients are affected by velocity and pore configurations. Thus, they must be experimentally determined for a given soil or aquifer material. This can readily be done in the laboratory by letting a salt solution enter a distilled-water-saturated sample of the material in a flume or column, observing salt-concentration changes at selected points in the material, and calculating D_L and/or D_T from Eq. (11.1). Using this technique, Harleman et al. (1963) found that D_L increased exponentially with v_x and with particle size of the material, and ranged from about 0.000 2 to 0.015 cm^2/s for sands of 0.45 to 1.4 mm particle size and v_x values of 0.008 to 0.12 cm/s. D_T ranged from 0.000 1 to 0.002 cm^2/s for sands with an average particle size $(D_{50\%})$ of 0.48 to 1.67 mm and various values of v_x (Shamir and Harleman, 1967a). Values of D_L and D_T could be related to intrinsic permeability and average particle size of the sand with correlation coefficients that were primarily dependent on particle shape and particle-size distribution. Bruch (1970) derived the following expressions for dispersion coefficients in sand with $D_{50\%} = 1.205$ mm, $D_{90\%} = 1.0$ mm, uniformity coefficient = 1.17, and porosity = 39 percent:

$$D_L = 1.8vN_R^{1.205} \tag{11.3}$$

and
$$D_T = 0.11vN_R^{0.7} \tag{11.4}$$

where v = kinematic viscosity of invading fluid (absolute viscosity divided by density of fluid)

N_R = Reynolds number, calculated as $v_x D_{50\%}/v$

These relations show that diffusion coefficients vary approximately directly with v_x. The ratio D_L/v_x theoretically is a constant, called the *dispersivity*, with the dimension of length. Reported values for dispersivities include 21.3 m longitudinally and 4.27 m transversely for a glacial outwash aquifer in Long Island, New York, consisting of beds of fine and coarse sand, gravel, and silt (Pinder, 1973), and 0.6 m longitudinally for the Bunter Sandstone aquifer near Mansfield, England (Oakes and Edworthy, 1976).

An approximate solution of Eq. (11.2) is (see, for example, DeWiest, 1969, and references therein)

$$C = \frac{1}{2}\operatorname{erfc}\left[\frac{x - v_x t}{2(D_L t)^{1/2}}\right] \tag{11.5}$$

where erfc is the complementary error function and the other symbols are the same as in Eq. (11.1). Equation (11.5) yields the characteristic sigmoid breakthrough curve of a salt solution invading a distilled-water-saturated sand column at a constant flow rate (Figure 11.2). Initially, the salt concentration at the outflow end of the sand column will be zero because the salt solution has not yet traversed the entire column of sand. Then, as the vanguard of the salt solution arrives, C at the outflow begins to increase and eventually reaches a value of 1 when all distilled water in the sand column has been displaced by the salt solution.

The vertical line in Figure 11.2 represents the breakthrough of the salt solution at the outflow end if there were no dispersion and the salt solution displaced the distilled water in the sand as "piston" flow with a sharp interface between the salt and fresh water. In that case, the salt solution would reach the outflow end of the sand at full strength, causing C to jump suddenly from zero to 1. This would occur when the volume of salt solution that has invaded the sand equals the pore volume of the sand. The sigmoid breakthrough curve and the piston-flow breakthrough line theoretically intersect at $C = 0.5$, which is also the inflection point of the symmetrical, sigmoid breakthrough curve if other attenuation reactions like adsorption, diffusion, and precipitation do not occur. Shapes of breakthrough curves are affected by many factors, including particle sizes of the medium and ionic diffusion (Biggar and Nielsen, 1960; Nielsen and Biggar, 1962). These authors also studied breakthrough curves for unsaturated media.

If polluted water invades a moving body of groundwater from a point source (septic tanks, leachate from waste disposal areas, etc.), the contaminated water moves laterally in a relatively narrow band that increases in width due to transverse dispersion (Figure 11.3). The concentration distribution of contaminants perpendicular to the flow in the resulting "plume" approaches a normal gaussian curve which becomes wider and flatter with increasing distance from the point source (Bear, 1969, and references therein). The rate of divergence of the plume depends on D_T and the groundwater velocity, and may vary from only a few degrees in granular materials to as much as 20 degrees or more in fractured rock.

The equations describing hydrodynamic dispersion have been extended to include molecular diffusion, adsorption of solutes by the medium, and decay of radioactive materials (Banks and Ali, 1964; Hoopes and Harleman, 1967; Marino, 1974; Ogata, 1970; and Tagamets and Sternberg, 1974). Dispersion equations are not readily solved analytically. Thus, numerical models have been used to obtain solutions (Chhatwal et al., 1973; Lawson, 1971; Peaceman and Rachford, 1962; Pickens and Lennox, 1976; Shamir and Harleman, 1966 and 1967b; and Tagamets and Sternberg, 1974). Dispersion and transport theory were used by Bredehoeft and Pinder (1965), Gershon and Nir (1969), Hoopes and Harleman (1967), and Reddell and Sunada (1970) to predict underground movement of solutes, tracers, or contaminants. In addition to the analytic treatment of dispersion, a statistical approach has been developed which considers dispersion a random phenomenon and uses probability theory to predict concentration patterns of foreign liquids invading porous media (Bear, 1969, and references therein).

Attenuation reactions other than dispersion are filtering of suspended solids (including microorganisms), biologic decomposition of organic compounds, nitrification of ammonium, denitrification of nitrate, retention and die-off of bacteria and viruses, ion exchange, precipitation of dissolved chemicals, salt sieving through dense clay layers, and decay of radioactive materials. Runnels (1976) lists and describes 11 physical-chemical processes that have potential for purifying wastes in the underground environment. Molecular diffusion is so slow that normally it is not significant, except where groundwater is stagnant or nearly so. Simultaneous attenuation reactions have been included in transport models to predict, for example, quality changes as water passes through a root zone (Dutt et al., 1972). In contrast to hydrodynamic dispersion, which produces the same dilution for all contaminants, other attenuation reactions differ for the various substances so that the concentration of some contaminants is reduced much more than that of others. Specific attenuation reactions are discussed with more detail in the sections of this chapter on actual sources of pollution.

The reduction in the concentration of contaminants by dispersion and other attenuation reactions eventually may enable normal use of affected groundwater, after it has moved a sufficient distance from the pollution source. The magnitude of this distance depends on many factors, including type of soil and aquifer materials, groundwater velocity, type of contamination, and intended use of groundwater. "Safe" distances may vary from several tens of meters for septic-tank outflow to thousands of meters for brines or other wastes from point sources that are attenuated only by lateral dispersion, and for radioactive wastes with long half-lives.

Fine-textured soils and aquifers are more effective in removing pollutants than gravel, fractured rock, or other material with large pores. Thus, porous aquifers that crop out at the surface or are exposed by gravel pits or other excavations, or aquifers that are covered only by thin layers of soil or coarse-textured material, are much more vulnerable to contamination by surface pollution sources than aquifers that are protected by a relatively thick zone of fine-textured material. The capability of soil and aquifer materials to remove contaminants sometimes is deliberately used to purify low-quality water, as in land treatment of sewage effluent and other wastewater (see Section 11.4).

11.2 SALTWATER INTRUSION

Invasion of saline water into fresh groundwater due to groundwater withdrawal commonly occurs in coastal aquifers, where seawater moves inland if groundwater levels decline. Salt water can also move upward into fresh groundwater in aquifers (coastal or inland) underlain by saline water. When water is pumped from wells that are too close to the freshwater-saltwater interface, the salt water beneath the well will rise and enter the well. This phenomenon is called *upconing*.

Seawater Intrusion

Because seawater is heavier than fresh water, it will form a saltwater wedge in coastal aquifers that drain into the ocean (Figure 11.4). Assuming that fresh water moves horizontally to the ocean and that the saltwater-freshwater interface is abrupt and originates at the shoreline, the freshwater pressure at any point of the interface can be expressed as $(h + z)\rho_f$, where h is the height of the water table above sea level, z is the distance of the interface below sea level, and ρ_f is the density of the fresh water (Figure 11.4). This pressure must be the same as the saltwater pressure on the other side of the interface, which is $z\rho_s$ where ρ_s is the density of the salt water. Equating the two expressions and solving for z yields

$$z = \frac{\rho_f}{\rho_s - \rho_f} h \tag{11.6}$$

Since $\rho_f = 1$ g/cm^3 and $\rho_s = 1.025$ g/cm^3 for seawater, Eq. (11.6) shows that for coastal aquifers $z = 40h$. Thus, for every meter that the water table is above sea level, fresh water will extend below sea level for 40 m before salt water occurs.

If groundwater withdrawal from the coastal aquifer exceeds the safe yield and

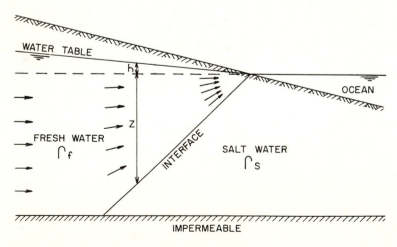

Figure 11.4 Freshwater-saltwater interface in coastal aquifer draining into ocean.

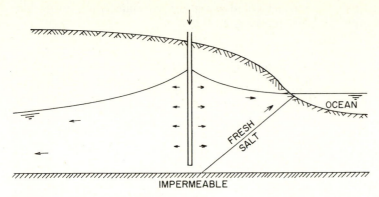

Figure 11.5 Freshwater barrier created by injection wells to prevent intrusion of seawater into a coastal aquifer with low water table.

water levels decline, the salt water will rise 40 m for every meter of drop in the water table. Thus, pumping from coastal aquifers must be managed very carefully to preserve the depth of the fresh-groundwater reservoir. Groundwater recharge for augmenting underground water supplies or for creating a freshwater ridge along the coast that forms a barrier against intruding seawater (Figure 11.5) are commonly practiced to protect fresh groundwater in coastal aquifers (see Section 8.3). Some of the fresh water used to maintain the barrier is lost to the ocean, but this amount is much less than the fresh groundwater that would be lost by uncontrolled seawater intrusion. The wedge-shaped body of fresh groundwater near the ocean is called the *Ghyben-Herzberg lens*, after the Dutch and German hydrologists who independently from each other derived Eq. (11.6) (Badon Ghyben, 1888; Herzberg, 1901).

In reality, conditions are not as simple as those assumed for Eq. (11.6). For example, the fresh water drains into the ocean over a certain area rather than at a point (Figure 11.6), and vertical flow components occur in the aquifer as the fresh water moves up along the freshwater-saltwater interface. Taking these factors into

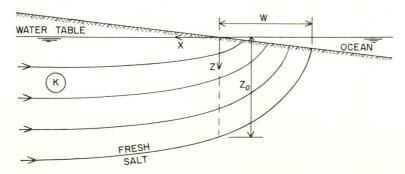

Figure 11.6 Geometry and symbols for Glover's solution of freshwater-saltwater interface.

account, Glover (1964) developed the following approximate equation for the shape of the freshwater-saltwater interface

$$z^2 - \frac{2qx}{(\rho_s - \rho_f)K} - \frac{q^2}{(\rho_s - \rho_f)^2 K^2} = 0 \tag{11.7}$$

where q = flow in aquifer per unit length of shoreline
$\quad K$ = hydraulic conductivity of aquifer
$\quad x, z$ = coordinate distances from shoreline (Figure 11.6)

For freshwater aquifers in contact with seawater, $\rho_s - \rho_f = 0.025$. Substituting $z = 0$ in Eq. (11.7) yields

$$W = \frac{q}{2(\rho_s - \rho_f)K} \tag{11.8}$$

for the width W of the bottom zone through which fresh water seeps into the ocean. The depth z_0 of the freshwater-saltwater interface beneath the shoreline is calculated as

$$z_0 = \frac{q}{(\rho_s - \rho_f)K} \tag{11.9}$$

by substituting $x = 0$ in Eq. (11.7). The interface position calculated with Eq. (11.7) is closer to the ocean than that obtained with Eq. (11.6).

Exact solutions for the shape of the saltwater front were obtained by Henry (1959; see also Pinder and Cooper, 1970) using conformal mapping and assuming a sharp interface. Pinder and Cooper (1970) developed a numerical model for predicting movement of saltwater fronts in coastal aquifers. The alternating-direction iterative procedure used in this model enabled inclusion of dispersion, transient flow, and nonhomogeneous or irregularly shaped aquifers in the solutions.

Upconing

When fresh groundwater is underlain by saline water, pumping a well in the freshwater zone causes the freshwater-saltwater interface to rise below the well (Figure 11.7). This "upconing" is in response to the pressure reduction on the interface due to drawdown of the water table around the well. If the bottom of the well is close to the saline water or the well discharge is relatively high, the saltwater cone may reach into the well, causing the well discharge to be a mixture of fresh and saline groundwater.

Assuming steady, horizontal flow of fresh water to the well, no lateral movement of salt water, and a sharp interface, the height z of the cone below the well center can be calculated in the same manner as the Ghyben-Herzberg lens, yielding

$$z = \frac{\rho_f s_w}{\rho_s - \rho_f} \tag{11.10}$$

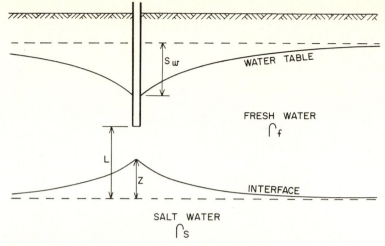

Figure 11.7 Geometry and symbols for upconing of salt water beneath a pumped well (dashed lines represent static positions of water table and interface).

where s_w = drawdown of water table at well. A more rigorous solution for upconing of saline water was developed by Bear and Dagan (1968), who presented the following equation for the rise of the cone below the center of the well (modified from the equation given by Schmorak and Mercado, 1969):

$$z_t = \frac{\rho_f Q}{2\pi(\rho_s - \rho_f)K_x L}\left(1 - \frac{2\rho_f nL}{2\rho_f nL + (\rho_s - \rho_f)K_z t}\right) \qquad (11.11)$$

where z_t = rise of cone center at time t
$\quad Q$ = well discharge
$\quad L$ = depth of freshwater-saltwater interface below well bottom prior to pumping
$\quad K_x$ = K of aquifer in horizontal direction
$\quad K_z$ = K of aquifer in vertical direction
$\quad n$ = porosity of aquifer
$\quad \rho_s$ = density of salt water
$\quad \rho_f$ = density of fresh water
$\quad t$ = time since start of pumping

For $t = \infty$, Eq. (11.11) becomes

$$z_\infty = \frac{\rho_f Q}{2\pi(\rho_s - \rho_f)K_x L} \qquad (11.12)$$

where z_∞ is the ultimate or equilibrium height of the saline-water cone below the well center. Values of z calculated with Eq. (11.11) agreed with field measurements up to some critical cone height, which generally was between $0.4L$ and $0.6L$ (Schmorak and Mercado, 1969). Similar results were obtained by Haubold (1975) using a Hele-Shaw model. When the cone height exceeded this critical value, z was no longer linear with Q and in some cases the cone reached the bottom of the well with a sudden jump, indicating conditions of instability. Saline water then entered the well, which in the studies by Schmorak and Mercado (1969) increased the salinity of the well discharge by 5 to 8 percent of the salt concentration of the saline water. Thus, where fresh groundwater is underlain by saline water, prediction of upconing is important for determining safe depths and pumping rates of wells (including "skimming" wells) that prevent entry of saline water into the well.

11.3 ROAD SALT

Another source of salt contamination of groundwater is the salt applied to snow- or ice-covered roads to provide " June driving conditions in January " (Field et al., 1974 and 1975). The bare-pavement policy of many highway departments in snowbelt states has resulted in greatly increased use of de-icing salts and less use of sand or other abrasives. The salts consist mostly of commercial rock salt and marine salt. Ferric ferrocyanide and sodium ferrocyanide are added to minimize caking of salt stocks. The sodium ferrocyanide is water-soluble and when exposed to sunlight it can generate cyanide in concentrations that are in excess of maximum limits for drinking water (Field et al., 1974, and references therein). Other additives include chromate and phosphate, which are used to reduce the corrosiveness of the salt. The chromate can produce excessive concentrations of hexavalent chromium in the melt water.

Highway salting rates generally range from 100 to 300 kg/km per application (Field et al., 1974, and references therein). In a winter season, roads may receive 10 000 kg of salt per lane per kilometer, which adds up to about 50 000 kg/km for typical highways. Chloride levels in road runoff during snow melt have been observed to range from 1 130 to 25 100 mg/l (Field et al., 1974, and references therein). Upon infiltration, this runoff and the leachate from exposed, year-around stockpiles of salt can seriously contaminate groundwater. Many such cases have been reported.

Field et al. (1974) alone cite and discuss over 20 references on the subject. In Massachusetts, for example, increases in the salt content of groundwater have been observed in more than 60 communities, forcing the abandonment of various wells. The city of Burlington, Massachusetts, suspended road salting when chloride contents in its wells began to increase to levels that eventually could exceed the maximum concentration of 250 mg/l for drinking water. In the town of Becket, the chloride content of water from a well increased to 1 360 mg/l due to salt storage upgradient from the well. For communities around Boston, road salting may eventually increase average concentrations of NaCl in groundwater from the natural 50 to 100 mg/l range to about 160 mg/l (Huling and Hollocker, 1972). Concentrations in excess of 59 mg/l are undesirable for heart patients and other persons restricted to a sodium intake of less than 1 g/day. Salt-contaminated groundwater below a salt storage pile at Reading, Massachusetts, moved at a rate of about 60 m/year and caused the chloride concentration to increase from less than 20 mg/l to 160 mg/l in water from a well located 330 m away (Pollock and Toler, 1973). Covering the salt pile reduced the rate of increase of the salt content. The state of New Hampshire before 1965 had already replaced more than 200 roadside wells, some of which yielded water with more than 3 500 mg/l of Cl. In Michigan, water from a well about 100 m from a salt-storage pile contained 4 400 mg/l of Cl. In Connecticut, taste and odors in domestic water supplies were traced to chlorides and sodium ferrocyanide from salt-storage areas (Field et al., 1974 and 1975, and references therein).

Percolation from salt piles in Monroe County, West Virginia, increased the chloride content of water from a well about 460 m away from 185 mg/l to 1 000 mg/l in 5 years (Wilmoth, 1972). Enlargement of the salt piles further increased the chloride level to 7 200 mg/l. This prompted removal of all salt piles, causing the chloride level in the well water to drop to 188 mg/l in 2 months. Chloride increases in wells at Peoria, Illinois, due to leachate from salt-storage facilities were reported by Walker and Wood (1973).

Groundwater damage by road salt is a widespread problem in snowbelt areas. The damage may be minimized by designing roads for reduced de-icing requirements and better collection and disposal of salty runoff, by designing salt-storage areas to minimize runoff and leachate infiltration, and by developing substitute materials for maintaining safe driving conditions.

11.4 SEWAGE AND SLUDGE

Sewage enters the ground intentionally from septic tanks, cesspools, and systems where sewage is applied to land for crop irrigation, groundwater recharge, or simply disposal. Unintentional entry of sewage into the underground environment include leakage from sewers and seepage from sewage lagoons (Miller et al., 1974), and seepage from streams or dry washes in which sewage effluent is discharged. Sewage sludge used as fertilizer for crops or disposed in landfills also is a potential source of groundwater contamination.

Characteristics and Composition of Sewage

The quality and thus the groundwater-pollution potential of sewage depend on the type treatment it has received before it enters the ground. Sewage that has not had any treatment is called *raw* sewage. Primary effluent is sewage that has received primary treatment, generally obtained by flowing raw sewage through a tank where everything that floats or sinks is removed. Secondary treatment is a biologic process in which bacteria are used to digest dissolved and suspended organic waste matter not removed by primary treatment. Secondary treatment is obtained by passing primary effluent through aerated tanks (activated sludge process) or by sprinkling it on deep beds of rocks (trickling filters). Tertiary treatment consists of various physical, chemical, and biochemical processes, including phosphate precipitation, ammonia stripping, denitrification, activated carbon adsorption, and chlorination or other disinfection. Conventional primary and secondary treatment each remove 40 to 50 percent of the organic waste matter expressed as BOD (biochemical oxygen demand), yielding a total BOD removal of 80 to 90 percent. Secondary effluent, however, still contains more than half of the N and P of the raw sewage. Most communities have, or will have, primary and secondary treatment. Tertiary treatment will be used where there are critical pollution problems in the receiving water or where sewage is to be reused.

Chemical characteristics of secondary sewage effluent presented by Weinberger et al. (1966) are shown in Table 11.1 as increases in concentration by one cycle of domestic water use, and as average concentrations in the effluent as such. Nitrogen increases may be higher than the values in Table 11.1. The California Department of Water Resources (1961), for example, lists a total-nitrogen increase of 20 to 40 mg/l. The same source also reports a boron increase of 0.1 to 0.4 mg/l in one cycle of domestic use.

Undisinfected secondary effluent harbors numerous microorganisms, including about 10^6 fecal coliforms and 1 to 500 virus units per 1. Disease-causing

Table 11.1 Average increases in concentration from tap water to secondary effluent, and average concentrations in secondary effluent

	Average increase, mg/l	Average concentration, mg/l
Organics	52	55
BOD	25	25
Sodium	70	135
Potassium	10	15
Ammonium-N	16	16
Calcium	15	60
Magnesium	7	25
Chloride	75	130
Nitrate-N	2	3
Nitrite-N	0.3	0.3
Bicarbonate (HCO_3)	100	300
Carbonate (CO_3)	0	0
Sulfate (SO_4)	30	100
Silica (SiO_3)	15	50
Phosphate (PO_4)	8	8
Hardness (as $CaCO_3$)	70	270
Alkalinity (as $CaCO_3$)	85	250
Total dissolved solids	320	730

Source: From Weinberger et al., 1966.

bacteria in sewage include bacteria of the Salmonella group (typhoid), Shigella (bacillary dysentery), Mycobacterium (tuberculosis), and Vibrio (cholera). Pathogenic viruses include enteroviruses, reoviruses, rotaviruses, adenoviruses, and hepatitis viruses. The latter have caused many disease outbreaks, but techniques to identify the hepatitis virus have not yet been developed. Viruses can cause a wide variety of diseases, including gastroenteritis, diarrhea, respiratory illness, paralysis, heart diseases, meningitis, liver disease, and various infections and rashes. They can also produce subclinical or latent infections. Other pathogens in sewage include protozoa, like *Endamoeba histolytica*, and helminth parasites such as ascaris and tapeworm ova (Bouwer and Chaney, 1974, and references therein).

In addition to the constituents in Table 11.1, sewage effluent contains a multitude of other chemicals, including metals, enzymes, hormones, biocides, phenols, PCBs, and other organic compounds. Sewage from industrialized areas tends to contain more metals and chemicals than that from residential areas. Even then, average metal concentrations usually are below the maximum limits for drinking and irrigation water, as shown by Mytelka et al. (1973) for sewage effluents in New York, New Jersey, and Connecticut, and by Blakeslee (1973) for effluents in Michigan (Table 11.2). Occasionally, however, excessive metal concentrations can occur which are cause for concern if not reduced by mixing. High metal concentrations normally result from industrial slug discharges and can be avoided by source control.

Most of the metals in raw sewage end up in the sludge, which is the digested solid material collected in the primary treatment phase. The digestion, which takes place anaerobically and at elevated temperatures in special digestion tanks, produces a rather stable product that can be dewatered or dried for disposal or used as fertilizer and soil conditioner. Nutrient and metal concentrations of dry sludge are shown in Table 11.3.

Table 11.2 Low, high, and median concentrations (mg/l) of metals in sewage effluent

	Mytelka et al.			Blakeslee		
	Low	High	Median	Low	High	Median
Cadmium	< 0.02	6.4	< 0.02	< 0.005	0.15	< 0.005
Cobalt	< 0.05	0.05	< 0.05			
Chromium	< 0.05	6.8	< 0.05	< 0.01	1.46	0.025
Copper	< 0.02	5.9	0.10	0.01	1.3	0.04
Mercury	< 0.000 1	0.125	0.000 9	< 0.000 2	0.007	0.002
Nickel	< 0.1	1.5	< 0.1	< 0.02	5.4	0.02
Lead	< 0.02	6.0	< 0.2	< 0.02	1.3	0.05
Zinc	< 0.02	20.0	0.12	0.03	4.7	0.19

Source: From Mytelka et al., 1973, and Blakeslee, 1973.

Table 11.3 Concentration in ppm of nutrients (ranges) and metals (mean values) in dry sewage sludge

Total N	35 000–64 000
Organic N	20 000–45 000
Phosphorus	8 000–39 000
Potassium	2 000–7 000
Cadmium	61
Copper	906
Mercury	14.5
Nickel	223
Lead	404
Zinc	2 420

Source: From a compilation by Dean and Smith, 1973.

Attenuation in Soils and Aquifers

When sewage moves through soil, the concentrations of a number of its constituents are greatly reduced, causing the wastewater to become "renovated" water. Most of the renovation takes place in the first 1 or 2 m of movement through soil where, for example, bacteria and other soil microorganisms reduce the BOD to essentially zero. BOD removal is more rapid and more complete under aerobic than under anaerobic conditions. Not all organic matter in sewage is biodegradable, however, so that the water moving down to the groundwater will contain some residual organic carbon. Concentrations of these refractory organics in the renovated water may be on the order of 5 mg/l as total organic carbon (Bouwer and Chaney, 1974, and references therein). The composition of the refractory organics has not been exhaustively studied, but carcinogens have been identified (Nupen and Hattingh, 1975). Presence of refractory organics and other potentially toxic substances (metals, for example) are the main reasons that large-scale use of renovated sewage effluent for drinking water is not yet encouraged (Ongerth et al., 1973). The upper level for total organic carbon (TOC) in reclaimed sewage water to be used for drinking has tentatively been set at 5 mg/l by the World Health Organization (Shuval, 1975, and references therein).

Another potential groundwater contaminant in sewage is nitrogen, which is primarily present in the organic and ammonium form. When the effluent moves through aerobic soil, organic nitrogen will be mineralized to ammonium which, along with the ammonium already present in the effluent, will then be converted to nitrate by nitrifying bacteria. This can produce nitrate levels in the renovated water and also in the underlying groundwater that are two to four times the maximum limit for drinking. Under anaerobic conditions, nitrification does not take place and mineralization of organic nitrogen stops at the ammonium level. Thus, the total nitrogen in the sewage effluent will then mostly be present as ammonium in the renovated water.

Nitrogen can be stored in the soil as adsorbed ammonium or as organic nitrogen in bacterial tissue or other organic matter. Removal of nitrogen from soil primarily takes place through uptake by plant roots and denitrification. The latter is an anaerobic bacterial process in which nitrate is reduced to free nitrogen gas and nitrogen oxides. Since denitrifying bacteria require organic carbon as an energy source (about 1 mg organic carbon for each milligram of nitrogen to be denitrified), denitrification occurs primarily in soils where organic matter is present and not in aquifers which usually are devoid of readily available organic carbon. Thus, once nitrate has reached groundwater, it tends to move with the groundwater without further attenuation other than dispersion. The same can be true for ammonium, which is not adsorbed in aquifers that do not contain clay.

Soil is an effective filter that removes bacteria and other relatively large microorganisms by straining and antagonistic effects of the native bacterial population. The much smaller viruses are removed by adsorption to clay, organic matter, and other negatively charged material in the soil. Since viruses consist of a nucleic-acid core with a protein coating, their physical-chemical behavior is similar to that of proteins. Thus, viruses are amphoteric. Their isoelectric point normally is at pH 5 or less. When the pH is below this point, viruses are positively charged. Above it, they are negatively charged. When positively charged, viruses are adsorbed like cations to clay and organic matter. When negatively charged, adsorption is less and, like anion adsorption, occurs as entrapment in adsorbed cations in the double layer, via cation bridges, or other mechanisms such as van der Waals forces. Virus adsorption in soil increases with decreasing pH, increasing salt content of the soil water, and increasing concentration of di- or trivalent cations relative to monovalent cations in the water (Bouwer and Chaney, 1974; Sproul, 1975; and references therein). Dissolved or suspended organic matter in the soil water competes for adsorption sites and, hence, decreases adsorption of viruses. Adsorbed viruses can be desorbed or mobilized again if water of a lower salt content passes through the soil (for example, rainwater after sewage effluent).

Pathogenic bacteria and viruses retained in the soil find themselves in a foreign, hostile environment. Thus, they do not multiply and eventually they die. Survival times for bacteria in soil normally range from a few weeks to a few months, and sometimes longer (Bouwer and Chaney, 1974, and references therein). Survival times for enteric viruses in soil may range from 1 to 6 months (Gerba et al., 1975a and b, and references therein).

The fecal-coliform test normally is used to indicate removal of pathogens. This is not always reliable, since some pathogens (particularly viruses) may survive fecal coliforms in the underground environment. Thus, testing for specific pathogenic organisms should be preferred, despite the fact that such tests are not as simple and fast as the fecal-coliform test. Virus concentrations are determined as plaque-forming units (PFUs) in monkey-kidney cells or other live-cell cultures. Specific virus types can then be identified serologically.

Many studies have reported complete absence or very low numbers of fecal coliforms or other microorganisms in sewage water after it has moved 1 or 2 m through soil. In other studies, appreciable numbers of microorganisms were found after much longer distances of underground movement, including a distance of 830 m in sand and gravel (Bouwer and Chaney, 1974; Gerba et al., 1975a; and references therein). Long underground survival distances for microorganisms, however, are usually associated with large pore sizes, as occur in uniform gravels and sands, structured clays, and fractured or cavernous rock, or with periods of high rainfall. The best protection against bacterial and viral contamination of groundwater by sewage is a relatively thick (1 to several meters) layer of medium- to fine-textured soil without pronounced structural features between the sewage source and the groundwater. Wells for potable water should be located as far as possible from the sewage source, and well water that may have the slightest chance of containing pathogenic organisms should be chlorinated or otherwise disinfected before potable use.

Phosphate in sewage moving through soil is adsorbed and precipitated, particularly if the soil contains iron and aluminum oxides. Initial adsorption generally fits a Langmuir adsorption isotherm. Retention of PO_4 in soil, however, increases with time because adsorbed PO_4 slowly precipitates with Ca, Al, or Fe. This regenerates adsorptive surfaces and causes the amount of P that can be stored in the soil to be much greater than that indicated by the initial, Langmuir-type adsorption (Ellis, 1973; Enfield and Bledsoe, 1975). Whether PO_4 precipitates in the soil with Ca or with Al and Fe depends on the pH. Above pH 6 to 7, PO_4 precipitates mostly as apatitelike Ca compounds, and below pH 6 to 7 as Al and Fe compounds (Lindsay and Moreno, 1960). Aluminum is particularly effective in retaining PO_4 in acid soils. Sandy soils with a low pH have the lowest capacity for removing phosphate from sewage effluent.

Metal ions in sewage effluent are bound by clay, hydrous oxides, and organic matter in the soil. Immobilization of some metals is favored by a high pH and aerobic conditions. Metals may react with organic compounds in the sewage to form chelates, which keep the metals in a mobile state and can cause deeper penetration of the metals. In fine soils, however, metal-organic complexes of high molecular weight may be physically filtered out (Bouwer and Chaney, 1974).

Other sewage constituents like boron and fluorine are also attenuated in soil. Fluoride can be adsorbed by clays or form insoluble compounds like fluorite (CaF_2) and fluorapatites. Boron is adsorbed by iron and aluminum oxides and clay minerals. In sandy soils, boron removal is insignificant (Bouwer and Chaney, 1974, and references therein).

Septic Tanks

Migration of city dwellers to suburbs and country has produced an almost explosive increase in the number of homes served by septic tanks, which now are the largest single contributor of sewage to groundwater in the United States. In 11 northeastern states alone, 23 percent of the population or 12 million people were served by septic tanks or other individual sewage disposal systems in the early 1970s (Miller et al., 1974). At an average flow of 0.15 to 0.3 m^3 per person per day, septic tanks thus are a significant source of " groundwater recharge." Septic tanks produce a poor-quality, primary-type of effluent. Because sludge accumulates and digests in the septic tank, the effluent contains relatively large concentrations of nitrogen (40 to 80 mg/l, mostly as NH_4) and phosphorus (10 to 30 mg/l, mostly as PO_4), while the BOD ranges from 20 to 450 mg/l (Sikora et al., 1976, and references therein).

The effluent of septic tanks is infiltrated into the soil with a drain field (Figure 11.8). Cesspools, consisting of a hole or buried chamber with porous walls (perforated concrete rings or blocks) to allow effluent water to seep into the ground, are commonly used in conjunction with or instead of septic tank systems. An unsaturated zone at least 1 or 2 m thick below the drain field is desirable to allow aerobic decomposition and other attenuation reactions. The design of septic-tank systems and the required separation distances between individual systems normally are regulated by local codes.

The most readily detectable effects of septic-tank drainage on underlying groundwater are increases in nitrate, chloride, and bacteria. The TDS content of the drainage also tends to be higher than that of the native groundwater. If the zone beneath the drain field is anaerobic, ammonium increases may be observed (Walker et al., 1973). Nitrate-nitrogen increases approaching the 10 mg/l upper limit for drinking water are quite common, and much higher levels have also been detected (Miller et al., 1974, and references therein). Nitrate-nitrogen concentrations of 0.2 to 13 mg/l (average 3.3 mg/l) in groundwater below Fresno, California, were attributed to septic-tank effluents and seepage from sewage lagoons (Schmidt, 1972). The highest nitrate levels were found in the upper 17 m of the aquifer below septic-tank areas. In Hall County, Nebraska, nitrate-nitrogen concentrations in groundwater ranged from a trace to 61 mg/l (average 3.2 mg/l), primarily due to contamination by septic tanks and feedlots (Piskin, 1973). Out of 511 wells tested, 52 domestic wells as well as 1 public well and 12 irrigation wells

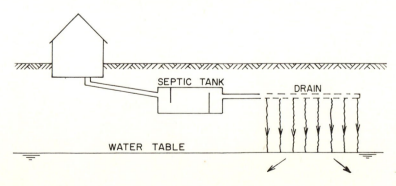

Figure 11.8 Schematic of septic tank and drain field.

yielded water with more than 10 mg/l of nitrate-nitrogen. The septic tanks also increased the chloride levels of the groundwater. Sodium enrichment was the best indicator of groundwater contamination by septic-tank effluent in southern Florida (Pitt et al., 1975). Shallow wells (about 3 m deep) also showed increased levels of ammonium, phosphorus, and fecal coliforms. Deeper wells showed less evidence of groundwater contamination. Agricultural drainage and storm runoff had greater effects on the groundwater than did septic tanks. Miller (1975) reported nitrate-nitrogen concentrations of as high as 31 mg/l in groundwater below septic tanks in Delaware. Fecal-coliform contamination of water from large-diameter, dug wells was attributed to high water tables which caused septic tanks to overflow, permitting effluent to flow over the land surface and seep into the wells.

The minimum separation between septic-tank systems and wells for potable water is specified in a number of local codes. In the State of New York, for example, a minimum distance of 30 m is required if the well is upgradient from the septic tank, and 60 m if downgradient.

Land-Treatment Systems

The capability of soil to remove suspended and dissolved constituents from sewage is utilized in land-treatment systems, where sewage (usually secondary effluent) is applied to land with sprinklers, irrigation furrows or borders, or infiltration basins. Sprinklers are suitable for irregular topographies, but they require more energy for pumping and cause more air pollution than surface-application techniques. The principle of cleaning wastes via the soil is very old indeed. As a matter of fact, soil is nature's system for decomposing and recycling waste produced by its biota. In our era of technology, treatment plants have become the acceptable way of purifying sewage. In recent years, however, land treatment is gaining interest as the need for preserving the quality of surface water becomes more critical and costs of purifying sewage with treatment plants continue to increase (Bouwer, 1976a).

There are three types of land-treatment systems: overland-flow, low-rate, and high-rate systems (Bouwer and Chaney, 1974). Overland-flow systems are used where surface soils are relatively impermeable and infiltration rates are low. In low-rate systems, sewage effluent is applied in amounts of 5 to 10 cm every 1 or 2 weeks to agricultural or forested land. In high-rate systems, applications range from 0.5 to 5 m per week. From a groundwater-quality standpoint, only low-rate and high-rate infiltration systems are of concern.

Prime objectives of low-rate systems are avoiding discharge of effluent into surface water, using nitrogen, phosphorus, and other "pollutants" in the sewage as fertilizer for crops, using the water in the effluent for irrigation of crops, and using the soil as a natural filter to obtain renovated deep-percolation water suitable for groundwater recharge. Nitrogen removal from the effluent water in low-rate systems is mainly accomplished via uptake by plant roots, which can absorb 50 to 500 kg/ha of nitrogen per year or season. The low figure applies to wheat and other grain crops, the high figure to certain frequently mowed grasses. Some nitrogen will also be removed by denitrification, which may amount to approximately 25 percent of the applied nitrogen (Bouwer and Chaney, 1974; Bouwer, 1976a; and references therein). Thus, if a crop takes up 100 kg of nitrogen per hectare and denitrification losses are 20 kg of nitrogen per hectare, 60 cm of sewage effluent containing 20 mg/l of nitrogen could be applied during the growing season without fear of contaminating underlying groundwater with nitrate. Assuming that the growing season is 6 months, 5 cm of effluent could then be applied every 2 weeks. Phosphorus uptake ranges from 10 kg/ha per year for crops like sugar beets, peanuts, and oats, to 70 kg/ha per year for forage crops (Tisdale and Nelson, 1975). Phosphorus not removed by crop roots, however, will largely be adsorbed or otherwise immobilized in most soils, except in sands with a low pH.

In addition to nitrogen and phosphorus, the soil-plant filter removes essentially all suspended solids, organic waste matter as expressed by the BOD, and bacteria and viruses (Bouwer and Chaney, 1974), so that the deep-percolation water from low-rate land-treatment systems is of relatively good quality, at least in humid regions. In arid regions, the TDS content of the deep-percolation water will be 3 to 10 times that of the sewage effluent if a salt balance in the root zone is to be maintained (see Section 11.7). Thus, unless the TDS content of the effluent is very low, the deep-percolation water from low-rate systems in arid regions will be too salty for most reuse purposes and must be handled in the same way as deep percolation from fields irrigated with normal water (see Section 11.7). Because of this concentration effect, nitrate contents in the deep-percolation water from sewage-irrigated fields in arid areas may also be high. This can produce nitrate concentrations in the underlying groundwater in excess of the maximum limit for drinking.

In high-rate systems, sewage effluent normally is put into infiltration basins, from where it infiltrates into the soil at rates of 0.5 to 5 m/week (depending on soil hydraulic conductivity). High-rate systems require relative permeable soils (sandy loams to loamy sands, for example), and most basins are not purposely vegetated. In humid areas, the main objective of high-rate systems may be disposal of sewage effluent without polluting surface water. In arid areas, renovating effluent and using it for groundwater recharge may be the primary purposes. The basins are regularly dried to maintain high infiltration rates (see Section 8.3.2) and to let oxygen enter the soil for decomposition of organic matter and nitrification of ammonium. Optimum flooding and drying schedules depend on wastewater characteristics, soil type, and climate, and may range from 8 h of flooding and 16 h of drying each day to flooding and drying periods of several weeks each (Bouwer and Chaney, 1974).

The amount of nitrogen that enters the soil with sewage effluent in high-rate systems could reach several tens of thousands of kilograms per hectare per year. This is much more than can be removed by plants, leaving denitrification in the soil as the only significant mechanism for nitrogen removal. By carefully controlling lengths of flooding and drying periods and infiltration rates, favorable conditions for denitrification can be created, enabling the removal of 30 to 80 percent of the nitrogen (Bouwer and Chaney, 1974; Bouwer et al., 1974; Bouwer, 1976a; Lance et al., 1976a). Since some of the nitrogen will remain in the ammonium form, nitrate contents in the deep-percolation water from high-rate systems can then be expected to be about equal to or less than the maximum limit for drinking water.

Under favorable soil and aquifer conditions, phosphates can be precipitated in high-rate systems to yield 80 to 90 percent phosphorus removal (Bouwer et al., 1974). In the same study, the BOD of the renovated water was essentially zero, but refractory organic carbon was present at concentrations of 4 to 7 mg/l TOC. Concentrations of several metals and fluoride in the renovated water were substantially lower than in the effluent, while boron concentrations were about the same.

High-rate land-treatment systems can produce renovated water free from fecal coliform bacteria and viruses. In a project near Phoenix, Arizona, most fecal coliforms were removed in the top meter of soil below the basins, but additional lateral movement of about 60 m below the water table was necessary before fecal coliforms were consistently absent (Bouwer et al., 1974). Viruses could not be detected in renovated water sampled from depths of 6 and 9 m below the infiltration basins (Gilbert et al., 1976a and b). The water table was at a depth of about 3 m in this project. In other land-treatment systems, however, viruses have been detected in underlying or adjacent groundwater (Mack et al., 1972; Schaub et al., 1975; Wellings et al., 1975). Deeper penetration of viruses frequently is associated with sands or other coarse-textured materials, structured clays, or humid areas where infiltration from heavy rains can remobilize previously adsorbed viruses in the soil. Lance et al. (1976b) found that rainfall following sewage infiltration mobilized previously adsorbed viruses only if the rain (simulated by demineralized water) occurred within 5 days after cessation of sewage infiltration. Addition of $CaCl_2$ to the demineralized water at concentrations of 1 to 3 mmol/l prevented desorption of the viruses. This means that the chances for virus contamination of groundwater below land-treatment systems can be reduced by applying $CaCl_2$ to the land if it threatens to rain during the beginning of a drying period.

Land treatment is used not only for sewage effluent, but also for effluent of agricultural processing plants, particularly of fruits and vegetables. Such effluents may have BODs of thousands of milligrams per liter (Bouwer and Chaney, 1974, and references therein), which requires low-rate and frequently intermitted applications to allow enough oxygen to enter the soil for decomposition of the organics. Landfill leachate and liquid waste and slurries from farm animals have also been applied to land (see Sections 11.5 and 11.7).

While the deep-percolation water from well-designed and well-managed land-treatment systems has a much better quality than the original sewage or other wastewater, the quality of this renovated water often is not as good as that of the native groundwater. To take advantage of land treatment without large-scale quality degradation of native groundwater, the spread of renovated water in the aquifer must be restricted. This can be achieved by locating the land-treatment system so that the deep-percolation or renovated water naturally drains into surface water, or by intercepting the renovated water with drains if the aquifer is shallow or with a series of wells if the aquifer is deep (Figure 11.9). After interception by drains or wells, the renovated water can be reused for irrigation, recreation, industrial purposes, and perhaps even for drinking (if proven safe and with additional treatment if necessary), or it can be discharged into surface water. To avoid movement of renovated water in the aquifer outside a system of infiltration areas and wells (Figure 11.9, bottom), the facility should be designed and managed so that the water table beneath the outer edges of the infiltration areas remains at the

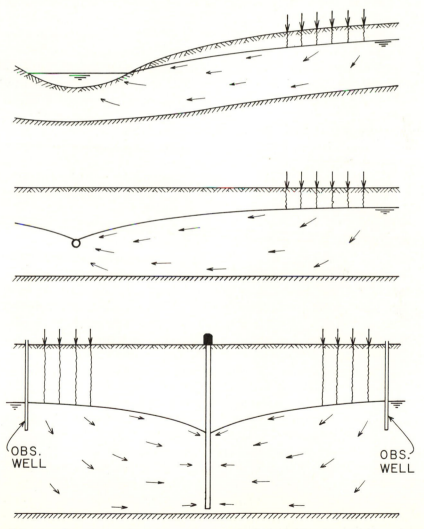

Figure 11.9 Removing renovated wastewater from aquifer by drainage into surface water (top) or by interception with drains (center) or wells (bottom).

same level as that in the aquifer adjacent to the system. This can be verified with observation wells installed at the perimeter of the system (Figure 11.9).

Systems like those in Figure 11.9 should be designed so that the renovated water will have had sufficient underground detention time and sufficient underground travel distance to be of acceptable quality when it leaves the aquifer. Desired times and distances depend on the quality of the input wastewater, the soil and aquifer materials, and the intended use of the renovated water. As a guide, an underground detention time of several weeks and a travel distance of about 100 m may be adequate for renovating sewage effluent in relatively fine textured soils and aquifers that are free from cracks or other large pores. Shuval (1975) reports a distance of 400 m between infiltration basins and wells and an underground detention time of about 400 days for a sewage reclamation project in Israel. Even then, carbon filtration and ozonation of the renovated water are planned before it will be released for potable use.

More detailed design criteria for the systems in Figure 11.9, including prediction of underground-flow systems and water-table heights, were presented by Bouwer (1974 and 1976b). Interception of renovated water with wells or drains generally is easier and less costly for high-rate than for low-rate systems, because the former require a much smaller infiltration area for the same effluent flow. Urban sewage production generally ranges from 0.3 to 0.4 m^3 per person per day, depending on industrial inputs of wastewater.

Sludge

Dry sewage sludge or sludge slurries containing about 5 percent solids are often applied to land as soil conditioner and fertilizer. Other disposal methods are incineration and disposal in landfills or oceans. Sludge is mostly relatively inert organic material with plant nutrients and metals (Table 11.3). Dry sludge contains 3.5 to 6.4 percent nitrogen, of which about two-thirds is tied up in the organic matter. When applied to soil, this organic matter gradually decomposes, releasing nitrogen in mineral form suitable for uptake by plant roots. The mineralization process may take several years. Pathogenic bacteria and viruses may also be present in sludge.

The amount of sludge that can be applied to soil without endangering underground water supplies normally is limited by its nitrogen content (Brown, 1975). Guidelines for sludge application to land, proposed by the U.S. Environmental Protection Agency (1976), call for applications whereby the nitrogen in the sludge will not exceed twice the nitrogen requirement of the crop. At these rates, the amounts of metals entering the soil probably are not excessive. Application of lime, however, is effective in inactivating zinc and other metals in the soil that may otherwise accumulate to levels toxic to plants (Dean and Smith, 1973). Penetration of heavy metals below root zones of sludge-fertilized crops has been observed in some instances (Kirkham, 1974 and 1976, and references therein).

11.5 SOLID WASTE

Landfills

Solid waste (mostly garbage and industrial waste) is disposed in landfills where it decomposes and produces a leachate that can contaminate underlying ground-water. Landfills range from unmanaged dumps, where refuse is piled up with little or no regard for environmental effects, to carefully designed and operated "sanitary" landfills. In the latter, garbage is dumped in rows or cells (Figure 11.10) that are covered with a layer of soil at the end of each day of dumping to keep away flies, rodents, birds, and other animals; to avoid nuisances like odor, blowing paper, and fire hazards; to provide escape routes for methane and other gases produced by the decomposing garbage; and to improve the appearance of the site (American Society of Civil Engineers, 1976; Brunner and Keller, 1972). The waste is compacted (usually with the same machinery used for placing the soil-cover layers) and the fills are relatively deep to minimize the required land area. When the fill has reached the desired height, it is topped by a thicker soil layer to reduce infiltration of rainfall and to enable use of the fill for landscaping, parking, etc. Unfortunately, only about 6 percent of the landfills in the United States can be classified as sanitary (Zanoni, 1972, and references therein). The others are plain landfills or dumps with varying degrees of management to minimize adverse environmental effects.

Leachate

The amount of leachate produced in a landfill depends on amount and distribu-tion of rainfall, hydraulic conductivity of cover soil (if any), evaporation from cover soil, and freezing and thawing. Studies in Illinois showed that somewhat less than one-half of the precipitation, which varied from 51 to 137 cm/year, infiltrated into landfills to produce leachate (Hughes et al., 1976). In arid areas, little or no

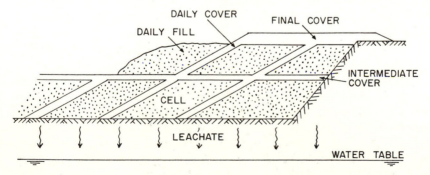

Figure 11.10 Cross section of sanitary landfill showing cells and daily and final soil covers. *(Modified from Brunner and Keller, 1972.)*

leachate is produced. If the soil below the fill is relatively impermeable, percolation of leachate to underlying groundwater is retarded. Landfills in high-water-table areas sometimes are surrounded by a perimeter trench to intercept the leachate. A good practice is to cover landfill bottoms with a clay layer or other artificial barrier (asphalt or plastic membrane, for example) to prevent movement of leachate to underlying groundwater. The leachate must then be collected by underdrains, after which it can be discharged into a sewer line for conveyance to a sewage plant or other treatment facility (Chian and DeWalle, 1976). Another possibility is to recycle the leachate through the landfill (Pohland and Kang, 1975) or to apply it to land for crop irrigation (Nordstedt et al., 1975). Fresh, compacted refuse is not yet saturated and must first absorb water before leachate can be produced. In one study, this absorption amounted to 10 to 14 cm of water per meter thickness of compacted garbage (Rovers and Farquhar, 1973). Hughes et al. (1976) estimated that it took 4 to 7 years for some Illinois landfills before leachate production began.

The chemical composition of landfill leachate depends on the nature of the refuse (particularly types of industrial waste present), on the leaching rate, and on the age of the fill (concentrations of the various constituents decrease with time and eventually reach relatively stable levels). Observed concentrations thus vary widely, as shown in Table 11.4, which applies primarily to domestic waste. The data indicate that landfill leachate is grossly polluted and a definite threat to groundwater quality. The hardness of leachate and contaminated groundwater is due to the CO_2 generated in the decomposition of the refuse. This CO_2 forms carbonic acid which dissolves calcium compounds in the soil material and causes increases in the hardness of underlying groundwater. In addition to the constituents shown in Table 11.4, leachate may contain all sorts of chemicals, depending on what is dumped in the landfill. If the refuse includes discarded electrical transformers, for example, the leachate is likely to contain the very persistent and carcinogenic PCBs (polychlorinated biphenyls). Chemical waste, dumped directly or in containers that may leak later, may add heavy metals, arsenic, pesticides, cyanide, and other toxicants to the leachate. In Iowa, for example, dumping arsenic-containing wastes produced arsenic concentrations of more than 175 mg/l in groundwater around the dump site (Lazar, 1975). The author also cited a case in New Jersey where leakage from drums with petrochemical wastes contaminated a major aquifer and resulted in the condemnation of about 150 private wells. Serious groundwater contamination was also detected near a large landfill in Delaware that received domestic and industrial wastes. The local aquifer serves over 40 000 residents and supplies water for industrial use. Costs of remedial measures may run into millions of dollars. Where sludge from sewage-treatment plants or septic tanks is also dumped, pathogenic microorganisms may be present in the leachate. A recent concern is the increased use of disposable diapers, which send large amounts of enteric bacteria and viruses to the landfill. Virus concentrations of 32 units per 100 g of municipal garbage, and virus recoveries in leachate for periods of as much as 20 weeks after disposal, have been reported (Gerba et al., 1975b).

Table 11.4 Chemical composition (mg/l) of landfill leachate

	Normal Range	Upper Limits
Calcium	240–2 330	4 080
Magnesium	64–410	15 600
Sodium	85–3 800	7 700
Potassium	28–1 700	3 770
Iron	0.1–1 700	5 500
Manganese		1 400
Zinc	0.03–135	1 000
Nickel	0.01–0.8	
Copper	0.1–9	9.9
Lead		5.0
Chloride	47–2 400	2 800
Sulfate	20–730	1 826
Orthophosphate	0.3–130	472
Total nitrogen	2.6–945	1 416
BOD	21 700–30 300	54 610
COD	100–51 000	89 520
pH (units)	3.7–8.5	8.5
Hardness ($CaCO_3$)	200–7 600	22 800
Alkalinity ($CaCO_3$)	730–9 500	20 850
Total residue	1 000–45 000	
TDS		42 276

Source: Data for normal range taken from American Society of Civil Engineers (1976) and Brunner and Keller (1972). Upper limits from Office of Solid Waste Management Programs (1973).

Wood and bark wastes stockpiled on land by wood-products industries produced a brown leachate that contained lignin-tannin (measured as tannic acid) and other organics. Underlying groundwater showed higher concentrations of lignin-tannin, iron, and manganese (maximum values 7.5, 13, and 106 mg/l, respectively) than unaffected groundwater (Sweet and Fetrow, 1975).

Contamination Plumes and Attenuation

Landfills are essentially point sources of pollution. When the leachate has reached the groundwater (some landfills in poorly selected sites already extend into the groundwater, as in old gravel pits or stone quarries), it travels laterally in the direction of general groundwater movement and forms a relatively narrow band or plume of contaminated water (Palmquist and Sendlein, 1975)—unless, of course, the groundwater is stagnant.

Contamination plumes can be detected with observation wells strategically placed downgradient from the landfill (Palmquist and Sendlein, 1975). The chloride ion normally is the best indicator for detecting the presence of leachate in groundwater. Geophysical techniques that have been used for detecting contamination plumes include temperature and resistivity surveys (Cartwright and McComas, 1968; Kelley, 1976; Klefstad et al., 1975; Stollar and Roux, 1975). The higher TDS content of the leachate plume reduced apparent resistivity to about one-fourth the value of unaffected areas (Cartwright and McComas, 1968). Contamination plumes have been detected hundreds of meters from the landfill. Anderson and Dornbusch (1967) reported a distance of 366 m in coarse, glacial alluvium for leachate from a landfill in an old gravel pit.

Attenuation reactions below landfills and in leachate plumes consist of dilution by dispersion (see Section 11.1), filtering of suspended solids, cation exchange, chemical precipitation, biological decomposition, and retention and die-off of pathogens (see Section 11.4). Cation exchange is important for the removal of metal ions, which are exchanged for Na^+ and other lesser-held cations on the exchange complex (clay, organic matter, and hydrous oxides of iron, manganese, and aluminum). Some metals can combine with organic compounds to form soluble chelates, allowing them to move through soils and aquifers. Precipitation is an important attenuation reaction for phosphate, fluoride, and heavy metals. Biodegradable materials in the leachate are decomposed by bacteria in the soil, eventually reducing the BOD to essentially zero. This decomposition is fastest and most complete under aerobic conditions, where the organics are mineralized to mostly H_2O, CO_2, NO_3, PO_4, and SO_4. Under anaerobic conditions, biodegradation is slower and less complete, often leaving CH_4, H_2S, NH_4, and organic acids as end-products.

Detailed studies of attenuation reactions in sand-and-gravel aquifers led Golwer et al. (1973) to conclude that if groundwater velocities are less than 1 m/day, deteriorated groundwater is "normalized" again at a distance of a few hundred meters from the landfill. This corresponds to underground detention times of 1 to 2 years. In the United States, minimum distances between landfills and water wells required by state and local agencies vary from 15 to 300 m, with most distances in the 30- to 150-m range (Zanoni, 1972). Contamination of wells at greater distances, however, has been observed. In the northeastern United States alone, 13 out of 60 selected wells with leachate contamination were more than 300 m from the landfill (Miller et al., 1974). About 100 contaminated wells were found in that study. Remedial measures included shutting down wells, abandoning landfills, and interception pumping of contaminated groundwater. In southern New Jersey, an industrial landfill was closed after leachate containing up to 18 mg/l of lead had moved 150 m toward municipal water supply wells 1 200 to 1 800 m away from the fill. For additional details, see Miller et al. (1974). On the other hand, unappreciable or unnoticeable effects of landfills on nearby groundwater have also been observed (Hughes et al., 1976).

Minimizing Groundwater Contamination

To minimize adverse effects of landfills on groundwater, flow of leachate to groundwater must be minimized by lining the bottom with impermeable material and collecting the leachate with drains or other facilities, and by covering the landfill with fine soil or impervious material as soon as it has reached the desired height. Landfills should preferably be located in fine-textured soil materials with a relatively deep water table. Many states require water-table depths of at least 0.6 to 9 m below the fill (Brunner and Keller, 1972), presumably to provide an aerobic zone. In view of the high BOD of the leachate, however, free oxygen cannot be expected in the unsaturated zone. The only benefit from unsaturated conditions would then be better attenuation by closer contact between leachate and solids. If the soil between a landfill and underlying water table is permeable, groundwater pollution may be more serious than when the landfill is in finer soil, even if for the latter the water table is higher and reaches into the fill (Apgar and Langmuir, 1971). Old gravel pits or limestone quarries as such are not suitable for landfill in humid areas because leachate can travel unrestricted and unattenuated through the large pores or openings in these materials.

In selecting landfill sites, patterns of underground water flow should be known so that the fill can be located where the contamination plume will do the least damage. In river-valley areas, for example, valley sites tend to be more favorable than upland sites because they are closer to the stream into which the groundwater drains (Sendlein and Palmquist, 1975).

11.6 CEMETERIES

Decomposing bodies in graveyards produce fluids that can leak to underlying groundwater if wooden coffins or other nonleakproof caskets are used. Groundwater in areas with high rainfall and high water tables is most vulnerable to this type of contamination. Historical cases range from a higher incidence of typhoid fever among people living near cemeteries in Berlin in the period 1863–1867, to a "sweetish taste and infected odor" of water from wells close to cemeteries in Paris, especially in hot summer periods (Mulder, 1954, and references therein).

An average human corpse contains about 10 kg protein, 5 kg fat, and 0.5 kg carbohydrate (van Haaren, 1951). Based on observations and an analysis of oxygen diffusion, van Haaren estimated that biooxidation of this material required about 10 years at a burial depth of 2.5 m in sandy soil under Dutch climatic conditions. The leaching rate, calculated as excess precipitation, was about 40 cm/year, yielding a leachate volume of 0.4 m^3/year for a typical single grave. Average chemical parameters of shallow groundwater below graves were:

Color (platinum-scale units)	75
Electrical conductivity	2.3 millimhos
COD (using $KMnO_4$)	95 mg/l
Chloride	500 mg/l
Sulfate	300 mg/l
Bicarbonate	450 mg/l

Table 11.5 Groundwater quality in relation to distance from a row of graves

Distance from graves (m):	0.5	1.5	2.5	3.5	4.5	5.5
Bacteria count (per ml):	6 000	8 000	8 000	3 600	1 200	180
NH_4 (mg/l):	6	0.75	—	—	—	—
NO_3 (mg/l):	4.8	0.1	—	—	—	—
COD (mg/l, using $KMnO_4$):	26.7	16.4	15.4	15.4	11.4	11.4

Source: From Schraps, 1972.

Individual samples were quite variable. One sample contained 45 mg/l ammonium. In another study, groundwater was sampled at a depth of 50 cm below grave level and at different distances downgradient from a row of graves in a West German cemetery (Schraps, 1972). Chemical analysis of the samples (Table 11.5) showed contamination in the immediate vicinity of the graves but rapid attenuation with distance. The graves were in unconsolidated alluvium underlain by siltstone. Schraps recommended that water tables in cemetery plots should be at least 2.5 m deep, which at the customary grave depth of 1.8 m would leave an unsaturated filter zone of 0.7 m. This should be adequate for groundwater protection in medium-textured soil materials (loams and sandy loams). Schraps cites studies on a Hamburg cemetery where there was no evidence of groundwater contamination under these conditions. Very permeable materials (sands and gravels) or soils so fine that anaerobic conditions prevail even if the filter zone is above the water table should be avoided. Serious contamination of groundwater can also result where cemeteries are underlain by fractured or cavernous rock. Careful studies are always necessary, however, before rash conclusions are drawn. For example, Brausz (1952) found that an increased bacterial content of water from a well in limestone near a cemetery was more likely caused by human activity in the nearby outcrop area of the limestone than by the cemetery itself. Groundwater contamination by cemeteries cannot be generalized, and each case must be considered individually.

Minimum distances between potable water wells and cemeteries required by law are 91.4 m (100 yards) in England (Goodman and Beckett, 1976), 100 m in France, and 50 m in Holland (Mulder, 1954). Mulder suggests, however, that larger distances—for example, 2 500 m—are better because biologic processes in the soil sometimes can lose their effectiveness. For aesthetic reasons alone, separation distances should always be as large as possible.

11.7 AGRICULTURE

Modern agriculture is based on extensive use of fertilizers and pesticides to obtain high crop yields. Some of the chemicals applied to farmland, however, move down with the deep-percolation water from the root zone and can contaminate underlying groundwater. Manure piles, feedlots, and similar concentrations of animal waste are other possible sources of groundwater contamination. Deep-percolation water from irrigated fields in arid regions tends to have a high salt content, which adversely affects underlying groundwater. With the exception of manure piles, agriculture is an area-wide or diffuse source of groundwater contamination. In humid areas, the major contaminant is nitrate, whereas TDS and nitrate are of most concern in arid, irrigated areas.

Humid Areas

Root uptake of nitrogen varies from about 50 kg/ha per crop season for barley to 200 kg/ha for alfalfa (Stewart et al., 1975). Frequently mowed grasses may absorb as much as 500 kg/ha per year. Some of this nitrogen is made available by mineralization of soil organic matter. For leguminous crops, considerable nitrogen is supplied by certain bacteria that live on the roots and can fix atmospheric nitrogen. Any deficit in the natural nitrogen supply must be made up by fertilizer to assure top yields. In the past, nitrogen fertilizer was cheap and it was often applied in greater amounts than needed. Also, fertilizers were applied in the fall or other convenient time that did not always coincide with the period of greatest nitrogen demand of the crop. These practices resulted in unnecessary leaching of nitrate from the root zone and contamination of groundwater. This is now rapidly changing as the cost of nitrogen fertilizer is rising, and the impact of improper fertilizer use on underlying groundwater is better understood.

Crop uptake of fertilizer nitrogen varies from about 40 to 80 percent of the amount applied (Stewart et al., 1975). Part of the 20 to 60 percent not taken up is returned to the atmosphere, primarily by denitrification and volatilization of ammonia. The rest mostly ends up as nitrate in the deep-percolation water and can move to underlying groundwater. Denitrification proceeds faster in warm soils than in cold soils. Also, it is more significant in wet than in dry soils and in fine than in coarse soils (fine soils tend to have more microanaerobic sites within an otherwise aerobic root zone). Denitrification also requires the availability of organic matter in the soil as energy source for the denitrifying bacteria. Because denitrification is affected by a number of factors, it is difficult to predict how much of the nitrogen will be denitrified and how much will go to the groundwater. As a guide, it can be assumed that for medium-textured soils and average climatic conditions, 10 to 30 percent of the nitrogen applied (or about half of the nitrogen not taken up by the crop) will be denitrified (Broadbent and Clark, 1965). The other 10 to 30 percent of the fertilizer nitrogen applied would then be leached out of the root zone as nitrate in the deep-percolation water.

Nitrate concentrations in deep-percolation water vary seasonally and with percolation rates. In The Netherlands, for example, concentrations in deep-percolation water from cropland (collected as outflow from tile drains) was highest in spring and lowest in fall (Kolenbrander, 1969). Average NO_3-N concentrations were higher for sandy soils than for clay soils, and ranged between 1 and 2 mg/l for grassland and between 4 and 10 mg/l for cropland. Nitrogen concentrations in drain effluent from cropland in New York varied from 3 to 51 mg/l, depending on fertilizer application rates (Zwerman et al., 1972). In Ohio, nitrogen concentrations in the outflow from tile drains were 1 to 24 mg/l, while TDS contents ranged from 250 to 450 mg/l (Logan and Schwab, 1976). These concentrations all pertain to nonirrigated fields in humid areas. Much higher concentrations can be expected below nonirrigated fields in semiarid areas, where so-called dry-land farming is practiced. Kreitler and Jones (1975), for example, reported an average NO_3-N concentration of 56 mg/l in the upper groundwater beneath fields in west-central Texas. The highest concentration was 690 mg/l! For additional data on nitrogen in deep-percolation water from agricultural land, reference is made to Viets (1975).

Arid Areas

Nitrogen content in deep-percolation water from irrigated fields in the western United States has been determined in water samples taken from vadose zones, drain discharges, or underlying groundwater. A review of these studies by Viets (1975) indicates that low NO_3-N concentrations were observed below grassland (2.8 mg/l in one case) and below a large irrigated region underlain by porous basalts in the Snake River valley of southern Idaho (3.2 mg/l). Most NO_3-N concentrations found in other studies were in the range of 15 to 50 mg/l with peak values of more than 100 mg/l below special, heavily fertilized crops like celery.

Agricultural drainage water from the San Joaquin Valley in California contained an average of 20 mg/l of NO_3-N (California Department of Water Resources, 1971). Total concentrations of other forms of nitrogen rarely exceeded 1 mg/l. Some areas yielded drainage water with 100 to 200 mg/l of NO_3-N. Drainage water from alluvial fans contained more nitrogen than that from basin soils which normally are of finer texture. Highest nitrogen concentrations were observed in drainage from soils in the Panoche group. Continuous flooding, as practiced in rice growing, produced lower nitrogen concentrations in the drainage water, probably due to dilution and increased denitrification. Average concentrations of other constituents in the San Joaquin Valley drainage flow were 0.09 mg/l of PO_4-P and 3 625 mg/l TDS. In Nebraska, deep-percolation water below an irrigated cornfield in sandy soil contained 25 mg/l of NO_3-N when fertilizer was applied with the irrigation water at a rate of 67 kg or nitrogen per hectare, and 30 mg/l of NO_3-N when 90 kg of nitrogen per hectare were applied as side-dressed anhydrous ammonia (Linderman et al., 1976).

Deep-percolation water from irrigated areas in arid regions generally has a high TDS content. This is a direct consequence of the need for maintaining a salt balance in the root zone, which is a prerequisite for permanent or long-term irrigated agriculture. Most irrigation waters contain 100 to 1 000 mg/l TDS (some go as high as 1 000 to 3 000 mg/l). Crop roots absorb only essentially pure water while leaving dissolved salts behind. If these salts are not leached out by regularly applying more irrigation water than needed for evapotranspiration (ET), salts accumulate in the root zone and the land eventually becomes too salty for agriculture. Considering only the amounts of salt that enter the root zone with irrigation water and leave it with deep-percolation water, a salt balance is maintained in the root zone if

$$C_i D_i = C_d D_d \qquad (11.13)$$

where C_i = TDS content of irrigation water
$\quad D_i$ = amount of irrigation water applied (expressed as a water depth)
$\quad C_d$ = TDS content of deep-percolation water
$\quad D_d$ = amount of deep-percolation water

This equation ignores other and usually minor components of the salt balance like fertilizer application, crop uptake of salts, precipitation or dissolution (weathering) of salt in the soil, and rainfall.

The terms C_i and C_d often are expressed in terms of electrical conductivity (EC) in millimhos (1 millimho corresponds to about 640 mg/l TDS; see Section 10.4). To determine D_d, it is customary to take C_d as EC of the saturation extract of a uniformly saline soil profile whereby crop yield is reduced to 50 percent of normal (see Section 10.5, Table 10.8, and Figure 10.1). This does not mean, however, that crop yields will be reduced by 50 percent in the field when the resulting value of D_d is used to calculate the extra irrigation requirement for salt leaching. In the field, salt concentrations of the soil water in the top of the root zone (where most of the roots occur) will be on the same order as C_i, especially when the field has just been irrigated. Further down, the salt concentration of the soil water becomes higher and reaches C_d at the bottom of the root zone. Most roots thus can absorb water with a salt content lower than C_d. This situation differs from the artificial uniform salt distributions used for obtaining the data in Table 10.8 and Figure 10.1. Experience has shown that irrigating on the basis of D_d calculated with C_d from Table 10.8 or Figure 10.1 for 50 percent yield reduction gives essentially normal crop yields (van Schilfgaarde et al., 1974).

By way of example, Figure 10.1 shows that EC of the saturation extract for 50 percent yield reduction of cotton is 16 millimhos. Assuming $C_i = 1$ millimho, Eq. (11.13) shows that $D_d = D_i/16$. Since $D_i = D_d + ET$, the amount of irrigation water to be applied should equal $(16/15)ET$. Thus, if $ET = 0.6$ cm/day and the field is irrigated every 2 weeks, 8.96 cm of water should be applied per irrigation. Of this amount, 8.4 cm will be evapotranspired and 0.56 cm will move down as deep-percolation water with an EC of 16 millimhos or a TDS content of about 10 240 mg/l. This deep-percolation rate should then produce a normal crop yield.

Most fields are irrigated with borders, basins, furrows, corrugations, or other surface techniques. With these systems, a uniform application is difficult to achieve and some parts of the fields will receive more water than others. This locally produces more deep-percolation water than the calculated D_d, which causes C_d to be lower than the theoretical value. Measured values of C_d/C_i in irrigated areas with a salt balance in the root zone thus can be expected to be lower than the theoretical values. For some fields in the San Joaquin Valley, California, for example, C_d/C_i at salt balance was estimated to range between 4 and 6 (Pillsbury et al., 1965). Since C_i was about 400 mg/l, C_d would be about 2 000 mg/l. For the Coachella Valley, California, C_d/C_i at salt balance was 2.7, which at a C_i value of 877 mg/l yields $C_d = 2\,400$ mg/l (Bower et al., 1969). Even under controlled conditions, C_d/C_i may be relatively low. Rhoades et al. (1973), for example, measured C_d with lysimeters planted to alfalfa and irrigated with different amounts of water to produce different leaching ratios D_d/D_i. Most of the C_d/C_i ratios were about 6 for $D_d/D_i = 0.1$, 3.7 for $D_d/D_i = 0.2$, and 3 for $D_d/D_i = 0.3$. These data apply to both calcareous and noncalcareous soils. Highest salt concentrations in the deep-percolation water occurred in the spring when crop growth and solubility of lime were maximum. A computer model for predicting the chemical composition of deep-percolation water was developed by Oster and Rhoades (1975). Input data for the model consist of the composition of the irrigation water, the leaching fraction D_d/D_i, the solubilities of aragonite and gypsum, and the measured partial pressure of CO_2.

Excessive deep percolation due to nonuniform water application in field irrigation systems can be taken into account with a *leaching efficiency*. In defining this term, the deep-percolation water is considered a hypothetical mixture of irrigation water that has passed unchanged through the larger pores of the root zone and of original soil water that has been directly displaced by the rest of the irrigation water. The leaching efficiency then is defined as the hypothetical fraction of the deep-percolation water that consists of displaced soil water (Bouwer, 1969, and references therein). The magnitude of the leaching efficiency depends on the heterogeneity, type, and structure of the soil and on the efficiency of the irrigation system. Field values normally range from 0.2 to 0.8. The leaching-efficiency concept yielded reasonably accurate predictions of C_d and D_d in irrigated areas where a salt balance was established (Bouwer, 1969).

In most surface-irrigation systems, 30 to 50 percent of the irrigation water becomes deep-percolation water—much more than would be needed to maintain salt balance in the root zone. Excessive deep percolation is undesirable because resulting irrigation water requirements and pumping costs are higher than necessary. Excessive deep percolation also leaches out valuable fertilizer, requires a more expensive tile-drainage system if the natural drainage is inadequate, and can leach more salts out of underlying formations. The latter is particularly undesirable where the deep-percolation water eventually drains to a stream.

With sprinkler- or drip-irrigation systems, the application rate is externally controlled and infiltration amounts are not affected by local soil differences. Thus, the water can be applied much more uniformly than in surface-irrigation systems and applications can be controlled so that D_d will be close to the desired value. If the water is applied frequently and in small amounts, an essentially steady-state condition of salt and water flow is created in the root zone (van Schilfgaarde et al., 1974). Salt concentrations in the water at the top of the root zone will then remain close to C_i. Since plant roots absorb most of the water from the top of the root zone, salt concentrations at the bottom of the root zone can then be much higher than for systems where water is applied less frequently and in larger amounts. If C_d can be taken higher, D_d will be correspondingly lower. As a result, irrigation-water requirements will be reduced, the amount of deep-percolation water will be less, and crop yields will not be affected. The high C_d values may also stimulate precipitation of salts like carbonate and gypsum in the soil, thus reducing the total salt load on the underlying groundwater. Low values of D_d also result in less dissolution of salts previously precipitated in the soil, in less weathering of soil minerals, and in less leaching of salt from underlying formations.

Deep-percolation water or return flow from irrigation systems in arid areas often is the main cause of quality degradation of underlying groundwater. With careful irrigation practices, the amount of deep-percolation water can be minimized and the TDS content maximized, which generally is preferable to a large amount of deep-percolation water with a relatively low salt content. Where water tables are so high that tile drainage is required, deep-percolation water is removed from the underground environment. This offers an opportunity to dispose of it in inland salt lakes, playas, evaporation ponds, or other places, where it can do relatively little harm. In irrigated river valleys, however, the deep-percolation water often returns to the stream where it contributes to downstream quality degradation of the water.

Animal Waste

Accumulations of animal waste are point sources of pollution that can contaminate groundwater via leachate that infiltrates directly into the soil below the waste, or via surface runoff that infiltrates elsewhere. Since manure piles and feedlots often are close to farmsteads, their leachate plumes can readily contaminate water from domestic wells on the premises. The main contaminant is nitrogen, which can be present in the leachate or runoff at concentrations of several hundred milligrams per liter, primarily as organic and ammonium nitrogen. This nitrogen is mostly converted to nitrate in the vadose zone.

In a Colorado study, NO_3-N concentrations in the upper groundwater below a feedlot ranged between 0 and 41 mg/l and averaged 13.4 mg/l (Viets, 1975, and references therein). These concentrations are about the same as those below irrigated fields. The amount of nitrogen in the vadose zone below the feedlots, however, was much higher than below irrigated fields. This suggests active denitrification in the lower vadose zone or delay in arrival of nitrate at the water table. Nitrate contamination of groundwater below feedlots may be more severe if the feedlots are regularly cleaned than when the manure is allowed to accumulate. In the latter case, the soil below the manure is anaerobic so that nitrification does not take place. Accumulation of ammonium will then cause the pH to rise to where ammonia volatilization becomes significant. Ammonium not volatilized can be adsorbed in the underlying soil (Porter et al., 1975). Groundwater contamination by feedlots tends to be more severe below coarse-textured soils with shallow water tables than below fine-textured soils with deep water tables.

Erickson et al. (1972) reduced the nitrogen content in deep-percolation water from field plots receiving animal-waste slurries by applying the slurry intermittently to maintain aerobic conditions in the upper soil layers. This caused most of the nitrogen in the leachate to be converted to nitrate, which was then denitrified in an anaerobic zone further down. This zone was artificially created by placing an impermeable barrier (asphalt, plastic membrane, or other) at a depth of about 2 m to establish a perched leachate mound from which oxygen was excluded. The system removed 96 to 99 percent of the nitrogen from a swine-manure slurry containing 310 to 660 mg/l nitrogen prior to infiltration.

Pesticides

A multitude of chemicals in the general category of pesticides are used in agriculture to control weeds (herbicides), insects (insecticides, miticides), nematodes (nematocides), fungi (fungicides), rodents (rodenticides), and other animals (Stewart et al., 1975). The chance for pesticide contamination is much greater for surface water than for groundwater. This is because pesticides tend to be adsorbed by soil particles, which can be carried down to streams by surface runoff from treated fields. Pesticide concentrations in surface runoff from agricultural fields normally range from a few tens to a few hundred ppt (parts per trillion, or 1 part per 10^{12} by weight). Concentrations of as much as 4000 ppt, however, have been measured in Mississippi River water (Edwards, 1970, and references therein).

Serious pesticide contamination of groundwater is rare (Kaufman, 1974; Stewart et al., 1975). Pesticides with a high molecular weight, such as chlorinated hydrocarbons, generally have very low water solubility and are adsorbed by clay and organic matter in the soil to be eventually degraded by bacteria. Sometimes, the intermediate degradation products are also toxic—like DDE, which is derived from DDT. Light-molecular-weight pesticides (various herbicides, for example) are partly returned to the atmosphere by volatilization. The portion that moves into the soil with infiltrating water can be adsorbed and decomposed (Weidner, 1974). Many herbicides are decomposed by anaerobic bacteria, so that they can still be degraded after they have moved deeper into the ground.

The rate of bacterial breakdown of chlorinated hydrocarbons increases with increasing temperature, water content, and organic-matter content of the soil. The half-life of the more persistent chlorinated hydrocarbons in soil generally is less than 4 years, with DDT and dieldrin having the longest half-lives (Edwards, 1970). Half-lives of 10 to 12 years, however, have also been reported (Nash and Woolson, 1967). Because of low solubility, adsorption, and biodegradation, chlorinated hydrocarbons do not penetrate soils very deeply and groundwater contamination may result only where the water table is close to the surface, where soils are coarse or cracked, or where soils are shallow and underlain by fractured or cavernous rock.

In a Texas study, DDT concentrations of 0.2 to 1.2 ppb (part per billion, or 1 part in 10^9) were observed in water from a seep near a field of pesticide application (Swoboda et al., 1971). The soil was a heavy clay that shrank and cracked upon drying. Toxaphene could not be detected in the seep water. Groundwater from wells 3 to 30 m deep in an intensively farmed area with predominantly sandy soils in South Carolina showed the following concentrations (ND means "not detected") of chlorinated-hydrocarbon insecticides (Achari et al., 1975):

	Range (ppt)	Average (ppt)
lindane	ND—21.1	1.19
aldrin	ND—44.8	7.11
DDT	6.4—161	37.7

In Ohio, deep-percolation water collected at a depth of 2.44 m from a lysimeter filled with silt loam showed only trace amounts of the chlorinated hydrocarbon herbicide 2,4,5-T (Edwards and Glass, 1971). Methoxychlor, which is a widely used insecticide (often as a replacement for DDT) because of its long residual action and low toxicity to humans and animals, could not be detected at all in the lysimeter percolate.

In other studies, LaFleur et al. (1973) sampled groundwater at a depth of 0.9 m in a cotton field that was treated with the chlorinated hydrocarbon insecticide toxaphene and the herbicide fluometuron at amounts of 10 times the normal dose. This produced initial concentrations of 0.03 ppm toxaphene and 0.44 ppm fluometuron in the shallow groundwater. These values decreased to 0.012 ppm and 0.12 ppm, respectively, after one year. With normal pesticide use, concentrations less than one-tenth of the reported values can be expected. The herbicide prometryne, which is less mobile in soil than fluometuron, could be detected in groundwater sampled at a depth of 1.1 m in sandy loam (LaFleur et al., 1975). Twenty months after its first appearance, however, it could no longer be detected. Tile-drain effluent in the irrigated San Joaquin Valley in California contained 0.093 to 0.418 ppb chlorinated hydrocarbons (Johnston et al., 1967). Surface runoff, however, had much higher concentrations and raised the level of chlorinated hydrocarbons to 2.55 ppb in drainage ditches that also received irrigation runoff. Water in these ditches also contained organophosphate pesticides (parathion and malathion, for example) at an average concentration of 0.654 ppb.

Most organic fungicides are biodegradable and persist only a few days or a few weeks in soil, except for the more persistent types like quintozene (Edwards, 1970). Inorganic fungicides contain copper, mercury, arsenic, or other toxic elements that can accumulate in the soil and persist for long times.

While pesticide contamination of groundwater from deep percolation of agricultural fields is generally limited to very shallow groundwater, direct entry of pesticides into deeper aquifers can occur if surface water is used for groundwater recharge with wells. Since most of the pesticides in surface water are associated with sediment which must be removed from the water anyway to avoid clogging of the well face, serious contamination of groundwater probably will not occur. Reported cases of groundwater contamination by pesticides normally are accidental or due to unusual circumstances. For example, contamination of a 12-m-deep well (5.5 m to water table) by chlorinated hydrocarbons was traced to the use of contaminated soil for backfilling around the casing (Lewallen, 1971). The soil was obtained from an area about 8 m from the well that was used for flushing and cleaning insecticide sprayers. Concentrations initially observed in the well water were 20.1 ppb toxaphene, 1.5 ppb DDT, and 0.5 ppb DDE. These concentrations decreased with time and were below drinking water limits after $1\frac{1}{2}$ years. In Minnesota, contamination of well water with arsenic was caused by burying arsenic-containing pesticides in the area. The contamination was detected in 1972 when 11 persons developed symptoms of arsenic poisoning, 35 years after disposal of the insecticides (Lazar, 1975). Chlordane, which is commonly used for termite control in homes, has entered improperly protected domestic wells via surface runoff from the treated areas.

While low levels of pesticides can be tolerated in drinking water (see standards in Table 10.3), complete absence should be preferred. Careful use of pesticides on the land is always necessary, shallow groundwater should be avoided for drinking, and biodegradability in soil is a must for any pesticide on the market or yet to be developed. Use of the very toxic, persistent, and possibly carcinogenic pesticides like DDT, dieldrin, aldrin, chlordane, heptachlor, and mirex has been severely restricted in recent years by the Environmental Protection Agency, starting with DDT in 1972.

11.8 MINING

Mining of mineral resources can result in contamination of groundwater by construction of shafts and tunnels that disrupt natural groundwater regimes and can allow atmospheric oxygen to enter the underground environment. Leachate from mine tailings and wastewater from ore-processing plants also are potential sources of groundwater contamination. A general reference on the effect of mining on water quality is the book edited by Hadley and Snow (1974). Technical and legal aspects of production and disposal of wastes in mining and processing of mineral resources are described by Williams (1975).

Coal

Much of the groundwater contamination due to coal mining in the eastern United States is caused by the oxidation of pyrite, which frequently occurs in the top 40 cm of coal seams and in the overlying black shales (Emrich and Merritt, 1969). Since coal with a high sulfur content is not desirable, the material from the top of the seams is usually left behind in the mines or in spoil banks, where the pyrite is oxidized by atmospheric oxygen to sulfuric acid and iron hydroxide. These products can leach down to the groundwater, where they increase iron and sulfate concentrations and decrease the pH. If contaminated water flows through limestone beds, the acids are neutralized and the pH is closer to 7.

In Pennsylvania alone, acid mine drainage is produced at a rate of 17 million m^3/day, or about 1.4 m^3 for each resident in the state (Merkel, 1972). In northwestern Pennsylvania, acid drainage from a coal mine contained 200 to 500 mg/l iron, 2 450 to 4 400 mg/l sulfate, and 0 to 2 mg/l chloride, while it had a pH of 2.6 to 3.3 (Emrich and Merritt, 1969). Groundwater in areas near mining activities had a lower pH and contained more iron and sulfate than comparable groundwater in areas without mining, as follows:

	Mining areas	Nonmining areas
pH	3.0–5.5	6.5–6.7
SO_4 (mg/l)	39–80	10–15
Fe (mg/l)	20–70	10–15

Resistivity techniques have been used to detect the extent of groundwater contamination by acid mine drainage (Merkel, 1972). Mine drainage also affects surface water, either by direct discharge of acid drainage or by seepage of contaminated groundwater into streams. Springs and flowing wells may also contribute contaminated groundwater to streams.

Strip mining of coal leaves large spoil areas of disturbed soil and rock materials. Coal from western states generally has a low sulfur content, so that acid drainage is not a problem. Spoil material, however, may contain soluble salt like the 2.4 and 23 kg of salt per cubic meter reported for two strip mines in Colorado (McWhorter et al., 1974 and 1975). A groundwater sample from a seepage face in the low-salt-content spoil area contained 3 400 mg/l TDS, while groundwater from an observation well in the area had 2 200 mg/l TDS. The pH of the water was above 7. The salt concentration in leachate from the spoil decreased with time and was reduced by 95 to 97 percent after the total leachate volume had reached a value of 10 times the spoil volume. Dominant ions in leachate and runoff from coal-mine strips in the upper Colorado River basin were Na, Ca, Mg, SO_4, and HCO_3.

Metals

Tailings and processing wastes from mining and milling metal ores can affect local groundwater quality. In northern Idaho, for example, leaching from tailings produced by 80 years of mining has increased metal concentrations in groundwater (Mink et al., 1972). The mines produced primarily silver, lead, and zinc, with copper and cadmium as important byproducts. Many of the old tailings were intermixed with the upper parts of the aquifer. Metal concentrations in affected groundwater included 37 mg/l zinc, 0.4 mg/l cadmium, and 1.6 mg/l lead, which compare with concentrations less than 0.1 mg/l in unaffected groundwater. Copper could not be detected in the contaminated water. Leaching of metals from tailings was stimulated by oxidation of sulfides and resulting decrease in pH (Galbraith et al., 1972). In certain areas of New Mexico, copper can be sufficiently immobilized in the vadose zone to greatly delay contamination of deep groundwater below tailings or waste ponds (Runnells, 1976). Molybdenum, which frequently occurs in waste and tailings from copper mines, was more mobile. Nevertheless, calculations for a given case showed that it would take some 1 000 years for the molybdenum to reach a depth of 30 m.

Uranium

Leachate from tailings and processing wastewater from uranium mines and mills contain uranium and its family of isotopes (including ^{226}Ra), plus elements like selenium, molybdenum, and vanadium that are commonly found in uranium ores. In northwestern New Mexico, mining and milling of uranium has adversely affected groundwater quality by seepage from tailings ponds, seepage from dry washes and ephemeral streams in which mine water is discharged, and by deep-well injection of processing wastewater (Kaufmann et al., 1975 and 1976). Effluents of operating mines generally contained 100 to 150 pCi/l ^{226}Ra. Natural background levels of local groundwater usually were below several picocuries per liter. Radium, selenium, nitrate, and uranium were the most valuable indicators of groundwater contamination by uranium mining.

Wastewater from one mine studied by Kaufmann et al. (1975 and 1976) contained 1 390 mg/l Na, 2 010 mg/l Cl, 105 mg/l nitrate, 13 200 mg/l TDS, 7 340 pCi/l natural U, 166 000 pCi/l ^{230}Th, and 292 pCi/l ^{226}Ra, while the pH was 2.5. From 1953 to 1973, about 6.5 million m^3 of this wastewater had entered the groundwater, 3.7 million m^3 by deep-well injection and 2.8 million m^3 by seepage from tailings ponds. Despite these large volumes and the potent nature of the wastewater, serious groundwater contamination had not yet been observed. Two on-site wells showed slightly increasing trends for TDS, chloride, and sulfate. Seven off-site wells within 4 km of tailings ponds showed no contamination and produced water with ^{226}Ra concentrations of 0.06 to 0.50 pCi/l, which is well below the maximum limit of 5 pCi/l proposed by the U.S. Environmental Protection Agency (see Section 10.5). Only 1 of the 71 groundwater samples collected in Kaufmann's studies showed a ^{226}Ra concentration in excess of 5 pCi/l. This water, however, came from a well in a restricted area downgradient from a tailings pond. At another mine, groundwater from monitoring wells at 0.5 and 0.6 km from a tailings pond contained 1.92 and 0.36 pCi/l ^{226}Ra, which was higher than the average natural background level of 0.16 pCi/l. Increased concentrations of other radionuclides, including uranium and polonium^{210}, were also observed for the monitoring wells. The most significant contaminant, however, was selenium, which reached concentrations of 3.4 mg/l (340 times the upper limit for drinking water; see Section 10.5) in water from a domestic well about 1.5 km from a tailings pond. Other wells in the area also showed high concentrations of Se, so that alternate water supplies have to be developed. Seepage of uranium-mine wastewater discharged in a dry stream bed produced NO_3-N concentrations of 24 mg/l in underlying groundwater.

Wastewater and tailings leachate from uranium mines are potential threats to groundwater quality. While in the New Mexico studies contamination seemed to be confined to restricted areas of the mines themselves and to stream beds receiving wastewater, extensive monitoring of groundwater quality is required to make sure that contaminated water does not advance toward existing wells or future well sites.

Phosphate

Mining and milling of rock phosphate can result in groundwater contamination by ^{226}Ra and other decay products of uranium, which usually occur in small concentrations in the rock. Untreated effluents from phosphoric acid plants in Florida, for example, contained about 30 times the maximum permissible concentration of ^{226}Ra (Rouse, 1974). Seepage from wastewater storage ponds has already contaminated limestone aquifers in Polk County, Florida, to the extent that shallow groundwater over an area of about $1\,000\ km^2$ may contain ^{226}Ra in excess of the U.S. Public Health Service Drinking Water Standard of 3 pCi/l (see Section 10.5). The highest concentration of ^{226}Ra observed in the local groundwater was 79 pCi/l.

11.9 DISPOSAL OF RADIOACTIVE WASTES

Safe storage of radioactive waste is a major problem with the operation of nuclear reactors for power generation and wherever radioactive materials are produced or used. Increasing use of nuclear power will result in increasing quantities of radioactive waste. By the year 2020, for example, high-level radioactive wastes may amount to $900\,000\ m^3$ as liquid or $70\,000\ m^3$ as solid (calcined or incorporated into glass or ceramic) with a total radioactivity of long-lived nuclides of about 8.7×10^{10} Ci (Gera and Jacobs, 1972). Most of the waste consists of ^{90}Sr and ^{137}Cs, which have half-lives of 28 to 30 years and will decay to safe levels in about $1\,000$ years. Longer containment is needed for the transuranic isotopes Pu and Am, which along with their radioactive daughters may require $500\,000$ years of isolation from the environment (Winograd, 1974).

Commercial radioactive wastes are now being held for ultimate disposal in federally operated subsurface and surface repositories, which should be ready sometime in the 1980s. Possible sites for long-term underground confinement of radioactive wastes include bedded salt and salt domes, brine aquifers, thick shale or clay sequences, tunnels or dry mines in granite or desert hills, vadose zones in arid regions, and river deltas. Surface storage may be in thick vaults or pyramids that are cooled with air or water (Winograd, 1974, and references therein). For the immediate future, disposal in deep mines may be the best solution (Kubo and Rose, 1973). Future options include disposal in mausolea or mines and in situ melt, possibly combined with chemical separation of long-lived nuclides. In situ melt is a technique where radioactive waste is poured into a subterranean cavity created by an underground nuclear explosion. Radioactive decay of the wastes then causes them to heat up, boil dry, and eventually melt with the surrounding rock into a glassy ball (Kubo and Rose, 1973). While the technique appears attractive, disadvantages are possible migration of radioactive material into the surrounding rock, breakage of feed and steam lines connecting the disposal site to the surface by faulting or earthquakes, and leaching of nuclides by underground water.

Safe storage of radioactive wastes can undoubtedly be achieved. Great care should be taken, however, in selecting proper sites, monitoring, and preventing accidents. In one case, radioactive wastes requiring 200 000 years of isolation had already leaked out of their containers 12 years after interment (*Ground Water Newsletter*, 1976).

Radionuclides with short half-lives normally are disposed of on or in the ground, relying on ion adsorption to retain the nuclides while they decay to harmless levels. At the Savannah River Plant in South Carolina, for example, solid waste has been buried in a trench 6 m deep while very-low-level waste went into seepage basins. Soils were mostly sandy, and groundwater depths were 18 to 36 m (Reichert, 1962). Radionuclides higher than background levels could be detected in groundwater below the seepage basin, but not below the trench. The radionuclides of strontium migrated faster than those of cerium, cesium, ruthenium, zirconium, and niobium, probably because the soil had a low pH and contained sandy streaks through which ^{90}Sr could rapidly move away from the basins. Concentrations of nonvolatile beta emitters (primarily ^{90}Sr) were in excess of 1 000 pCi/l for a distance of about 50 m on one side of a seepage basin. The highest level was 3 700 pCi/l. ^{90}Sr was still detectable in groundwater as far as 150 m from the basins. Plans for reducing the hazards of groundwater contamination included the use of bentonite or other sealing material to decrease infiltration, and to excavate and remove the layers of sandy material adjacent to the seepage basins.

At the Oak Ridge National Laboratory in Tennessee, one of the five burial grounds for radioactive wastes was found to be a major contributor of ^{90}Sr to the groundwater (Duguid, 1974). The site, which measured about 10 ha, had been used since 1951, primarily for disposal of contaminated chemicals and laboratory equipment. Disposal was stopped in 1959, after which the area was used for disposal of uncontaminated fill. This resulted in higher infiltration and leaching rates, which raised the underlying water table. The increased leaching and groundwater flow caused movement of 1 to 2 Ci of ^{90}Sr to a nearby creek. Concentrations of ^{90}Sr in the local groundwater ranged from 90 to 16 400 pCi/l. Concentrations in excess of 180 000 pCi/l were measured in overflow from the trench when it was completely filled with water. The leachate also contained ^{137}Cs, but this was largely adsorbed in the underlying shale.

Operations at the atomic energy facility near Richland, Washington, have resulted in the disposal of some 500 million m^3 low-level waste (cooling water) in ponds and swamps and of 30 million m^3 intermediate-level waste in cribs or trenches since 1944 (Phillips and Raymond, 1975). The cribs are inverted timber boxes placed 3 to 6 m below ground surface. They were filled with gravel to improve percolation and were covered with soil. The trenches are about 6 m deep. The waste liquids must percolate through 50 to 100 m of silts, sands, and gravels before reaching an unconfined aquifer of sedimentary material underlain by basalt. Waste disposal has raised local groundwater levels 3 to 23 m. The longer-lived and more hazardous nuclides like ^{95}Sr, ^{137}Cs, and ^{239}Pu were effectively adsorbed in the vadose zone by ion exchange. Other nuclides, such as ^{106}Ru, ^{99}Tc, and ^{3}H, were more mobile and reached the groundwater. More than 600 cased monitoring wells were installed in the area, primarily to follow the movement of the more mobile nuclides ^{106}Ru and ^{3}H. Based on gross beta radiation (calculated as ^{106}Ru), groundwater zones with activities of 100 to 1 000, 1 000 to 10 000, and > 10 000 pCi/l were distinguished. The values above 10 000 pCi/l, which were the only ones exceeding the concentration guide limit for ^{106}Ru, were found only immediately beneath some disposal sites. Groundwater movement was sufficiently slow in relation to decay rates of higher-level contaminants that there was little danger of groundwater contamination outside the study area. Tritium concentrations in the groundwater ranged from 1 000 to > 3 000 000 pCi/l, with the high values occurring below and adjacent to a few disposal sites. Tritium concentrations in groundwater outside the study area were expected to be below detection limits.

Surface and atmospheric nuclear explosions have had only minor effects on radionuclide contents of groundwater. Hazardous nuclides such as ^{90}Sr and ^{137}Cs are adsorbed in most soils and can enter groundwater only through coarse sands and gravels or through exposed fractured rock. Concentrations in groundwater of highly mobile nuclides like ^{3}H, ^{36}Cl, and ^{14}C are generally not hazardous.

Underground nuclear explosions must be carefully planned to avoid groundwater contamination with fission products that have condensed on the rock surfaces of the chimney rubble region (Levy, 1972). Other contamination sources from underground explosions are residual radioactivity from the exploded device and induced radioactivity in the surrounding rock by escaping neutrons. The increased porosity of the rubble chimney will draw surrounding groundwater to the chimney so that groundwater flow initially will be toward the site of the explosion. Since it may take months or years before natural groundwater flow is restored, this can significantly delay movement of radionuclides into adjacent aquifers (Levy, 1972). Because groundwater movement tends to be slow, it is usually sufficient to consider only radionuclides with half-lives of more than a few months or a year. Using theoretical relations to predict attenuation due to decay, ion exchange, and dispersion, Levy (1972) calculated for a hypothetical underground explosion that, after 1.6 km of underground movement, ^{90}Sr and ^{137}Cs will have completely decayed away and ^{3}H will be below the maximum permissible concentration (MPC), which is 3 μCi/l according to the Code of Federal Regulations. The same code specifies an MPC of 300 pCi/l for ^{90}Sr and 20 000 pCi/l for ^{137}Cs (Levy, 1972, and references therein).

Predictions for idealized conditions (uniform aquifers, uniform adsorption coefficients, etc.) must be used with great caution when applying them to actual situations. Thin layers of very permeable material in aquifers, fractured rock, or highly anisotropic materials could make theoretical predictions virtually useless and cause contaminants to arrive at certain points much sooner than anticipated.

It could also cause them to arrive where they are not expected at all or to travel in a much broader band than anticipated (Levy, 1972; Fisher, 1972). Extensive post-explosion monitoring is always necessary to make sure that radioactive contamination has decreased to acceptable levels at points of groundwater withdrawal.

11.10 UNDERGROUND STORAGE OF LIQUID WASTE

Waste fluids that are difficult to treat or dispose of are increasingly stored underground with injection wells. In contrast to land application of sewage effluent or similar wastewater where soil is used to purify water on a renewable basis, underground storage of waste liquids is a nonrenewable use of the earth's crust (Bouwer, 1976a). The underground storage space must be used carefully and sparingly, making sure that waste fluids will not move into fresh groundwater and that they remain isolated from the hydrologic cycle. Containment is not always assured even under seemingly favorable conditions, because the waste fluids may migrate over long distances and eventually find their way into freshwater aquifers.

Waste and Well Characteristics

Underground storage of waste liquids by deep-well injection is primarily practiced by chemical, petrochemical, and pharmaceutical industries, and, to a lesser extent and in decreasing order, by refineries, gas plants, and metal industries. The total number of industrial waste-injection wells in the United States recently was estimated to be at least 278, of which 34 were used rarely and 11 not at all (Warner and Orcutt, 1973). These numbers do not include oil-field brine return wells, of which there are more than 20 000 in Texas alone (Warner and Orcutt, 1973). Return of oil-field brines to the same formation from where they came up with the oil is frequently practiced in oil production. Municipal sewage and storm runoff are occasionally injected underground—for example, in saline, cavernous limestone aquifers (Garcia-Bengochea et al., 1973).

Injection wells are constructed in basically the same way as pumped wells. Casings and screens should resist corrosion by the waste liquids, and the wells should be carefully grouted to prevent migration of fluid along the casing to freshwater aquifers or to ground surface. Most of the underground formations used for waste storage are saline aquifers of sands, sandstones, and carbonate rock of Permian age or older. Well depths generally are between 300 and 1 800 m. Injection rates are less than 260 m^3/day for 43 percent of the industrial wells and less than 2 160 m^3/day for 85 percent of the wells (Warner and Orcutt, 1973). Injection pressures at the well head are zero (gravity feed) for 21 percent of the wells, between 0 and 10 kg/cm^2 for 19 percent, 10 and 20 kg/cm^2 for 21 percent, 20 and 40 kg/cm^2 for 17 percent, and 40 and 100 kg/cm^2 for 21 percent of the wells. A detailed survey of injection wells, showing type of waste liquid, well characteristics, and geology of target formation was published by the U.S. Environmental Protection Agency (1974).

The receiving formation should be separated from freshwater aquifers or from the surface environment by (essentially) impermeable layers. Storage of waste liquid then is predicated on elastic deformation of the receiving strata by the injection pressure, or on inconsequential displacement of fluids already present in the target formation (Wolff et al., 1975).

Injected liquids should be low in suspended solids to avoid clogging of the hole wall and the aquifer. The liquid should not chemically react with underground materials to form precipitates, to open pathways through confining layers, or to cause clay to deflocculate. Also, the liquid should not create favorable conditions for bacterial growth that could reduce hydraulic conductivity. Finally, the target zone should be sufficiently permeable or elastic to accept the waste liquid at the desired rate.

Quality Changes

Since the main objectives of underground disposal are permanent storage and isolation from the hydrologic cycle, quality changes in the waste liquid as it migrates through the receiving formation are of minor importance. Physical, chemical, and biological processes, however, do take place and may somewhat improve the quality of the waste liquid (Kaufman et al., 1973; Kharaka, 1973; DiTommaso and Elkan, 1973; Donaldson and Johansen, 1973; Peek and Heath, 1973). On the other hand, quality degradation like increases in iron (Vecchioli and Ku, 1972) and other ions mobilized in the receiving formation may also occur. Salt sieving in clay layers acting as semipermeable membranes can reduce the salt content of waste liquids moving through such layers into freshwater aquifers (Kharaka, 1973).

Hydraulic Fracturing

To prevent migration of injected liquids into freshwater aquifers, injection pressures must be less than the pressure whereby hydraulic fracturing of the receiving formation begins to occur. Such fractures could propagate into confining layers and provide pathways for waste liquids to move into overlying aquifers.

Hydraulic fracturing is a common technique for increasing the production of oil wells. It was generally assumed that fracturing was initiated when hydraulic pressures exceeded the pressure due to the weight of the overburden material, and that the fractures developed laterally and advanced parallel to bedding planes. Subsequent analyses and measurements have indicated, however, that fracturing already begins at lower injection pressures and that the fractures are vertical and extend radially from the well (Hubbert and Willis, 1957). The increase in well productivity then is caused by improved communication between permeable strata due to fracturing of the slowly permeable layers separating them.

Hubbert and Willis (1957) distinguished three mutually perpendicular stresses in underground formations: the major, intermediate, and minor principal stresses. In tectonically relaxed areas, the major principal stress is vertical and equal to the overburden pressure. The two other principal stresses are horizontal and, if the medium is isotropic, equal. The liquid pressure in the bore hole necessary to initiate fracturing theoretically is equal to the minor principal stress plus the tensile strength of the rock. In reality, however, additional liquid pressure is generally required to overcome distortion of the stress field in the rock around the well due to the drilling action. Also, as the fractures propagate into the formation, additional liquid pressure in the bore hole is required to overcome friction losses in the fractures themselves. When injection is stopped and liquid is prevented from leaving the well, the pressure of the liquid will drop to the level required to keep the fractures open. This pressure, which is called the *instantaneous shut-in* pressure, represents the minor principal stress of the target formation (Wolff et al., 1975). Where hydraulic fracturing is used to increase well productivity, liquids or gels are injected into the formation at high pressures. Glass beads or other granular "propping" agents are sometimes mixed with the injected liquid to keep the new fractures open when liquid pressures decrease to normal values.

The effectiveness of hydraulic fracturing for increasing injection rates of waste-disposal wells was demonstrated for a disposal well near Denver, Colorado, where pressure increases resulting from a threefold increase in injection rates doubled the transmissivity of the receiving formation, which was a fractured Precambrian gneiss (Pickett, 1968). Hydraulic fracturing of consolidated rock combined with mixing the waste in a cement slurry or other grouting material that hardens in place after injection is a technique for disposing of hazardous wastes (including radioactive materials) that must remain immobile once stored underground (Sun, 1973).

If hydraulic fracturing is to be avoided (for example, to prevent fracturing of confining layers separating the receiving formation from overlying freshwater aquifers), liquid pressures inside the well should be kept at values that are considerably less than the overburden pressure (major principal stress) at any depth. In tectonically relaxed areas, such injection pressures should not exceed two-thirds of the overburden pressure. For other formations, lower values may be desirable. Hubbert and Willis (1957), for example, show a minor principal stress in dolomite limestone that is equal to about one-half the overburden pressure. If it is desirable to use injection pressures that are higher than safe, nonfracturing values, state-of-stress measurements should be made for the receiving formation to make sure that hydraulic fracturing is not occurring (Wolff et al., 1975).

Where the injected waste liquid is saline and the receiving formation is separated from a freshwater aquifer by a clay layer acting as a semipermeable membrane, additional pressures can develop in the receiving formation by inmigration of fresh water from the aquifer through the clay layer due to osmotic effects (Hanshaw, 1972; see also Section 10.3). This has two consequences: (1) injection pressures at the well head must be higher if the same injection rate is to be maintained, and (2) the buildup of pressure in the receiving formation, which may continue for tens or hundreds of years after injection is stopped, could eventually cause hydraulic fracturing of the receiving formation and of the confining clay layer, even if safe injection pressures were used initially.

Environmental, Legal, and Economic Constraints

In addition to movement through induced fractures, injected waste liquids or saline formation water displaced by injected wastes may migrate into freshwater aquifers through fault zones or other weak spots in confining layers, through abandoned and improperly sealed wells, or along the casing of the injection well itself if it is inadequately grouted and sealed or if the casing is corroded and leaking. Sometimes, movement of waste or saline liquids into fresh groundwater can occur at a considerable distance from the injection wells. Some of these escape paths and resulting return of the fluid to the hydrologic cycle are difficult to foresee when hydrogeologic investigations are carried out to determine the suitability of a certain site for waste disposal by well injection.

Another adverse environmental effect of well injection is increased seismicity, which may result from entry of pressurized waste liquids into stressed faults, causing the fault planes to move. The number of earthquakes in the Denver, Colorado, area, for example, was directly related to the calculated pressure of waste liquids injected into an underground formation (Pickett, 1968). The area already was seismically active before injection ever started (Simon, 1969), but injection of waste liquid probably increased the severity of the seismicity (Healy et al., 1968; Harrison and Longley, 1970). It has also been postulated that fluid injection (mainly brines) for secondary oil recovery in the Inglewood field caused, or contributed to, surface cracking in the Baldwin Hills area of Los Angeles County, California, and resulting failure of the Baldwin Hills reservoir (Hamilton and Meehan, 1971; Castle et al., 1973).

Well injection of waste fluids normally is regulated by the states. Before a permit can be obtained, intensive hydrogeologic investigations may be required to make sure that the injection well will not endanger freshwater supplies. Contingency plans showing how the waste liquid will be handled in case of failure of the injection well may also have to be submitted. Injection of radioactive waste is subject to federal control and regulations.

Underground storage of waste liquids also is legally constrained by the property rights of adjacent landowners or those having mineral rights. Such rights traditionally extend to the center of the earth and legal action on grounds like nuisance, negligence, liability, and trespassing may result if such rights are violated by the injection of waste liquids (Walker and Cox, 1973).

Well injection of waste liquids is not cheap. If the cost of preinvestigations is added to that of constructing the wells and installing pumps and pipelines to transport the wastewater to the well, several hundred thousands of dollars may have been invested before the first liter of waste can be injected. In addition, monitoring of local groundwater quality during and after injection may be necessary, which adds to the cost. Depending on the volume of waste liquid injected over the life of the well, costs of underground storage may range from less than $1 per cubic meter to several tens of dollars per cubic meter of waste fluid. This high cost is another reason why well injection is used primarily as a last-resort solution to a waste problem.

11.11 LAGOONS AND EVAPORATION PONDS

Lagoons are commonly used in the treatment and/or disposal of industrial wastes. If the lagoons are not lined, they will leak and the resulting seepage may contaminate underlying groundwater. Hexavalent chromium, for example, was detected in groundwater below an unlined lagoon in which chromium-laden process water was disposed of (Yare, 1975). The lagoon was located in coastal plain sediments in southern New Jersey. The contamination zone advanced at a rate of 18.3 cm/day and had moved a distance of 915 m from the source in 14 years. Resistivity surveys were carried out to determine the lateral and vertical extent of the contaminated zone. This minimized the number of observation wells that needed to be drilled to obtain samples of the contaminated groundwater (Berk and Yare, 1977). In an old coal strip-mine area of Ohio, test holes adjacent to a disposal pit for spent pickling liquor from the steel industry yielded groundwater with a pH of around 4 and an iron concentration that sometimes exceeded 1 000 mg/l (Pettyjohn, 1975). In California, discharge of industrial waste and sewage in the normally dry Mojave River near Barstow has produced a contamination plume in groundwater that advances 30 to 45 cm/day. The plume, which has traveled a distance of 6.4 km since 1910, has forced abandonment of several domestic wells and is threatening others (Hughes, 1975).

On Long Island, New York, contamination of groundwater by hexavalent chromium from disposal basins receiving wastewater from an aircraft plant where chromium was used in anodizing aluminum and aluminum alloys was first detected in 1942, when water from a private well near a disposal pit was found to contain 0.1 mg/l Cr (Pinder, 1973, and references therein). Test wells, installed in 1945 at about 100 m from the plant, in 1948 showed 1 to 3.5 mg/l Cr, along with Cd, Cu, and Al. Similar metal contamination was detected in a shallow domestic well 450 m south of the disposal pond. Additional test wells drilled in 1949 and 1950 indicated that the contamination plume at that time was about 1 200 m long and 260 m wide (maximum), and had reached Massapequa Creek in which the aquifer drained. The rate of advance of the plume had been about 0.43 m/day. In 1953, 22 new test wells were installed and more wells were added in 1958 and 1962, at which time the plume had become about 1 300 m long and 300 m wide (maximum). Meanwhile, peak chromium concentrations in well samples decreased from 40 mg/l in 1940 to 14 mg/l in 1962, as observed 900 m from the disposal ponds. The aquifer was about 25 to 43 m thick and consisted of glacial outwash materials with a porosity of 35 percent and K values of 173 and 648 m/day, as determined on two samples. The behavior of the plume was accurately simulated with a finite-element model, which indicated that contamination of Massapequa

Creek by chromium-laden water from the aquifer would continue about 7 years after disposal of metal in the ponds has been halted (Pinder, 1973).

Industrial brines, including brines from oil fields, are commonly disposed in evaporation ponds which should be lined to exclude seepage. Unfortunately, some pits leak and contaminate underlying groundwater. In southwest Arkansas, for example, seepage from an oil-field brine-disposal pit contaminated the underlying aquifer over an area of 3 km^2 (Fryberger, 1975). The situation was aggravated by leakage from an abandoned, corroded oil well that was used for injection of some of the brines. The TDS content of the brine was 84 750 mg/l, mostly as NaCl. Because of the high density of the brine, groundwater levels in the polluted zone were depressed about 0.5 m. Chloride levels in the contaminated zone ranged from a few hundred to almost 50 000 mg/l, with the high values occurring primarily in the lower part of the aquifer near the evaporation pit. The contamination was first detected when a farmer lost his rice crop because the water from his irrigation well had become salty. The contaminated groundwater is slowly moving to the Red River about 7 km away. After discontinuation of brine disposal, it would take at least 250 years before all contaminated water will have moved out of the aquifer. Changing the disposal method to deep-well injection and letting the salt slug move naturally to the river was cheaper than other measures for rehabilitating the aquifer, such as containment of affected groundwater with a bentonite wall, acceleration of movement to the river by groundwater recharge near the source, or pumping the salty water out of the aquifer for desalting or blending with other water.

11.12 OIL LEAKS AND SPILLS

Gasoline and other petroleum products can enter soils and aquifers from leaking pipelines or storage tanks and from accidents involving tank trucks or railroad cars. Most groundwater contamination cases are caused by underground tanks from gasoline stations. In Maryland alone, 60 cases of groundwater contamination with petroleum products were reported in the period 1969–1970 (Matis, 1970). Numerous other cases, mostly of local significance, were reported by McKee et al. (1972).

The best way to detect gasoline or other petroleum products in well water is by smell and taste. This " technique " has threshold values of 0.005 mg/l for gasoline and 0.01 mg/l for fuel oil and is more sensitive than gas chromatography (Matis, 1971, and references therein). The main problem of petroleum contamination of groundwater is taste. Toxicity is not a problem because the water is already undrinkable due to taste and odor well before concentrations reach toxic levels. The solubility of modern gasolines in water ranges from 20 to 80 mg/l and averages about 50 mg/l (McKee et al., 1972). Free gasoline or oil may actually float on the groundwater in severe cases. In some instances, gasoline-contaminated groundwater has seeped into basements. Resulting gasoline vapors then produce serious fire hazards, and explosions have been reported for two homes (Rhindress, 1971). Early detection of gasoline or other petroleum-product leaks and immediate action to stop the leak and contain the contamination are the best ways to minimize undesirable effects (Hall and Quam, 1976). Remedial action may include pumping of wells to remove the oil from the aquifer, digging of collection pits or interceptor trenches, and excavating contaminated soil. Holzer (1976) reported fuel-oil contamination of groundwater beneath oil-products storage tanks in New England. The affected area was about 1.6 ha. The oil-bearing zone floated on the groundwater and was less than 0.45 m thick. To avoid drainage of affected groundwater into a river by an existing drain, a new tile drain was installed upgradient from the old drain to intercept the water for processing in a separation system. In one year, 22.3 m³ of oil were collected this way.

A well-documented case of groundwater contamination by a leaking gasoline pipeline is that of Forest Lawn Memorial Park near Los Angeles, California (McKee et al., 1972; Williams and Wilder, 1971). The contamination was first detected by a peculiar taste and odor of water from a drinking fountain supplied by a local well. When pumping of another well in the area was stopped, free gasoline floated on the water in that well and could be removed by bailing. By that time, an area of about 1.5 ha was affected. The gasoline was concentrated near the top of the groundwater. The thickness of the free-gasoline layer in the aquifer ranged between 15 and 45 cm. The total amount of gasoline that had leaked from the pipeline was estimated at 400 to 1 000 m³. To prevent the spread of gasoline in the aquifer, the pipeline was drained and filled with water. A nearby well field was closed to reduce velocity of groundwater, and 70 wells were drilled in the affected area to remove the gasoline from the aquifer. These wells were then pumped with small jet pumps placed just below the " water " level to collect as much free gasoline as possible. In 3 years of cleanup, over 200 m³ of gasoline were removed. Although gasoline was still left in the aquifer, most wells no longer showed measurable amounts of gasoline. According to the theory of two-phase flow, gasoline in aquifers cannot be completely replaced by water and some gasoline will remain entrapped in the aquifer. Extensive laboratory studies by McKee et al. (1972) of the simultaneous movement of air, water, and gasoline in soil columns showed that gasoline tends to cling to soil particles (referred to as pellicular gasoline), particularly in the dewatered zone above a falling water table. Even after passing 844 pore volumes of water through a gasoline-contaminated soil column, gasoline could still be tasted in the outflow. Final removal of gasoline may thus have to rely on bacterial decomposition by species like *Pseudomonas* and *Arthrobacter*, which are particularly active under aerobic conditions (McKee et al., 1972).

11.13 URBAN RUNOFF AND POLLUTED SURFACE WATER

Many streams receive municipal and industrial wastewater. Seepage of such water into underlying groundwater may adversely affect groundwater quality. This type of groundwater contamination can be minimized by adequate treatment of wastewater before discharging it into streams. Another potential source of groundwater contamination is urban runoff (Whipple et al., 1974), which may infiltrate directly into the ground through or adjacent to pavements, after it has reached a stream, or via recharge pits or "dry wells" constructed for disposal of storm runoff.

The chemical composition of urban runoff depends on the land use (residential, industrial, etc.) and on rainfall characteristics. Analyses of urban runoff thus show wide variation, as indicated in Table 11.6, which applies to runoff from a 440-ha area in Durham, North Carolina (Colston and Tafuri, 1975). Average population density in the area was 2.4 per ha. Land use was 59 percent residential, 19 percent commercial, 12 percent public and institutional, and 10 percent idle. The data in the table show that urban runoff is a potential pollutant that should be kept out of aquifers. In other studies, urban runoff sampled at the edge of a road bed in Oklahoma City contained 3.6 to 8.5 mg/l of lead (average 5.5 mg/l), primarily from vehicular emissions (Newton et al., 1974). An average lead concentration of 0.48 mg/l was measured in urban runoff from Durham, North Carolina (Bryan, 1974). BOD values of 4 to 15 mg/l and nitrate and phosphate concentrations of several milligrams per liter were reported by Whipple et al. (1974). Wear from rubber tires may contribute zinc and cadmium to urban runoff. For a discussion of de-icing salts in runoff and melt water from streets, reference is

Table 11.6 Pollutant concentrations in urban runoff from an area in Durham, North Carolina

Pollutant	Mean mg/l	Standard deviation	Range (mg/l)	
			Low	High
COD	170	135	20	1 042
TOC	42	35	5.5	384
Total solids	1 440	1 270	194	8 620
Volatile solids	205	124	33	1 170
Total suspended solids	1 223	1 213	27	7 340
Volatile suspended solids	122	100	5	970
Kjeldahl nitrogen as N	0.96	1.8	0.1	11.6
Total phosphorus as P	0.82	1.0	0.2	16
Fecal coliform per ml	230	240	1	2 000
Aluminum	16	8.15	6	35.7
Calcium	4.8	5.6	1.1	31
Cobalt	0.16	0.11	0.04	0.47
Chromium	0.23	0.10	0.06	0.47
Copper	0.15	0.09	0.04	0.50
Iron	12	9.1	1.3	58.7
Lead	0.46	0.38	0.1	2.86
Magnesium	10	4.0	3.6	24
Manganese	0.67	0.42	0.12	3.2
Nickel	0.15	0.05	0.09	0.29
Zinc	0.36	0.37	0.09	4.6
Alkalinity	56	30	24	124

Source: From Colston and Tafuri, 1975.

made to Section 11.3. Control of urban runoff and its effect on groundwater quality may be expensive.

In addition to the material presented in this chapter, general discussions of groundwater contamination have been prepared by Todd and McNulty (1976) and Lehr et al. (1976). The latter also describe laws, regulations, and institutional requirements for control of groundwater pollution. Waste, wastewaters, and residues are unavoidable by-products of modern society. They cannot be legislated away, and they must be disposed of, stored, recycled, or rendered harmless in one way or another. Usually, there are no good solutions but only various alternatives from which the one that is economically the most attractive and environmentally the least undesirable must be selected. Groundwater, though inconspicuous and out of sight, is a very important part of the environment that must be protected for the benefit of present and future generations.

PROBLEMS

11.1 Derive Eq. (11.10).

11.2 A freshwater ridge 4 km in length is to be maintained by recharge wells parallel to a coastline to prevent saltwater intrusion into a coastal aquifer with $K = 4$ m/day. If the freshwater-saltwater interface must be at a depth of at least 50 m below sea level at a distance of 200 m inland from the shore, what is the minimum amount of injected fresh water that will be lost to the ocean [use Eq. (11.7)]? What will be the width of the zone on the ocean bottom through which fresh water will seep into the ocean?

11.3 A well is drilled in the coastal aquifer of Problem 11.2 so that the well bottom is 20 m above the freshwater-saltwater interface. To prevent salt water from entering the well, upconing should be less than 10 m. What is the maximum permissible discharge of the well [Use Eq. (11.12)]?

11.4 Sewage effluent with a TDS content of 900 mg/l is to be renovated by a rapid-infiltration groundwater recharge system in an area where rainfall is 20 cm/year and potential evapotranspiration 180 cm/year. If the renovated water should contain no more than 1 000 mg/l TDS, what is the minimum annual recharge rate?

11.5 Wet sludge (98 percent water) with a nitrogen content of 4 percent on a dry-weight basis is to be disposed of on a field with a corn crop that normally requires 70 kg/ha of fertilizer nitrogen per year. How many cubic meters of the wet sludge can be applied per hectare without creating undue soil or groundwater pollution problems?

11.6 Corn is irrigated every 2 weeks in an arid area with an evapotranspiration rate of 0.8 cm/day. The irrigation water has a TDS content of 640 mg/l. How much water should be applied per irrigation to maintain a salt balance in the root zone (evaluate C_d from Figure 10.1)? How much deep-percolation water will be produced per irrigation, what is its salt concentration, and what is the salt load on the underlying groundwater in kilograms per hectare per irrigation? Repeat for irrigation water with a TDS content of 1 493 mg/l.

11.7 Assuming that the irrigation system in Problem 11.6 is inefficient and that 4 cm more water are applied than needed (i.e., 17.07 cm and 20.8 cm for the two waters), what will be the amount and salt concentration of the deep-percolation water and the salt load on the underlying groundwater for both irrigation waters? Compare the values of C_d and salt load with those for the efficient irrigation system of Problem 11.6. What system is better?

11.8 Compare pesticide and radioactivity levels in groundwater reported in Sections 11.7, 11.8, and 11.9 with maximum limits for drinking water given in Section 10.5.

11.9 A new industrial-waste lagoon leaks at a rate of 0.455 cm/day. The lagoon is located in a desert area with the water table at a depth of 100 m and a dry vadose zone with a volumetric water content of

0.05. The soil of the vadose zone is a sandy loam. The relation between unsaturated hydraulic conductivity and water content of this soil is the same as the one shown in Figure 7.26. How long will it take before contaminated seepage water can be expected to reach the groundwater?

11.10 Waste liquids with a density of 1 g/cm^3 are to be injected into a saline sandstone aquifer at a depth of 1 000 m. To avoid fracturing, injection pressures at the target depth should not exceed 50 percent of the overburden pressure. If the average weight of the overburden is 2.2 g/cm^3 and the friction loss in the piping system inside the well is 10 m pressure head, what is the maximum injection pressure head at the well head?

REFERENCES

Achari, R. G., S. S. Sandhu, and W. J. Warren, 1975. Chlorinated hydrocarbon residues in ground water. *Bull. Environm. Contam. and Toxicol.*, Springer Verlag, New York, **13**(1): 94–96.

Am. Soc. Civ. Eng., 1976. *Sanitary Landfill.* Prepared by Solid Waste Manag. Comm., ASCE Manuals and Reports on Engineering Practice No. 39, 50 pp.

Anderson, J. R., and J. N. Dornbusch, 1967. Influence of sanitary landfill on ground-water quality. *J. Am. Water Works Assoc.* **59**: 457–470.

Apgar, M. A., and D. Langmuir, 1971. Ground-water pollution potential of a landfill above the water table. *Ground Water* **9**(6): 76–94.

Bachmat, Y., and J. Bear, 1964. The general equations of hydrodynamic dispersion in homogeneous, isotropic, porous mediums. *J. Geophys. Res.* **69**: 2561–2567.

Badon Ghyben, W., 1888. Nota in verband met de voorgenomen putboring nabij Amsterdam. *Tijdschrift Koninklijk Instituut van Ingenieurs*, The Hague, Netherlands, pp. 8–22.

Banks, R. B., and I. Ali, 1964. Dispersion and adsorption in porous media flow. *J. Hydraul. Div., Am. Soc. Civ. Eng.* **90**(HY5): 13–31.

Bear, J., 1969. Hydrodynamic dispersion. In *Flow through Porous Media*, R. J. M. DeWiest (ed.), Academic Press, New York and London, pp. 109–199.

Bear, J., and G. Dagan, 1968. Solving the problem of local interface upconing in a coastal aquifer by the method of small perturbations. *J. Hydraul. Res.* **6**(1): 16–44.

Berk, W. J., and B. S. Yare, 1977. An integrated approach to delineating contaminated groundwater. *Ground Water* **15**(2): 138–145.

Biggar, J. W., and D. R. Nielsen, 1960. Diffusion effects in porous materials. *J. Geophys. Res.* **65**: 2887–2895.

Blakeslee, P. A., 1973. Monitoring considerations for municipal wastewater effluent and sludge application to the land. *Proc. Conf. on Recycling Municipal Sludges and Effluent on Land*, U.S. Environmental Protection Agency, U.S. Dept. of Agriculture and Nat. Assoc. State Univ. and Land-Grant Coll., Washington, D.C., pp. 183–198.

Bouwer, H. 1969. Salt balance, irrigation efficiency, and drainage design. *J. Irrig. Drain. Div., Am. Soc. Civ. Eng.* **95**(IR1): 153–170.

Bouwer, H. 1974. Design and operation of land treatment systems for minimum contamination of groundwater. *Ground Water* **12**(3): 140–147.

Bouwer, H. 1976a. Use of the earth's crust for treatment or storage of sewage effluent and other waste fluids. *Crit. Reviews in Envir. Contr.*, Chem. Rubber Comp. Press, **6**(2): 111–130.

Bouwer, H. 1976b. Zoning aquifers for tertiary treatment of wastewater. *Ground Water* **14**(6): 386–395.

Bouwer, H., and R. L. Chaney, 1974. Land treatment of wastewater. In *Adv. in Agron. 26*, N. C. Brady (ed.), Academic Press, Inc., New York, pp. 133–176.

Bouwer, H., J. C. Lance, and M. S. Riggs, 1974. High-rate land treatment II: water quality and economic aspects of the Flushing Meadows project. *J. Water Poll. Contr. Fed.* **46**: 844–859.

Bower, C. A., J. R. Spencer, and L. O. Weeks, 1969. Salt and water balance, Coachella Valley, California. *J. Irrig. Drain. Div., Am. Soc. Civ. Eng.* **95**(IR1): 55–64.

Brausz, F. W. 1952. Bedeutet die Nachbarschaft eines Friedhofes in jedem Fall eine Gefährdung der Wasserversorgung? *Gesund. Ing.* **73**(5/6): 86–89.

Bredehoeft, J. D., and G. F. Pinder, 1973. Mass transport in flowing groundwater. *Water Resour. Res.* **9:** 194–210.

Broadbent, F. E., and F. Clark, 1965. Denitrification. In *Soil Nitrogen*, W. V. Bartholomew and F. Clark (eds.), Agronomy Monograph No. 10, Am. Soc. Agron., Madison, Wis. pp. 344–359.

Brown, R. E., 1975. Significance of trace metals and nitrates in sludge soils. *J. Water Poll. Contr. Fed.* **47:** 2863–2875.

Bruch, J. C., 1970. Two-dimensional dispersion experiments in a porous medium. *Water Resour. Res.* **6:** 791–800.

Bruch, J. C., and R. L. Street, 1967. Two-dimensional dispersion. *J. Sanit. Eng. Div., Am. Soc. Civ. Eng.* **93** (SA6): 17–39.

Brunner, D. R., and D. J. Keller, 1972. Sanitary landfill design and operation. *U.S. Environmental Protection Agency Report SW-65ts*, 59 pp.

Bryan, E. H., 1974. Concentrations of lead in urban stormwater. *J. Water Poll. Contr. Fed.* **46:** 2419–2421.

California Department of Water Resources, 1961. Feasibility of reclamation of water from wastes in the Los Angeles Metropolitan Area. *Bull. 80, Calif. Dept. of Water Resources, Sacramento.*

California Department of Water Resources, 1971. Nutrients from tile drainage systems. *Report 13030 ELY5/71-3, California Dept. of Water Resources*, Sacramento, 90 pp.

Cartwright, K., and M. R. McComas, 1968. Geophysical surveys in the vicinity of sanitary landfills. *Ground Water* **6**(5): 23–30.

Castle, R. O., R. F. Yerkes, and T. L. Youd, 1973. Ground rupture in the Baldwin Hills—an alternative explanation. *Bull. Assoc. Eng. Geol.* **10**(1): 21–46.

Chhatwal, S. S., R. L. Cox, D. W. Green, and B. Ghandi, 1973. Experimental and mathematical modeling of liquid-liquid miscible displacement in porous media. *Water Resour. Res.* **9:** 1369–1377.

Chian, E. S. K., and F. B. DeWalle, 1976. Sanitary landfill leachates and their treatment. *J. Env. Eng. Div., Am. Soc. Civ. Eng.* **102**(EE2): 411–431.

Colston, N. V., and A. N. Tafuri, 1975. Urban land runoff considerations. In *Urbanization and Water Control*, W. Whipple, Jr. (ed.), Am. Water Resour. Assoc., Minneapolis, Minn., pp. 120–128.

Dean, R. B., and J. E. Smith, Jr., 1973. The properties of sludges. *Proc. Conf. on Recycling Municipal Sludges and Effluent on Land*, U.S. Environmental Protection Agency, U.S. Dept. of Agriculture, and Nat. Assoc. State Univ. and Land-Grant Coll., Washington, D.C., pp. 39–47.

DeWiest, R. J. M., 1969. Fundamental principles of ground-water flow. In *Flow through Porous Media*, R. J. M. DeWiest (ed.), Academic Press, New York and London, pp. 1–52.

DiTommaso, A., and G. H. Elkan, 1973. Role of bacteria in decomposition of injected liquid waste at Wilmington, North Carolina. In *Proc. 2nd Int. Symp. on Underground Waste Management and Artificial Recharge, 26 to 30 September, New Orleans, La.*, Vol. 1, J. Braunstein (ed.), Am. Assoc. of Petrol. Geol., U.S. Geol. Survey, and Int. Assoc. of Hydrol. Sci., Tulsa, Okla, pp. 585–599.

Donaldson, E. C., and R. T. Johansen, 1973. History of a two-well industrial-waste disposal system. In *Proc. 2nd Int. Symp. on Underground Waste Management and Artificial Recharge, 26 to 30 September, New Orleans, La.*, Vol. 1, Braunstein (ed.), Am. Assoc. of Petrol. Geol., U.S. Geol. Survey, and Int. Assoc. of Hydrol. Sci., Tulsa, Okla. pp. 603–621.

Duguid, J. O. 1974. Groundwater transport of radionuclides from buried waste: a case study at Oak Ridge National Laboratory, *ORNL Report WASH-1332*, 18 pp.

Dutt, G. R., M. J. Shaffer, and W. J. Moore, 1972. Computer simulation model of dynamic bio-physiochemical processes in soils. *Agric. Exp. Sta., Univ. of Arizona, Tucson, Techn. Bull. 196*, 101 pp.

Edwards, C. A., 1970. *Persistent Pesticides in the Environment.* Chem. Rubber Comp. Press, Cleveland, Ohio, 78 pp.

Edwards, W. M., and B. L. Glass, 1971. Methoxychlor and 2,4,5-T in lysimeter percolation and runoff water. *Bull. Environ. Contam. Toxicol.*, Springer Verlag, New York, **6**(1), 81–84.

Ellis, B. G., 1973. The soil as a chemical filter. In *Recycling Treated Municipal Wastewater and Sludge through Forest and Cropland*, W. E. Sopper and L. T. Kardos (eds.), Pennsylvania University Press, University Park, Pa., pp. 46–70.

Emrich, G. H., and G. L. Merritt, 1969. Effects of mine drainage on ground water. *Ground Water* 7(3): 27–32.

Enfield, C. G., and B. E. Bledsoe, 1975. Fate of wastewater phosphorus in soil. *J. Irrig. Drain. Div., Am. Soc. Civ. Eng.* **101**(IR3): 145–155.

Erickson, A. E., J. M. Tiedje, B. G. Ellis, and C. M. Hansen, 1972. Initial observations of several medium-sized Barriered Landscape Water Renovation Systems for animal wastes. *Proc. Cornell Agric. Waste Management Conference Syracuse, New York.* Cornell Univ., pp. 405–410.

Field, R., E. J. Struzeski, Jr., H. E. Masters, and A. N. Tafuri, 1974. Water pollution and associated effects from street salting. *J. Env. Eng. Div., Am. Soc. Civ. Eng.* **100**(EE2): 459–477.

Field, R., E. J. Struzeski, H. E. Masters, and A. N. Tafuri, 1975. Water pollution and associated effects from street salting. In *Water Pollution Control in Low Density Areas*, W. J. Jewell and R. Swand (eds.), Univ. Press of New England, Hanover, N.H., pp. 317–340.

Fisher, H. L., 1972. Prediction of the dosage to man from the fallout of nuclear devices. 6. Transport of nuclear debris by surface and groundwater. *UCLR Report No. 50163, Part 6*, Lawrence Livermore Laboratory, Univ. of Calif. 22 pp.

Fryberger, J. S., 1975. Investigation and rehabilitation of a brine-contaminated aquifer. *Ground Water* 13(2): 155–160.

Fungaroli, A. A. 1971. Pollution of subsurface water by sanitary landfills. *U.S. Environmental Protection Agency Report SW-12g*, 186 pp.

Galbraith, J. H., R. E. Williams, and P. L. Siems, 1972. Migration and leaching of metals from old mine tailings deposits. *Ground Water* 10(3): 33–44.

Garcia-Bengochea, J. I., C. R. Sproul, R. O. Vernon, and H. J. Woodard, 1973. Artificial recharge of treated wastewaters and rainfall runoff into deep saline aquifers of Peninsula of Florida. In *Proc. 2nd Int. Symp. on Underground Waste Management and Artificial Recharge, 26 to 30 September, New Orleans, La.*, Vol. 1, J. Braunstein (ed.), Am. Assoc. of Petrol. Geol., U.S. Geol. Survey, and Int. Assoc. of Hydrol. Sci., Tulsa, Okla. pp. 505–525.

Gera, F., and D. G. Jacobs, 1972. Considerations in the long-term management of high-level radioactive wastes. *Report ORNL 4762, Oak Ridge Nat. Lab.*, Oak Ridge, Tenn., 151 pp.

Gerba, C. P., C. Wallis, and J. L. Melnick, 1975a. Fate of wastewater bacteria and viruses in soil. *J. Irrig. Drain. Div., Am. Soc. Civ. Eng.* **101**(IR3): 157–174.

Gerba, C. P., C. Wallis, and J. L. Melnick, 1975b. Viruses in water: the problem, some solutions. *Env. Sci. and Technol.* **9**: 1122–1126.

Gershon, N. D., and A. Nir, 1969. Effects of boundary conditions of models on tracer distribution in flow through porous media. *Water Resour. Res.* **5**: 830–839.

Gilbert, R. G., R. C. Rice, H. Bouwer, C. P. Gerba, C. Wallis, and J. L. Melnick, 1976a. Wastewater renovation and reuse: virus removal by soil filtration. *Science* **192**: 1004–1005.

Gilbert, R. G., C. P. Gerba, R. C. Rice, H. Bouwer, C. Wallis, and J. L. Melnick, 1976b. Virus and bacteria removal from wastewater by land treatment. *Appl. Environ. Microbiol.* **32**: 333–338.

Glover, R. E. 1964. The pattern of fresh-water flow in a coastal aquifer. In *Sea Water in Coastal Aquifers, U.S. Geol. Survey Water Supply Paper 1613-C*, pp. C32–C35.

Golwer, A., K. H. Knoll, G. Matthess, W. Schneider, and K. H. Wällhauser, 1973. Biochemical processes under anaerobic and aerobic conditions in groundwater contaminated by solid waste. *Proc. Symp. Hydrochem. and Biogeochem, Tokyo, Sept. 1970.* Clarke, Washington, D.C., pp. 344–357.

Goodman, A. H., and M. J. Beckett, 1976. Legislative aspects of groundwater quality. *Intern. Conf. on Groundwater Quality, Measurement, Prediction, and Protection.* Water Research Center, Reading, England, Paper No. 20, 14 pp.

Ground Water Newsletter 5(5), 1976. N. P. Gillies (ed.), Where have all the radionuclides gone?

Hadley, R. F., and D. T. Snow (eds.), 1974. *Water Resources Problems Related to Mining.* Amer. Water Works Assoc., Minneapolis, Minn., 236 pp.

Hall, P. L., and H. Quam, 1976. Countermeasures to control oil spills in western Canada. *Ground Water* 14(3), 163–169.

Hamilton, D. H., and R. L. Meehan, 1971. Ground rupture in the Baldwin Hills. *Science* **172**: 333–344.

Hanshaw, B. B., 1972. Natural-membrane phenomena and subsurface waste emplacement. In *Underground Waste Management and Environmental Implications*, T. D. Cook (ed.), *Amer. Assoc. Petrol. Geol. Memoir* **18**: 308–317.

Harleman, D. R. F., P. F. Mehlhorn, and R. R. Rumer, Jr., 1963. Dispersion-permeability correlation in porous media. *J. Hydraul. Div., Am. Soc. Civ. Eng.* **89**(HY2): 67–85.

Harrison, J. C., and W. W. Longley, 1970. Denver earthquakes. *Science* **169**: 211.

Haubold, R. G., 1975. Approximation for steady interface beneath a well pumping fresh water overlying salt water. *Ground Water* **13**(3): 254–259.

Healy, J. H., W. W. Rubey, D. T. Griggs, and C. B. Raleigh, 1968. The Denver earthquakes. *Science* **161**: 1301–1310.

Henry, H. R. 1959. Salt intrusion into fresh-water aquifers. *J. Geophys. Res.* **64**, 1911–1919.

Herzberg, B., 1901. Die Wasserversorgung einiger Nordseebader. *Z. Gasbeleuchtung und Wasserversorgung* **44**, 815–819, 842–844.

Holzer, T. L., 1976. Application of ground-water flow theory to a subsurface oil spill. *Ground Water* **14**(13): 138–145.

Hoopes, J. A., and D. R. F. Harleman, 1967. Wastewater recharge and dispersion in porous media. *J. Hydraul. Div., Am. Soc. Civ. Eng.* **93**(HY5), 51–71.

Hubbert, M. K., and E. G. Willis, 1957. Mechanics of hydraulic fracturing. *Trans. Amer. Inst. Mech. Engin. 210*. [Reprinted in *Underground Waste Management and Environmental Implications*, T. D. Cook (ed.), *Am. Assoc. Petrol. Geol. Memoir*, **18**: 239–257].

Hughes, J. L., 1975. Evaluation of ground-water degradation resulting from waste disposal to alluvium near Barstow, California. *U.S. Geol. Survey Prof. Paper 878*, 33 pp.

Hughes, G. M., J. A. Schleicher, and K. Cartwright, 1976. Supplement to the final report on the hydrogeology of solid waste disposal sites in northeastern Illinois. *Envir. Geol. Note No. 80*, Illinois State Geological Survey, Urbana, 24 pp.

Huling, E. E., and T. C. Hollocker, 1972. Groundwater contamination by road salt: steady state concentrations in East Central Massachusetts. *Science* **176**: 288–290.

Johnston, W. R., F. T. Ittahadieh, K. R. Craig, and A. F. Pillsbury, 1967. Insecticides in tile drainage effluent. *Water Resour. Res.* **3**: 525–537.

Kaufman, W. J., 1974. Chemical pollution of ground waters. *J. Am. Water Works Assoc.* **66**: 152–159.

Kaufman, M. I., D. A. Goolsby, and G. L. Faulkner, 1973. Injection of acidic industrial waste into a saline carbonate aquifer: Geochemical aspects. In *Proc. 2nd Int. Symp. on Underground Waste Management and Artificial Recharge, 26 to 30 September, New Orleans, La.*, Vol. 1, J. Braunstein (ed.), Am. Assoc. of Petrol. Geol., U.S. Geol. Survey, and Int. Assoc. of Hydrol. Sci., Tulsa, Okla. pp. 526–541.

Kaufmann, R. F., G. G. Eadie, and C. R. Russell, 1975. Ground-water quality impacts of uranium mining and milling in the Grants Mineral Belt, New Mexico. *Techn. Note ORP/LV-75-4*, U.S. Environmental Protection Agency, Office of Rad. Progr. Las Vegas, Nev., 90 pp.

Kaufmann, R. F., G. G. Eadie, and C. R. Russell, 1976. Effects of uranium mining and milling on ground water in the Grants Mineral Belt, New Mexico. *Ground Water* **14**(5): 296–308.

Kelley, W. E., 1976. Geoelectric sounding for delineating ground-water contamination. *Ground Water* **14**(1): 6–10.

Kharaka, Y. K., 1973. Retention of dissolved constituents of waste by geologic membranes. In *Proc. 2nd Int. Symp. of Underground Waste Management and Artificial Recharge, 26 to 30 September, New Orleans, La.*, Vol. 1, J. Braunstein (ed.), Am. Assoc. of Petrol. Geol., U.S. Geol. Survey, and Int. Assoc. of Hydrol. Sci., Tulsa, Okla. pp. 420–435.

Kirkham, M. B., 1974. Disposal of sludge on land: effect on soils, plants and ground water. *Compost Sci.* **15**(2): 6–10.

Kirkham, M. B., 1976. Trace elements in sludge on land: effect on plant, soils, and ground water. *Proc. Eighth Annual Waste Management Conference*, Cornell University, Ithaca, N.Y., pp. 209–247.

Klefstad, G., L. V. A. Sendlein, and R. C. Palmquist, 1975. Limitations of the electrical resistivity method in landfill investigations. *Ground Water* **13**(5): 418–427.

Kolenbrander, G. J., 1969. Nitrate content and nitrogen loss in drainwater. *Netherlands J. Agric. Sci.* **17**: 246–252.

Kreitler, C. W., and D. C. Jones, 1975. Natural soil nitrate: the cause of the nitrate contamination of ground water in Runnels County, Texas. *Ground Water* **13**(1): 53–61.

Kubo, A. S., and D. J. Rose, 1973. Disposal of nuclear wastes. *Science* **182**: 1205–1211.

LaFleur, K. S., G. A. Wojeck, and W. R. McGaskill, 1973. Movement of toxaphene and fluometuron through Dunbar soil to underlying ground water. *J. Environ. Qual.* **2**: 515–518.

LaFleur, K. S., W. R. McGaskill, and D. S. Adams, 1975. Movement of prometryne through Congaree soil into ground water. *J. Environ. Qual.* **4**: 132–133.

Lance, J. C., F. D. Whisler, and R. C. Rice, 1976a. Maximizing denitrification during soil filtration of sewage water. *J. Environ. Qual.* **5**: 102–107.

Lance, J. C., C. P. Gerba, and J. L. Melnick, 1976b. Virus movement in soil columns flooded with secondary sewage effluent. *Appl. and Envir. Microbiol.* **32**: 520–526.

Lawson, D. W., 1971. Improvements in the finite difference solution of two-dimensional dispersion problems. *Water Resour. Res.* **7**: 721–725.

Lazar, E. C. 1975. Summary of damage incidents from improper land disposal. *Paper presented at Nat. Conf. on Management and Disposal of Residues from the Treatment of Industrial Wastewaters.* Washington, D.C., Environmental Protection Agency, Office of Solid Waste Man. Progr., 15 pp.

Lehr, J. H., W. A. Pettyjohn, T. W. Bennett, J. R. Hanson, and L. E. Sturtz, 1976. A manual of laws, regulations, and institutions for control of ground water pollution. *U.S. Environmental Protection Agency Report 440/9-76-006*, Washington, D.C., 416 pp.

Levy, H. B. 1972. On evaluating the hazards of groundwater contamination by radioactivity from an underground nuclear explosion. *UCLR Report No. 51278*, Lawrence Livermore Laboratory, Univ. of Calif., 20 pp.

Lewallen, M. J., 1971. Pesticide contamination of a shallow bored well in the southeastern Coastal Plains. *Ground Water* **9**(6): 45–48.

Linderman, C. L., L. N. Mielke, and G. E. Schuman, 1976. Deep percolation in a furrow-irrigated sandy soil. *Trans. Am. Soc. Agric. Eng.* **19**(2): 250–253.

Lindsay, W. L., and E. C. Moreno, 1960. Phosphate phase equilibria in soils. *Proc. Soil Sci. Soc. Am.* **24**: 177–182.

Logan, T. J., and G. O. Schwab, 1976. Nutrient and sediment characteristics of tile effluent in Ohio. *J. Soil Water Conserv.* **31**(1): 24–27.

Mack, W. N., L. Yue-Shoung, and D. B. Coohon, 1972. Isolation of poliomyelites virus from a contaminated well. *Health Serv. Rep.* **87**: 271–274.

Marino, M. A., 1974. Distribution of contaminants in porous media flow. *Water Resour. Res.*, **10**: 1013–1018.

Matis, J. R., 1971. Petroleum contamination of ground water in Maryland. *Ground Water* **9**(6): 57–61.

McKee, J. E., F. B. Laverty, and R. Hertel, 1972. Gasoline in groundwater. *J. Water Poll. Contr. Fed.* **44**: 293–302.

McWhorter, D. B., R. K. Skogerboe, and G. V. Skogerboe, 1974. Potential of mine and mill spoils for water quality degradation. In *Water Resources Problems Related to Mining*, R. F. Hadley and D. T. Snow (eds.), Amer. Water Resour. Assoc., Minneapolis, Minn. pp. 123–137.

McWhorter, D. B., R. K. Skogerboe, and G. V. Skogerboe, 1975. Water quality control in mine spoils Upper Colorado River Basin. *Environmental Protection Agency Report 670/2-75-048*, Cincinnati, Ohio, 100 pp.

Merkel, R. H., 1972. The use of resistivity techniques to delineate acid mine drainage in ground water. *Ground Water* **10**(5): 38–42.

Miller, J. C., 1975. Nitrate contamination of the water-table aquifer by septic tank systems in the Coastal Plain of Delaware. In *Water Pollution Control in Low Density Areas*, W. J. Jewell and R. Swand (eds.), Univ. Press of New England, Hanover, N.H., pp. 121–133.

Miller, D. W., F. A. DeLuca, and T. L. Tessier, 1974. Ground water contamination in the northeast States. *U.S. Environmental Protection Agency Report 660/2-74-056*, 325 pp.

Mink, L. L., R. E. Williams, and A. T. Wallace, 1972. Effect of early day mining operations on present day water quality. *Ground Water* **10**(1): 17–26.

Mulder, R. D., 1954. Verontreiniging van grondwater. *Meded. No. 2, Raad Bijstand van Samenwerkende Waterleidinglaboratoria.* Keuringsinstituut voor Waterleidingartikelen, The Hague, pp. 19–22.

Mytelka, A. I., J. S. Czachor, W. R. Guggino, and H. Golub, 1973. Heavy metals in wastewater and treatment plant effluents. *J. Water Poll. Contr. Fed.* **45:** 1859–1864.

Nash, R. G., and E. A. Woolson, 1967. Persistence of chlorinated hydrocarbon insecticides in soils. *Science* **157:** 924–927.

Newton, C. D., W. W. Shephard, and M. S. Coleman, 1974. Street runoff as a source of lead pollution. *J. Water Poll. Contr. Fed.* **46:** 999–1000.

Nielsen, D. R., and J. W. Biggar, 1962. Miscible displacement: III. Theoretical considerations. *Proc. Soil Sci. Soc. Amer.* **26:** 216–221.

Nordstedt, R. A., L. B. Baldwin, and L. M. Rhodes, 1975. Land disposal of effluent from a sanitary landfill. *J. Water Poll. Contr. Fed.* **47:** 1961–1970.

Nupen, E. M., and W. H. J. Hatting, 1975. Health aspects of reusing wastewater for potable purposes. In *Proc. Workshop on Research Needs for the Potable Reuse of Municipal Wastewater*, K. D. Linstedt and E. R. Bennett (eds.), *Environmental Protection Agency Report 600/9-75-007*, pp. 109–119.

Oakes, D. B., and K. J. Edworthy, 1976. Field measurements of dispersion coefficients in the United Kingdom. *International Conference on Groundwater Quality, Measurement, Prediction, and Protection.* Water Research Center, Reading, England, Paper No. 12, 16 pp.

Office of Solid Waste Management Program, 1973. An environmental assessment of potential gas and leachate problems at land disposal sites. *U.S. Environmental Protection Agency Publ. SW-110,* 33 pp.

Ogata, A., 1970. Theory of dispersion in a granular medium. *U.S. Geol. Survey Prof. Paper 411-I,* pp. 11–134.

Ongerth, H. J., D. P. Spath, J. Crook, and A. E. Greenberg, 1973. Public health aspects of organics in water. *J. Am. Water Works Assoc.* **65:** 495–498.

Oster, J. D., and J. D. Rhoades, 1975. Calculated drainage water composition and salt burdens resulting from irrigation with river waters in the western United States. *J. Environ. Qual.* **4:** 73–79.

Palmquist, R., and L. V. A. Sendlein, 1975. The configuration of contamination enclaves from refuse disposal sites on floodplains. *Ground Water* **13**(2): 167–181.

Peaceman, D. W., and H. H. Rachford, 1962. Numerical calculation of multidimensional miscible displacement. *J. Soc. Petrol. Eng.* **2**(4): 327–339.

Peek, H. M., and R. C. Heath, 1973. Feasibility study of liquid-waste injection into aquifers containing salt water, Wilmington, North Carolina. In *Proc. 2nd Int. Symp. on Underground Waste Management and Artificial Recharge, 26 to 30 September, New Orleans, La.,* Vol. 2, J. Braunstein (ed.), Am. Assoc. of Petrol. Geol., U.S. Geol. Survey, and Int. Assoc. of Hydrol. Sci., Tulsa, Okla., pp. 851–875.

Pettyjohn, W. A., 1975. Pickling liquors, strip mines, and ground-water pollution. *Ground Water* **13**(1): 4–10.

Pettyjohn, W. A., 1976. Monitoring cyclic fluctuations in ground-water quality. *Ground Water* **14**(6): 472–480.

Phillips, S. J., and J. R. Raymond, 1975. Monitoring and characterization of radionuclide transport in the hydrogeologic system. *Reprt No. BNWL-SA-5494, Part I,* Battelle Pacific Northwest Laboratories, Richland, Wash., 79 pp.

Pickens, J. F., and W. C. Lennox, 1976. Numerical simulation of waste movement in steady ground-water flow systems. *Water Resour. Res.* **12:** 171–180.

Pickett, G. R., 1968. Properties of the Rocky Mountain Arsenal disposal reservoir and their relation to Derby earthquakes. In *Geophysical and Geological Studies of the Relationships between the Denver Earthquakes and the Rocky Mountain Arsenal Well,* vol. 63, Quarterly of the Colorado School of Mines, J. C. Hollisher and R. J. Weimer (eds.), Golden, Colo., pp. 73–100.

Pillsbury, A. F., W. R. Johnston, F. Ittihadieh, and R. M. Daum, 1965. Salinity of tile drainage effluent. *Water Resour. Res.* **1:** 531–535.

Pinder, G. F., 1973. A Galerkin finite-element simulation of groundwater contamination on Long Island, New York. *Water Resour. Res.* **9:** 1657–1669.

Pinder, G. F., and H. H. Cooper, Jr., 1970. A numerical technique for calculating the transient position of the saltwater front. *Water Resour. Res.* **6:** 875–882.

Piskin, P., 1973. Evaluation of nitrate content in ground water in Hall County, Nebraska. *Ground Water* **11**(6): 4–13.

Pitt, W. A., Jr., H. C. Mattraw, Jr., and H. Klein, 1975. Ground-water quality in selected areas serviced by septic tanks, Dade County, Florida. *U.S. Geol. Survey Open File Report 75-607*, 82 pp.

Pohland, F. G., and S. J. Kang, 1975. Sanitary landfill stabilization with leachate recycle and residual treatment. *Am. Inst. Chem. Engrs. Symp. Series No. 145, Water-1974* **71**: 308–318.

Pollock, S. J., and L. G. Toler, 1973. Effects of highway de-icing salts on groundwater and water supplies in Massachusetts. *Highway Research Board* **425**: 17–22.

Porter, L. K., F. G. Viets, Jr., T. M. McCalla et al., 1975. Pollution abatement from cattle feedlots in northeastern Colorado and Nebraska. *Environmental Protection Agency Report 660/2-75-015*.

Reddell, D. L., and D. K. Sunada, 1970. Numerical simulation of dispersion in groundwater aquifers. *Hydrology Paper 41*, Colorado State Univ., Fort Collins, Colo., 79 pp.

Reichert, S. O., 1962. Radionuclides in groundwater at the Savannah River plant waste disposal facilities. *J. Geophys. Res.* **67**: 4363–4374.

Rhindress, D., 1971. Remarks in Bull Session No. 2: Chemical Contamination of Ground Water. *Ground Water* **9**(6): 63.

Rhoades, J. D., R. D. Ingvalson, J. M. Tucker, and M. Clark, 1973. Salts in irrigation drainage waters: I. Effects of irrigation water composition, leaching fraction, and time of year on the salt compositions of irrigation drainage waters. *Proc. Soil Sci. Soc. Am.* **37**: 770–774.

Rouse, J. V., 1974. Radiochemical pollution from phosphate rock mining and milling. In *Water Resources Problems Related to Mining*, R. F. Hadley and D. T. Snow (eds.), Amer. Water Resour. Assoc., Minneapolis, Minn. pp. 65–71.

Rovers, F. A., and G. J. Farquhar, 1973. Infiltration and landfill behavior. *J. Env. Eng. Div., Am. Soc. Civ. Eng.* **99**(EE5): 671–690.

Runnells, D. B., 1976, Wastewaters in the vadose zone of arid regions: geochemical interactions. *Ground Water* **14**(6): 374–385.

Schaub, S. A., E. P. Meier, J. R. Kolmer, and C. A. Sorber, 1975. Land application of wastewater: the fate of viruses, bacteria, and heavy metals at a rapid infiltration site. *Report No. TR7504*. U.S. Army Medical Research and Development Command, Washington, D.C., 57 pp.

Scheidegger, A. E., 1961. General theory of dispersion in porous media. *J. Geophys. Res.* **66**: 3273–3278.

Schmidt, K. D., 1972. Nitrate in ground water of the Fresno-Clovis metropolitan area, California. *Ground Water* **10**(1): 50–61.

Schmorak, S., and A. Mercado, 1969. Upconing of fresh water-sea water interface below pumping wells, field study. *Water Resour. Res.* **5**: 1290–1311.

Schraps, W. G., 1972. Die Bedeutung der Filtereigenschaften des Bodens für die Anlage von Friedhofen. *Mittcilungen Deutsche Bodenkundl. Gesellschaft* **16**: 225–229.

Sendlein, L. V. A., and R. C. Palmquist, 1975. A topographic-hydrogeologic model for solid waste landfill siting. *Ground Water* **13**(3): 260–268.

Shamir, U. Y., and D. R. F. Harleman, 1966. Numerical and analytical solutions of dispersion problems in homogeneous and layered aquifers. Rep. 89, Dept. Civ. Eng. Hydrodyn. Lab., Massachusetts Institute of Technology, Cambridge, Mass., 206 pp.

Shamir, U. Y., and D. R. F. Harleman, 1967a. Dispersion in layered porous media. *J. Hydraul. Div., Am. Soc. Civ. Eng.* **93**(HY5): 237–260.

Shamir, U. Y., and D. R. F. Harleman, 1967b. Numerical solutions for dispersion in porous mediums. *Water Resour. Res.* **3**: 557–581.

Shuval, H. I., 1975. Evaluation of the health aspects of reusing wastewater for potable purposes in Israel. In *Proc. Workshop on Research Needs for the Potable Reuse of Municipal Wastewater*, K. G. Linstedt and E. R. Bennett (eds.), *Environmental Protection Agency Report 600/9-75-007*, pp. 97–108.

Sikora, L. J., M. G. Bent, R. B. Corey, and D. R. Keeney, 1976. Septic tank nitrogen and phosphorus removal test system. *Ground Water* **14**: 309–314.

Simon, R. B., 1969. Seismicity of Colorado: Consistency of recent earthquakes with those of historical record. *Science* **165**: 897–899.

Sproul, O. J., 1975. Virus movement into groundwater from septic tank systems. In *Water Pollution*

Control in Low Density Areas, W. J. Jewell and R. Swand (eds.), Univ. Press of New England, Hanover, N.H., pp. 135–144.

Stewart, B. A., F. G. Viets, Jr., G. L. Hutchinson, and W. D. Kemper, 1967. Nitrate and other water pollutants under fields and feedlots. *Env. Sci. Technol.* **1**: 736–739.

Stewart, B. A., D. A. Woolhiser, W. H. Wischmeier, J. H. Caro, and M. H. Frere, 1975. Control of water pollution from cropland. Joint report by Agric. Res. Service, U.S. Dept. of Agriculture, and Environmental Protection Agency Report No. ARS-H-5-1, U.S. Dept. of Agriculture, 111 pp.

Stollar, R. L., and P. Roux, 1975. Earth resistivity surveys—a method for defining ground-water contamination. *Ground Water* **13**(2): 145–150.

Sun, R. J., 1973. Hydraulic fracturing as a tool for disposal of wastes in shale. In *Proc. 2nd Int. Symp. on Underground Waste Management and Artificial Recharge, 26 to 30 September, New Orleans, La.,* vol. 1, J. Braunstein (ed.), Am. Assoc. of Petrol. Geol. U.S. Geol. Survey, and Int. Assoc. of Hydrol. Sci., Tulsa, Okla., pp. 219–227.

Sweet, H. R., and R. H. Fetrow, 1975. Ground-water pollution by wood waste disposal. *Ground Water* **13**(2): 227–231.

Swoboda, A. R., G. W. Thomas, F. B. Cady, R. W. Baird, and W. G. Knisel, 1971. Distribution of DDT and toxaphene in Houston black clay on three watersheds. *Env. Sci. Technol.* **5**: 141–145.

Tagamets, T., and Y. M. Sternberg, 1974. A predictor-corrector method for solving the convection-dispersion equation for adsorption in porous media. *Water Resour. Res.* **10**: 1003–1011.

Tisdale, S. L., and W. L. Nelson, 1975. *Soil Fertility and Fertilizers* 3d ed., MacMillan, New York, 694 pp.

Todd, D. K., and D. E. O. McNulty, 1976. *Polluted Groundwater.* Water Information Center, Inc., Port Washington, N.Y., 179 pp.

U.S. Environmental Protection Agency, 1974. Compilation of Industrial and Municipal Injection Wells in the United States, *Report 520/9-74-20,* Environmental Protection Agency, Washington, D.C.

U.S. Environmental Protection Agency, 1976. Municipal sludge management. *Federal Register* **41**(108): 22531–22536.

van Haaren, F. W. J., 1951. Kerkhoven als bronnen van waterverontreiniging (Cemeteries as sources of groundwater contamination). *Water* **35**(16): 167–172.

van Schilfgaarde, J., L. Bernstein, J. D. Rhoades, and S. L. Rawlins, 1974. Irrigation management for salt control. *J. Irrig. Drain. Div., Am. Soc. Civ. Eng.* **100**(IR3): 321–338.

Vecchioli, J., and H. F. H. Ku, 1972. Preliminary results of injecting highly treated sewage-plant effluent into a deep sand aquifer at Bay Park, New York. *U.S. Geol. Survey Professional Paper 751-A,* Washington, D.C.

Viets, F. G., 1975. The environmental impact of fertilizers. *Critical Reviews in Environmental Control,* Chem. Rubber Corp. Press, Inc., **5**: 423–453.

Walker, W. R., and W. E. Cox, 1973. Legal and institutional consideration of deep-well waste disposal. In *Proc. 2nd Int. Symp. on Underground Waste Management and Artificial Recharge, 26 to 30 September, New Orleans, La.,* vol. 1, J. Braunstein (ed.), Am. Assoc. of Petrol. Geol., U.S. Geol. Survey, and Int. Assoc. of Hydrol. Sci., Tulsa, Okla., pp. 3–19.

Walker, W. H., and F. O. Wood, 1973. Road salt use and the environment. *Highways Research Board* **425**: 67–76.

Walker, W. G., J. Bouma, D. R. Keeney, and F. R. Magdoff, 1973. Nitrogen transformations during subsurface disposal of septic tank effluent in sands: I. Soil transformations. *J. Env. Quality* **2**: 475–480.

Warner, D. L., and D. H. Orcutt, 1973. Industrial wastewater-injection wells in the United States—Status of use and regulation, 1973. In *Proc. 2nd Int. Symp. on Underground Waste Management and Artificial Recharge, 26 to 30 September, New Orleans, La.,* vol. 2, J. Braunstein (ed.), Am. Assoc. of Petrol. Geol., U.S. Geol. Survey, and Int. Assoc. of Hydrol. Sci., Tulsa, Okla., pp. 687–697.

Weidner, C. W., 1974. "Degradation in groundwater and mobility of herbicides." MSc thesis, Agronomy Dept., Univ. Nebraska, Lincoln, 69 pp.

Weinberger, L. W., D. G. Stephan, and F. M. Middleton, 1966. Solving our water problems—water renovation and reuse. *Ann. New York Acad. Sci.* **136**: 131–154.

Wellings, F. M., A. L. Lewis, C. W. Mountain, and L. V. Pierce, 1975. Demonstration of virus in groundwater after effluent discharge onto soil. *Appl. Microbiol.* **29**: 751–757.

Whipple, W., J. V. Hunter, and S. L. Yu, 1974. Unrecorded pollution from urban runoff. *J. Water Poll. Contr. Fed.* **46**: 873–885.

Williams, R. E., 1975. *Waste Production and Disposal in Mining, Milling and Metallurgical Industries.* Miller Freeman Publications, Inc., San Francisco, 489 pp.

Williams, D. E., and D. G. Wilder, 1971. Gasoline pollution of a ground-water reservoir—a case history. *Ground Water* **9**(6): 50–54.

Wilmoth, B. M., 1972. Salty ground water and meteoric flushing of contaminated aquifers in West Virginia. *Ground Water* **10**(1): 99–105.

Winograd, I. J., 1974. Radioactive waste storage in the arid zone. *EOS Trans. Am. Geophys. Un.* **55**: 884–894.

Wolff, R. G., J. D. Bredehoeft, W. S. Keys, and E. Shuter, 1975. Stress determination by hydraulic fracturing in subsurface waste injection. *J. Am. Water Works Assoc.* **67**: 519–523.

Yare, B. S. 1975. The use of a specialized drilling and ground-water sampling technique for delineation of hexavalent chromium contamination in an unconfined aquifer, southern New Jersey Coastal Plain. *Ground Water* **13**(2): 151–154.

Zanoni, A. E., 1972. Ground-water pollution and sanitary landfills—a critical review. *Ground Water* **10**(1): 3–13.

Zwerman, P. J., T. Greweling, S. D. Klausner, and D. J. Lathwell, 1972. Nitrogen and phosphorus content of water from tile drains at two levels of management and fertilization. *Proc. Soil Sci. Soc. Am.* **36**: 134–137.

TWELVE

GROUNDWATER RIGHTS

For legal purposes, groundwater is divided into percolating water and underground streams or reservoirs. Percolating waters are those that "seep, drip or ooze through the interstices of the earth and below the surface thereof" (Smith, 1959). Underground streams and ponds, which also include subflows of surface streams, are associated with clearly defined pathways or open spaces. All underground water legally is presumed to be percolating water unless proven otherwise, with the burden of proof on those claiming it to be different. For underground streams, this proof involves presentation of convincing evidence that there is a bed with distinct banks through which flows a distinct stream of water. In a case in New Mexico (*Keeney v. Carillo*, 2 New Mexico 480, 1883), this was successfully argued for a surface stream that went underground and emerged 3.2 km downgradient in the same valley, by showing that obstruction of the stream flow upgradient from the point of disappearance reduced the discharge in the stream after its reappearance (Smith, 1959). Often, however, the legal distinction between percolating groundwater and underground streams is of little hydrologic significance, as so eloquently expressed by Coogan (1975) who stated:

> The law—a formal set of rules by which society is ordered—seems to the physical scientist a strangely confusing and confused tool with which to define, even in a social context, the parameters and limits of a physical continuum. For example, on the basis of attorney's briefs, bolstered even by expert testimony, judges have legally defined "subterranean streams" and erected criteria for recognizing such streams that sound more like the rhetoric of Humpty Dumpty than a description of a body of water one could scoop up in a bucket, or upon which one could float a rubber raft.

The objective of legally classifying underground water as different from the true, percolating groundwater is, of course, to get around some of the legal restrictions and consequences of the various laws and doctrines pertaining to true ground-water, such as the English rule (see discussions after next paragraph). Under-ground streams, for example, can be subject to the same rules as surface streams.

Historically, the first water rights dealt with the uses of surface water, which in the United States are primarily governed by the riparian doctrine in the humid, eastern states, and by the prior-appropriation doctrine in the arid, western states (Clark, 1972; Davis, 1976; Linsley and Franzini, 1964; Smith, 1959; Walton, 1970; and references therein). The riparian doctrine is a reasonable-use concept, allow-ing owners of land adjacent to streams to withdraw reasonable portions of the stream flow for beneficial use without jeopardizing downstream water users. In times of low flow, all water users share in reducing their withdrawals. Nobody really owns the water. Riparian landowners only have the right to beneficial use of the water. The prior-appropriation doctrine is a first-in-time, first-in-right concept that was the basis for water rights for miners and irrigation farmers settling in the West. In times of water shortage, those holding the oldest water rights remained entitled to their full shares while junior appropriators had to curtail their diver-sions. Domestic use of water generally has preference over agricultural use, which in turn has preference over industrial use.

Development of groundwater law, which is handled by the states, has been sporadic and uneven (Clark, 1975). Coogan (1975) characterized the status of groundwater law in Ohio as one where "there have been only a few issues actually decided," whereas Chalmers (1974) implied that the Colorado groundwater doc-trine is one of "modified confusion." There are four doctrines that govern with-drawal and use of groundwater and the consequences thereof: the English rule, the American rule, the correlative-rights rule, and the prior-appropriation doctrine (Chalmers, 1974; Clark and Clyde, 1972; Fleming, 1977; Smith, 1959; Walton, 1970).

The *English* rule, also called the *common-law* or *absolute* rule, is based on the theory that percolating groundwater is part and parcel of the land above it and thus belongs to the owner of that land under the legal maxim that property rights include everything that is above or below the land ("cujus est solem ejus est usque ad coelum et ad infernos," which literally translates to "his is his alone and is from the heavens to the depths of the earth"; see Smith, 1959, and Coogan, 1975). Under this doctrine, landowners can withdraw as much groundwater from below their land as they wish, without liability for harm to adjacent landowners as long as malicious intent is not demonstrated. Owners can sell and transfer water from their property to other locations, they can install deeper wells and dry up wells of their neighbors, all without liability (Smith, 1959). Because of the absolute owner-ship, the English rule is also referred to as the law of the biggest pump, or the might-makes-right rule (Coogan, 1975, and references therein). The English rule is based on nineteenth-century English conditions and is followed in most eastern states. In Ohio, for example, the English rule was established in 1861 by the holding of the Ohio Supreme Court that Jacob Brown was not liable when he dug a hole on his property that stopped the flow of a spring on Joseph Frazier's land (*Frazier v. Brown;* see Coogan, 1975; Hanson, 1965; and references therein). The

English rule also was the initial doctrine for percolating groundwater in several western states. In *Mosier v. Caldwell* (7 Nevada 363, 1872), for example, the court held that percolating groundwater belonged to the owner of the land above it and that diversion of such water did not make the owner responsible for damage to others, even if this groundwater was the source of someone else's spring, as in *Mosier v. Caldwell*.

One reason for the dearth of court cases on groundwater in states governed by the English rule undoubtedly is that the absolute and unyielding character of the rule discourages litigation. Before condemning the English rule, however, it should be realized that it originated and was used in humid areas where water generally was plentiful, and that nineteenth century groundwater hydrology was rather primitive to say the least. In the absence of any knowledge regarding the origin, movement, and occurrence of groundwater, it was logical to assign complete ownership of this "occult and secret substance" (Hanson, 1965) from the bowels of the earth to the owner of the land above it. As competition for groundwater increases also in the humid states and as modern groundwater hydrology enables much better evaluation of sources, occurrence, movement, and quantitative aspects of groundwater, plaintiffs undoubtedly will argue increasingly for softening or abandonment of the English rule (Coogan, 1975; Hanson, 1965). Precedents for more reasonable-use type of interpretations have already been set in various English-rule states (see examples of court cases at end of chapter).

The *American* rule or *reasonable-use* rule is like the English rule, but the use is more restricted to a reasonable-type use on the owners' land. Within those restrictions, however, owners have the right to withdraw as much groundwater as they wish, even if it would affect the groundwater supply of neighboring owners (Fleming, 1977). The question of what constitutes a reasonable use normally is settled in court. If property owners sell or transfer water away from their lands, neighboring landowners must show resulting harm to sustain legal action.

The *correlative-rights* doctrine is a modification of the American rule and was originated in California to provide an equitable distribution of withdrawal rights where percolating groundwater is in limited supply. Landowners then are restricted to withdrawing groundwater amounts in proportion to the land area they own over the groundwater supply (Fleming, 1977). As under the American rule, water can be withdrawn only for the beneficial use of the lands from which it is taken. If there is more than enough groundwater to meet the needs for beneficial use on overlying lands, the surplus may be appropriated for export to other lands. The overlying-land owners and the appropriators of surplus groundwater then have co-equal rights to the groundwater. An actionable wrong results when sale or transfer of water to other lands harms groundwater rights of adjacent landowners. Prior to 1902, percolating groundwater in California was governed by the English rule. In *Katz v. Wilkinshaw* (141 California 116, 70 Pac. 663, 1902; 74 Pac. 766, 1903), however, the court held that the English rule could never be of any value as a rule of property because it offers "so little protection" and "so much temptation to others to capture the water." This case, which dealt with a diversion of groundwater by the defendant that was injurous to the use of artesian groundwater by the plaintiff, was the start of the correlative-rights doctrine. Despite increased management of groundwater basins on a district basis and a trend toward greater state involvement, the correlative-rights doctrine offers no protection against groundwater depletion (Chalmers, 1974, and references therein).

The *prior appropriation doctrine* is not predicated on ownership of the land, but on who first started to withdraw groundwater. This creates a chronological hierarchy of appropriators that protects those with the most seniority (Fleming, 1977). The quantity to which an appropriator is entitled is determined by his record of historical use. Appropriators may transfer water to any site where it can be beneficially used. When there is not enough groundwater, junior appropriators must curtail their pumping and may be required to compensate senior appropriators for added pumping costs due to lower groundwater levels. In some states, all water, surface water as well as percolating groundwater and underground streams, by statute belongs to the public and is subject to appropriation (Nevada and New Mexico, for example). In other states, only surface water and water in underground streams are governed by prior appropriation, whereas percolating groundwater is governed by one of the other three doctrines.

State control of groundwater pumping normally is achieved through a system of permits for drilling a well and of licensing drillers. Areas of serious groundwater overdraft can be declared " critical areas " by the states. In Arizona, for example, 10 such areas have been designated. Drilling of new irrigation wells in critical areas must first be approved by the state land department. This permit system, however, does not apply to domestic wells or replacement irrigation wells (Fleming, 1977). Groundwater withdrawal may also be restricted where land subsidence causes serious problems, such as flooding by high tides or river stages, or where groundwater overdraft produces other undesirable effects like saltwater intrusion.

Conflicting water uses are resolved in local or state courts of law, where rulings provide interesting insight into the interpretation and administration of groundwater law and local codes and regulations, and into what is legal and what, at least to the layperson, seems right. Conflicts arise primarily from interference between wells of different owners, between wells and springs, and between wells or springs and surface excavations or mining. A second group of cases has to do with conflicting uses of interconnected surface and underground water supplies. A third category encompasses cases arising from damage to property due to a decline or rise in groundwater level (Coogan, 1975). Such damage may consist of subsidence and damage to structures or foundations, loss of natural subirrigation water for orchards or other deep-rooted crops, flooded basements or fields, soggy soils, loss of soil-bearing capacity, and landslides or other instability. Examples of court cases and resulting holdings and interpretations of the various groundwater doctrines are given in the following paragraphs.

An early abandonment of the harsh English rule took place in New Jersey, where a court held that a city could not export and sell water from 20 artesian wells that intercepted groundwater which normally drained into a stream, without liability for damage to a riparian landowner whose use of water from the stream would be impaired by the reduction in base flow due to the wells (*Meeker v. East*

Orange, New Jersey L. 623, 74 A. 379, 1909; Coogan, 1975). This case also is an example of conflicting uses of interrelated stream water and groundwater.

The English rule was also modified in Delaware, where groundwater that was pumped to fill the pool of a new swim club caused the well of a neighboring landowner to go dry (41 Delaware Ch. 26, 187 A. 2d 417, 1963). In litigation, the Court of Chancery applied the test of "objective reasonableness" and ordered the plaintiff to either deepen his well with the cost split between plaintiff and defendant, or to connect his home water system to a commercial water supply company. Plaintiff was to assume the risk in case deepening the well did not produce more water (Coogan, 1975, and references therein).

A third example of abandonment of the English rule in favor of a more reasonable-use type of doctrine is the case of the city of Columbia, Missouri. This city acquired 7 ha of land for 5 new wells that would pump about 44 000 m^3/day from a shallow aquifer for municipal use, without beneficial use to the overlying land, which comprised a much greater area than the land purchased by the city. A farmer owning 2 400 ha in the area tried to obtain a court declaration that the city was without right to extract the water, because without such declaration, the city's pumping would deprive the landowners from the reasonable use of the groundwater under their land (*Higday v. Nickolaus*, Missouri 469 S.W. 2d 859, 1971). While the city argued that Missouri had recognized the English rule of groundwater rights, the court eventually held that the city could not deprive the landowners of the beneficial use of the normal groundwater by acquiring "minuscule plots" and using "powerful pumps" to draw water into "wells on its own land for merchandising groundwater stored in plaintiff's land." The court then directed the city to withdraw groundwater only to the extent that it would not interfere with the plaintiff's beneficial use of the water (Coogan, 1975, and references therein).

A growing problem in arid regions is the increasing need for water by expanding municipalities in areas where surface water is not available and groundwater below land around the city belongs to private landowners who use it for crop irrigation. One approach, used by the city of Tucson, Arizona, for example, has been for the city to buy the land, retire it from farming, and pump the underlying groundwater to the city for municipal use. In litigation, the court restricted the amount of groundwater that could be withdrawn in such a manner to the amount that was historically used for irrigation (*Jarvis v. Arizona State Land Department*, 106 Arizona 506, 479 P. 2d 169, 1970). This historical use was further interpreted by the court in subsequent action as being the difference between the amount of groundwater pumped out of the aquifer and the amount seeping back to the aquifer as deep-percolation water from the irrigated fields. This difference, which is essentially equal to the evapotranspiration or consumptive use of the crop, was set at one-half the amount of groundwater pumped from the wells. The court also ruled that this pumpage should be based on the average pumping over a number of years, and not on a year of unusually high pumping (Fleming, 1977, and references therein). Transfer of groundwater from beneath retired farmland to mining industries, however, was not permitted by the court inasmuch as industrial water requirements have lower priority than municipal or agricultural needs in critical water areas.

In Arizona, water from underground streams is governed by prior appropriation and is subject to control by the state through a permit system. Thus, when a
junior appropriator in 1919 sank wells to capture the underflow of the Santa Cruz
River and caused the water level in two wells of a senior appropriator (installed in
1913) further downstream to drop almost 6.7 m, the court held that the senior
appropriator had a right to have the water level in his well remain the same. The
junior appropriator had to stop pumping or supply the senior appropriator with
the amount of water he formerly was able to pump (*Pima Farms Co. v. Proctor*, 30
Arizona 96, 245 P. 2d 369, 1926; Martin and Erickson, 1976).

Use of percolating groundwater in Arizona originally was governed by the
English rule, but later by the American rule. Thus, while an early case held that
percolating groundwater is an integral part of the soil and that, therefore, the
owner of the overlying land also is the absolute owner of the groundwater
(*Howard v. Perrin*, 8 Arizona 347, 76 p. 460, 1904), a later decision held that
withdrawal of groundwater was restricted to reasonable use on the lands above it,
and that it was unreasonable to transport water away from the land from which it
is pumped (*Bristor v. Cheatham*, 75 Arizona 227, 225 P. 2d 173. 1953; see also
Chalmers, 1974; Clark, 1975; Fleming, 1977; Martin and Erickson, 1976).

Loss of artesian head due to installation of another well was the subject of a
case in Murray City, Utah, where private owners of free-flowing wells took the
city to court because a new municipal well would lower piezometric levels in their
wells, forcing them to pump or obtain water elsewhere. While early Utah law held
that prior appropriators who receive their water from artesian wells have a vested
right in the artesian pressure head, the court in this case held that free-flowing
wells were a wasteful diversion of groundwater that prevented full beneficial use of
the groundwater resource. The city thus was allowed to use the new well without
having to compensate the private well owners (*Wayman v. Murray City Corp.*, 23
Utah 2d 97, 458 P. 2d 861, 1969). There are also cases, however, in which junior
appropriators had to curtail their pumping where it would lower water levels in
pumped wells of senior appropriators (*Pima Farms Co. v. Proctor*, 30 Arizona 96,
245 P 2d 369, 1926; *Baker v. Ore-Ida Foods, Inc.*, 95 Idaho 575, 513 P. 2d 627,
1973; Martin and Erickson, 1976).

Where surface water and groundwater are hydraulically connected and con-
junctively used, conflicts may arise between appropriators of stream water and
those of groundwater. Senior appropriators may change their point of diversion
from a stream to a well and vice versa. Junior appropriators of groundwater may
have to curtail their pumping if it interferes with the rights of senior appropriators
of stream water. A case in point is *Langenegger v. Carlsbad Irrigation District* (82
New Mexico 416, 483 P. 2d 297, 1971), where a senior appropriator of stream
water wished to change his point of water diversion to an aquifer that drained into
the stream (Martin and Erickson, 1976). The aquifer was a more reliable source of
water than the stream, whose base flow was reduced by pumping from the aquifer
by junior appropriators. Despite the fact that the groundwater was already
heavily appropriated and junior well owners objected to the drilling of new wells,
the court ruled that the senior appropriator could rely on all sources tributary to
the waters to which he was entitled, and that he could pursue those waters to their

source and make his diversion there. Thus, he was allowed to drill the well and obtain his water from the aquifer. Where the connection between the surface water and the groundwater could not be established, however, the owner of surface-water rights could not change his point of diversion to an aquifer (*Kelly v. Carlsbad Irrigation District*, 76 New Mexico 466, 415 P. 2d 849, 1966). In other cases, owners of wells pumping groundwater from an aquifer connected to a river could be forced to reduce their pumping if it interfered with the rights of senior appropriators on the river (*Kuiper v. Well Owners Conserv. Assoc.*, 176 Colorado 119, 490 P. 2d 268, 1971; *Fellhauer v. People*, 167 Colorado 320, 447 p. 2d 986, 1968).

Artificial recharge of groundwater may interfere with other uses of underground resources, like gravel digging. In Alameda County, California, groundwater recharge by a water district to replenish a depleted aquifer caused water to rise in gravel pits, prompting the owners of the pits to pump the water out and discharge it into a stream which carried it to San Francisco Bay. While one party thus was putting water underground, the other was pumping it out again. Ironically, it was the groundwater depletion in the first place that enabled gravel operators to dig deep without getting water in their pits. The issue was taken to court, which held that groundwater recharge could be practiced without interfering pumping from the gravel pits as long as groundwater levels were below the natural groundwater levels that prevailed before groundwater depletion started (*Niles Sand and Gravel Co. v. Alameda County Water District*, 37 California 3d 924, 112 Cal. Rptr. 846, 1974; Martin and Erickson, 1976).

Other questions in groundwater recharge are: who owns the recharged groundwater and can an aquifer be used as a facility for transmitting water? These questions arise when a party wants to recharge the aquifer at one point and pump the recharged water out at another point without other well owners in between availing themselves of the water. In a court case on this question, it was ruled that aquifers may be used for storage and transmission of water in the same manner as surface facilities, and that others have no right to use that water (*City of Los Angeles v. City of San Fernando*, 123 Cal. Rptr. 1, 537 P. 2d 1250, 1975; Martin and Erickson, 1976).

Saving water by lining irrigation canals to reduce seepage losses does not always create a right to the water saved. Once others have established a prior right to the seepage or other return flow, the party salvaging that water cannot claim the water saved (*Dannenbrink v. Burger*, 23 Cal. App. 587, 138 Pac. 751, 1913; *Comstock v. Ramsey*, 55 Colorado 244, 133 Pac. 1107, 1913). In Arizona, a farmer lined his irrigation ditch with concrete and used the seepage water thus saved to irrigate more land. While he diverted no more water from the river than before, the court ruled that he could not use the water saved to irrigate additional land and that he should return that water to the stream for the benefit of downstream users with prior rights (*Salt River Valley Water Users Association v. Kovacovich*, 3 Arizona App. 28, 411 P. 2d 201, 1966; Martin and Erickson, 1976).

Reducing seepage losses from surface streams by removing phreatophytes from floodplains does not necessarily entitle the remover of the phreatophytes to the water saved. In one case, for example, farmers that cleared phreatophytes along a portion of the Arkansas River could not divert an amount of river water equal to that previously used by the phreatophytes. The court's argument was that the phreatophytes took water from senior appropriators on the river and that farmers who removed these phreatophytes to take the water that these phreatophytes used in fact took water from the senior appropriators (*Southeastern Colorado Water Conservancy District v. Shelton Farms, Inc.*, 529 P. 2d 1321, 1974; Martin and Erickson, 1976).

The legal status of hot geothermal water or steam for energy production and heating is nebulous as states are debating whether it should be considered a water or a mineral. Classification problems particularly arise on public and other lands where surface rights and mineral rights are not in the same hand, and each party claims a right to the geothermal resource. A recent ruling of the Superior Court of Sonoma County, California, held that geothermal water is a mineral and thus belongs to the owner of the mineral rights (*Water Newsletter*, 1976). The argument was that since geothermal water is hot, corrosive, toxic, and unsuitable for potable or irrigation use, it is more a mineral resource than a water resource. There may, however, be exceptions to this as some geothermal waters may be of sufficient quality to enable their use for purposes other than power production or heating (see Section 10.4 and 10.8). Some states are considering classifying geothermal water as neither water nor mineral, but as a special, energy-type resource.

State water laws are not applicable to federal lands. This was established by the U.S. Supreme Court on June 7, 1976, in a unanimous ruling that water rights on federal reserved lands are the jurisdiction and property of the Federal Government (*Cappeart et al. v. United States et al.*, No. 74-1107; see also *Water Well Journal Staff*, 1977). The case in question was the pumping of groundwater by a private landowner from wells about 4 km away from Devil's Hole, which is a small pool on federal land in Death Valley, Nevada. The pool is formed by blockage of groundwater flow in cavernous limestone, due to a fault. The landowner had obtained permission from the state of Nevada to withdraw groundwater under the prior-appropriation doctrine, and began pumping in 1968. The United States then asserted that the groundwater was hydrologically connected to the water in Devil's Hole and that pumping from some of the wells had lowered the water level in the pool, thus threatening the survival of several hundred desert pupfish. This is a rare and unique, minnow-type fish that was common to the region in Pleistocene times and apparently survived in Devil's Hole from original stock. In 1971, the United States sought an injunction to limit pumping from six specific wells to that needed for domestic use only. The District Court in 1974 then enjoined the owners from pumping that would lower the water level in the pool more than 0.9 m below a copper nail in the rock that serves as a benchmark. The recent normal water level in the pool had been about 0.5 m below the nail.

The argument for federal supremacy over local water rights was that " when the Federal Government withdraws its land from the public domain and reserves it for a Federal purpose, the government, by implication, reserves appurtenant water when unappropriated to the extent needed to accomplish the purpose of the reservation. In so doing, the United States acquires a reserved right in unappropriated water which rests on the date of the reservation and is superior to the rights of future appropriators." Thus, the federal government's rights to the water were not bound by Nevada water laws and the federal government did not have to apply for a state permit to appropriate groundwater when the federal reservation was formed (*United States v. Cappaert*, 508 F 2d 313, 1974). This case, which is known as the pupfish decision, may have far-reaching implications on groundwater rights situations involving federal or Indian lands (*Water Well Journal Staff*, 1977; Bird, 1976). The pupfish case basically is an extension to groundwater of the Winters decision (*Winters v. United States*, 207 U.S. 567, 1908), which held that the federal government was considered to have reserved rights to water on federal lands necessary for such lands (see also Chalmers, 1974).

PROBLEMS

12.1 Obtain a feeling for the complex issues and the jargon of groundwater law by studying the references of this chapter, particularly those by Chalmers, Clark, Coogan, and Smith.

12.2 Select a state and contact the natural-resources or land department to find out about applicable groundwater law and how to obtain permission to drill a new well and withdraw a certain flow of water.

REFERENCES

Bird, J. W., 1976. Your water—how secure? *J. Irrig. Drain. Div., Am. Soc. Civ. Eng.* **102**(IR3): 363–367.

Chalmers, J. R., 1974. Southwestern groundwater law: A textural and bibliographic interpretation. *Arid Lands Resource Information Paper Number 4*, Univ. of Arizona, Tucson, 228 pp.

Clark, R. E., 1972. Introduction to water law of the Western States. In *Waters and Water Rights*, vol. 5, R. E. Clark (ed.), The Allen Smith Company, Indianapolis, Indiana, pp. 1–37.

Clark, R. E., 1975. Arizona ground water law: the need for legislation. *Ariz. Law Rev.* **16**(4): 799–819.

Clark, R. E., and E. W. Clyde, 1972. Western ground-water law. In *Water and Water Rights*, vol. 5, R. E. Clark (ed.), The Allen Smith Company, Indianapolis, Indiana, pp. 407–474.

Coogan, A. H., 1975. Problems of groundwater rights in Ohio. *Akron Law Rev.* **9**(1): 34–115.

Davis, C., 1976. Introduction to water law of the eastern United States. In *Waters and Water Rights*, vol. 7, R. E. Clark (ed.), The Allen Smith Company, Indianapolis, Indiana, pp. 1–26.

Fleming, R., 1977. Arizona groundwater law: historical development. *Arizona Water Resources Project Information Bull. No. 15*, 4 pp.

Hanson, J. R., 1965. Ohio's secret substance. *Ground Water* **3**(2): 28–31.

Linsley, R. K., and J. B. Franzini, 1964. *Water-Resources Engineering*. McGraw-Hill Book Co., New York, 690 pp.

Martin, W. A., and G. H. Erickson, 1976. Summary of selected court cases in water conservation and groundwater litigation. *WCL Report No. 10*, U.S. Water Conservation Lab., Agric. Res. Serv., U.S. Dept. of Agric., Phoenix, Ariz., 24 pp.

Smith, C. H., 1959. *Smith's Review of Mining Law and Water Rights*. West Publishing Co., St. Paul, Minn., 150 pp.

Walton, W. C., 1970. *Groundwater Resource Evaluation*. McGraw-Hill Book Co., New York, 664 pp.

Water Newsletter, 1976. N. P. Gillies (ed.), Geothermal steam belongs to the owner of mineral rights. Vol. 18, no. 14.

Water Well Journal Staff, 1977. U.S. Supreme Court creates new ground water law. *Water Well J.* **31**(2): 52–53.

ANSWERS TO CALCULATION PROBLEMS

2.1 25 m

2.3 0.37 mm, 0.66 mm, 1.94, and sand for sample A; 0.014 mm, 0.27 mm, 25, and loamy sand for sample B; 0.001 mm, 0.005 5 mm, 9, and silty clay loam for sample C

2.5 12.9 percent, 0.24 (volume fraction), 0.3, 0.429, 80 percent, and 1.85, respectively

2.7 0.17

2.8 0.000 1

2.9 0.42 and 0.33

3.3 1 096 years, 5 000 persons

3.4 10 m^3/day

3.5 0.5 m^3/day per meter length of stream, 4.5 m^3/day per meter length of stream, 90 percent

3.6 0.005, 0.05, and 0.25 m for the 1-m piezometer; 0.025, 0.25, and 1.25 m for the 5-m piezometer; and 0.05, 0.5, and 2.5 m for the 10-m piezometer

3.7 The t values for the eight z values are 0.010 7, 0.046, 0.113, 0.222, 0.386, 0.633, 1.008, and 1.619 days. From the slope of the z-vs.-t curve, dz/dt is evaluated as 20 m/day at $z = 2$ m, 7.5 m/day at $z = 4$ m, and 3.33 m/day at $z = 6$ m. Multiplying these values by $f = 0.1$ yields values of v that agree with v calculated with Eq. (3.15)

3.8 4.67 m and 2.5 m/day

3.9 $K = 8 \times 10^{-6}$ m/day; 62.5 km^2

3.10 -0.4 m of water

3.11 1 152 m/day, 57.6, hardly

3.12 0.016 6 m/day; 0.014 8 m^3/day

4.2 $Q = 12 279$ m^3/day

4.4 2.99 m

4.5 $h_w = 42.99$ m

4.6 0.000 814 m^{-2} days

4.7 45.4, 44.2, 43.7, and 41.9 m

4.8 18 m and 682 m^2/day

4.9 The s values are 0.017, 0.073, 0.178, 0.44, 0.65, 0.94, and 1.14 m, respectively. These values agree with s calculated with the Theis solution because s is small compared to H

4.10 219, 186, 136, 103, and 81 m^2/day

5.1 1 day for the 50- and 100-m wells, 3.7 days for the 100- and 200-m wells, and 2.5 days for the 50- and 200-m wells. This shows that the observation wells, or at least one of them, should be relatively close to the pumped well in order to obtain a constant s difference in relatively short time (the distance should not be less than 1.5H, however, to avoid vertical-flow components)

5.2 $T = 2024$ m^2/day and $S = 0.159$

5.3 $T = 2000$ m^2/day and $S = 0.161$

5.4 $T = 2000$ m^2/day and $S = 0.15$

5.5 $T = 1987$ m^2/day and $S = 0.156$

5.6 $r/B = 3$; $T_A = 1194$ m^2/day and $S_A = 0.000096$, $T_Y = 973$ m^2/day and $S_Y = 0.107$ (these results agree with $T = 1000$ m^2/day, $S_A = 0.0001$, and $S_Y = 0.1$ used to calculate the s-vs.-t curve in Figure 5.7); $s = 0.0051$ m calculated and 0.0046 m on graph; $t_{wt} = 1.04$ days, which is about where the curves join

5.7 $K = 31$ m/day

5.8 $K = 1.34$ m/day

5.9 $K = 0.5$ m/day

5.10 $K = 0.52$ m/day

6.1 19.8 and 44.3 m, and 20 and 46.2 m, respectively

6.4 Total water production cost is $15 per 1000 m^3, of which 9.3 percent is for amortization and maintenance

7.3 An infinite number; however, the effect of the remote wells on drawdown in the real system is insignificant

7.4 840 m^3/day

7.6 -75.3 and -17.2 cm, respectively

7.7 27 percent

7.8 500 m; 1000 m; 9.69 m and 2.92 m

8.1 $S_i = 17.5$ cm/day$^{1/2}$; best fit is obtained when $A = 30.5$ cm/day or $A = 2K/3$; $C = 47$ cm/day and $\alpha = 1.02$ for Kostiakov's equation; β in Horton's equation varies from about 140 days^{-1} for small t to 5 days^{-1} for large t

8.2 0.08333 days or 2 h, 1.67 cm, and 0.04458 days or 1.1 h.

8.3 2.88 cm infiltration and 2.12 cm runoff

8.4 0.30 cm/day

8.5 θ is 24.4 percent from 0 to 40 cm depth, 25.8 percent from 40 to 80 cm, 27.2 percent from 80 to 120 cm, and 28.6 percent from 120 to 160 cm depth. Corresponding f values are 0.156, 0.142, 0.128, and 0.114

8.6 0.39 cm/day

8.7 $I_s/K = 1.63$, water loss is 6520 m^3/day per kilometer

8.8 15120 m^3

8.9 5.5 days, 2 m, and 0.47 × 10^6 m^3 per hectare per year

8.10 $v_{min} = 0.14$ m/day, infiltration periods are 7.3 days, maximum hydraulic loading is 48.6 m/year or 0.486 × 10^6 m^3 per hectare per year

8.11 45 m and 63 m^3/day

8.12 559 m^3/year, 143 m^3/year, and 416000 m^3/km per year

9.1 For the dense sand, $C_c = 0.010$ and $C_u = 0.0063$; for the loose sand, $C_c = 0.16$ and $C_u = 0.077$; and for blue clay, $C_c = 0.40$ and $C_u = 0.21$

9.2 0.036 m and 0.036 m

9.3 0.026 m and 0.037 m

9.4 0.489 m and 0.487 m

9.5 Total subsidence is 0.024 6 m. Storage coefficient is 0.006 1 if water comes from compression of clay and aquifer, 0.001 if water comes from compression of aquifer alone, and 7.5×10^{-6} if water comes from decompression of aquifer water alone

9.7 0.067 4 m

9.8 0.5 cm for well and 5 m for basin

9.9 0.8 m

9.10 2.5 kg/cm^2

10.3 About 75°C per 100 m vs. 2.5°C per 100 m for normal conditions

11.2 2462 m^3/day and 3.1 m

11.3 126 m^3/day

11.4 16 m

11.5 175 m^3

11.6 13.07 cm per irrigation, 1.87 cm deep-percolation water with salt concentration of 4 480 mg/l, and salt load of 838 kg/ha per irrigation for $C_i = 640$ mg/l. For $C_i = 1 493$ mg/l, the answers are 16.8 cm, 5.6 cm, 4 480 mg/l, and 2 508 kg/ha per irrigation, respectively

11.7 5.87 cm, 1 861 mg/l, and 1 092 kg/ha for $C_i = 640$ mg/l; 9.6 cm, 3 235 mg/l, and 3 105 kg/ha for $C_i = 1 493$ mg/l

11.9 9.6 years

11.10 110 m

INDEX